The Papillomaviruses

Robert L. Garcea   Daniel DiMaio
Editors

The Papillomaviruses

 Springer

Robert L. Garcea
UCDHSC at Fitzsimons
RC1-North, Room 4100
PO Box 6511, Mail Stop 8302
Aurora, Colorado 80045
USA
Bob.Garcea@uchsc.edu

Daniel DiMaio
Department of Genetics
Yale University School of Medicine
Sterling Hall of Medicine, I-141
333 Cedar Street
New Haven, CT 06510
USA
U.S. Mail:
P.O.Box 208005
New Haven, CT 06520-8005
daniel.dimaio@yale.edu

Library of Congress Control Number: 2006929455

ISBN-10: 0-387-36522-2   e-ISBN-10: 0-387-36523-0
ISBN-13: 978-0-387-36522-0   e-ISBN-13: 978-0-387-36523-7

Printed on acid-free paper.

9 8 7 6 5 4 3 2 1

springer.com

PREFACE

Our understanding of the biology of papillomaviruses has increased dramatically over the past 25 years. Before then, the lack of a facile tissue culture system had left this virus family behind the other "DNA tumor viruses," such as SV40 and murine polyomavirus. However, with the advent of molecular cloning and other new techniques, coupled with the pivotal stimulus given by connecting HPV infection and cervical cancer, the field has flourished. This progress has culminated in the development of the first true anticancer vaccine, using virus-like particles of the HPV16 and 18 capsid proteins. All investigators in the field should take pride in their contribution to a wonderful scientific journey with such a spectacular humanitarian outcome. Their basic studies, technological innovations, and clinical investigation serve as a model of the positive impact of science in society.

The obvious difficulty in assembling a book about any rapidly moving scientific field is that many observations may be out of date by the time of publication. Given the rate of progress in papillomavirus research, this text may suffer in this regard. However, the release of the HPV vaccine completes a major era, which should be marked by a summary of the work to date. In this summary, we have emphasized the biology of papillomaviruses as well as their clinical importance. Many of the lessons learned have obvious relevance to other viruses and cancers, and they illuminate new pathways to oncogenesis. These lessons stand alongside the vaccine as monuments of progress.

We thank all the contributors to this volume and our other colleagues in the field for their dedication and efforts. The field could not have moved forward so rapidly without such a committed, collaborative, and interactive group of investigators. We hope that the vaccine is not perceived as the end of research into papillomavirus biology, but rather the end of the beginning. We look forward to another 25 years in which new basic science questions are answered, and therapeutic and prophylactic HPV vaccines and other treatment modalities reach everyone in need worldwide.

Ciao bene
Bob Garcea
Dan DiMaio

CONTENTS

LIST OF CONTRIBUTORS

M. Teresa Aguado
Vaccine Research and Development
Global Programme for Vaccine and
Immunization
World Health Organization
Geneva, Switzerland

Hans-Ulrich Bernard
Department of Molecular Biology and
Biochemistry
University of California
Irvine, California, USA

Thomas R. Broker
University of Alabama at
Birmingham
Department of Biochemistry and
Molecular Genetics
Birmingham, Alabama, USA

Xiaojiang Chen
Department of Molecular and
Computational Biology
University of Southern California
Los Angeles, California, USA

Louise T. Chow
University of Alabama at
Birmingham
Department of Biochemistry and
Molecular Genetics
Birmingham, Alabama, USA

Christopher P. Crum
Division of Women's and Perinatal
Pathology
Department of Pathology
Brigham and Women's Hospital and
Harvard Medical School
Boston, Massachusetts, USA

Daniel DiMaio
Departments of Genetics and
Therapeutic Radiology
Yale University School of Medicine
New Haven, Connecticut, USA

Erin M. Egelkrout
Fred Hutchinson Cancer Research
Center
Seattle, Washington, USA

Denise A. Galloway
Fred Hutchinson Cancer Research
Center
Seattle, Washington, USA

Robert L. Garcea
Section of Pediatric Oncology
University of Colorado School of
Medicine

Aurora, Colorado, USA

Anne E. Griep
Department of Anatomy
University of Wisconsin, Madison
Madison, Wisconsin, USA

Peter Howley
Department of Pathology
Harvard Medical School
Boston, Massachusetts, USA

W. Martin Kast
Norris Comprehensive Cancer Center
Department of Molecular
Microbiology & Immunology
University of Southern California
Los Angeles, California, USA

George Klein
Microbiology and Tumor Biology
Center (MTC)
Karolinska Institutet
Stockholm, Sweden

Laimonis A. Laimins
Department of
Microbiology-Immunology
Feinberg School of Medicine
Northwestern University
Chicago, Illinois, USA

Paul F. Lambert
McArdle Laboratory
University of Wisconsin
Madison, Wisconsin, USA

Choogho Lee
Department of
Microbiology-Immunology
Feinberg School of Medicine
Northwestern University
Chicago, Illinois, USA

Douglas R. Lowy
Laboratory of Cellular Oncology
National Cancer Institute
National Institutes of Health
Bethesda, Maryland, USA

Gretchen Eiben Lyons
Norris Comprehensive Cancer Center
Department of Molecular
Microbiology & Immunology
University of Southern California
Los Angeles, California, USA

Karl Münger
Brigham & Women's Hospital
Channing Labs
Boston, Massachusetts, USA

Michael I. Nishimura
Department of Surgery
University of Chicago
Chicago, Illinois, USA

Sonia Regina Pagliusi
World Health Organization
Department of Immunization,
Vaccines and Biologicals
Geneva, Switzerland

Donald Maxwell Parkin
Clinical Trials Service Unit/
Epidemiological Studies Unit
University of Oxford
Headington, United Kingdom

Martin Sapp
Institute for Medical Microbiology
and Hygiene
University of Mainz
Mainz, Germany

Ralph M. Richart
Department of Pathology
Columbia University College of
Physicians and Surgeons

New York, New York, USA

John Schiller
Laboratory of Cellular Oncology
National Cancer Institute
National Institutes of Health
Bethesda, Maryland, USA

Hans-Christoph Selinka
Institute for Medical Microbiology
and Hygiene
University of Mainz

Mainz, Germany

Arne Stenlund
Cold Spring Harbor Laboratory
Cold Spring Harbor
New York, New York, USA

Rolf E. Streeck
Institute for Medical Microbiology
and Hygiene
University of Mainz
Mainz, Germany

1
Principles of Human Tumor Virology

George Klein[1] and Daniel DiMaio[2]

[1] *Microbiology and Tumor Biology Center (MTC), Karolinska Institutet, S-171 77 Stockholm, Sweden*
[2] *Departments of Genetics and Therapeutic Radiology, Yale University School of Medicine, New Haven, CT 06520-8005*

1.1. Introduction

Viruses containing either RNA or DNA genomes can cause tumors, and studies of tumor viruses have provided important insights into basic molecular and cellular processes. Retroviruses are the only class of RNA viruses that are known to cause tumors in animals, whereas several different families of DNA viruses can cause tumors. Four DNA viruses are known to cause or contribute to the development of cancer in humans: Epstein-Barr virus, high-risk types of human papillomaviruses, hepatitis B virus, and human herpesvirus-8, which is also known as Kaposi's sarcoma herpesvirus (KSHV). Two RNA viruses are etiologically involved in human cancer: human T lymphotropic virus type 1 and hepatitis C virus. The identification of human tumor viruses has generated optimism that it will be possible to develop improved methods to prevent and treat certain cancers in humans. Vaccination and other public health measures that prevent infection or transmission can be expected to reduce the incidence of the cancer. Antiviral treatments that inhibit virus replication are likely to reduce viral load in infected people and impair transmission to other individuals. It may also be possible to generate therapeutic cell-mediated immune responses against virally encoded proteins expressed in tumor cells. In those cancers where continued viral gene expression is required to maintain the survival or proliferation of cancer cells, it may be beneficial to inhibit viral gene expression or the action of viral proteins. Because viral genes and gene products are absent from uninfected cells, it is possible that such anticancer treatments will be highly specific and relatively nontoxic.

1.2. Precedents in Animals

Although it has been known for almost 100 years that viruses can cause tumors in animals, only in recent decades have human tumor viruses been identified. During the past century, views on the role of viruses in the etiology of human cancer have varied widely. In 1911, Peyton Rous discovered that chicken sarcomas could be transmitted with cell-free filtrates (Rous, 1911). The great excitement caused by this finding soon subsided when similar filtrates of mouse and rat tumors failed to induce tumors in experimental animals. This failure led to the premature conclusion that viruses may cause tumors in birds, but not in mammals.

Tumor viruses in mammals were first described in the 1930s. Richard Shope found that benign warts in rabbits could be transmitted by cell free filtrates (Shope, 1933), and John Bittner discovered the milk factor, later identified as the mouse mammary tumor virus (MMTV), which contributes to the development of breast cancer in mice (Bittner, 1936). However, neither of these discoveries changed the prevailing skepticism about the role of viruses in human cancer. Warts were benign papillomas, not cancers, even though rabbit papillomas occasionally progressed to carcinomas. Bittner's agent was viewed as a "maternal influence" that contributed to the high cancer incidence of selectively inbred mouse strains. Although MMTV could increase the risk of mammary cancer development, it was neither sufficient nor necessary, and hormonal and genetic factors were shown to be involved as well.

The view that viruses were not involved in cancer in mammals changed in the 1950s with the discovery of the murine leukemia virus by Ludwik Gross (Gross, 1951) and the polyoma virus by Sarah Stewart and Bernice Eddy (Stewart et al., 1958). Gross found that cell-free filtrates prepared from the "spontaneous" leukemias of the high leukemic AKR strain could transmit the disease to the low leukemic C3H strain. These filtrates were later shown to contain a leukemogenic retrovirus. Stewart and Eddy showed that Gross's preparations also contained a second virus, polyomavirus, which caused parotid tumors. These results were soon confirmed, and new tumor viruses were isolated from many different rodent, feline, simian, and fowl tumors. For the next decade, there was widespread belief that viruses played a role in many if not most cancers. Even cancers induced by chemicals and radiation were thought to be due to the activation of latent viruses. In the 1970s, cellular oncogenes were discovered, and the view again shifted away from viruses as an important cause of human cancer. More recent studies indicate that most human cancer is not virally induced, although as outlined in the next section, viruses are important carcinogens for a restricted group of human tumors.

1.3. Human Tumor Viruses

Although many viruses were identified that caused tumors in animals, the discovery of human tumor viruses was slow and at times disappointing. Human tumor viruses simply did not follow the pattern set by experiments in animals, where tumor formation often followed viral infection rapidly and efficiently. There are several difficulties in identifying human tumor viruses. The process of viral tumorigenesis in humans is inefficient, and typically only a small percentage of infected individuals actually develop cancer. This inefficiency reflects effective immuno-surveillance by the host and possibly other factors as well. Moreover, virus infection is not sufficient for tumor formation; additional events must also occur. These events may include somatic mutation, immuno-suppression, exposure to another carcinogen, or inheritance of genes that confer high susceptibility to infection or carcinogenesis. Also, viral tumorigenesis in humans is a slow process, with incubation periods measured not in days or weeks but in years or even decades. Finally, human virally induced tumors often do not produce infectious viruses. Several of these features obviously hindered epidemiological studies.

Despite these difficulties, six viruses are now widely recognized as playing an etiologic role in human cancers. Four of these viruses, human papillo-maviruses (HPV), Epstein–Barr virus (EBV), hepatitis B virus (HBV), and human herpesvirus-8 (HHV-8), contain DNA genomes, and the remaining two, human T lymphotropic virus 1 (HTLV-1) and hepatitis C virus (HCV), contain RNA genomes.

1.3.1. Human Papillomaviruses

The HPVs are small, non-enveloped DNA viruses that cause benign epithelial papillomas or warts in their natural hosts, with particular HPV types causing specific types of papillomas or lesions at particular anatomic sites. The main medical importance of the HPVs is the causal role played by certain high-risk HPV types, primarily HPV16 and HPV18, in a variety of carcinomas. The most important and best documented HPV-associated cancer is cervical carcinoma (Dürst et al., 1983; Bosch and Muñoz, 2002), but HPV is also thought to play an etiologic role in other anogenital cancers, skin cancers in immunosuppressed individuals, and some head-and-neck tumors (e.g., Gillison, 2004).

1.3.2. Hepatitis Viruses

Hepatitis B virus and hepatitis C virus are genetically unrelated viruses. HBV is a small DNA virus that utilizes an unusual replication strategy involving an RNA intermediate (Summers and Mason, 1982), whereas HCV contains an RNA genome related to flaviviruses (Choo et al., 1989; Miller and Purcell, 1990). Infections by both viruses are prevalent in humans and can cause acute and chronic liver disease. Importantly, chronic HBV- and HCV-induced liver disease can progress to primary hepatocellular carcinoma (Beasley, 1988; Di Bisceglie, 1997).

1.3.3. Epstein–Barr Virus

Epstein–Barr virus is a herpesvirus that infects virtually all adult humans (Henle et al., 1969). Unlike the viruses mentioned above, EBV is a large, complex virus that encodes more than 100 proteins. In the developed world, EBV causes infectious mononucleosis, a benign and self-limiting prolifer-ation of B lymphocytes, in about one-half of primarily infected adolescents or adults, while the rest and all young children undergo silent seroconversion. EBV is also believed to contribute to the origin of high incidence Burkitt's lymphoma and nasopharyngeal carcinoma, a malignant cancer of epithelial cells (Henle et al., 1969; de The et al., 1975). EBV-driven B cell lymphoproliferative disease can occur in immunosuppressed persons, such as transplant recipients (Nalesnik, 1998) and patients with congenital immunodeficiencies such as X-linked lymphoproliferative disease (Purtilo et al., 1992). EBV also has been implicated in some forms of Hodgkin's disease and gastric carcinoma (e.g., Weiss et al., 1989).

1.3.4. Human Herpesvirus-8

HHV-8, also known as Kaposi sarcoma herpesvirus, is distantly related to EBV. This virus was first identified in the tumor DNA of a patient with Kaposi's sarcoma (Chang et al., 1994), a tumor of endothelial cells. HHV-8 is believed to play an etiologic role in this tumor, as well as in Castleman's disease and body cavity lymphoma (Sarid et al., 1999). Kaposi's sarcoma was a relatively rare tumor until the AIDS epidemic, when it became one of the most common causes of cancer death in AIDS patients (Cesarman et al., 1995). Following the introduction of highly effective antiretroviral therapy, the incidence of Kaposi sarcoma has decreased dramatically.

1.3.5. Human T Lymphotropic Virus Type I

HTLV-1 is a complex retrovirus that causes a relatively rare tumor, adult T cell leukemia/lymphoma, in the Far East and the Caribbean basin (Gallo et al., 1983), as well as some non-neoplastic diseases.

1.3.6. SV40 as a Potential Human Tumor Virus

Simian virus 40 (SV40) has also been discussed in relation to several human tumors. SV40 is a small DNA virus that causes tumors in experimental animals. The natural host of SV40 is rhesus monkeys, and there is inconclusive evidence regarding the ability of this virus to establish infections in humans. SV40 DNA has been reported to be present in some mesotheliomas, osteosarcomas, childhood brain tumors, and non-Hodgkin's B cell lymphomas and to play an etiologic role in these cancers (reviewed in Shah, 2004). However, these reports remain controversial.

1.3.7. Viral Tumors in the Developing World

In summary, more than 10% of all cancers in humans is strongly associated with infection by tumor viruses (Table 1.1) (Parkin et al., 1999). In addition, HIV infection and its associated immunosuppression predisposes individuals to cancer development. The great majority of tumors associated with virus infection occur in the developing world, where they are a leading cause of cancer death. The preponderance of such tumors in the developing world is due to several factors.

TABLE 1.1. Annual Number of Viral-Associated Cases Worldwide*

Agent	Cancer	No. of cases
Human papillomaviruses	Cervical cancer	492,800
	Anogenital carcinoma	46,700
	Oral and pharyngeal carcinoma	14,500
Hepatitis B viruses	Hepatocellular carcinoma	535,640
Hepatitis C viruses		
Kaposi's sarcoma herpes virus	Kaposi's sarcoma	66,000
Epstein–Barr virus	Nasopharyngeal carcinoma	78,100
	Hodgkin's disease	28,600
	Burkitt's lymphoma	6,700
Human T lymphotropic virus	Adult T cell leukemia/lymphoma	3,000

* Compiled by D.M. Parkin from 2000 statistics.

In some cases, infection by the tumor virus is more prevalent in the developing world. For example, there is a high correlation between the areas of the world with a high prevalence of chronic HBV infection and hepatocellular carcinoma (Beasley, 1988). In addition, ineffective screening programs to identify lesions at a precancerous, treatable stage may contribute to the high incidence of virally associated cancer in the developing world. The introduction of effective Pap smear screening programs for cervical precursor lesions caused a dramatic drop in the incidence of cervical cancer in the developed world (Gustafsson et al., 1997). Poor nutrition and general health status resulting in impaired immune function and other social, behavioral, and possibly genetic factors may also contribute to the high prevalence of virally associated tumors in the developing world.

1.4.   Mechanisms of Viral Tumorigenesis

Molecular studies of tumor viruses in cultured cells and animals have provided important insights into many fundamental cellular processes (Nevins, 2001). These studies resulted in the discovery of mRNA splicing, transcriptional enhancers, oncogenes, and tumor suppressor proteins, and elucidated important aspects of signal transduction, immune regulation, and cell cycle control. The earliest efforts at restriction mapping and genetic engineering involved tumor viruses (Danna and Nathans, 1971; Jackson et al., 1972), and viruses were the first organisms to yield to the genomics effort (Sanger et al., 1977; Reddy et al., 1978; Fiers et al., 1978). These studies revealed that DNA tumor viruses carry their own oncogenes as essential parts of the viral genome. In contrast, most oncogenic retroviruses induce tumors under natural conditions by activating the expression of a normal cellular gene in its native chromosomal location. Retroviruses can also transform cells by transducing modified cellular genes.

Viruses have evolved to replicate their own genome and spread to new hosts, not to induce cancer. Virus-induced tumor development can thus be regarded as a biological accident or a by-product of the biochemistry of virus propagation.

In order to synthesize large amounts of viral DNA during productive infection of their host cells, most DNA viruses must first mobilize the cells to enter a state conducive to high-level DNA replication by inducing expression of proteins required for cell-cycle progression (Nevins, 2001). Thus, many DNA viruses stimulate DNA synthesis in their target cells. If lytic infection is aborted but this proliferative stimulus is sustained, growth transformation may result. For this to occur, viral genomes must persist in the infected cells, either by stable integration into the cellular DNA or by extrachromosomal DNA replication.

In the case of the HPVs (and other small DNA tumor viruses), the viral oncogene products inactivate both the Rb and the p53 tumor suppressor pathways (Nevins, 2001). Strikingly, these pathways are often crippled during the development of nonviral tumors as well. Inactivation of these tumor suppressor pathways not only provides a proliferative stimulus, but also elicits genetic instability, in part by decreasing the likelihood of growth arrest and/or apoptosis in response to DNA damage (zur Hausen, 2002). The ensuing mutations in cellular growth control genes undoubtedly play a role in further carcinogenic progression. However, despite these accumulated mutations, the proliferation of HPV-induced cervical cancer cells requires continued expression of the viral E6 and E7 oncogenes (von Knebel Doeberitz et al., 1988; Hwang et al., 1993). The mechanisms of herpesvirus transformation are more complex and appear to involve modulation of cellular signaling and cell cycle control pathways by viral proteins persistently expressed in tumor cells. In both EBV-positive (endemic) and EBV-negative (sporadic) Burkitt's lymphoma cells, the tumor is driven by the constitutive expression of the c-*myc* proto-oncogene activated by juxtaposition to an immunoglobulin gene by chromosomal translocation. Nevertheless, it has been reported that inhibition of the EBV-encoded EBNA1 protein can induce apoptosis in some EBV-positive Burkitt's lymphoma cell lines (Hammerschmidt and Sugden, 2004). HTLV-1 encodes an essential viral regulatory protein, Tax, which appears responsible for its transforming activity (Ross et al., 1996; Nerenberg et al., 1987). The mechanisms of hepatitis B and C virus tumorigenesis are not clear but are thought to involve repeated mutagenic cycles of liver damage and regeneration.

1.5.   Immunity and Tumor Viruses

Viruses have adopted many strategies to evade the immune response (Vossen et al., 2002). For example, many viruses down-regulate or inactivate the products of the major histocompatibility antigen class I locus to avoid immune clearance. In addition, herpesviruses such as EBV can enter latent states with limited viral gene expression to hide from the immune system (Klein, 1989). Nevertheless, virally transformed tumor cells often express viral proteins that can serve as tumor rejection antigens. Recognition of such neoantigens on virally induced cancer cells plays an important role in preventing viral tumorigenesis. Accordingly, immunosuppression can permit the malignant proliferation of virally transformed cells that would be otherwise rejected. It is therefore not surprising that transplant recipients, congenitally immunodeficient persons, and AIDS patients are prone to develop EBV-carrying B-cell malignancies and other virally induced tumors including Kaposi sarcoma, and HPV-associated skin and cervical carcinomas (e.g., Beral et al., 1991). Strikingly, reconstitution of EBV-specific immunity can lead to regression of EBV-associated B-cell proliferations (Heslop et al., 1994; O'Reilly et al., 1997). These considerations suggest that manipulations that increase cellular immunity to viral antigens in cancer cells may be therapeutically useful.

1.6.   Prospects for Prevention and Treatment

The identification of a human tumor virus immediately suggests strategies for tumor prevention and control. Public heath measures can be instituted to protect the population from exposure or to identify carriers or people at elevated risk of cancer. Successful examples of this approach are the elimination of HBV and HCV from the blood supply and the use of Pap screening to identify women with HPV-induced cervical dysplasia. In a more recent example, maternal-to-infant transmission of HTLV-1 is reduced if carrier mothers refrain from breast-feeding (Hino et al., 1997). Another well-established modality to control viral infection is vaccination. An effective hepatitis B vaccine is already reducing the incidence of chronic hepatitis B virus infection and hepatocellular carcinoma (Huang and Lin, 2000), and on the basis of clinical trials demonstrating protection against persistent high-risk HPV infection and the development of precancerous lesions, HPV vaccines were recently approved for distribution (Koutsky et al., 2002; Harper et al., 2004; Villa et al., 2005). Vaccines such as these that are targeted at virus particles or structural proteins are referred to as prophylactic vaccines and rely on the production of neutralizing antibodies that prevent viral infection. These vaccines are expected to be of greatest benefit to people who have not yet been exposed to the targeted virus. Prophylactic vaccines may also be used in infected people to reduce the titer of infectious virus in tissues and body fluids, the number of infected cells, or the concentration of viral gene products in infected cells. Such effects may well reduce the likelihood of malignant progression or transmission of the virus to a new host. Antiviral drugs that interfere with virus replication may provide similar benefits. However, prophylactic vaccines are unlikely to be beneficial to patients who already have precancerous changes or cancer itself. For such people, therapeutic vaccines are being developed that induce the generation of cytotoxic T lymphocytes that recognize and kill tumor cells that express viral proteins (Schreckenberger and Kaufmann, 2004). Since the expression of many cellular genes is under the control of viral proteins in cancer cells, some cellular proteins that are induced by the virus may also serve as tumor rejection antigens.

Tumor virus genomes and proteins in the cancer cells are also well-defined therapeutic targets. It may be possible to use antiviral agents that inhibit the replication or expression of viral genomes in cancer cells or that interfere with function of viral proteins required for cancer cell survival or growth. Such agents might include RNAi, micro RNA, ribozymes, or other nucleases that directly bind to and degrade viral DNA or RNA, as well as small molecules that interfere with expression or activity of viral proteins. These approaches are expected to be useful in virally induced cancers that continuously require viral gene products for proliferation or survival. Because viral genes and gene products are present in infected cells only, it may be possible to develop highly specific antiviral drugs that prevent or inhibit cancer growth with minimal effects on normal uninfected tissue.

1.7. Conclusions

Studies of tumor viruses have provided important insights into fundamental cellular processes such as the storage and expression of genetic information, cell-cycle control, signal transduction, immune regulation, and carcinogenesis. The identification of a limited number of viruses that contribute to the development of specific human cancers provides new clinical opportunities as well. If we combine various vaccination strategies with specific antiviral approaches, it is not unrealistic to foresee a future in which the incidence of these cancers is greatly reduced and their treatment is effective and relatively nontoxic.

References

Beasley, R.P. (1988). Hepatitis B virus. The major etiology of hepatocellular carcinoma. *Cancer* 61:1942–1956.

Beral, V., Teterman, T., Berkelman, R., and Jaffe, H. (1991). AIDS-associated non-Hodgkin lymphoma. *Lancet* 337:805–809.

Bittner, J.J. (1936). Some possible effects of nursing on the mammary tumor incidence in mice. *Science* 84:162.

Bosch, F.X., and Muñoz, N. (2002). The viral etiology of cervical cancer. *Virus Res.* 89:183–190.

Cesarman, E., Chang, Y., Moore, P.S., Said, J.W., and Knowles, D.M. (1995). Kaposi's sarcoma-associated herpesvirus-like DNA sequences are present in AIDS-related body cavity based lymphoma. *N. Engl. J. Med.* 332:1186–1191.

Chang, Y., Cesarman, E., Pessin, M.S., Lee, F., Culpepper, J., Knowles, D.M., and Moore, P.S. (1994). Identification of herpesvirus-like DNA sequences in AIDS-associated Kaposi's sarcoma. *Science* 266:1865–1869.

Choo, Q.L., Kuo, G., Weiner, A.J., Overby, L.R., Bradley, D.W., and Houghton, M. (1989). Isolation of a cDNA clone derived from a blood-borne non-A, non-B viral hepatitis genome. *Science* 244:359–362.

Danna, K., and Nathans, D. (1971). Specific cleavage of simian virus 40 DNA by restriction endonuclease of Hemophilius influenzae. *Proc. Natl. Acad. Sci. USA* 68:2913–2917.

de The, G., Day, N., Geser, A., Ho, J.H., Simons, M.J., Sohier, R., Tukei, P., and Vonka, V. (1975). Epidemiology of the Epstein-Barr virus infection and associated tumors in man. *Bibl. Haematol.* 43:216–220.

Di Bisceglie, A.M. (1997). Hepatitis C and hepatocellular carcinoma. *Hepatology* 26:34S–38S.

Dürst, M., Gissmann, L., Ikenberg, H., and zur Hausen, H. (1983). A papillomavirus DNA from a cervical carcinoma and its prevalence in cancer biopsy samples from different geographic regions. *Proc. Natl. Acad. Sci. USA* 80:3812–3815.

Fiers, W., Contreras, R., Haegemann, G., Rogiers, R., Van de Voorde, A., Van Heuverswyn, H., Van Herreweghe, J., Volckaert, G., and Ysebaert, M. (1978). Complete nucleotide sequence of SV40 DNA. *Nature* 273:113–120.

Gallo, R.C., Kalyanaraman, V.S., Sarngadharan, M.G., Sliski, A., Vonderheid, E.C., Maeda, M., Nakao, Y., Yamada, K., Ito, Y., Gutensohn, N., Murphy, S., Bunn, Jr., P.A., Catovsky, D., Graves, M.F., Blayney, D.W., Blattner, W., Jarrett, W.F., zur Hausen, H., Seligmann, M., Brouet, J.C., Haynes, B.F., Jegasothy, B.V., Jaffe, E., Cossman, J., Broder, S., Fisher, R.I., Golde, D.W., and Robert-Guroff, M. (1983). Association of the human type C retrovirus with a subset of adult T-cell cancers. *Cancer Res.* 43:3892–3899.

Gillison, M.L. (2004). Human papillomavirus-associated head and neck cancer is a distinct epidemiologic, clinical, and molecular entity. *Semin Oncol.* 31:744–54.

Gross, L. (1951). Spontaneous leukemia developing in C3H mice following inoculation in infancy with AK leukemic extracts of AK embryos. *Proc. Soc. Exp. Biol. Med.* 76:27–32.

Gustafsson, L., Ponten, J., Zack, M., and Adami, H.-O. (1997). International incidence rates of invasive cervical cancer after introduction of cytological screening. *Cancer Causes and Control* 8:755–763.

Hammerschmidt, W., and Sugden, B. (2004). Epstein-Barr virus sustains Burkitt's lymphomas and Hodgkin's disease. *TRENDS Mol. Med.* 10:331–336.

Harper, D.M., Franco, E.L., Wheeler, C., Ferris, D.G., Jenkins, D., Schuind, A., Zahaf, T., Innis, B., Naud, P., de Carvalho, N.S., Rotelli-Martins, C.M., Teixeira, J., Blatter, M.M., Korn, A.P., Quint, W., and Dubin, G. (2004). Efficacy of a bivalent L1 virus-like particle vaccine in prevention of infection with human papillomavirus types 16 and 18 in young women: A randomized controlled trial. *Lancet* 364:1757–1765.

Henle, G., Henle, W., Clifford, P., Diehl, V., Kafuko, G.W., Kirya, B.G., Klein, G., Morrow, R.H., Munube, G.M., Pike, P., Tukei, P.M., and Ziegler, J.L. (1969). Antibodies to EB virus in Burkitt's lymphoma and control groups. *J. Natl. Cancer Inst.* 43:1147–1157.

Heslop, H.E., Brenner, M.K., and Rooney, C.M. (1994). Donor T cells to treat EBV-associated lymphoma. *N. Engl. J. Med.* 330:1185–1191.

Hino, S., Katamine, S., Miyata, H., Tsuji, Y., Yamabe, T., and Miyamoto, T. (1997). Primary prevention of HTLV-1 in Japan. *Leukemia* 11:57–59.

Huang, K., and Lin, S. (2000). Nationwide vaccination: A success story in Taiwan. *Vaccine* 18:S35–38.

Hwang, E.S., Riese II, D.J., Settleman, J., Nilson, L.A., Honig, J., Flynn, S., and DiMaio, D. (1993). Inhibition of cervical carcinoma cell line proliferation by the introduction of a bovine papillomavirus regulatory gene. *J. Virol.* 67:3720–3729.

Jackson, D.A., Symons, R.H., and Berg, P. (1972). Biochemical method for inserting new genetic information into DNA of Simian Virus 40: Circular SV40 DNA molecules containing lambda phage genes and the galactose operon of *Escherichia coli*. *Proc. Natl. Acad. Sci. USA* 69:2904–2909.

Klein, G. (1989). Viral latency and transformation: The strategy of Epstein-Barr virus. *Cell* 58:5–8.

Koutsky, L.A., Ault, K.A., Wheeler, C.M., Brown, D.R., Barr, E., Alvarez, F.B., Chiacchierini, L.M., and Jansen, K.U. (2002). A controlled trial of a human papillomavirus type 16 vaccine. *N. Engl. J. Med.* 347:1645–1651.

Miller, R.H., and Purcell, R.H. (1990). Hepatitis C virus shares amino acid sequence similarity with pestiviruses and flaviviruses as well as members of two plant virus supergroups. *Proc. Natl. Acad. Sci. USA* 87:2057–2061.

Nalesnik, M.A. (1998). Clinical and pathological features of post-transplant lymphoproliferative disorders (PTLD). *Springer Semin. Immunopathol.* 20:325–342.

Nerenberg, M., Hinrichs, S.M., Reynolds, R.K., Khoury, G., and Jay, G. (1987). The *tat* gene of human T-lymphotropic virus type 1 induces mesenchymal tumors in transgenic mice. *Science* 237:1324–1329.

Nevins, J.R. (2001). Cell Transformation by Viruses. In Field's *Virology*. Chapter 10, pp. 245–265. Lippincott, Williams & Wilkins.

O'Reilly, R.J., Small, T.N., Papdopoulous, E., Lucas, K., Lacerda, J., and Koulova, L. (1997). Biology and adoptive cell therapy of Epstein-Barr virus-associated lymphoproliferative disorders in recipients of marrow allografts. *Immunol. Rev.* 157:195–216.

Parkin, D.M., Pisano, P., Muñoz, N., and Ferlay, J. (1999). The global health burden of infection associated cancers. *Cancer Surv.* 33:5–33.

Purtilo, D.T., Strobach, R.S., Okano, M., and Davis, J.R. (1992). Epstein-Barr virus-associated lymphoproliferative disorders. *Lab. Invest.* 67:5–23.

Reddy, V.B., Thimmappaya, B., Dhar, R., Subramanian, K.N., Zain, B.S., Pan, J., Ghosh, P.K., Celma, M.L., and Weissman, S.M. (1978). The genome of simian virus 40. *Science* 200:494–502.

Ross, T.M., Pettiford, S.M., and Green, P.L. (1996). The *tax* gene of human T-cell leukemia virus type 2 is essential for transformation of human T lymphocytes. *J. Virol.* 70:5194–5202.

Rous, P. (1911). A sarcoma of fowl transmissible by an agent from the tumor cells. *J. Exp. Med.* 13:397–411.

Sanger, F., Air, G.M., Barrell, B.G., Brown, N.L., Coulson, A.R., Fiddes, C.A., Hutchison, C.A., Slocombe, P.M., and Smith, M. (1977). Nucleotide sequence of bacteriophage phi X174 DNA. *Nature* 265:687–695.

Sarid, R., Olsen, S.J., and Moore, P.S. (1999). Kaposi's sarcoma-associated herpesvirus: Epidemiology, virology, and molecular biology. *Adv. Virus Res.* 52:139–232.

Schreckenberger, C., and Kaufmann, A.M. (2004). Vaccination strategies for the treatment and prevention of cervical cancer. *Curr. Opin. Oncol.* 16:485–491.

Shah, K.V. (2004). Simian Virus 40 and human disease. *J. Infect. Dis.* 190:2061–2064.

Shope, R.E. (1933). Infectious papillomatosis of rabbits; with a note on the histopathology. *J. Exp. Med.* 68:607–624.

Stewart, S.-E. Eddy, B.-E., Borgear, N. (1958). Neoplasms in mice inoculated with a tumor agent carried in tissue culture. *J. Natl. Cancer Inst.* 20:1223–1243.

Summers, J., and Mason, W.S. (1982). Replication of the genome of a hepatitis B-like virus by reverse transcription of an RNA intermediate. *Cell* 29:403–415.

Villa, L.L., Costa, R.L., Petta, C.A., Andrade, R.P., Ault, K.A., Giuliano, A.R., Wheeler, C.M., Koutsky, L.A., Malm, C., Lehtinen, M., Skjeldestad, F.E., Olsson, S.E., Steinwall, M., Brown, D.R., Kurman, R.J., Ronnett, B.M., Stoler, M.H., Ferenczy, A., Harper, D.M., Tamms, G.M., Yu, J., Lupinacci, L., Railkar, R., Taddeo, F.J., Jansen K.U., Esser, M.T., Sings, H.L., Saah, A.J., and Barr, E. (2005). Prophylactic quadrivalent human papillomavirus (types 6, 11, 16, and 18) L1 virus-like particle vaccine in young women: a randomized double-blind placebo-controlled multicentre phase II efficacy trial. *Lancet Oncol.* 6:271–278.

von Knebel Doeberitz, M., Oltersdorf, T., Schwarz, E., and Gissmann, L. (1988). Correlation of modified human papilloma virus early gene expression with altered growth properties in C4-1 cervical carcinoma cells. *Cancer Res.* 48: 3780–3786.

Vossen, M.T.M., Westerhout, E.M., Soderberg-Naucler, C., and Wiertz, E.J.H.J. (2002). Viral immune evasion: A masterpiece of evolution. *Immunogenetics* 54: 527–542.

Weiss, L.M., Mohared, L.A., Warnke, R.A., and Sklar, J. (1989). Detection of Epstein-Barr viral genomes in Reed-Sternberg cells of Hodgkin's disease. *N. Engl. J. Med.* 320:502–506.

zur Hausen, H. (2002). Papillomaviruses and cancer: From basic studies to clinical application. *Nature Rev.* 2:342–350.

2
History of Papillomavirus Research

Douglas R. Lowy

*Laboratory of Cellular Oncology, National Cancer Institute, National Institutes
of Health, Bethesda, MD 20892*

2.1. Introduction

Papillomavirus research has passed through several phases. The field began
slowly with the experimental transmission of human and animal warts prior
to 1930. Greater interest in these viruses was stimulated in the 1930s by the
demonstration that filtered extracts from cutaneous papillomas of wild cottontail
rabbits could induce lesions with malignant potential in cottontail and domestic
rabbits. Although investigations in the 1930s and 1940s were limited to in vivo
studies in outbred rabbits, many principles of papillomavirus biology were established by observations made during this period. The availability of infectious
extracts permitted the reproducible induction of lesions whose natural history
could be followed or be experimentally modified. Interest in papillomaviruses
diminished during the 1950s and 1960s. This change was attributable to several
factors, including the inability of papillomaviruses to propagate in culture at a
time when the life-cycle and transforming activity of other oncogenic viruses
could be studied in vitro, permitting rapid advances in molecular understanding
of these processes in the more tractable systems. In addition, human papillomaviruses (HPV) were believed to have limited medical importance because the
conditions they induced were thought to be limited to benign lesions with little or
no potential for malignant progression. The advent of molecular cloning during
the 1970s led to a resurgence in papillomavirus research. As in other areas of
biology, this technical revolution was critical to progress in the investigation
of papillomaviruses. The unlimited availability of wild-type and mutant viral
genomes made it possible to study the function of viral genes and their products,
to use viral sequences as molecular probes to detect papillomavirus sequences in
tissue, and to identify and molecularly clone new viral genotypes. Application
of these molecular techniques led to the identification of HPV as the necessary
infectious cause of a major public health problem, cervical cancer. These studies
also provided insight into the pathogenesis of HPV-induced disease, established
new paradigms for cellular transformation by viral genes, and identified candidate
antigens for protection against papillomavirus infection.

There have been excellent reviews relevant to the history of papillomavirus research. They include those for tumor viruses in general (Gross, 1983); for DNA tumor viruses (Grodzicker and Hopkins, 1981); for animal papillomaviruses (Lancaster and Olson, 1982); for cottontail rabbit papillomavirus (Breitburd et al., 1997; Kreider and Bartlett, 1981; Syverton, 1952); for HPV (Rowson and Mahy, 1967); for HPV in epidermodysplasia verruciformis (Orth, 1986); and for HPV in genital neoplasia (zur Hausen and de Villiers, 1994). This chapter highlights research advances from the 1930s though the early 1990s. Prior to the mid-1970s, most observations were first made in animal papillomavirus systems. Since then, the analysis of HPVs and the diseases they induce has produced many key observations, although experimental animal papillomavirus systems have also continued to yield important insights.

2.2.   The 1930s and 1940s: Biology of the Shope Papillomavirus and Other Animal Papillomaviruses

Papillomaviruses were the second class of viruses, after retroviruses, shown to induce malignant tumors (Shope, 1933). As such, the Shope papillomavirus, which is now designated the cottontail rabbit papillomavirus (CRPV), became an important experimental model of viral tumorigenesis. Not only did infectious extracts induce benign papillomas in cottontail and domestic rabbits, but some of the benign lesions progressed to squamous cell cancers, the first demonstration that a mammalian virus could cause a malignant solid tumor (Rous and Beard, 1935; Syverton, 1952; Syverton and Berry, 1935). It was also found that some papillomas regressed, while others persisted without progression. A causal relationship between persistence of a papilloma and the risk of malignant progression was inferred, as the carcinoma developed at the site of the papilloma and malignant tumors still expressed viral antigens. In addition, if a lesion regressed, the site of the former lesion was no longer at risk for carcinoma development. However, it was not clear whether the virus in the papilloma played a specific role in progression or if the virus was required for maintenance of the carcinoma. Another poorly understood feature of the malignant tumors was that the virus was "masked," which meant that infectious virus could not be recovered from them (Kidd and Rous, 1940). Such uncertainties should not obscure the fact that the observations made with CRPV established many salient biological characteristics of papillomavirus infection that were subsequently found to be relevant to HPV infection and carcinogenesis.

In the decade following the description of CRPV, prominent themes of CRPV research involved efforts to understand the basis for the outcome of infection and to modify the frequency with which the virus induced papillomas or carcinomas. The combination of CRPV and coal tar greatly increased the rate at which carcinomas developed (Rous, 1938; 1944), thus establishing the concept that certain environmental exposures could promote the likelihood of progression. It was also found that the development of papillomas was associated with the induction of serum neutralizing antibodies and with concomitant resistance to CRPV challenge at other cutaneous sites (Kidd et al., 1936; Shope, 1933). Furthermore, systemic immunization with papilloma suspensions could, without establishing cutaneous infection, induce serum-neutralizing antibodies and protect against high dose cutaneous viral challenge (Shope, 1937). These early studies therefore laid the foundation for the belief that induction of humoral immunity could form the basis of a preventive vaccine against papillomaviruses. These results also showed that distinct processes mediated resistance to de novo infection and regression of established papillomas, as neutralizing antibodies did not influence regression. Importantly, spontaneous regression appeared to be an immune phenomenon in that it usually affected most or all papillomas on the rabbit.

Early research also established the exquisitely specific host range of papillomaviruses. CRPV induced papillomas in rabbits only when nongenital skin was inoculated; genital skin and various mucous membranes were resistant (Parsons and Kidd, 1936a; Shope, 1933). When a second papillomavirus, the rabbit oral papillomavirus (ROPV), was isolated from spontaneous papillomas in the mouth of domestic rabbits (Parsons and Kidd, 1936b, 1943), it was shown to induce papillomas only to the oral mucosa and to lack oncogenic potential. CRPV and ROPV were shown to be distinct viruses, as animals infected by one virus developed resistance to the homologous virus while remaining susceptible to the heterologous one. Furthermore, species other than rabbits were resistant to infection by CRPV and ROPV. These findings indicated that these viruses are remarkably restricted in the range of host species that they infect and that papillomaviruses may exploit distinct ecological niches by infecting various subsets of stratified squamous epithelia. These observations also showed that more than one virus may infect a single host species, and that protection against infection by one papillomavirus may not confer protection against another.

Experimental studies with the canine oral papillomavirus (COPV) also emphasized the narrow host range of papillomaviruses. Experimental transmission of oral papillomas in dogs had been reported in the late 19th century (M'Fadyean and Hobday, 1898) with material obtained from an outbreak in foxhound puppies (Penberthy, 1898). The filterable nature of the etiological agent was demonstrated in the early 1930s (DeMonbreun and Goodpasture, 1932; Findlay, 1930). The papillomas induced by COPV uniformly regressed spontaneously, dogs were the only species susceptible to COPV infection, and inoculation of epithelial sites in the dog other than the oral mucous membranes failed to induce papillomas. COPV and ROPV were clearly distinct viruses, as neither induced oral lesions in the other host species.

Bovine papillomavirus type 1 (BPV1) was also identified during this period (Creech, 1929; Magalhaes, 1920). In its natural host, BPV1 induced fibropapillomas (consisting of both dermal fibroblasts and epidermal epithelial cells), in contrast to the strictly epithelial lesions (papillomas) characteristic of other papillomaviruses, including HPV. Although BPV1 produced the characteristic epithelial changes seen with other papillomaviruses, its ability to induce morphologic changes in the underlying dermis was a reflection of the wider host range of BPV1. However, the significance of this feature does not appear to have been appreciated until the early 1950s, with the observation that BPV1 could induce a benign fibroblastic tumor in horses (Olson and Cook, 1951). The experimental lesions closely resembled those of equine sarcoid, a naturally occurring condition of horses subsequently found to contain BPV DNA (Lancaster et al., 1977).

2.3. The 1950s and 1960s: Cell Differentiation and Virus Replication

The development of tissue culture techniques in the 1940s and 1950s did not lead to the successful in vitro propagation of papillomaviruses. By contrast, the life-cycle and transforming activity of polyoma virus and SV40 virus, which had been discovered in the late 1950s (Stewart et al., 1958; Sweet and Hilleman, 1960), could be studied in monolayer cultures. The latter viruses, therefore, became the most popular DNA tumor viruses to study, leading to remarkable advances in understanding their molecular biology and impact on cells. Interest in papillomaviruses waned in the 1950s and 1960s largely because they continued to be less tractable to tissue culture analysis and because human papillomaviruses were not thought to be agents of medically important disease. During this period, papillomaviruses were classified as belonging to the same virus family as polyoma and SV40 (the *papovaviridae*), as the papillomavirus capsid and genome were both structurally similar to, but larger than, those of polyoma and SV40 (Crawford, 1969; Stone et al., 1959; Williams et al., 1961).

Despite the shift in focus to other tumor viruses, there were some notable advances in papillomavirus biology in the 1950s and 1960s. The development of fluorescent antibody microscopy and improvements in electron microscopy made it possible to examine viral structural antigens and virus particles in papillomas. The analysis of CRPV papillomas showed that structural antigen and particles were limited to the nuclei of differentiated keratinocytes in the upper layers of the lesion and that infectious virus was also found in these layers (Moore et al., 1959; Noyes, 1959; Noyes and Mellors, 1957). The chronic nature of papillomas had implied that the virus would be present in the basal cells of lesions, as most epithelial cell division occurs in basal cells. Therefore, the lack of virus particles in the basal cells of papillomas led to the inference that the virus in the basal cells was present in an immature form. The results therefore implied that papillomavirus replication was closely tied to the differentiation process of stratified squamous epithelial cells, an insight which provided a likely explanation for the inability to propagate papillomaviruses in monolayer cultures. In fact, suitable multilayer differentiated culture systems were not developed and applied to the in vitro study of papillomaviruses until many years later (McCance et al., 1988; Dollard et al., 1992; Meyers et al., 1992).

These observations with virions were followed by the demonstration that DNA from CRPV papillomas could induce papillomas that resembled those induced by CRPV virions (Ito and Evans, 1961). This technique was also used to establish that the viral DNA was present in a transplantable carcinoma (VX7) that had been induced by CRPV, as DNA extracts from the carcinoma could induce papillomas, although the extracts did not contain infectious virus (Ito and Evans, 1965). The data, therefore, indicated that a non-encapsidated form of the viral genome was present in the transplantable tumor, and provided the first example of an oncogene transferable as naked DNA.

During this period, additional evidence was developed that regression in rabbit papillomas was an immunologically mediated event that required more than neutralizing antibodies or was independent of them. The frequency of regression was decreased when immunosuppression was induced with methylprednisolone (McMichael, 1967). Conversely, the rate of regression was increased when rabbits with papillomas were immunized systemically with minced papilloma tissue containing intact cells, although regression could not be induced by passive transfer of serum to rabbits with persistent papillomas (Evans et al., 1962). In addition, the resistance of regressor rabbits to papilloma formation was shown to be qualitatively different from that of naïve, virus-immune, and papilloma-bearing rabbits. This difference was shown first by making short-term cultures of skin explants from rabbits with each type of history, infecting the explants with CRPV, and then transferring them as an autograft to the rabbit from which each had been taken. Autographs did not form papillomas in regressor rabbits, while they did form papillomas in naïve, virus-immune, and papilloma-bearing rabbits (Kreider, 1963). Analogous results were obtained when regressor rabbits or rabbits with persistent papillomas were inoculated with extracts containing CRPV DNA. Although the CRPV DNA could produce papillomas in rabbits with persistent papillomas and neutralizing antibodies, the viral DNA was ineffective in regressor rabbits (Evans and Ito, 1966). In addition to confirming that the presence of neutralizing antibodies was not sufficient to induce regression, these studies supported the conclusion that regression was mediated by nonhumoral, cellular immunity.

2.4.   The 1970s to the Early 1990s: Viral Genetics and the Emergence of HPV as a Medically Important Virus

Molecular cloning and related techniques developed in the mid-1970s partially overcame the experimental limitations to studying papillomaviruses, leading to renewed interest in these viruses and to a wealth of new information about them. During the late 1970s and early 1980s, papillomavirus research followed two main themes: experimentally oriented studies of animal papillomaviruses, especially BPV1, and more clinically oriented studies of HPVs. By the second half of the 1980s, clinical and experimental aspects of HPVs became predominant, following the recognition of their medical importance.

BPV1 had two attractive properties. First, there was a readily renewable source of virions, as BPV1 could be serially propagated in vivo from extracts of the large lesions it induced in cattle. Second, building on earlier studies of BPV1 with primary cell cultures (Black et al., 1963; Thomas et al., 1964), BPV1 was found to induce morphologic transformation and focus formation of established tissue culture cell lines, such as the mouse C127 and NIH 3T3 cell lines (Dvoretzky et al., 1980). In contrast, CRPV and the HPVs known in 1980 did not display this activity. The ability of BPV1 to transform cultured nonepithelial cells is related to its wider in vivo host range. As noted earlier, BPV1 is the prototype for a class of animal papillomaviruses that induce fibropapillomas in their natural host. This broad host range at the cellular level presumably endows BPV1 with the ability to induce nonepithelial lesions in heterologous hosts (Friedmann et al., 1963; Olson and Cook, 1951).

The availability of a bio-assay for BPV1 in established cells, combined with the ability to molecularly clone and mutate the viral genome, made it possible to study the genetics of this virus. The experimental importance of BPV1 attracted the interest of molecular biologists and resulted in BPV1 being the first papillomavirus genome to be completely sequenced (Chen et al., 1982), with those of HPV1 (Danos et al., 1982, 1983) HPV6 (Schwarz et al., 1983), and CRPV (Danos et al., 1984; Giri et al., 1985) completed shortly thereafter. These data showed that distinct papillomaviruses share a similar genetic organization and possess considerable sequence homology. When combined with transcription analysis (Heilman et al., 1982; Engel et al., 1983), the viral genome could be divided into three segments: a noncoding region, a region coding for the nonstructural ("early" [E]) genes, and a region coding for the two viral capsid proteins ("late" [L]). In contrast to earlier expectations, the organization of the papillomavirus genome was distinct from that of SV40 and polyoma, and there was almost no sequence homology between papillomaviruses and polyomaviruses. For example, the papillomavirus E and L genes are transcribed from the same strand, while those of SV40 and polyoma are transcribed from opposite strands. Such differences eventually led to papillomaviruses being designated a separate virus family.

The genetics of BPV1 was studied largely by examining two parameters: rodent cell transformation and the generation of episomal viral DNA. These latter studies were stimulated by the surprising finding that cells transformed by BPV1 contained multiple episomal copies of the viral genome (Law et al., 1981), in contrast to cells transformed by SV40 or adenoviruses, which contained exclusively integrated viral DNA. Cells transformed by BPV1 appeared to be the in vitro counterpart to the underlying fibroblastic portion of BPV1 fibropapillomas in the natural host or the nonepithelial lesions in heterologous hosts. As in nonepithelial lesions in the animal, the L genes were not expressed in tissue culture and did not contribute to transformation or to viral DNA replication. Genetic analysis identified two genes, E5 and E6, with a direct role in morphologic transformation (Groff and Lancaster, 1986; Schiller et al., 1984; Schiller et al., 1986; Yang et al., 1985). E5 was found to transform cells by activating receptor tyrosine kinases (Martin et al., 1989) via the direct activation of PDGF β-receptors (Petti et al., 1991), with a possible contribution by its binding to a component of the vacuolar H^+-ATPase (Goldstein et al., 1991). The E2 gene was found to regulate the expression of other viral E genes, by trans-activation following the binding of E2 protein to cognate binding sites located mainly in the viral noncoding region (Androphy et al., 1987; Spalholz et al., 1985). E1 was shown to be an ATP-dependent helicase required for the replication of viral DNA (Clertant and Seif, 1984; Lusky and Botchan, 1985; Yang et al., 1993; Seo et al., 1993), with E2 having a key ancillary role in viral DNA replication (Mohr et al., 1990; Ustav and Stenlund, 1991; Yang et al., 1991).

Infection of the esophagus by BPV4 represented an important animal papillomavirus model developed in the 1970s (Jarrett et al., 1978). As with CRPV, BPV4 causes benign tumors that sometimes progress to invasive cancer. Unlike CRPV, the BPV4 lesions were located at mucosal, not cutaneous surfaces. The development of esophageal cancer in the infected cattle was related to their eating bracken fern, which contained a carcinogen. This was a natural situation where a nonviral environmental exposure contributed to the carcinogenic progression of a papillomavirus infection. The esophageal cancers did not contain the viral genome, in contrast to the benign lesions (Campo et al., 1985). This form of progression therefore operates via a hit-and-run mechanism, which is unusual for viral-induced cancers in general, and for other papillomavirus-associated cancers in particular.

The study of HPVs also reached a major turning point. Transmission of warts from one individual to another had been achieved in the late 19th century, and the viral etiology of genital and nongenital warts was shown in the first part of the 20th century (Ciuffo, 1907; Serra, 1924; Ullmann, 1923; Variot, 1894). Furthermore, nongenital warts could be induced experimentally by extracts from genital warts and from laryngeal papillomas, suggesting that the responsible viruses might be identical. A particularly noteworthy advance came from the recognition, based on molecular analysis, that there were several HPV genotypes. In one situation, it was recognized that several HPV genotypes caused nongenital cutaneous warts (Gissmann et al., 1977; Orth et al., 1977). In another, it was noted that distinct papillomavirus genotypes were associated with the skin lesions of epidermodysplasia verruciformis (EV) (Orth et al., 1978b), a rare susceptibility to widespread cutaneous warts. Still other HPV types were responsible for genital warts (Gissmann and zur Hausen, 1980; Orth et al., 1978a). These findings implied that there were a large number of HPV genotypes, as has been borne out by subsequent studies.

Studies on EV conducted by Orth and colleagues uncovered a treasure trove of interesting properties of HPVs (Orth, 1978b, 1979, 1986). Some skin lesions in EV progress to squamous cell carcinoma. Analysis of the malignant tumors indicated that the vast majority of them contained either of two closely related HPV genotypes (HPV5 or HPV8), although many HPV genotypes were present in benign EV lesions. These findings represented the first clear evidence that HPV was involved in human cancer. A second important conclusion was that some HPV genotypes had greater malignant potential than others. Although the benign EV lesions were present on both covered and exposed areas of the body, almost all the malignant tumors were found on exposed areas. This observation implied that while HPV infection with HPV5 or HPV8 might be necessary for skin cancer, progression to invasive cancer also required exposure of the lesion to ultraviolet light, a carcinogen known to be active as an initiator and a promoter in skin cancer.

The study of the genital-mucosal HPVs, primarily by zur Hausen and colleagues, led to findings of even greater medical importance. It was noted in the mid-1970s that the histologic appearance of cervical dysplasia, the cellular precursor to cervical cancer, resembled that of viral papillomas (Meisels and Fortin, 1976). This observation, combined with other evidence, led to the suggestion that HPV infection may play a role in cervical carcinogenesis (Meisels and Morin, 1981; zur Hausen et al., 1981). The first genital-mucosal HPV types to be isolated were HPV6 and HPV11 (de Villiers et al., 1981; Gissmann et al., 1983). Although these viruses were present in a high proportion of benign genital warts, related sequences were found in only a minority (~10%) of cervical cancers. A major breakthrough in revealing a role of HPV in cervical cancer came from the identification and molecular cloning of HPV16 (Durst et al., 1983) and HPV18 (Boshart et al., 1984) DNA, two genotypes only distantly related to HPV6/11. It was soon determined that the majority of cervical cancers contained DNA that hybridized under stringent conditions to probes from HPV16 or HPV18. Under less stringent conditions, an even higher proportion of cervical cancers hybridized to these viral DNAs. The findings therefore implicated HPV in a common human cancer. The potential public health implications of these findings gave new prominence to HPV research and attracted many new investigators, particularly medical epidemiologists, to the field. Although some epidemiological studies in the mid-1980s identified major differences in the incidence of HPV16 in cervical cancers, compared with low grade cervical lesions or normal cervices (McCance et al., 1985), others did not. These variable results, which resulted from a combination of false-positive and false-negative HPV DNA data, were overcome by the development of PCR primers for the reliable detection of small quantities of genital-mucosal HPV DNA in cervical smears (Resnick et al., 1990; van den Brule et al., 1990), leading to cervical HPV infection being recognized as the predominant risk factor for cervical cancer and its precursor lesions (Schiffman, 1992), although HPV infection is not sufficient for cervical cancer. Further studies indicated that virtually all cases of cervical cancer were initiated by infection with high-risk HPV types (Walboomers et al., 1999).

A series of observations about the oncogenes of HPV16 and 18 and the biochemical activities of their encoded proteins also had an enormous impact on the field. It was shown initially in cervical cancer cell lines that the viral DNA was often integrated, in a deleted form, into the host genome, in contrast to the episomal nature of the viral genome in benign lesions (Schwarz et al., 1985). Although integration was relatively random with respect to the host DNA, integration was not random with respect to the viral genome. In the tumors, the noncoding region of the genome along with the E6 and E7 genes were preferentially retained, and both genes were expressed in the tumors (Schwarz et al., 1985; Smotkin and Wettstein, 1986; Schneider-Gadicke and Schwarz, 1986). These results provided a strong clue that E6 and E7 were major transforming genes of the virus. The expression of E6 and E7 also appeared to be necessary for cervical cancer cell lines to maintain their transformed phenotype (Thierry and Yaniv, 1987; von Knebel Doeberitz et al., 1988; Hwang et al., 1993).

HPV6 and HPV11 were almost never found alone in cervical cancer, which gave rise to the notion that these genital-mucosal HPVs were "low-risk" types, while those HPVs that were found regularly in cervical cancer, such as HPV16 and HPV18, were "high-risk" types. Although low-risk and high-risk HPVs were able to stimulate human keratinocyte growth, the high-risk HPVs possessed additional biological properties, including the ability to immortalize primary human keratinocytes (Durst et al., 1987; Pirisi et al., 1987; Schlegel et al., 1988; Woodworth et al., 1988, 1989; Kaur and McDougall, 1989). Furthermore, efficient keratinocyte immortalization could be attributed to a cooperative activity between high-risk E6 and high-risk E7 (Hawley-Nelson et al., 1989; Munger et al., 1989). Consistent with the notion that HPV infection is necessary, but not sufficient, for malignant progression, human keratinocytes immortalized by HPV16 DNA were not tumorigenic in experimental animals, but malignant transformation could be induced by continued passage or transfer of an activated *ras* oncogene (DiPaolo et al., 1989; Pecoraro et al., 1991). In transgenic mice, co-expression of HPV-16 E6 and E7 in epithelial cells induced papillomas, epithelial hyperplasia, and carcinoma (Lambert et al., 1993; Arbeit et al., 1994).

Insight into the mechanisms underlying these properties of high-risk E6 and E7 genes came from identification of biochemical activities displayed by their protein products. First, it was shown that high-risk E7, but not low-risk E7, bound and inactivated the pRb tumor suppressor protein (Dyson et al., 1989; Munger et al., 1989; Gage et al., 1990; Chellappan et al., 1992), a property originally identified in the adenovirus E1A protein and also in the SV40 large T antigen. It was then determined that high-risk E6, but not low-risk E6, formed a complex with the p53 tumor suppressor protein (Werness et al., 1990), which is also targeted by the other small DNA tumor viruses. p53 activity was reduced because E6 binding led to p53 degradation via a ubiquitin-dependent process involving the formation of a trimolecular complex that in addition to E6 and p53 included E6AP, a newly identified protein that was found to be a ubiquitin ligase (Scheffner et al., 1990, 1993). These results were correlated with the finding that the p53 and Rb genes were wild type, rather than mutated, in human cervical cancer cell lines associated with high-risk HPV, while they were frequently mutated in HPV-negative lines (Scheffner et al., 1991; Wrede et al., 1991).

Thus, by the early 1990s, HPVs had been identified as the etiolologic agent for cervical cancer, and the continued expression of E6 and E7 in cancer cells was recognized, as were some of the main biochemical activities of the viral oncoproteins. The combination of the medical importance of HPVs and their mechanisms of cell transformation attracted widespread interest, from basic scientists to clinicians. It is noteworthy that the achievements of papillomavirus research from the 1970s through the early 1990s occurred despite the inability to propagate the virus in culture. Research during this period set the stage for many subsequent important advances, including the use of HPV DNA testing in cervical cancer screening, the identification of HPV in normal individuals and in a wider spectrum of cancers, the efforts to interfere with HPV infection by vaccination and other modalities, the recognition that most viral genes, including E6 and E7, are multifunctional, and the development of in vitro assays for papillomavirus replication. Many of these advances are described in detail elsewhere in this volume.

References

Androphy, A.J., Lowy, D.R., and Schiller, J.T. (1987). Bovine papillomavirus E2 trans-activating gene product binds to specific sites in papillomavirus DNA. *Nature* 325:70–73.

Arbeit, J.M., Munger, K., Howley, P.M., and Hanahan, D. (1994). Progressive squamous epithelial neoplasia in K14-human papillomavirus type 16 transgenic mice. *J. Virol.* 68:4358–4368.

Black, P.H., Hartley, J.W., Rowe, W.P., and Huebner, R.J. (1963). Transformation of bovine tissue culture cells by bovine papilloma virus. *Nature* 199:1016–1018.

Boshart, M., Gissmann, L., Ikenberg, H., Kleinheinz, A., Scheurlen, W., and zur Hausen, H. (1984). A new type of papillomavirus DNA, its presence in genital cancer biopsies and in cell lines derived from cervical cancer. *EMBO J.* 3(5):1151–1157.

Breitburd, F., Salmon, J., and Orth, G. (1997). The rabbit viral skin papillomas and carcinomas: a model for the immunogenetics of HPV-associated carcinogenesis. *Clin. Dermatol.* 15(2):237–247.

Campo, M.S., Moar, M.H., Sartirana, M.L., Kennedy, I.M., and Jarrett, W.F. (1985). The presence of bovine papillomavirus type 4 DNA is not required for the progression to, or the maintenance of, the malignant state in cancers of the alimentary canal in cattle. *EMBO J.* 4(7):1819–1825.

Chellappan, S., Kraus, V.B., Kroger, B., Munger, K., Howley, P.M., Phelps, W.C., and Nevins, J.R. (1992). Adenovirus E1A, simian virus 40 tumor antigen, and human papillomavirus E7 protein share the capacity to disrupt the interaction between transcription factor E2F and the retinoblastoma gene product. *Proc. Natl. Acad. Sci. U S A* 89:4549–4553.

Chen, E.Y., Howley, P.M., Levinson, A.D., and Seeburg, P.H. (1982). The primary structure and genetic organization of the bovine papillomavirus type 1 genome. *Nature* 299(5883):529–534.

Ciuffo, G. (1907). Innesto positivo con filtrate di verruca vulgare. *Giornale Italiano delle Malattie Veneree* 42:12–17.

Clertant, P., and Seif, I. (1984). A common function for polyoma virus large-T and papillomavirus E1 proteins? *Nature* 311(5983):276–279.

Crawford, L.V. (1969). Nucleic acids of tumor viruses. *Adv Virus Res* 14:89–152.

Creech, G.T. (1929). Experimental studies of the etiology of common warts in cattle. *J. Agric. Res.* 39:723–737.

Danos, O., Engel, L.W., Chen, E.Y., Yaniv, M., and Howley, P.M. (1983). Comparative analysis of the human type 1a and bovine type 1 papillomavirus genomes. *J. Virol.* 46(2):557–566.

Danos, O., Giri, I., Thierry, F., and Yaniv, M. (1984). Papillomavirus genomes: sequences and consequences. *J. Invest. Dermatol.* 83(1 Suppl.):7s–11s.

Danos, O., Katinka, M., and Yaniv, M. (1982). Human papillomavirus 1a complete DNA sequence: a novel type of genome organization among papovaviridae. *Embo. J.* 1(2):231–236.

de Villiers, E.-M., Gissmann, L., and zur Hausen, H. (1981). Molecular cloning of viral DNA from human genital warts. *J. Virol.* 40:932–935.

DeMonbreun, W.A., and Goodpasture, E.W. (1932). Infectious oral papillomatosis of dogs. *Am. J. Pathol.* 8:43–55.

DiPaolo, J.A., Woodworth, C.D., Popescu, N.C., Notario, V., and Doniger, J. (1989). Induction of human cervical squamous cell carcinoma by sequential transfection with human papillomavirus 16 DNA and viral Harvey ras. *Oncogene* 4(4):395–399.

Dollard, S.C., Wilson, J.L., Demeter, L.M., Bonnez, W., Reichman, R.C., Broker, T.R., and Chow, L.T. (1992). Production of human papillomavirus and modulation of the infectious program in epithelial raft cultures. *Genes Dev.* 6:1131–1142.

Durst, M., Gissmann, L., Ikenberg, H., and zur Hausen, H. (1983). A papillomavirus DNA from a cervical carcinoma and its prevalence in cancer biopsy samples from different geographic regions. *Proc. Natl. Acad. Sci. U S A* 80(12):3812–3815.

Durst, M., Dzarlieva-Petrusevska, R.T., Boukamp, P., Fusenig, N.E., and Gissmann, L. (1987). Molecular and cytogenetic analysis of immortalized human primary keratinocytes obtained after transfection with human papillomavirus type 16 DNA. *Oncogene* 1:251–256.

Dvoretzky, I., Shober, R., Chattopadhyay, S.K., and Lowy, D.R. (1980). Focus assay in mouse cells for bovine papillomavirus. *Virology* 103:369–375.

Dyson, N., Howley, P.M., Munger, K., and Harlow, E. (1989). The human papillomavirus-16 E7 oncoprotein is able to bind the retinoblastoma gene product. *Science* 243:934–937.

Engel, L.W., Heilman, C.A., and Howley, P.M. (1983). Transcriptional organization of bovine papillomavirus type 1. *J. Virol.* 47:516–528.

Evans, C.A., Gorman, L.R., Ito, Y., and Weiser, R.S. (1962). Antitumor immunity in the Shope papilloma–carcinoma complex of rabbits. I. Papilloma regression induced by homologous and autologous tissue vaccines. *J. Nat. Cancer Inst.* 29:277–285.

Evans, C.A., and Ito, Y. (1966). Antitumor immunity in the Shope papilloma–carcinoma complex of rabbits. III. Response to reinfection with viral nucleic acid. *J. Nat. Cancer Inst.* 36:1161–1166.

Findlay, G.M. (1930). Warts. Chapter XVIII. *A System of Baceriology in Relation to Medicine*. Vol. 7, pp. 252–258. London: Great Britain Medical Research Council.

Friedmann, J.C., Levy, J.P., Lasneret, J., Thomas, M., Boiron, M., and Bernard, J. (1963). Induction of subcutaneous fibromas in the golden hamster by inoculation of bovine papilloma acellular extracts. *C R Hebd Seances Acad Sci* 257:2328–2331.

Gage, J.R., Meyers, C., and Wettstein, F.O. (1990). The E7 proteins of the nononcogenic human papillomavirus type 6b (HPV-6b) and of the oncogenic HPV-16 differ in retinoblastoma protein binding and other properties. *J. Virol.* 64:723–730.

Giri, I., Danos, O., and Yaniv, M. (1985). Genomic structure of the cottontail rabbit (Shope) papillomavirus. *Proc. Natl. Acad. Sci. U S A* 82(6):1580–1584.

Gissmann, L., Pfister, H., and Zur Hausen, H. (1977). Human papilloma viruses (HPV): characterization of four different isolates. *Virology* 76(2):569–580.

Gissmann, L., Wolnik, L., Ikenberg, H., Koldovsky, U., Schnurch, H.G., and zur Hausen, H. (1983). Human papillomavirus types 6 and 11 DNA sequences in genital and laryngeal papillomas and in some cervical cancers. *Proc. Natl. Acad. Sci. U S A* 80(2):560–563.

Gissmann, L., and zur Hausen, H. (1980). Partial characterization of viral DNA from human genital warts (Condylomata acuminata). *Int. J. Cancer* 25(5):605–609.

Goldstein, D.J., Finbow, M.E., Andresson, T., McLean, P., Smith, K., Bubb, V., and Schlegel, R. (1991). Bovine papillomavirus E5 oncoprotein binds to the 16K component of vacuolar H(+)-ATPases. *Nature* 352(6333):347–349.

Grodzicker, T., and Hopkins, N. (1981). Origins of contemporary DNA tumor virus research. In J. Tooze (ed.): *DNA Tumor Viruses, Molecular Biology of Tumor Viruses*, 2nd ed. Pp. 1–59. Cold Spring Harbor Laboratory.

Groff, D.E., and Lancaster, W.D. (1986). Genetic analysis fo the 3' early region transformation and replication functions of bovine papillomavirus type 1. *Virology* 150:221–230.

Gross, L. (1983). Papillomas, warts, and related neooplasms in rabbits, dogs, horses, cattle, hamsters and in man. *Oncogenic Viruses*, 3rd ed., 2 vols. Oxford: Pergamon Press, Vol. 1, pp. 48–76.

Hawley-Nelson, P., Vousden, K.H., Hubbert, N.L., Lowy, D.R., and Schiller, J.T. (1989). HPV16 E6 and E7 proteins cooperate to immortalize primary human foreskin keratinocytes. *EMBO J.* 8:3905–3910.

Heilman, C.A., Engel, L., Lowy, D.R., and Howley, P.M. (1982). Virus-specific transcription in bovine papillomavirus-transformed mouse cells. *Virology* 119:22–34.

Hwang, E.S., Riese II, D.J., Settleman, J., Nilson, L.A., Honig, J., Flynn, S., and DiMaio, D. (1993). Inhibition of cervical carcinoma cell line proliferation by the introduction of a bovine papillomavirus regulatory gene. *J. Virol.* 67:3720–3729.

Ito, Y., and Evans, C.A. (1961). Induction of tumors in domestic rabbits with nucleic acid preparations from partially purified Shope papilloma virus and from extracts of papillomas of domestic and cottontail rabbits. *J. Exp. Med.* 114:485–500.

Ito, Y., and Evans, C.A. (1965). Tumorigenic nucleic acid extracts from tissues of a transplantable carcinoma, Vx7. *J. Natl. Cancer Inst.* 34:431–437.

Jarrett, W.F., McNeil, P.E., Grimshaw, W.T., Selman, I.E., and McIntyre, W.I. (1978). High incidence area of cattle cancer with a possible interaction between an environmental carcinogen and a papilloma virus. *Nature* 274(5668):215–217.

Kaur, P. and McDougall, J.K. (1989). HPV-18 immortalization of human keratinocytes. *Virology* 173:302–310.

Kidd, J.G., Beard, J.W., and Rous, P. (1936). Serological reactions with a virus causing rabbit papillomas which become cancerous. I. Tests of the blood of animals carrying the papillomas. *J. Exp. Med.* 64:63–77.

Kidd, J.G., and Rous, P. (1940). A transplantable rabbit carcinoma originating in a virus-induced papilloma and containing the virus in masked or altered form. *J. Exp. Med.* 71:813–838.

Kreider, J.W. (1963). Studies on the mechanism responsible for the spontaneous regression of the shope rabbit papilloma. *Cancer Res.* 23:1593–1599.

Kreider, J.W., and Bartlett, G.L. (1981). The Shope papilloma–carcinoma complex of rabbits: a model system of neoplastic progression and spontaneous regression. *Adv. Cancer Res.* 35:81–110.

Lambert, P.F., Pan, H., Pitot, H.C., Liem, A., Jackson, M., and Griep, A.E. (1993). Epidermal cancer associated with expression of human papillomavirus type 16 E6 and E7 oncogenes in the skin of transgenic mice. *Proc. Natl. Acad. Sci. U S A* 90(12):5583–5587.

Lancaster, W.D., and Olson, C. (1982). Animal papillomaviruses. *Microbiol. Rev.* 46:191–207.

Lancaster, W.D., Olson, C., and Meinke, W. (1977). Bovine papilloma virus: presence of virus-specific DNA sequences in naturally occurring equine tumors. *Proc. Natl. Acad. Sci. U S A* 74(2):524–528.

Law, M.F., Lowy, D.R., Dvoretzky, I., and Howley, P.M. (1981). Mouse cells transformed by bovine papillomavirus contain only extrachromosomal viral DNA sequences. *Proc. Natl. Acad. Sci. U S A* 78:2727–2731.

Lusky, M., and Botchan, M.R. (1985). Genetic analysis of bovine papillomavirus type 1 trans-acting replication factors. *J. Virol.* 53(3):955–965.

M'Fadyean, J., and Hobday, F. (1898). Note on experimental transmission of warts in the dog. *J. Comp. Pathol. Ther.* 11:341–344.

Magalhaes, O. (1920). Verruga dos bovideos. *Braseil-Medico* 34:430–431.

Martin, P., Vass, W.C., Schiller, J.T., Lowy, D.R., and J., V.T. (1989). The bovine papillomavirus E5 transforming protein can stimulate the transforming activity of EGF and CSF-1 receptors. *Cell* 59(1):21–32.

McCance, D.J., Campion, M.J., Clarkson, P.K., Chesters, P.M., Jenkins, D., and Singer, A. (1985). Prevalence of human papillomavirus type 16 DNA sequences in cervical intraepithelial neoplasia and invasive carcinoma of the cervix. *Br. J. Obstet. Gynaecol.* 92(11):1101–1105.

McCance, D.J., Kopan, R., Fuchs, E., and Laimins, L.A. (1988). Human papillomavirus type 16 alters epithelial cell differentiation *in vitro. Proc. Natl. Acad. Sci. U S A* 85:7169–7173.

McMichael, H. (1967). Inhibition by methylprednisolone of regression of the Shope rabbit papilloma. *J. Natl. Cancer Inst.* 39(1):55–65.

Meisels, A. and Fortin, R. (1976). Condylomatous lesions of the cervix and vagina. I. Cytologic patterns. *Acta Cytol.* 20:505–509.

Meisels, A. and Morin, C. (1981). Human papillomavirus and cancer of the uterine cervix. *Gynecol. Oncol.* 12:S111–S123.

Meyers, C., Frattini, M.G., Hudson, J.B., and Laimins, L.A. (1992). Biosynthesis of human papillomavirus from a continuous cell line upon epithelial differentiation. *Science* 257(5072):971–973.

Mohr, I.J., Clark, R., Sun, S., Androphy, E.J., MacPherson, P., and Botchan, M.R. (1990). Targeting the E1 replication protein to the papillomavirus origin of replication by complex formation with the E2 transactivator. *Science* 250:1694–1699.

Moore, D.H., Stone, R.S., Shope, R.E., and Gelber, D. (1959). Ultrastructure and site of formation of rabbit papilloma virus. *Proc. Soc. Exp. Biol. Med.* 101(3):575–578.

Munger, K., Phelps, W.C., Bubb, V., Howley, P.M., and Schlegel, R. (1989). The E6 and E7 genes of the human papillomavirus type 16 together are necessary and sufficient for transformation of primary human keratinocytes. *J. Virol.* 63:4417–4421.

Munger, K., Werness, B.A., Dyson, N., Phelps, W.C., Harlow, E., and Howley, P.M. (1989). Complex formation of human papillomavirus E7 proteins with the retinoblastoma tumor suppressor gene product. *EMBO J.* 8:4099–4105.

Noyes, W.F. (1959). Studies on the Shope rabbit papilloma virus. II. The location of infective virus in papillomas of the cottontail rabbit. *J. Exp. Med.* 109(4):423–428.

Noyes, W.F., and Mellors, R.C. (1957). Fluorescent antibody detection of the antigens of the Shope papilloma virus in papillomas of the wild and domestic rabbit. *J. Exp. Med.* 106(4):555–562.

Olson, C., Jr., and Cook, R.H. (1951). Cutaneous sarcoma-like lesions of the horse caused by the agent of bovine papilloma. *Proc. Soc. Exp. Biol. Med.* 77(2):281–284.

Orth, G. (1986). Epidermodysplasia verruciformis: a model for understanding the oncogenicity of human papillomaviruses. *Ciba Found Symp.* 120:157–174.

Orth, G., Favre, M., and Croissant, O. (1977). Characterization of a new type of human papillomavirus that causes skin warts. *J. Virol.* 24(1):108–120.

Orth, G., Favre, M., Jablonska, S., Brylak, K., and Croissant, O. (1978a). Viral sequences related to a human skin papillomavirus in genital warts. *Nature* 275(5678):334–336.

Orth, G., Jablonska, S., Favre, M., Croissant, O., Jarzabek-Chorzelska, M., and Rzesa, G. (1978b). Characterization of two types of human papillomaviruses in lesions of epidermodysplasia verruciformis. *Proc. Natl. Acad. Sci. U S A* 75(3):1537–1541.

Orth, G., Jablonska, S., Jarzabek-Chorzelska, M., Obalek, S., Rzesa, G., Favre, M., and Croissant, O. (1979). Characteristics of the lesions and risk of malignant conversion associated with the type of human papillomavirus involved in epidermodysplasia verruciformis. *Cancer Res.* 39(3):1074–1082.

Parsons, R.J., and Kidd, J.G. (1936a). Tissue affinity of Shope papilloma virus. *Proc. Soc. Exp. Biol. Med.* 35:438–441.

Parsons, R.J., and Kidd, J.G. (1936b). A virus causing oral papillomatosis in rabbits. *Proc. Soc. Exp. Biol. Med.* 35:441–443.

Parsons, R.J., and Kidd, J.G. (1943). Oral papillomatosis of rabbits: a virus disease. *J. Exp. Med.* 77:233–250.

Pecoraro, G., Lee., M., Morgan, D., and Defendi, V. (1991). Evolution of *in vitro* transformation and tumorigenesis of HPV16 and HPV18 immortalized primary cervical epithelial cells. *Am. J. Pathol.* 138:1–8.

Penberthy, J. (1898). Contagious warty tumours in dogs. *J. Comp. Pathol. Ther.* 11:363–365.

Petti, L., Nilson, L.A., and DiMaio, D. (1991). Activation of the platelet-derived growth factor receptor by the bovine papillomavirus E5 transforming protein. *EMBO J.* 10(4):845–855.

Pirisi, L., Yasumoto, S., Feller, M., Doniger, J., and DiPaolo, J.A. (1987). Transformation of human fibroblasts and keratinocytes with human papillomavirus type 16 DNA. *J. Virol.* 61:1061–1066.

Resnick, R.M., Cornelissen, M.T., Wright, D.K., Eichinger, G.H., Fox, H.S., ter Schegget, J., and Manos, M.M. (1990). Detection and typing of human papillomavirus in archival cervical cancer specimens by DNA amplification with consensus primers. *J. Natl. Cancer Inst.* 82(18):1477–1484.

Rous, P., and Beard, J. (1935). The progression to carcinoma of virus induced rabbit papillomas (Shope). *J. Exp. Med.* 2:523–545.

Rous, P., and Friedewald, W.F. (1944). The effect of chemical carcinogens on virus induced rabbit papillomas. *J.Exp.Med.* 79:511–538.

Rous P., and Kidd, J.G. (1938). The carcinogenic effect of a papilloma virus on the tarred skin of rabbits. I Description of the phenomenon. *J.Exp.Med.* 67:399–422.

Rowson, K.E.K., and Mahy, B.W.J. (1967). Human papova (wart) virus. *Bacteriol. Rev.* 31:110–131.

Scheffner, M., Huibregtse, J.M., Vierstra, R.D., and Howley, P.M. (1993). The HPV-16 E6 and E6-AP complex functions as a ubiquitin–protein ligase in the ubiquitination of p53. *Cell* 75(3):495–505.

Scheffner, M., Munger, K., Byrne, J.C., and Howley, P.M. (1991). The state of the p53 and retinoblastoma genes in human cervical carcinoma cell lines. *Proc. Natl. Acad. Sci. U S A* 88:5523–5527.

Scheffner, M., Werness, B.A., Huibregtse, J.M., Levine, A.J., and Howley, P.M. (1990). The E6 oncoprotein encoded by human papillomavirus types 16 and 18 promotes degradation of p53. *Cell* 63:1129–1136.

Schiffman, M.H. (1992). Recent progress in defining the epidemiology of human papillomavirus infection and cervical neoplasia. *J. Natl. Cancer Inst.* 84:394–398.

Schiller, J.T., Vass, W.C., and Lowy, D.R. (1984). Identification of a second transforming region in bovine papillomavrirus DNA. *Proc. Natl. Acad. Sci. U S A* 81:7880–7884.

Schiller, J.T., Vass, W.C., Vousden, K.H., and Lowy, D.R. (1986). E5 open reading frame of bovine papillomavirus type 1 encodes a transforming gene. *J. Virol.* 57:1–6.

Schlegel, R., Phelps, W.C., Zhang, Y.L., and Barbosa, M. (1988). Quantitative keratinocyte assay detects two biological activities of human papillomavirus DNA and identifies viral types associated with cervical carcinoma. *EMBO J.* 7(10):3181–3187.

Schneider-Gadicke, A., and Schwarz, E. (1986). Different human cervical carcinoma cell lines show similar transcription patterns of human papillomavirus type 18 early genes. *EMBO J.* 5:2285–2292.

Schwarz, E., Durst, M., Demankowski, C., Lattermann, O., Zech, R., Wolfsperger, E., Suhai, S., and zur Hausen, H. (1983). DNA sequence and genome organization of genital human papillomavirus type 6b. *EMBO J.* 2(12):2341–2348.

Schwarz, E., Freese, U.K., Gissmann, L., Mayer, W., Roggenbuck, B., Stremlau, A., and zur Hausen, H. (1985). Structure and transcription of human papillomavirus sequences in cervical carcinoma cells. *Nature* 314(6006):111–114.

Seo, Y.S., Muller, F., Lusky, M., and Hurwitz, J. (1993). Bovine papilloma virus (BPV)-encoded E1 protein contains multiple activities required for BPV DNA replication. *Proc. Natl. Acad. Sci. U S A* 90:702–706.

Serra, A. (1924). Studi sul virus della verruca, del papilloma, del condiloma acuminato. *Giornale Italiano delle Malattie Veneree e delle Pelle* 65:1808–1814.

Shope, R.E. (1933). Infectious papillomatosis of rabbits (with a note on the histopathology by E. W. Hurst). *J. Exp. Med.* 58:607–624.

Shope, R.E. (1937). Immunization of rabbits to infectious papillomatosis. *J. Exp. Med.* 65:219–231.

Smotkin, D., and Wettstein, F.O. (1986). Transcription of human papillomavirus type 16 early genes in a cervical cancer and a cancer-derived cell line and identification of the E7 protein. *Proc. Natl. Acad. Sci. U S A* 83:4680–4686.

Spalholz, B.A., Yang, Y.C., and Howley, P.M. (1985). Transactivation of a bovine papilloma virus transcriptional regulatory element by the E2 gene product. *Cell* 42:183–191.

Stewart, S.E., Eddy, B.E., and Borgese, N. (1958). Neoplasms in mice inoculated with a tumor agent carried in tissue culture. *J. Natl. Cancer Inst.* 20(6):1223–1243.

Stone, R.S., Shope, R.E., and Moore, D.H. (1959). Electron microscope study of the development of the papilloma virus in the skin of the rabbit. *J. Exp. Med.* 110:543—546.

Sweet, B.H., and Hilleman, M.R. (1960). The vacuolating virus, S.V. 40. *Proc. Soc. Exp. Biol. Med.* 105:420–427.

Syverton, J.T. (1952). The pathogenesis of the rabbit papilloma-to-carcinoma sequence. *Ann. N Y Acad. Sci.* 54(6):1126–1140.

Syverton, J.T., and Berry, G.P. (1935). Carcinoma in the cottontail rabbit following spontaneous virus papilloma (Shope). *Proc. Soc. Exp. Biol. Med.* 33:399–400.

Thierry, F., and Yaniv, M. (1987). The BPV1-E2 transactivating protein can be either an activator or a repressor of the HPV18 regulatory region. *EMBO J.* 6:3391–3397.

Thomas, M., Boiron, M., Tanzer, J., Levy, J.P., and Bernard, J. (1964). *In vitro* transformation of mice cells by bovine papilloma virus. *Nature* 202:709–710.

Ullmann, E.V. (1923). On the aetiology of the laryngeal papilloma. *Acta Oto-Laryng* 5:317–334.

Ustav, M., and Stenlund, A. (1991). Transient replication of BPV-1 requires two viral polypeptides encoded by the E1 and E2 open reading frames. *EMBO J.* 10:449–457.

van den Brule, A.J., Snijders, P.J., Gordijn, R.L., Bleker, O.P., Meijer, C.J., and Walboomers, J.M. (1990). General primer-mediated polymerase chain reaction permits the detection of sequenced and still unsequenced human papillomavirus genotypes in cervical scrapes and carcinomas. *Int. J. Cancer* 45(4):644–649.

Variot, G. (1894). Un cas d'incoluation experimentale des verrues de l'enfant a l'homme. *J. Clin. Ther. Infant* 2:529–531.

von Knebel Doeberitz, M., Oltersdorf, T., Schwarz, E., and Gissmann, L. (1988). Correlation of modified human papilloma virus early gene expression with altered growth properties in C4-1 cervical carcinoma cells. *Cancer Res.* 48(13):3780–3786.

Walboomers, J.M., Jacobs, M.V., Manos, M.M., Bosch, F.X., Kummer, J.A., Shah, K.V., Snijders, P.J., Peto, J., Meijer, C.J., and Munoz, N. (1999). Human papillomavirus is a necessary cause of invasive cervical cancer worldwide. *J. Pathol.* 189:12–19.

Werness, B.A., Levine, A.J., and Howley, P.M. (1990). Association of human papillomavirus types 16 and 18 E6 proteins with p53. *Science* 248:76–79.

Williams, M.G., Howatson, A.F., and Almeida, J.D. (1961). Morphological characterization of the virus of the human common wart (verruca vulgaris). *Nature* 189:895–897.

Woodworth, C.D., Bowden, P.E., Doniger, J., Pirisi, L., Barnes, W., Lancaster, W.D., and DiPaolo, J.A. (1988). Characterization of normal human exocervical epithelial cells immortalized *in vitro* by papillomavirus types 16 and 18 DNA. *Cancer Res.* 48:4620–4628.

Woodworth, C.D., Doniger, J., and DiPaolo, J.D. (1989). Immortalization of human foreskin keratinocytes by various human papillomavirus DNAs corresponds to their association with cervical carcinoma. *J. Virol.* 63:159–164.

Wrede, D., Tidy, J.A., Crook, T., Lane, D., and Vousden, K.H. (1991). Expression of RB and p53 proteins in HPV-positive and HPV-negative cervical carcinoma cell lines. *Mol. Carcinog* 4(3):171–175.

Yang, Y.C., Okayama, H., and Howley, P.M. (1985). Bovine papillomavirus contains multiple transforming genes. *Proc. Natl. Acad. Sci. U S A* 82(4):1030–1034.

Yang, L., Li, R., Mohr, I.J., Clark, R., and Botchan, M.R. (1991). Activation of BPV-1 replication *in vitro* by the transcription factor E2. *Nature* 353:628–632.

Yang, L., Mohr, I., Fouts, E., Lim, D.A., Nohaile, M., and Botchan, M. (1993). The E1 protein of bovine papilloma virus 1 is an ATP-dependent DNA helicase. *Proc. Natl. Acad. Sci. U S A* 90:5086–5090.

zur Hausen, H., de Villiers, E.M., and Gissmann, L. (1981). Papillomavirus infections and human genital cancer. *Gynecol. Oncol.* 12:S124–S128.

zur Hausen, H., and de Villiers, E.M. (1994). Human papillomaviruses. *Annu. Rev. Microbiol.* 48:427–447.

3
Phylogeny and Typing of Papillomaviruses

Hans-Ulrich Bernard
Department of Molecular Biology and Biochemistry, University of California, Irvine, CA 92697

3.1. Introduction

Papillomaviruses have circular double-stranded DNA genomes with sizes close to 8 kb. Characterization of these genomes has always been the foundation for papillomavirus nomenclature and taxonomy, as there are neither reliable cell culture systems nor serological markers for these viruses. Traditionally, papillomavirus DNA isolates that were significantly different from any other isolate were called "types," and all papillomaviruses together were lumped with the polyomaviruses in one taxonomic family, the papovaviridae. The affiliation of papillomaviruses with polyomaviruses has been discontinued due to the molecular findings of the last 25 years. Phylogenetic comparisons of the increasingly large number of papillomavirus isolates were the basis to establish a taxonomy with multiple layers, which are now officially recognized by the International Committee on Taxonomy of Viruses. Papillomaviruses form a family of their own, papillomaviridae. Within this family, each major phylogenetic branch formed by remotely related papillomavirus types is referred to as a "genus," while groups of closely related types are considered as "species." Papillomavirus isolates are recognized as separate "types" when the nucleotide sequence of their L1 gene differs from every other type by at least 10 percent. The term "subtype," frequently erroneously confused with the term "type," refers to the very rare isolates that differ by 2 to 10 percent from the sequence of any type. Independent isolates of the same type show minor nucleotide differences and are called "variants." Papillomavirus genomes evolve very slowly, and the phylogenetic tree of all known animal and human papillomavirus types appears to represent a evolutionary history as ancient as that of their hosts, namely, more than 100 million years in the case of all vertebrate papillomaviruses and several hundred thousand years in the case of all the variants of human papillomavirus types. On the taxonomic level of genera, there is only limited correlation between taxonomic classification and molecular biological and pathogenic properties, but on lower taxonomic levels, the papillomavirus types that form one species have similar biological and pathogenic properties. Taxonomy forms the foundation for HPV detection and typing by DNA diagnosis, which is based in basic research on the polymerase chain reaction with consensus primers, and in clinical studies on several commercial kits based on various hybridization and amplification reactions.

TABLE 3.1. Principles of the Classification of Papillomaviruses: The Taxa, Their Definition, Examples, and Explanations

Taxon	Definition	Example	Explanations
Family	The highest taxon to classify most viruses	Papillomaviridae	All papillomviruses have similar genome organization, no other viruses with significant homologies to papillomaviruses known
Genus	Formerly called "major branches" and "supergroups," 40–55% nucleotide sequence diversity to members of other genera	Alpha-papillomaviruses	While genera are defined just by sequence analysis; members of one genus often share molecular, but not necessarily clinical properties
Species	Fomerly called "minor branches" and "groups," 30–40% nucleotide sequence diversity to members of other species	Human papillomavirus species 9 (which includes HPV types 16, 31, 33, 35, 52, 58, 67)	Members of one species often share molecular and clinical properties; the term "species" was introduced at this taxonomic level to conform with ICTV regulations; under evolutionary aspects, HPV types behave like species
Type	The traditional phylogenetic taxon, 10–25% nucleotide sequence diversity to other types	HPV-16	Papillomavirus types are the taxonomic level to which most molecular, etiological and epidemiological studies refer
Subtype	2–10% nucleotide sequence diversity to other subtypes	HPV-44/55, HPV-68A and B	For unknown reasons, very few HPV isolates fulfill this definition
Phylogenetic groups of variants	Branches with variants in intratype phylogenetic studies	Af1, Af2, AA, As, E, NA branches	Within particular HPV types, related variants (i.e., a branch) may have unique molecular and clinical properties
Variant	Any isolate that has up to 2% nucleotide differences from the original HPV type isolate, the reference or prototype clone	Unregulated ad hoc abbreviations	Phylogenetically interesting as they coevolved with ethnic groups, possibly no clinical relevance of individual variants

3.2.   Papillomaviridae: A Taxon Separate from the Polyomaviruses

Traditionally, papillomaviruses had been placed with the polyomaviruses such as the simian virus 40 (SV40) in the family papovaviridae based on the properties that could be observed in the 1970s, namely, electronmicroscopically similar viral capsids that lack envelopes and double-stranded circular DNA genomes. Research during the 1980s revealed that polyoma- and papillomaviruses have different genome sizes (5 kb vs. 8 kb) and different genome organizations, since early and late genes face one another in the polyomaviruses but are oriented in the same direction in the papillomavirues. Lastly, there are no homologies between the proteins of papilloma- and polyomaviruses, with the exception of a small domain in the E1 gene and the T-antigen gene, respectively (Clertant and Seif, 1984), which is associated with the helicase activity of these proteins. While it was therefore obvious since the mid-1980s that papillomaviruses and polyomaviruses are unrelated and should be considered as two separate families, formal recognition of the family "papillomaviridae" by the International Committee on Taxonomy of Viruses (ICTV) was confirmed only recently (de Villiers et al., 2004). The taxonomy of papillomaviruses is summarized in Table 3.1, which also contains definitions and arguments for the hierarchy of taxa that will be explained in the following paragraphs of this chapter.

3.3.   Novel Attempts to Improve the Nomenclature of Groups of Papillomavirus (PV) Types

3.3.1.   The Genus

PV genomes were traditionally described as "types" (see below). There are today more than 118 formally described animal and human PV types and observations of hundreds of additional PV genomes that would qualify to become types after isolation and formal description of their complete genomes. If all mammals and possibly other vertebrates would be studied as intensively as humans for infection by PVs, the number of PV types would become without doubt virtually unlimited. It was therefore desirable to establish classification schemes to place new PV types into a taxonomic order. Since the early 1990s, efforts have been made to base PV taxonomy on phylogenetic algorithms that compare nucleotide sequences. This research revealed that the analysis of whole PV genomes, individual PV genes, or even PV gene segments led to very similar phylogenetic trees (Chan et al., 1992, 1995; Van Ranst et al., 1992; Farmer et al., 1995; Matsukura and Sugase, 2001; de Villiers et al., 2004). From this one has to conclude that PV genomes do not recombine and that each homologous and sufficiently informative nucleotide segment suffices to document the evolutionary history of each virus and to establish its relationship with all other PV types.

When one compares the nucleotide sequences of a large number of PV types, one observes hierarchical branching patterns as those shown in Figure 3.1. In order to describe these patterns, previous publications used such terms as "major branches" and "minor branches" or "supergroups" and "groups" (Chan et al., 1995; Farmer et al., 1995). These phylogenetic trees were initially considered confusing, as they led to taxonomic affiliations that did not necessarily unite PV types with similar biological or pathogenic properties. For example, one major branch was formed by all HPV types that were originally described from genital or mucosal lesions, and therefore called "genital HPVs."

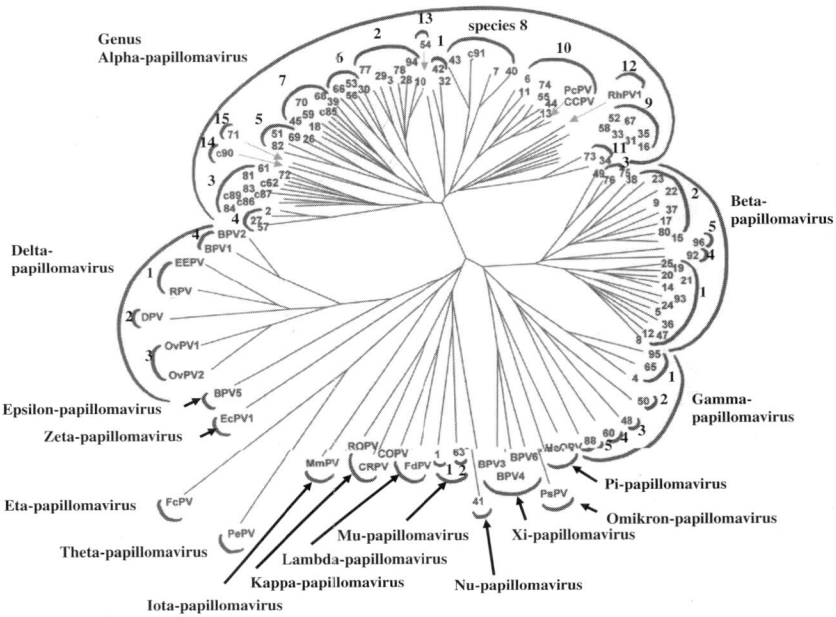

FIGURE 3.1. Phylogenetic tree of 118 papillomavirus types. All numbers refer to HPV types, c-numbers to candidate types, i.e., HPV genomes isolated as PCR amplicons. All abbreviations including letters refer to animal papillomaviruses. BPV, bovine PV; COPV, canine oral PV; CRPV, cottontail rabbit PV; DPV, deer PV; EcPV, *Equus caballus* or horse PV; EEPV, European elk PV; FcPV, *Fringilla coelebs* or chaffinch PV; FdPV, *Felix domesticus* or cat PV; HaOPV, hamster oral PV; MnPV, *Mastomys natalensis* or African rat PV; OvPV ovine or sheep PV; PePV, *Psittacus erythacus* or African grey parrot PV; PsPV, *Phocoena spinipinnis* or porpoise PV; RPV, reindeer PV; ROPV, rabbit oral PV. The outermost semicircular symbols identify papillomavirus genera, for example, the genus alpha-papillomavirus. The inner semicircular symbols identify papillomavirus species. For example, the HPV species 8 within the genus alpha-papillomaviruses lumps the HPV types 7, 40, 43, and c and 91. Reprinted from "Classification of papillomaviruses," by de Villiers, E.M., Fauquet, C., Broker, T.R., Bernard, H.U., and zur Hausen, *H. Virology*, 324, 17–27 (2004), with permission from Elsevier.

Unexpectedly, this branch included together with all those HPV types that are infecting genital epithelia also HPV types that normally cause nongenital cutaneous lesions, such as HPV-2 and -27, the most prevalent HPV types in common warts (Chan et al., 1997a). To make matters even more complicated, HPV types that cause pathologically similar cutaneous lesions, like HPV-4, were found to be completely unrelated to HPV-2 and -27 based on phylogeny. As another example for misleading terms, HPV types originally isolated from patients with epidermodysplasia verruciformis (EV-HPVs) have subsequently been documented to be widespread in the skin of completely healthy individuals (Steger et al., 1990; Boxman et al., 1999).

These examples show that a phenotypically based taxonomy, namely, one that creates taxa of PV types that give rise to similar pathology, would not reflect the true relationship among these PVs. A nomenclature that uses terms like "genital" HPVs should also be avoided as it mistakenly implies that there exists a correlation between the histological target of an HPV type and its taxonomic status. In addition to these biological and clinical inconsistencies, it has to be noted that terms like "branch" and "group" had been introduced since they were operationally useful in phylogenetic PV research. However, these terms are generally not used in the taxonomy of biological organisms, where taxa below the level of families are described by the terms "genus" and "species."

In order to overcome these inconsistencies, the working group on PV taxonomy of the ICTV (consisting of E.M. de Villiers, H. zur Hausen, H. U. Bernard, and T. Broker) agreed with the ICTV on the term "genus" for higher-order associations that were originally referred to as "supergroups" or "major branches." It was decided that Greek letters should be used to name such genera, as shown in Figure 3.1. A complete list of all PV types and their has taxonomic order has recently been published (de Villiers et al., 2004), and a short form is shown in Table 3.2. To give examples of the consequences of this agreement, all "genital" PVs are referred to as alpha-papillomaviruses, and those PV types most often associated with epidermodysplasia verruciformis as beta-PVs. The working group and the ICTV took a cue to introduce the Greek alphabet for this taxonomy from the generally accepted taxonomy of herpesviruses. The parties involved in this decision are aware that it will be problematic to assure that these genus names (and the species designations discussed in the next paragraph) will be widely used, but it adopted the view that the taxonomy of PVs required a formal solution replacing the somewhat vernacular lingo used in the past.

3.3.2. Papillomavirus "Species"

Traditionally, novel genomic PV isolates have been described as "types." As will be discussed below, today a PV type is defined as a complete PV genome whose L1 gene nucleotide sequence differs by at least 10 percent from that of any previously described PV type. According to this definition the genomes of PV types are as remotely related to one another as those of many distinct host species, for example, humans and the four living ape species. In addition, PV

TABLE 3.2. The Most Frequently Studied Papillomavirus Types, Their Biological and Clinical Properties, and Phylogenetic Associations on the Level Of "Genus" and "Species" as Recently Recognized by the International Council on the Taxonomy of Viruses (ICTV)

Family papillomaviruses (*papillomaviridae*) Genus	Species	Type (s)	Properties
Alpha-papillomaviruses	4	HPV-2, HPV-27, HPV-57	Common skin warts, frequently in genital warts of children
	5	HPV-26, HPV-51, HPV-69, HPV82	High-risk malignant and benign mucosal lesions
	6	HPV-53, HPV-30, HPV-56, HPV-66	High-risk malignant and benign mucosal lesions
	7	HPV-18, HPV-39, HPV-45, HPV-59, HPV-68, HPV-70	High-risk malignant mucosal lesions, some (esp. HPV-18) more frequent in adeno- than in squamous carcinoma of the cervix
	8	HPV-7, HPV-40, HPV-43	Low-risk mucosal and cutaneous lesions, HPV-7 known as butcher's wart virus, often in lesions of HIV infected patients
	9	HPV-16, HPV-31, HPV-33, HPV-35, HPV-52, HPV-58, HPV-67	High-risk malignant mucosal lesions, some (esp. HPV-16) more frequent in squamous than in adenocarcinoma of the cervix, HPV-16 most prevalent HPV type in cervical malignancies
	10	HPV-6, HPV-11, HPV-13, HPV-44, HPV-74	Benign mucosal lesions; HPV-6 and HPV-11 in male and female genital warts, condylomata acuminata of cervix, laryngeal papillomas; some of these lesions can progress malignantly
Beta-papillomaviruses	1	HPV-5, HPV-8 (selected from a very type-rich genus)	Cutaneous benign and malignant lesions in EV and immune-suppressed patients
Gamma-papillomaviruses	1	HPV-4, HPV-65 (selected from a very type-rich genus)	Cutaneous benign lesions
Delta-papillomaviruses	4	Bovine papillomavirus-1 (BPV-1) (selected from a type-rich genus)	Fibropapillomas in cattle, sarcoids in horses; an important cell culture model
Kappa-papillomaviruses	1	Cottontail rabbit papillomavirus (CRPV)	Cutaneous lesions; an important animal model
Mu-papillomaviruses	1,2	HPV-1, HPV-63	Cutaneous lesions, frequently in foot warts
Nu-papillomaviruses	1	HPV-41 (unrelated to any other HPV type)	Cutaneous lesions
Xi-papillomaviruses	1	BPV-3, BPV-4, BPV-4	Papillomas of the alimentary canal; BPV-4 is an important model for multi-step carcinogenesis and vaccination research

genome sequences are, just like their hosts, phylogenetically separate entities, and PV genomes intermediate to any two types do not exist. Lastly, there is evidence that even closely related PV types became separate taxa over periods as long as those that gave rise to related mammalian hosts, namely several hundred thousand to a few million years. As a consequence of these three arguments, one may be inclined to consider a "PV type" the same as a "PV species."

The ICTV is in charge of formalizing definitions of "virus species" and all taxa above the species level. Unfortunately, the reasoning analyzed in the last paragraph was not acceptable to the ICTV, as its rules require that a virus species is defined by biological and pathogenic (and not genomic) characteristics that set it apart from any other virus species. This rule could not be fulfilled by many related PV types. For example, no biological properties distinguish between HPV-2 and -27, which cause common warts, and HPV-6 and -11, which cause genital warts, laryngeal papillomas, and cervical condylomata acuminata. As discussed above, major phylogenetic branches, which are now identified as genera, unite PV types that often do not share biological properties. Fortunately, however, HPV types like the two pairs HPV-2/27 and HPV-6/11 are part of the same minor phylogenetic branches. Therefore, in order to resolve the disagreement over the relationship between "PV types" and "PV species," a system was created where these minor branches of related PV types are recognized as PV species and identified by numbers. A complete list of these PV species has recently been published (de Villiers et al., 2004). It can be derived from Figure 3.1 and is summarized in a shortened form in Table 3.2.

Among the PV types that are now considered to form one species, the term "type species" has been coined to identify the most thoroughly studied PV type of each species. While the biological and pathogenic properties of many rare HPV types are not yet sufficiently known, the properties of the type species are those that define a species by ICTV criteria.

The ICTV definitions of "genera" and "species" have several advantages. They integrate PV taxonomy into the same system that governs the taxonomy of all viruses. They are at the same time biologically meaningful and solidly based on phylogeny. Importantly, they leave the decision to describe new PV types in the hands of PV researchers and outside the "jurisdiction" of the ICTV, a clear advantage at a time when many new PV types are isolated every year, whose description by the present rules does not require major bureaucratic formalities and the involvement of multiple parties. Unfortunately, the ICTV genus and species definitions have also clear disadvantages: Most importantly, it does not seem to be very likely that the majority of molecular and clinical PV researchers will adopt the unfamiliar system of Greek letters and numbers in publications and oral presentations. For the phylogenetically interested PV researcher, it may also be important to remember that based on phylogenetic distances and evolutionary time scales the genomes of PV types certainly behave like the genomes of organisms that have species status elsewhere in biology.

3.4.   Papillomavirus Types, the Natural Taxonomic Entities in Molecular and Clinical Investigations

A PV type is today defined as a complete PV genome whose L1 nucleotide sequence differs from that of any other one by more than 10 percent. Many PV genomes, in particular those of HPV types, have been isolated numerous times from populations living in different parts of the world, and each isolate could be clearly identified as one of the original types. Since these independent isolates showed minor genomic variation, the nucleotide sequence of the original isolates is normally referred to as "prototype" or "reference genome."

Genomes intermediate to two PV types or recombinants of PV types have never been found. Under natural conditions, PV types are specific for one host, although multiple species of hoofed animals in agricultural settings can be infected by the same PV types (Trenfield et al., 1990; Chambers et al., 2003). It is not yet clear whether there are molecular mechanisms that limit the spread of PVs between different host species, as no rigorous laboratory studies have attempted a systematic transfer of PV types between numerous potential mammalian hosts.

Most PV researchers concentrate on those HPV types that cause malignancies threatening human lives or other medically important neoplasias, most notably cervical cancer and genital and laryngeal warts. HPV taxonomy may have a major impact on these fields: In order to give an example, about 50 percent of all cervical carcinomas and its precursor lesions contain HPV-16, a molecularly thoroughly investigated HPV type. HPV-16 is the type species of the HPV species 9, which unifies HPV-16 with HPV-31, -33, -35, -52, -58, and -67 (de Villiers et al., 2004). Each of these HPV types is likely to be a "high-risk" type with an etiological potential similar to HPV-16 (Munoz et al., 2003). While it will be impossible to study each of these types with the same intensity as HPV-16, their close relationship may allow us to extrapolate that they will have the same or similar molecular and pathogenic properties. Once medical strategies will be in place to combat HPV-16 infections, one may find that the same strategies will be useful to target all seven related HPV types.

3.5.   Papillomavirus Subtypes

The term "subtype" was originally used to identify HPV isolates that were apparently closely related, for example based on strong cross-hybridization in Southern blots, but differed in their restriction patterns. Today, subtypes are narrowly defined as PV genomes that differ from another type by 2 to 10 percent of their nucleotide sequence. The outcome of this definition was surprising, as it became clear that there are very few PV genomes that fall in this category. There are only four isolates in the literature that qualify as being subtypes. HPV-67 was found to be a subtype of HPV-34, HPV-46 of HPV-20, and HPV-55 of HPV-44. Also, the pigmy chimpanzee-PV and the chimpanzee (Bonobo) PV are related to one another as subtypes (de Villiers et al., 2004). The type listed first in each pair became eliminated from the nomenclature based on the prior isolation and description of the second type in each pair. The reasons for the near-absence of subtypes are not understood but are likely intrinsic to the question of the evolution of PV types.

Several isolates that were originally described as subtypes rather fall into the category of genomic "variants" (see below) or were isolates that had previously been described as types. For example, the HPV-2 subtype HPV-2c was found to be identical to HPV-27, while other HPV subtypes were variants of the HPV-2 reference genome (Chan et al., 1997b).

The term "subtype" is sometimes confused with the term "type" by scientists who are not familiar with HPV taxonomy, and this misuse should be avoided, as it creates misunderstandings.

3.6. Papillomavirus Variants

In contrast to the rare subtypes, repeated isolation of genomes of each HPV type showed that the worldwide human population is a rich source of "variants" of these HPV types. Variants are defined as HPV genomes less than 2 percent different from the original isolate of the respective HPV type, the prototype. The genomic diversity is often less in conserved parts of genes, but higher in noncoding regions, particularly the long control region (LCR).

Numerous variants of all commonly diagnosed PV types have been found in humans (Deau et al., 1993; Ho et al., 1993; Ong et al., 1993; Heinzel et al., 1995; Stewart et al., 1996; Chan et al., 1997b; Yamada et al., 1997; Calleja-Macias et al., 2004) and some animals (Chambers et al., 2003). If one defines variation as nucleotide exchanges and deletions which occur within a short genomic segment, for example, 400 bp of the LCR, most studies typically identified 20 to 60 variants of each HPV types, and variation elsewhere in the genome is typically linked to variation of the LCR. When analyzed by phylogenetic algorithms, the variants of each HPV type typically form two to six phylogenetic branches. Certain variants of each HPV type likely evolved in human populations of particular geographic regions, since specific variants of many HPV types are prevalent in some ethnic groups and absent or rare elsewhere (Ho et al., 1993; Ong et al., 1993; Stewart et al., 1996). This is particularly obvious when one compares European, East Asian, and African people.

The distribution patterns of variants in today's human population probably likely have two origins, namely, ancient bottlenecks during the evolution and spread of human ethnic groups, and infections after travel and mixing of populations in more recent times. In some cases variants of HPV types spread like chromosomally inherited genes and the viral load of a population reflects their ethnic composition. For example, in European populations one finds nearly exclusively European variants, and in African countries African variants. In many countries with populations of mixed ethnicity, patterns of the distribution of variants have emerged that are not always plausible. For example, the population of Brazil, which originated genetically in nearly equal proportions from Europeans, Native Americans, and Africans, has a high prevalence of European and Native American but a very low prevalence of African HPV-16 and HPV-18 variants (Ho et al., 1993; Ong et al., 1993). A genetically predominantly European population in Northern Mexico contained 70 percent HPV-16 variants of Native American origin (Calleja-Macias et al., 2004), while a tribe of Native Americans in Argentina contained European variants more frequently than Native American ones (Picconi et al., 2003).

Variants of HPV-16 may have differing biological and pathogenic properties (Villa et al., 2000). Such observations led to epidemiological studies that asked the question whether some variants of the same HPV type may be more carcinogenic than others (Xi et al., 1998; da Costa et al., 2002). There is some support for the possibility that non-European variants are more frequently associated with malignantly progressing lesions. A final answer to this question is still lacking, as in some studies European variants were often compared with all non-European variants, a phylogenetically heterogenous group of four different HPV-16 variant branches. Yet other investigations were restricted to European nations that carried only one group of variants, namely those that occurred traditionally in Europeans (Nindl et al., 1999; Kammer et al., 2002). In spite of the limitations of this research, intratype molecular diversity should be kept in mind as a potential explanation for the higher incidence of cervical cancer in developing countries in addition to variables originating from public health and life style questions. One can hypothesize that human populations in Latin America, Africa, and parts of Asia, which have a much higher prevalence of cervical cancer than people in other parts of the world, are exposed to genomically different variants of HPV types.

3.7.   Papillomavirus Typing in Research and Clinical Practice

The first PV types were isolated from tissues with a very high viral load, namely, warts of rabbits (leading to the isolation of DNA of the cottontail rabbit PV, CRPV), cattle (bovine PV-1, BPV-1), and human flat warts (HPV-1). Using the DNA from these viruses as probes, the cutting edge technologies of the late 1970s and early 1980s, such as Southern blot hybridization and hybridization to phage lambda libraries, helped to identify and clone novel HPV types by low stringency hybridization. Differences in the intensity of hybridization and varying restriction patterns in Southern blots led to the earliest efforts to determine relationships between different PV types (Pfister, 1987; de Villiers, 1989, 1997). As an increasing number of HPV types were sequenced, it became clear that the nucleotide sequence differences among any two types always exceeded 10 percent. This amount of sequence dissimilarity was eventually used to define novel PV types, and the defining region was restricted to the L1 gene. It was a very fortunate coincidence that this arbitrary definition identified natural taxonomic entities. As discussed above, PV genomes that diverge from one another by more than 10 percent are common, while genomes with distances of 10 percent down to 2 percent are very rare.

During the last few years, the detection of novel and known PV types in research and clinical studies was normally achieved by a combination of hybridization and amplification steps. Typical experiments to detect novel HPV types use "consensus primers," which are modeled according to highly conserved segments of the L1 gene, to amplify an HPV genome segment with the polymerase chain reaction (PCR). Distinction between known or novel types is achieved by sequencing the amplicon and comparing it against sequence databases. The most successful primers for the identification of novel genital HPV types are MY09/11 (for a review and references see Bernard, 1994) and GP5/6 (Jacobs et al., 1995), and for cutaneous HPV types FAP59 and FAP64 (Antonsson et al., 2000).

With the aim to promote DNA diagnosis for the detection of known HPV types in the clinical practice, several companies have developed kits, which are based on some variations of this general scheme. The line blot assay, developed in collaboration with Roche Molecular Systems Inc., uses a combination of PCR amplification with subsequent hybridization against probes of common genital HPV types that are fixed to a polymer surface (Gravitt et al., 1998). Digene Diagnostics, Inc., developed the hybrid capture technique, which is based on the formation of hybrids between HPV DNA that may be present in clinical specimens and complementary unlabeled HPV RNA probes. The RNA-DNA hybrids are captured and immobilized by antihybrid antibodies. Immobilized hybrids are reacted with a monoclonal antibody reagent that is conjugated to alkaline phosphatase, and the complexes are detected via a chemiluminescent substrate reaction (Ferris et al., 1998). The Nuclisens–NASBA technique combines the ability to work under isothermal conditions with the detection of RNA rather than DNA, which may favor the detection of cells that actively express HPV genomes (Smits et al., 1995).

3.8.   General Considerations Regarding the Evolution of Papillomaviruses

There are numerous data that papillomaviruses (PVs) evolve in linkage to their hosts. For example, PVs that were isolated from related hosts, such as humans and monkeys or different species of hoofed animals, are related to one another. Among the alpha-PVs, PVs from apes are closely related to one another and related to specific HPV types such as HPV-6 (de Villiers et al., 2004). PVs from some monkey species form host-specific groups, reminiscent of HPV-species and remotely related to certain alpha-HPV species (Chan et al., 1997a). Yet deeper branches of the phylogenetic tree of PVs (Fig. 1) show that hosts remotely related to humans are infected by PV types remotely related to HPVs. Some of these viruses have diverged significantly from the typical composition of PV genomes, such as by replacement of the E6 and E7 oncogenes (Scobie et al., 1997; Terai et al., 2002). The only two known bird PVs form a separate remote branch without close linkage to any mammalian PVs. These affiliations suggest that the evolution of novel PV taxa often started with evolutionary splits of their host species.

The processes that give rise to the large number of different PV types, even in single host species like humans, certain monkeys, and cattle are not known. It seems plausible that genetic drift, genetic bottlenecks, and a lack of selection are likely sources to create diversity. "Genetic drift" refers to the fact that mutations resulting in exchange, deletion, or insertion of nucleotides can accumulate and lead to novel genomes as long as these mutations are not deleterious to the viral biology. The published data of the relationship between variants of HPV types confirm that this process is continuously taking place. While there is a continuum of such variants, over long periods of time this process leads to discontinuities between taxa when intermediate genomes become extinct as a result of extinction of most individuals of the host population or host species. The small number of host individuals that survives such an extinction contains only a subset of viral genomes and are therefore considered an evolutionary "bottleneck."

This process could have led to evolution of novel taxa without any selective pressure. Selection of PV types could, in principle, result from immune evasion, or from evolution into novel ecological niches. While details of the immune response against PV infected cells are still little understood, the absence of inflammation in response to PV infection suggests an absence of ongoing evolutionary processes to evade immune recognition. The large numbers of different HPV types that are not biologically distinct does not suggest that each PV type evolved into a novel niche (Chan et al., 1995; Antonsson et al., 2000, 2003; Antonsson and Hansson, 2002).

3.9. Evolution of Papillomaviruses: The Time Scale

Molecular changes of PV genomes occur clearly at a very low rate and have never been observed in vitro nor during the last 20 years, when PV genome sequences have been available. Therefore, the rate of PV genome changes has to be estimated from comparisons of naturally occurring genomes and assumptions about their origin. There are several observations that all lead to estimates that PV genomes change at the same slow rates as the genomes of their vertebrate hosts, namely, roughly 1 percent nucleotide exchange per 100,000 to 1,000,000 years. 1) All ethnic groups of world, including reproductively isolated people such as certain Amazonian Indians (Ho et al., 1993, L. L. Villa, pers. comm.) contain apparently the same complement of high-risk, genital HPV types. This would mean that humankind was never without these viruses, and cervical cancer has always been with us. It is believed that the different ethnic groups of *Homo sapiens* came into being during the last 200,000 to 1,000,000 years, the time span that would have been available to generate 2 percent genomic diversity among the different HPV variants (Bernard, 1994). 2) The chimpanzee and pigmy chimpanzee PVs are (subtypes of) a PV type within the same species (No. 10), which also contains HPV-6 and -11, suggesting that over a period of a few million years, which led to the divergence of humans and apes, PV types diversified to a level that make ape PVs still members of HPV species (de Villiers et al., 2004). 3) All PVs from different species of monkeys (Chan et al., 1997a) form branches within the genital HPVs, which are probably the equivalent of species, but not termed this way, as this phylogeny was based on short amplicons. This would mean that it would take 10–25 million years to give rise to the phylogenetic assemblage of PV genomes that qualify as closely related species. 4) The extremely remote relationship of bird PVs to all mammalian PVs and idiosyncrasies of their genome organization (Terai et al., 2002) suggests that these viruses split from the mammalian PVs at a time when reptilian ancestors gave rise to the separate lineages of mammals and birds. The combination of these observations suggests that the phylogenetic tree of PVs represents a period of time as ancient as that of their hosts, and the genomic organization of PVs was apparently maintained for more than 100 million years.

There are, however, also splits between PV taxa independent of the speciation of their host as shown by Colobus monkey PVs in alpha- as well as in beta-PVs (Chan et al., 1997c). PV diversification can apparently sometimes also occur independent of host evolution, and one can conclude that separation of the human PV genera, such as the alpha- and beta-PVs, must have been ancient and predated the origin of primates.

So far, no PVs have been detected in lower vertebrates or nonvertebrates, where one might expect to find yet more divergent PVs or PV-related genomes. As a consequence, the origin of the papillomaviridae is unknown. The E1 helicase domain mentioned above is clearly a homology. The same homology was detected in a virus-like episome in planaria and in adeno-associated viruses (Rebrikov et al., 2002). Its origin provides interesting evolutionary speculation.

3.10.   Pathogenicity Versus Latency

Infections with PVs are clearly the cause of common and genital warts and a central causal factor in the etiology of cervical cancer. While this concept is well confirmed, it is nevertheless erroneous to characterize PVs as aggressive neoplastic agents. Among the arguments for this statement is the observation of a huge number of HPV types in the skin of asymptomatic individuals (Antonsson et al., 2000, 2003; Antonsson and Hansson, 2002) and the wide variety of HPV types in healthy individuals that give rise to epidermodysplasia in high-risk patients (Steger et al., 1990; de Villiers, 1989). Even the so-called "high-risk" HPV types such as HPV-16 most often give rise to neoplasia only in the context of the transformation zone of the cervix, while they lead to subclinical infection elsewhere, for example throughout the vagina and at penile epithelia. All of these observations lead to a concept that PVs evolved toward a life cycle that does not necessarily lead to clinically detectable neoplasia but rather to a commensalic "episome"-like life cycle.

References

Antonsson, A., Forslund, O., Ekberg, H., Sterner, G, and Hansson, B.G. (2000). The ubiquity and impressive genomic diversity of human skin papillomaviruses suggest a commensalic nature of these viruses. *J. Virol.* 74:11636–11641.

Antonsson, A., and Hansson, B.G. (2002). Healthy skin of many species harbours papillomaviruses, which are closely related to their human counterparts. *J. Virol.* 76:12537–12542.

Antonsson, A., Erfurt, C., Hazard, K., Holmgren, V., Simon, M., Kataoka, A., Hossain, S., Hakangard., C, and Hansson, B.G. (2003). Prevalence and type spectrum of human papillomaviruses in healthy skin samples collected in three continents. *J. Gen. Virol.* 84:1881–1886.

Bernard, H.U. (1994). Coevolution of papillomaviruses and human populations. *Trends Microbiol.* 2:140–143.

Bernard, H.U., Chan, S.Y., Manos, M.M., Ong, C.K., Villa, L.L., Delius, H., Bauer, H.M., Peyton, C., and Wheeler, C.M. (1994). Assessment of known and novel human papillomaviruses by polymerase chain reaction, restriction digest, nucleotide sequence, and phylogenetic algorithms. *J. Inf. Dis.* 170:1077–1085.

Boxman, I.L., Mulder, L.H., Russell, A., Bouwes-Bavinck, J.N., Green, A., and Ter Schegget, J. (1999). Human papillomavirus type 5 is commonly present in immunosuppressed and immunocompetent individuals. *Br. J. Dermatol.* 141:246–249.

Calleja-Macias, I.E., Kalantari, M., Huh, J., Ortiz-Lopez, R., Rojas-Martines, A., Gonzales-Guerrero, J.F., Williamson, A.L., Hagmar, B., Wiley, D.J., Villarreal, L., Bernard, H.U., and Barrera-Saldana, H.A. (2004). High prevalence of specific variants of human papillomavirus-16, 18, 31, and 35 in a Mexican population. *Virology*, 319:315–323.

Chambers, G., Ellsmore, V.A., O'Brien, P.M., Reid, S.W., Love, S., Campo, M.S., and Nasir, L. (2003). Sequence variants of bovine papillomavirus E5 detected in equine sarcoids. *Virus Res.* 96:141–145.

Chan, S.Y., Bernard, H.U., Ong, C.K., Chan, S.P., Hofmann, B., and Delius H. (1992). Phylogenetic analysis of 48 papillomavirus types and 28 subtypes and variants: a show case for the molecular evolution of DNA viruses. *J. Virol.* 66:5714–5725.

Chan, S.Y., Delius, H., Halpern, A.L., and Bernard, H.U. (1995). Analysis of genomic sequences of 95 papillomavirus types: uniting typing, phylogeny, and taxonomy. *J. Virol.* 69:3074–3083.

Chan, S.Y., Bernard, H.U., Ratterree, M., Birkebak, T.A., Faras, A.J., and Ostrow, R.S. (1997a). Genomic diversity and evolution of papillomaviruses in Rhesus monkeys in Rhesus monkeys. *J. Virol.* 71:4938–4943.

Chan, S.Y., Chew, S.H., Egawa, K., Grussendorf-Conen, E.I., Honda, Y., Ruebben, A., Tan, K.C., and Bernard, H.U. (1997b). Phylogenetic analysis of the human papillomavirus type 2 (HPV-2), HPV-27, and HPV-57 group, which is associated with common warts. *Virology* 239:296–302.

Chan, S.Y., Ostrow, R.S., Faras, A.J., and Bernard, H.U. (1997c). Genital papillomaviruses (PVs) and epidermodysplasia verruciformis PVs occur in the same monkey species: implications for PV evolution. *Virology* 228:213–217.

Clertant, P, and Seif, I. (1984). A common function for polyoma virus large-T and papillomavirus E1 proteins? *Nature* 311:276–279.

Da Costa, M.M., Hogeboom, C.J., Holly, E.A., and Palefsky, J.M. (2002). Increased risk of high-grade anal neoplasia associated with a human papillomavirus type 16 E6 sequence variant. *J. Infect. Dis.* 185:1229–1937.

Deau, M.C., Favre, M., Jablonska, S., Rueda, L.A., and Orth, G. (1993). Genetic heterogeneity of oncogenic human papillomavirus type 5 (HPV5) and phylogeny of HPV5 variants associated with epidermodysplasia verruciformis. *J. Clin. Microbiol.* 31:2918–2926.

de Villiers, E.M. (1989). Heterogeneity of the human papillomavirus group. *J. Virol.* 63:4898–4903.

de Villiers, E-M. (1997). Papillomavirus and HPV typing. *Clin. Dermatol.* 15:199–206.

de Villiers, E-M. (1998). Human papillomavirus infections in skin cancer. *Biomed. Pharmacother.* 52:26–33.

de Villiers, E.M., Fauquet, C., Broker, T.R., Bernard, H.U., and zur Hausen, H. (2004). Classification of papillomaviruses. *Virology* 324:17–27.

Farmer, A.D., Calef, C.E., Millman, K., and Myers, G.L. (1995). The human papillomavirus database. *J. Biomed. Sci.* 2:90–104 (http://hpv-web.lanl.gov/stdgen/virus/hpv/compendium/htdocs/).

Ferris, D.G., Wright, T.C. Jr., Litaker, M.S., Richart, R.M., Lorincz, A.T., Sun, X.W., and Woodward, L. (1998). Comparison of two tests for detecting carcinogenic HPV in women with Papanicolaou smear reports of ASCUS and LSIL. *J. Fam. Pract.* 46:136–141.

Gravitt, P.E., Peyton, C.L., Apple, R.J., and Wheeler, C.M. (1998). Genotyping of 27 human papillomavirus types by using L1 consensus PCR products by a single-hybridization, reverse line blot detection method. *J. Clin. Microbiol.* 36:3020–3027.

Heinzel, P.A., Chan, S.Y., Ho, L., O'Connor, M., Balaram, P., Campo, M.S., Fujinaga, K., Kiviat, N., Kuypers, J., Pfister, H., Steinberg, B.M., Tay, S.K., Villa, L.L., and Bernard, H.U. (1995). Variation of human papillomavirus type 6 (HPV-6) and HPV-11 genomes sampled throughout the world. *J. Clin. Microb.* 33:1746–1754.

Ho, L., Chan, S.Y., Burk, R.D., Das, B.C., Fujinaga, K., Icenogle, J.P., Kahn, T., Kiviat, N., Lancaster, W., Mavromara, P., Labropoulou, V., Mitrani-Rosenbaum, S., Norrild, B., Pillai, M.R., Stoerker, J., Syrjaenen, K., Syrjaenen, S., Tay, S.K., Villa, L.L., Wheeler, C.M., Williamson, A.L., and Bernard, H.U. (1993). The genetic drift of human papillomavirus type 16 is a means of reconstructing prehistoric viral spread and movement of ancient human populations. *J. Virol.* 67:6413–6414.

Jacobs, M.V., de Roda Husman, A.M., van den Brule, A.J., Snijders, P.J., Meijer, C.J., and Walboomers, J.M. (1995). Group-specific differentiation between high- and low-risk human papillomavirus genotypes by general primer-mediated PCR and two cocktails of oligonucleotide probes. *J. Clin. Microbiol.* 33:901–905.

Kammer, C., Tommasino, M., Syrjanen, S., Delius, H., Hebling, U., Warthorst, U., Pfister, H., and Zehbe, I. (2002). Variants of the long control region and the E6 oncogene in European human papillomavirus type 16 isolates: implications for cervical disease. *Br. J. Cancer* 86:269–273.

Matsukura, T., and Sugase, M. (2001). Relationships between 80 human papillomavirus genotypes and different grades of cervical intraepithelial neoplasia: association and causality. *Virology* 283:139–147.

Munoz, N., Bosch, F.X., de Sanjosé, S., Herrero, R., Castellsagué, X., Shah, K.V., Snijders, P.J.F., and Meijer, C.J.L.M. (2003) Epidemiological classification of human papillomavirus types associated with cervical cancer. *N. Engl. J. Med.* 348:518–527.

Nindl, I., Rindfleisch, K., Lotz, B., Schneider, A., and Durst, M. (1999). Uniform distribution of HPV 16 E6 and E7 variants in patients with normal histology, cervical intra-epithelial neoplasia and cervical cancer. *Int. J. Cancer* 82:203–207.

Ong, C.K., Chan, S.Y., Campo, M.S., Fujinaga, K., Mavromara, P., Labropoulou, V., Pfister, H., Tay, S.K., ter Meulen, J., Villa, L.L., and Bernard, H.U. (1993). Evolution of human papillomavirus type 18: an ancient phylogenetic root in Africa and intratype diversity reflect coevolution with human ethnic groups. *J. Virol.* 67: 6424–6431.

Pfister, H. (1987). Relationship of papillomaviruses to anogenital cancer. *Obstet. Gynecol. Clin. North Am.* 14:349–361.

Picconi, M.A., Alonio, L.V., Sichero, L., Mbayed, V., Villa, L.L., Gronda, J., Campos, R., and Teyssie, A. (2003). Human papillomavirus type-16 variants in Quechua aboriginals from Argentina. *J. Med. Virol.* 69:546–552.

Rebrikov, D.V., Bogdanova, E.A., Bulina, M.E., and Lukyanov, S.A. (2002). A new planarian extrachromosomal virus-like element revealed by subtractive hybridization. *Mol. Biol.* 36:813–820.

Scobie, L., Jackson, M.E., and Campo, M.S. (1997). The role of exogenous p53 and E6 oncoproteins in *in vitro* transformation by bovine papillomavirus type 4 (BPV-4): significance of the absence of an E6 ORF in the BPV-4 genome. *J. Gen. Virol.* 78:3001–3008.

Smits, H.L., van Gemen, B., Schukkink, R., van der Velden, J., Tjong-A-Hung, S.P., Jebbink, M.F., and ter Schegget. J. (1995). Application of the NASBA nucleic acid amplification method for the detection of human papillomavirus type 16 E6–E7 transcripts. *J. Virol. Methods* 54:75–81.

Steger, G., Olszewsky, M., Stockfleth, E., and Pfister, H. (1990). Prevalence of antibodies to human papillomavirus type 8 in human sera. *J. Virol.* 64:4399–4406.

Stewart, A.C., Eriksson, A.M., Manos, M.M., Munoz, N., Bosch, F.X., Peto, J., and Wheeler, C.M. (1996). Intratype variation in 12 human papillomavirus types: a worldwide perspective. *J. Virol.* 70:3127–3136.

Terai, M., DeSalle, R., and Burk, R.D. (2002). Lack of canonical E6 and E7 open reading frames in bird papillomaviruses: Fringilla coelebs papillomavirus and Psittacus erithacus timneh papillomavirus. *J. Virol.* 76:10020–10023.

Trenfield, K., Spradbrow, P.B., and Vanselow, B.A. (1990). Detection of papillomavirus DNA in precancerous lesions of the ears of sheep. *Vet. Microbiol.* 25:103–116.

Van Ranst, M., Kaplan, J.B., and Burk, R.D. (1992). Phylogenetic classification of human papillomaviruses: correlation with clinical manifestations. *J. Gen. Virol.* 73:2653–2660.

Villa, L.L., Sichero, L., Rahal, P., Caballero, O., Ferenczy, A., Rohan, T., and Franco, E.L. (2000). Molecular variants of human papillomavirus types 16 and 18 preferentially associated with cervical neoplasia. *J. Gen. Virol.* 81:2959–2968.

Xi, L.F., Critchlow, C.W., Wheeler, C.M., Koutsky, L.A., Galloway, D.A., Kuypers, J., Hughes, J.P., Hawes, S.E., Surawicz, C., Goldbaum, G., Holmes, K.K., and Kiviat, N.B. (1998). Risk of anal carcinoma *in situ* in relation to human papillomavirus type 16 variants. *Cancer Res.* 58:3839–3844.

Yamada, T., Manos, M.M., Peto, J., Greer, C.E., Munoz, N., Bosch, F.X., and Wheeler, C.M. (1997). Human papillomavirus type 16 sequence variation in cervical cancers: a worldwide perspective. *J. Virol.* 71:2463–2472.

4

The Differentiation-Dependent Life Cycle of Human Papillomaviruses in Keratinocytes

Choogho Lee and Laimonis A. Laimins
Department of Microbiology-Immunology, Feinberg School of Medicine, Northwestern University, Chicago, IL 60611

4.1. Introduction

Human papillomaviruses (HPVs) are small DNA viruses that induce a variety of hyperproliferative lesions in epithelial tissues. More than 100 different HPV types have been identified and classified on the basis of their nucleotide sequence (Howley and Lowy, 2001). One group of HPVs induces warts on hands or feet and exhibits a high tropism for cutaneous tissues. Another group of HPVs primarily targets genital and oral mucosa. Among these mucosotropic HPVs, types 6 and 11 induce benign lesions and are called "low-risk" HPVs due to their lack of association with cancers. In contrast, HPV types 16, 18, 31, and 45 are referred to as "high-risk" types, since they induce lesions that can ultimately progress to cervical and other cancers (Laimins, 1993; Lowy et al., 1994; zur Hausen, 2002). Despite the differences in oncogenic potential of these various HPV types, it is likely that they share common mechanisms during their productive life cycles.

4.2. Life Cycle of HPVs

The productive life cycle of HPVs is closely linked to the differentiation program of the host epithelial tissue (Fig. 4.1) (Howley and Lowy, 2001). The cellular receptor for viral entry is still unknown although heparin sulfate seems to facilitate virus attachment in some cell types (Joyce et al., 1999; Giroglou et al., 2001). The targets of initial infection are thought to be basal epithelial cells, which become exposed as a result of microwounds of the stratified epithelia (Howley and Lowy, 2001). Following virus binding and entry, virions migrate to the nucleus and establish their genomes as multi-copy extrachromosomal plasmids, which are maintained at approximately 20 to 100 copies per infected basal cell (Stubenrauch and Laimins, 1999). The replication of viral genomes

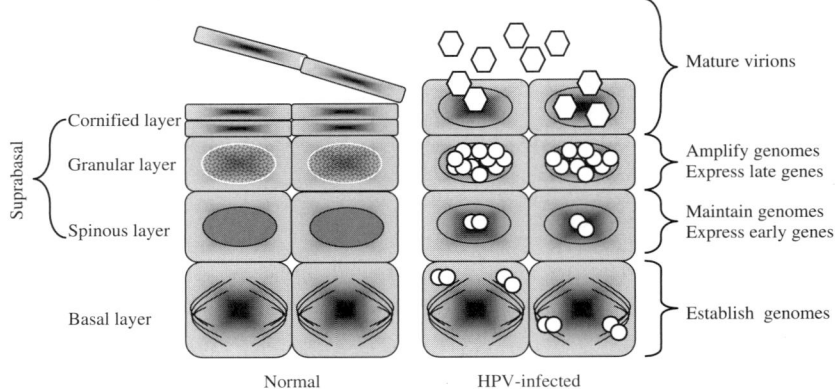

FIGURE 4.1. Tight linkage of the HPV life cycle to the differentiation program of the host epithelial tissue. Drawings show normal (**left**) and HPV-infected epithelia (**right**) with various differentiation layers associated with different stages of the HPV life cycle.

occurs in the S-phase of the cell cycle through the action of the viral E1 and E2 proteins in cooperation with cellular replication proteins (Lambert, 1991; Ustav and Stenlund, 1991). Following viral genome replication and cell division, one of the daughter cells migrates away from the basal layer and starts a program of differentiation. Normal uninfected keratinocytes exit the cell cycle after leaving the basal layer, and the nuclei are degraded in many of these differentiating cells. In contrast, HPV-infected cells undergo differentiation but remain active in the cell cycle (Stubenrauch and Laimins, 1999). This allows highly differentiated suprabasal cells to re-enter S-phase and support high-level productive viral DNA replication. The ability of differentiated cells to undergo cell cycle progression is mediated largely through the action of the viral E7 protein (Chen et al., 1995; Flores et al., 1999; Halbert et al., 1992). Differentiation-dependent amplification of viral genomes in suprabasal cells coincides with activation of the late viral promoter (Chen et al., 1995; Flores et al., 1999; Halbert et al., 1992; Hummel et al., 1992). The late viral transcripts encode the capsid proteins L1 and L2 as well as two additional mediators of late viral functions, E1^E4, and E5 (Hummel et al., 1992). Progeny virions are assembled in highly differentiated cells and then released to the extracellular environment (Frattini et al., 1996; Howley and Lowy, 2001; Meyers et al., 1992).

4.3.   Methods to Study the HPV Life Cycle

Study of the papillomavirus life cycle has been hampered by the inability of these viruses to replicate to high titer in the laboratory. Initial studies on the life cycle involved descriptive studies of the distribution of viral genomes and gene products in naturally occurring lesions in animals and humans. Cells transformed in culture by bovine papillomavirus were regarded as a model of the early phase of infection, while the outer epithelial layers of productively infected warts represented the late phase. In recent years, considerable progress has been made in studying the HPV life cycle by using various tissue culture models that include cell lines that stably maintain viral genomes as plasmids and methods to induce epithelial differentiation, which appears to be required for production of progeny virions. In addition, HPV genomes carrying specific mutations have been used to assess the role of viral gene products on the virus life cycle in differentiated epithelial cells. These studies have been supplemented by the analysis of individual viral genes and proteins in cellular and biochemical assays that mirror certain aspects of the virus life cycle, such as cell transformation, transcriptional regulation, and viral DNA replication.

A particularly useful cultured epithelial cell system is the organotypic raft culture system initially developed by Asselineau and Prunieras (1984). In this system, undifferentiated keratinocytes are first layered on top of a fibroblast and collagen plug that is then placed on a metal grid at the air–liquid interface (Fig. 4.2). Diffusion of nutrients through the plug allows growth of these cells at this air–liquid interface for about 2 weeks, which induces the differentiation and stratification of the epithelial cells. If the starting cells contain HPV genomes, the resulting differentiated cultures faithfully duplicate the morphology

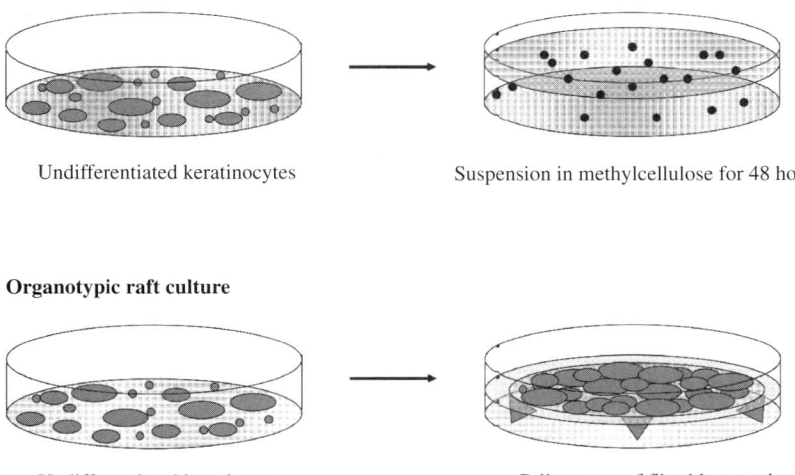

Semisolid media

Undifferentiated keratinocytes   Suspension in methylcellulose for 48 hours

Organotypic raft culture

Undifferentiated keratinocytes   Cells on top of fibroblasts and
                                                    collagen/grid

FIGURE 4.2. Methods to induce the epithelial differentiation to study late viral functions. **Top**: In suspension culture, undifferentiated keratinocytes maintaining HPV genomes are suspended in semisolid media supplemented with 1.6 % methylcellulose for 48 h. **Bottom**. In organotypic raft culture, undifferentiated keratinocytes maintaining HPV genomes are layered on top of a fibroblast and collagen plug that is then placed on a metal grid at the air–liquid interface for 2 weeks.

of HPV-induced lesions in vivo (McCance et al., 1988). Furthermore, epithelial differentiation in this system is accompanied by activation of late viral functions, including the focal synthesis of progeny virions in cells in the most highly differentiated layers. These virions are infectious but the titers are low. In some studies, differentiation and virus production is stimulated by agents such as protein kinase C activators (Ozbun and Meyers, 1998b; Sen et al., 2002). A second system for differentiation that utilizes suspension of cells in 1.6 % methylcellulose is useful for studying other aspects of the HPVs life cycle (Flores et al., 1999; Ruesch et al., 1998) (Fig. 4.2). The advantage of suspension culture is that cells progress through differentiation in a coordinate fashion within 48 hours. However, the methylcellulose system obviously does not reproduce the morphologies seen in vivo or in organotypic cultures. Immortalized cell lines derived from biopsy materials that stably maintain viral genomes as plasmids have been studied in both systems. These cell lines include W12 cells, which contain up to 500 copies of the HPV16 genome, and CIN 612 cells, which contain approximately 100 copies of HPV31 DNA (Bedell et al., 1991; Stanley et al., 1989). Both of these cell lines produce HPV virions following differentiation in organotypic raft cultures (Flores et al., 1999; Meyers et al., 1992).

The first genetic technique that allowed the analysis of the full papillomavirus life cycle including production of infectious progeny was the introduction of specifically altered cottontail rabbit papillomavirus (CRPV) DNA into rabbit skin (Brandsma and Xiao, 1993; Brandsma et al., 1991). A major technical advance for the study of the HPV life cycle was the development of genetic methods to analyze HPV functions in cell culture (Fig. 4.3) (Frattini et al., 1996).

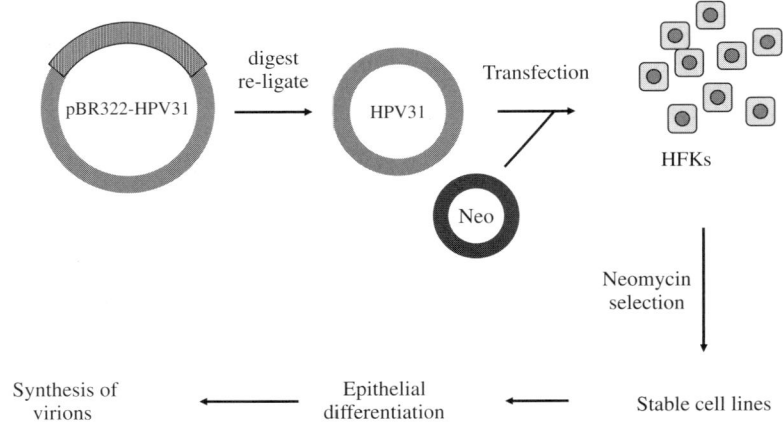

FIGURE 4.3. Genetic approach to study the life cycle of the HPVs in tissue culture. Wild-type or mutant HPV genome is released from a bacterial plasmid, recircularized and cotransfected with a neomycin resistance gene into human foreskin keratinocytes (HFKs). Transfected cells are subjected to G418 selection, and resistant colonies are pooled to generate stable cell lines maintaining episomal HPV genomes. Epithelial differentiation is induced to study the late stages of the HPV life cycle.

To perform this analysis, mutations are first introduced into the complete HPV genome contained in a bacterial plasmid. The viral DNA is then excised from vector sequences, circularized, and transfected into normal human keratinocytes along with drug-selectable markers. Following selection and expansion of transfected cells in monolayer culture, effects on the viral life cycle can be studied in either organotypic raft cultures or following suspension in methylcellulose. This method allows analysis of early HPV functions such as establishment and stable replication of viral genomes as well as late functions such as vegetative viral DNA synthesis and late transcription in conditions that mimic the natural situation (Frattini et al., 1996, 1997; Meyers et al., 1997). Such an approach is important because the activities displayed by individual viral proteins in heterologous overexpression assays or in in vitro assays do not necessarily faithfully mimic the actual function of the protein expressed at physiologic levels in the context of other viral proteins in differentiated host epithelial cells. The effects of mutations in HPV types 11, 6, 16, 18 and 31 have been examined using these methods (Flores et al., 1999; Frattini et al., 1996, 1997; Meyers and Laimins, 1994; Meyers et al., 1997; Oh et al., 2004; Thomas et al., 1999).

4.4.   Genome Organization and Gene Products of HPVs

The genetic organizations of all HPV types are similar (Fig. 4.4). The circular HPV DNA genomes are approximately 8 kb in size and consist of early- and late-coding regions as well as sequences that regulate transcription and replication. The major regulatory region is referred to as the upstream regulatory region (URR) or long control region (LCR). HPVs encode six to eight open reading frames that contribute to the productive viral life cycle.

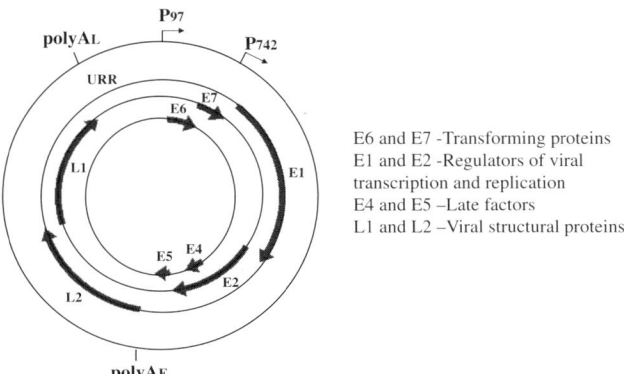

E6 and E7 -Transforming proteins
E1 and E2 -Regulators of viral transcription and replication
E4 and E5 –Late factors
L1 and L2 –Viral structural proteins

FIGURE 4.4. Genomic organization and gene products of high-risk HPV type 31. The early (P_{97}) and late (P_{742}) viral promoters are shown with arrows. URR stands for the upstream regulatory region. Eight major open reading frames are also indicated with their biological functions.

The early viral gene products, E6 and E7, play critical roles in the immortalization and transformation of epithelial cells by high-risk HPVs. These proteins bind to and inactivate a number of cellular proteins that inhibit cell cycle progression, creating a cellular environment favorable for viral DNA replication. The E6 protein forms a complex with the tumor suppressor protein p53 and induces its rapid proteosomal degradation (Huibregtse et al., 1991; Scheffner et al., 1990, 1993; Werness et al., 1990). E7 binds to the retinoblastoma (pRb) family of proteins, resulting in the degradation of Rb family members, and thereby facilitates cell-cycle progression by alleviating pRb-mediated repression of genes required for entry into S phase (Cheng et al., 1995; Dyson et al., 1989; Martin et al., 1998; Munger et al., 1989). As discussed in a subsequent section, the E6 and E7 proteins are also important for the maintenance of viral genomes as well as for differentiation-dependent late viral functions (Oh et al., 2004; Thomas et al., 1999).

The E1 and E2 proteins regulate viral replication and transcription. Binding of E1 and E2 proteins to their respective recognition sequences in the URR is required for the recruitment of DNA polymerase and other cellular DNA replication proteins to the viral origin of DNA replication (Conger et al., 1999; Frattini and Laimins, 1994; Mohr et al., 1990). The ATPase and helicase activities of E1 protein are also necessary for unwinding of supercoiled DNA to allow efficient viral replication (Seo et al., 1993; Yang et al., 1993), but the binding affinity of the monomeric form of the E1 protein for its recognition sequence is very weak. However, the formation of a complex of the E1 protein with the full-length E2 protein recruits E1 to its recognition sequence and allows it to assemble into hexameric structures with higher affinity for DNA and increased helicase activity (Sanders and Stenlund, 1998; Sedman and Stenlund, 1998).

The dimeric E2 protein is required for transcriptional control of viral gene expression as well as for viral DNA replication (Laimins, 1998). Four E2 binding sites consisting of conserved palindromic sequences ($ACCN_6GGT$) are found in the URRs of genital papillomaviruses (Laimins, 1998). Through its association with these binding sites, the E2 protein regulates transcription initiated from the early viral promoter, which is located upstream of the E6 open reading frame, and contributes to copy number control (Klumpp and Laimins, 1999). However, the transactivation function of the E2 protein does not appear to be required for the HPV31 life cycle (Stubenrauch et al., 1998a).

L1 and L2 are the major and minor capsid proteins, respectively, and are expressed during the late phase of virus life cycle only in the most differentiated epithelial cells where the assembly of HPV virions takes place (Howley and Lowy, 2001). Assembly of L1 and L2 into icosahedral capsids occurs around a single copy of the viral genome associated with cellular histones prior to cell lysis.

The most highly expressed HPV protein during productive infection is the E1^E4 fusion protein, which consists of the first five amino acids of E1 fused to E4 coding sequences. The E1^E4 proteins are synthesized in the late phase of the viral life cycle from spliced transcripts that initiate at the late promoter.

These transcripts utilize a major splice donor located 15 nucleotides inside the E1 open reading frame and a splice acceptor just upstream of the E4 open reading frame (Doorbar et al., 1996; Nasseri et al., 1987). Because the E4 ORF lacks an ATG, the E1 sequences provide the initiation codon for translation. In the differentiation-dependent life cycle, E1^E4 synthesis precedes that of the capsid proteins, L1 and L2, but occurs concurrently with genome amplification (Palefsky et al., 1991; Doorbar et al., 1997; Middleton et al., 2003; Peh et al., 2002; Ruesch and Laimins, 1998). After translation, the E1^E4 proteins are processed into several proteolytically cleaved forms that associate into multimeric structures in the cytoplasm and bind to cytokeratins (Sterling et al., 1993; Pray et al., 1995; Roberts et al., 1993, 1997; Wang et al., 2004). In addition, yeast-two hybrid assays showed that E1^E4 binds a novel DEAD box-containing RNA helicase (Doorbar et al., 2000).

Genetic analysis of E1^E4 in the context of complete viral genomes revealed a role in the virus life cycle. Cottontail rabbit papillomaviruses (CRPV) that contained translation termination mutations in the E4 ORF displayed impaired production of progeny virions, indicating an important role for this protein in the productive stage of the life cycle (Peh et al., 2004). Mutations that abrogated HPV31 E4 synthesis had similar effects (Wilson et al., 2005). Cells containing these mutant HPV31 genomes were found to be impaired in differentiation-dependent genome amplification and had significantly reduced levels of late viral transcripts.

Overexpression of E1^E4 in heterologous expression systems can result in the collapse of cytokeratin networks leading to the hypothesis that E1^E4 may induce cytokeratin collapse to facilitate viral egress (Doorbar et al., 1991; Roberts et al., 1994). However, only a limited amount of keratin collapse was observed in natural high-risk HPV infections, suggesting that E1^E4 may play other roles in the viral life cycle (Pray et al., 1995; Doorbar et al., 1996). Overexpression of HPV16 E1^E4 in transient transfection assays also induces G2-M arrest, leading to a model in which the protein facilitates late viral functions by allowing cells to remain in S-phase by blocking progression into mitosis (Davy et al., 2002; Nakahara et al., 2002; Knight et al., 2004). This model is consistent with the recent observation that cyclin B/cdk1 complexes bound to E1^E4 were sequestered in the cytoplasm (Davy et al., 2005). Elucidation of the role of E1^E4 in the life cycle is likely to yield important insights into the control of late viral functions.

The E5 protein is a small hydrophobic protein that is localized to intracellular membranes (Conrad et al., 1993). The BPV1 E5 protein is expressed in both basal keratinocytes and in highly differentiated cells (Burnett et al., 1992), implying that it plays roles in both the early and the late phase of the virus life cycle. However, the absence of a recognizable E5 gene from some HPV genomes suggests that it is not absolutely required for productive virus infection. Removal of the E5 gene from the CRPV genome inhibited but did not abolish papilloma induction in rabbits (Brandsma et al., 1992). The HPV E5 proteins enhance epidermal growth factor (EGF) receptor signaling, apparently through upregulation of EGF receptor recycling (Conrad et al., 1993; Straight et al., 1993). E5 also associates with the vacuolar H^+-ATPase and blocks acidification of endosomes, activities which might stimulate receptor recycling (Straight et al., 1995; Conrad et al., 1993).

4.5. Control of HPV Transcription

Viral gene expression in the high-risk HPV types is mediated by two major promoters and a series of minor promoters (Fig. 4.4). The major early viral promoter, called p97 or p99 in HPV31 and HPV16 and p105 in HPV18, is constitutively active in undifferentiated basal cells as well as most differentiated suprabasal cells (Hummel et al., 1992; Rohlfs et al., 1991). These promoters direct transcription of early viral E6, E7, E1, E2, E1^E4, and E5 genes. In the high-risk HPV types, differential expression of E6 and E7 is mediated through a single primary transcript that is alternatively spliced in the E6 gene. The low-risk genital papillomaviruses have two early promoters: one initiates transcription upstream of the E6 open reading frame while the second is within the E7 gene (DiLorenzo and Steinberg, 1995; Smotkin et al., 1989). Most early transcripts terminate at the early polyadenylation (poly A) sequence located downstream of the E5 gene, and inhibitory sequences in the L2 and L1 genes inhibit extension of transcripts that pass through the early poly A sequence in undifferentiated cells (Collier et al., 2002; Oberg et al., 2003). Epithelial differentiation abolishes the inhibitory effect of these elements. HPV early gene expression has been assessed soon after infection of epithelial cells with HPV31 isolated from organotypic cultures. These studies revealed a complex pattern of alternatively spliced transcripts, with E1 and E2 mRNAs being the earliest detectable species (Ozbun, 2002).

Upon cellular differentiation, a late viral promoter is activated and drives expression of late RNA with a heterogeneous set of start sites located in the E7 gene (del Mar Pena and Laimins, 2001; Grassmann et al., 1996). The late viral promoter is called p742 in HPV31 and p680 in HPV16 (Grassmann et al., 1996; Hummel et al., 1992) and directs two set of transcripts; one set terminates at the early poly A site while the second set passes through the early poly A site and terminates at the late poly A site located downstream of L1. The first group of late transcripts encodes E1^E4, E5, and E1/E2 and the second group of late transcripts encodes the capsid proteins, L1 and L2. Changes in the levels of cellular polyadenylation factors such as CstF64 direct the differentiation-dependent read-through of the early polyadenylation sequence (Terhune et al., 1999, 2001). Additional promoters have also been described (Ozbun and Meyers, 1998b).

A variety of cellular and viral transcription factors regulate the activity of the early and late promoters. For the early promoter, cellular transcription factors bind to sequences in the URR and regulate expression of viral genes in a keratinocyte-specific manner (Sen et al., 2002). This cell type-specificity is likely mediated by the cooperative action of multiple constitutively expressed keratinocyte-specific transcription factors (Chang et al., 1991). The spectrum of factors that bind to the URR sequences vary between HPV types but all HPV genomes contain binding sites for Ap-1 (Kyo et al., 1995, 1997), Sp-1 (Apt et al., 1996) and TFIID (Tan et al., 1994). Other transcription factors that bind the URR sequences include YY1 (Bauknecht et al., 1995), KRF-1 (Mack and Laimins, 1991), AP-2 (Beger et al., 2001), Oct-1 (Chong et al., 1991), TEF-2 (Chong et al., 1991), TEF-1 (Ishiji et al., 1992), NF-1 (Kyo et al., 1995) and the glucocorticoid receptor (Mittal et al., 1993). The composition of cellular transcription complexes binding to the URR varies during different stages of the HPV life cycle (Sen et al., 2004). In the case of HPV6, that CCAAT displacement protein (CDP) appears to negatively regulate differentiation-specific viral promoters during the early phase of the viral life cycle (Pattison et al., 1997).

Little is known about the regulation of late transcription, which is activated upon epithelial differentiation. Changes in chromatin configuration occur in the vicinity of the late HPV 31 promoter, and this is likely to contribute to activation of late viral gene transcription (del Mar Pena and Laimins, 2001). In addition, recent studies using inhibitors of DNA replication have demonstrated that amplification of viral genomes is not necessary for activation of p742 in HPV31 and that differentiation alone is sufficient for its activation (Spink and Laimins, 2005). However, during normal infection, genome amplification does increase template numbers, which also contributes to increased levels of late viral gene expression. The cellular factors that regulate induction of late transcription are currently unknown. In addition, two promoters in the URR appear to be down-regulated during epithelial differentiation (Ozbun and Meyers, 1999).

All early HPVs transcripts are synthesized as polycistronic mRNAs that use common start and polyadenylation sites, and alternative splicing is used to regulate expression of viral genes such as E6 and E7 (Fig. 4.5). The E6 protein is translated from unspliced transcripts, whereas the E7 protein is translated from transcripts that have removed the E6 intron, resulting in the generation of an alternative deleted form of E6 called E6* (Howley and Lowy, 2001). Many early and late transcripts utilize the E1^E4 splice donor and acceptors while those messages encoding E2 use alternate acceptor sequences (Stubenrauch and Laimins, 1999). It is likely that splicing of viral transcripts is regulated in a differentiation-specific manner through the action of splicing enhancers and silencers. The identification of these elements is a difficult but important undertaking.

The regulation of HPV transcription through posttranscriptional regulatory mechanisms such as splicing and polyadenylation enables HPVs to fine-tune the expression of early and late viral genes in a differentiation-specific manner. Differential promoter usage may also affect translation. Since HPV translation occurs by a leaky scanning mechanism in which ribosomes bind at the 5′ end of messages and scan transcripts in a linear fashion for initiation codons, open reading frames located proximal to 5′ ends of mRNA are preferentially translated (Remm et al., 1999). Most early viral transcripts contain E4 and E5 open reading frames, which are the third and fourth open reading frames on these polycistronic messages and hence translated at a low level. In contrast, the E4 and E5 proteins

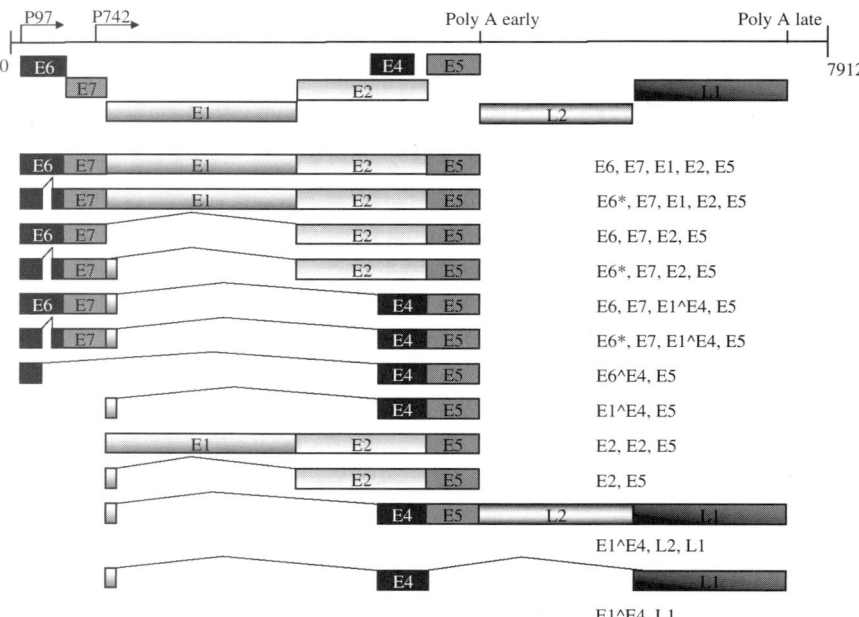

FIGURE 4.5. Early and late viral transcripts of high-risk HPV type 31. The HPV 31 genome is shown in a linear fashion. The early (P_{97}) and late (P_{742}) viral promoters are designated with arrows and the early and late polyadenylation sites are identified as Poly A early and Poly A late, respectively. Splice sites are shown by thin lines connecting open reading frames boxes. Potential gene products encoded by each transcript are indicated at the right of the figure.

are the products of the first and second open reading frames on late viral transcripts, allowing their efficient translation in differentiated cells.

The E5 protein may also play a role in controlling the switch to late transcription. When the E5 gene is mutated in the context of the complete HPV31 or HPV 16 genomes, the induction of late viral function is impaired, implicating it as a regulator of these activities (Fehrmann et al., 2003; Genther et al., 2003). However, the HPV31 E5 mutant had no effect on the levels of the EGFR or on its phosphorylation status (Fehrmann et al., 2003). Additional studies are needed to further clarify the biological activity of the HPV E5 proteins during the virus life cycle.

4.6.   Differentiation-Dependent Mechanisms to Regulate HPV DNA Replication

The replication of HPV genomes during the productive life cycle can be divided into three phases: establishment, maintenance, and amplification. During the establishment phase following infection, one HPV genome migrates to the nucleus and initiates several rounds of replication to reach 20 to 50 copies per cell. Following establishment, viral genomes replicate on an average of once per cell cycle during S-phase, with the maintenance of a constant copy number for periods ranging up to decades through the regulation of E1 and E2 expression by the E2 protein (Steger and Corbach, 1997; Stubenrauch et al., 1998b). The levels of the E2 proteins regulate early viral transcription. At low genome copy number and levels of E2 expression, E2 activates expression from the early promoter which drives expression of E1 and E2, whereas at high levels, it represses transcription (Stubenrauch et al., 1998b). In this way, expression of the viral replication proteins and viral copy number is believed to be maintained at a stable, optimal level in undifferentiated cells. Upon differentiation, transcription of the replication genes switches from the E2-regulated early promoter to the more active E2-independent late promoter (Klumpp and Laimins, 1999). High-level expression of the E1 and E2 replication proteins from this differentiation-specific promoter drives amplification of viral genomes to thousands of copies per cell. Analysis of viral transcript levels during vegetative viral DNA synthesis implies that E1 expression may be the major regulator of viral genome amplification (Ozbun and Meyers, 1998a). In addition, a spliced E8/E2C protein that inhibits transient replication of viral DNA is required for maintenance of episomal HPV31 DNA in keratinocytes (Stubenrauch et al., 2000). It is likely that additional uncharacterized mechanisms regulate viral genome replication.

4.7.   Roles of the HPV E6 and E7 Proteins in the Viral Life Cycle

The genetic system described above has been used to examine the functions of the E6 and E7 proteins in the viral life cycle. E6 and E7 proteins are encoded by all genital papillomaviruses, suggesting that they provide important functions in viral propagation. In addition, they function as oncoproteins in the high-risk types HPVs. Genetic studies indicate that the E6 and E7 proteins play essential roles in the life cycle of CRPV and the HPVs (Brandsma et al., 1991; Flores et al., 2000; Thomas et al., 1999). High-risk HPV31 genomes that contain translation termination mutations in either E6 or E7 are unable to replicate as plasmids in the stable replication assays (Thomas et al., 1999). The mechanism by which E6 and E7 contribute to the replication and maintenance of viral genomes as plasmids is not yet fully understood.

Both the E6 and the E7 proteins bind and alter the activity of cell cycle-regulatory proteins (Huibregtse and Beaudenon, 1996; Kubbutat and V.K.H., 1996). Inactivation of the retinoblastoma (pRb) family of proteins is the best characterized function of the E7 protein (Dyson et al., 1989). This interaction sequesters pRb protein away from E2F-pRb complexes and destabilizes the protein, resulting in the constitutive transcription of E2F-responsive S phase-specific genes (Edmonds and Vousden, 1989; Weintraub et al., 1995).

The low-risk E7 proteins bind to Rb family members with an affinity that is about tenfold lower than that seen with high-risk E7 proteins (Heck et al., 1992). Therefore, high-affinity Rb binding is not required for productive virus growth. This conclusion is consistent with the ability of a pRb binding-deficient CRPV E7 mutant to induce papillomas in rabbits (Defeo-Jones et al., 1993).

E7 proteins also bind to histone deacetylases (HDACs), which remove acetyl groups from histones that are in nucleosomes, leading to the formation of heterochromatin that prevents transcription factor binding (Brehm et al., 1999; Longworth and Laimins, 2004). High-risk E7 proteins bind to HDACs through C-terminal sequences that are independent of the Rb binding site (Brehm et al., 1999). The binding to HDACs is indirect and mediated through the Mi2β protein, a component of the mammalian NuRD chromatin remodeling complex. Both pRb and HDAC binding activities are important for the ability of high-risk E7 proteins to immortalize keratinocytes (Longworth and Laimins, 2004). Mutation of the HDAC binding site in E7 leads to an inability to stably maintain viral plasmid genomes, while mutation of the Rb binding domain alters the ability of the E7 protein to maintain differentiating cells active in the cell cycle (Cheng et al., 1995; Flores et al., 2000; Longworth and Laimins, 2004; Thomas et al., 1999).

Central to the ability of the E6 protein to immortalize cells is formation of a complex with p53 and E6AP, resulting in rapid p53 ubiquitination and degradation (Huibregtse et al., 1991; Scheffner et al., 1990, 1993; Werness et al., 1990). E6 can also activate expression of hTERT, the catalytic subunit of telomerase (Klingelhutz et al., 1996; Oh et al., 2001) as well as bind to a series of PDZ domain-containing cellular proteins. It is not clear which, if any, of these activities is most important for the viral life cycle, because the low-risk E6 proteins fail to degrade p53, activate telomerase or bind PDZ proteins, yet they propagate efficiently.

The binding of E6 to p53 appears to play a role in the replication of high-risk genomes, because E6 mutant HPV31 genomes defective for p53 binding integrated into the host genomes (Thomas et al., 1999). However, E6/p53 binding is required for plasmid maintenance only if the corresponding E7 protein binds Rb with high-affinity (Park and Androphy, 2002). When the Rb binding domain of E7 was mutated to that found in low-risk E7 proteins, the E6 protein/p53 interaction was no longer required for plasmid maintenance. Modulation of the activities of cell cycle regulators such as p53, Rb, HDACs, and other unknown cellular proteins is required to facilitate growth of HPV-infected cells and the maintenance of extrachromosomal viral DNAs. It is speculated that the balance of the activities of these cellular proteins is important and that loss of one function influences the action of other factors.

The ability of the E6 protein to bind factors other than p53 is likely to be important in the viral life cycle. Additional cellular proteins that bind to the E6 protein include those involved in calcium signaling (Das et al., 2000), cell adhesion (Tong and Howley, 1997), transcriptional control (Degenhardt and Silverstein, 2001b; Kumar et al., 2002; Patel et al., 1999; Ronco et al., 1998),

DNA synthesis (Kuhne and Banks, 1998), apoptosis (Filippova et al., 2002; Thomas and Banks, 1998), cell cycle control (Gao et al., 2000), DNA repair (Iftner et al., 2002; Srivenugopal and Ali-Osman, 2002), and small G protein signaling (Gao et al., 1999). As noted above, another set of E6 binding proteins possess multiple copies of a protein–protein interaction domain called PDZ (PSD-95/disc large/ZO-1). PDZ proteins are primarily found at the membrane–cytoskeleton interfaces of cell–cell contact and form multiprotein signaling complexes at the inner surface of the membrane. The primary functions of PDZ proteins involve the regulation of cell growth, cell polarity, and cell adhesion in response to cell contact (Craven and Bredt, 1998; Fanning and Anderson, 1999). In overexpression assays, the E6 protein targets these PDZ proteins for degradation via ubiquitin-proteasome pathway (Gardiol et al., 1999; Glaunsinger et al., 2000; Lee et al., 2000; Nakagawa and Huibregtse, 2000; Pim et al., 2000; Thomas et al., 2002). Mutation of the PDZ-binding motif at the C-terminus of high-risk E6 proteins in the context of the complete HPV 31 genome results in reduced cellular growth rates and low plasmid copy number, which in turn results in reduced virion production upon differentiation (Lee and Laimins, 2004). The PDZ binding motif is therefore important in the life cycle of high-risk HPV types.

Several other E6-binding partners contain seven amino acid motifs which form part of an α-helical structure that forms a recognition sequence for E6 (Be et al., 2001). α-Helix binding partners include E6AP, E6BP (Chen et al., 1995), a calcium-binding protein of the CREC family, paxillin (Tong and Howley, 1997; Vande Pol et al., 1998), a focal adhesion protein, E6-TP1 (Gao et al., 1999), a putative Rap1 GAP protein, and IRF3 (Ronco et al., 1998), a transcriptional regulator involved in interferon response. The role of these interactions in the productive life cycle of HPVs remains to be elucidated.

A series of binding partners have also been identified for the low-risk E6 proteins, which do not bind p53. These include zyxin (Degenhardt and Silverstein, 2001a), GPS2 (Degenhardt and Silverstein, 2001b), Bak (Thomas and Banks, 1998), and MCM7 (Kukimoto et al., 1998). As in the case of HPV 31, mutation of E6 in the context of the complete HPV 11 genome leads to loss of the ability to maintain viral plasmids (Oh et al., 2004). Of the identified binding partners of low-risk E6 protein, the association with the cellular replication protein MCM7 is most likely to play a role in mediating plasmid maintenance. It is possible that as-yet-unknown binding partners of E6 that are common to both high- and low-risk types play essential roles in virus replication.

4.8.   Effects of HPVs on Epithelial Differentiation

In normal epithelia, the basal layer contains stem cells as well as transitory amplifying cells, which are the primary generators of cells in stratified epithelia (Zegers et al., 2003). Once transitory amplifying cells divide, one of the daughter cells exits the cell cycle and migrates away from the basal layer to begin differentiation. The presence of HPV gene products maintains differentiating cells in the cell cycle and blocks the degradation of nuclei (Stubenrauch and Laimins, 1999). Examination of the expression of markers of epithelial cell differentiation such as involucrin, keratin 10, transglutaminase, and filaggrin indicate that HPVs exert modest effects on cell differentiation during the productive stage of the viral life cycle (Ruesch and Laimins, 1998). During the productive life cycle, there is a close correlation between the expression of filaggrin and the induction of late viral functions (Ruesch and Laimins, 1998). HPV-positive cells activate late viral functions in differentiating cells that have re-entered S-phase, and fail to express the intermediate filament associated protein, filaggrin. In contrast, the cells that do not amplify viral DNA nor express late genes remain in G_1 and synthesize filaggrin. Since viral late gene expression occurs only upon differentiation, it is likely that differentiation-specific cellular factors regulate expression of the viral genome. Elucidation of these factors should help in understanding how the productive phase of the viral life cycle is controlled.

4.9.   Conclusion

The development of genetic methods to study HPVs functions in the productive life cycle has provided an opportunity to understand the action of viral proteins in the proper context under physiological conditions. Further application of these methods will help us to continue to expand our knowledge of how infection by only certain HPV types leads to malignant transformation and to characterize the mechanisms that regulate viral pathogenesis.

Acknowledgements. This work was supported by grants from the National Cancer Institute and National Institute of Allergy and Infectious Diseases to L.A.L.

References

Apt, D., Watts, R.M., Suske, G., and Bernard, H.U. (1996). High Sp1/Sp3 ratios in epithelial cells during epithelial differentiation and cellular transformation correlate with the activation of the HPV-16 promoter. *Virology* 224:281–291.

Asselineau, D., and Prunieras, M. (1984). Reconstruction of 'simplified' skin: control of fabrication. *Br. J. Dermatol.* 111(Suppl 27):219–222.

Bauknecht, T., Jundt, F., Herr, I., Oehler, T., Delius, H., Shi, Y., Angel, P., and Zur Hausen, H. (1995). A switch region determines the cell type-specific positive or negative action of YY1 on the activity of the human papillomavirus type 18 promoter. *J. Virol.* 69:1–12.

Be, X., Hong, Y., Wei, J., Androphy, E.J., Chen, J.J., and Baleja, J.D. (2001). Solution structure determination and mutational analysis of the papillomavirus E6 interacting peptide of E6AP. *Biochemistry* 40:1293–1299.

Bedell, M.A., Hudson, J.B., Golub, T.R., Turyk, M.E., Hosken, M., Wilbanks, G.D., and Laimins, L.A. (1991). Amplification of human papillomavirus genomes *in vitro* is dependent on epithelial differentiation. *J. Virol.* 65:2254–2260.

Beger, M., Butz, K., Denk, C., Williams, T., Hurst, H.C., and Hoppe-Seyler, F. (2001). Expression pattern of AP-2 transcription factors in cervical cancer cells and analysis of their influence on human papillomavirus oncogene transcription. *J. Mol. Med.* 79:314–320.

Brandsma, J.L., and Xiao, W. (1993). Infectious virus replication in papillomas induced by molecularly cloned cottontail rabbit papillomavirus DNA. *J. Virol.* 67:567–571.

Brandsma, J.L., Yang, Z.H., Barthold, S.W., and Johnson, E.A. (1991). Use of a rapid, efficient inoculation method to induce papillomas by cottontail rabbit papillomavirus DNA shows that the E7 gene is required. *Proc. Natl. Acad. Sci. U S A* 88:4816–4820.

Brandsma, J.L., Yang, Z.H., DiMaio, D., Barthold, S.W., Johnson, E., and Xiao, W. (1992). The putative E5 open reading frame of cottontail rabbit papillomavirus is dispensable for papilloma formation in domestic rabbits. *J. Virol.* 66:6204–6207.

Brehm, A., Nielsen, S.J., Miska, E.A., McCance, D.J., Reid, J.L., Bannister, A.J., and Kouzarides, T. (1999). The E7 oncoprotein associates with Mi2 and histone deacetylase activity to promote cell growth. *EMBO J.* 18:2449–2458.

Burnett, S., Jareborg, N., and DiMaio, D. (1992). Localization of bovine papillomavirus type 1 E5 protein to transformed basal keratinocytes and permissive differentiated cells in fibropapilloma tissue. *Proc. Natl. Acad. Sci. U S A* 89:5665–5669.

Chang, F., Syrjanen, S., Kellokoski, J., and Syrjanen, K. (1991). Human papillomavirus (HPV) infections and their associations with oral disease. *J. Oral Pathol. Med.* 20:305–317.

Chen, J.J., Reid, C.E., Band, V., and Androphy, E.J. (1995). Interaction of papillomavirus E6 oncoproteins with a putative calcium-binding protein. *Science* 269:529–531.

Cheng, S., Schmidt-Grimminger, D.C., Murant, T., Broker, T.R., and Chow, L.T. (1995). Differentiation-dependent up-regulation of the human papillomavirus E7 gene reactivates cellular DNA replication in suprabasal differentiated keratinocytes. *Gene Dev.* 9:2335–2349.

Chong, T., Apt, D., Gloss, B., Isa, M., and Bernard, H.U. (1991). The enhancer of human papillomavirus type 16: binding sites for the ubiquitous transcription factors oct-1, NFA, TEF-2, NF1, and AP-1 participate in epithelial cell-specific transcription. *J. Virol.* 65:5933–5943.

Collier, B., Oberg, D., Zhao, X., and Schwartz, S. (2002). Specific inactivation of inhibitory sequences in the 5' end of the human papillomavirus type 16 L1 open reading frame results in production of high levels of L1 protein in human epithelial cells. *J. Virol.* 76:2739–2752.

Conger, K.L., Liu, J.S., Kuo, S.R., Chow, L.T., and Wang, T.S. (1999). Human papillomavirus DNA replication. Interactions between the viral E1 protein and two subunits of human DNA polymerase α/primase. *J. Biol. Chem.* 274:2696–2705.

Conrad, M., Bubb, V.J., and Schlegel, R. (1993). The human papillomavirus type 6 and 16 E5 proteins are membrane-associated proteins which associate with the 16-kilodalton pore-forming protein. *J. Virol.* 67:6170–6178.

Craven, S.E., and Bredt, D.S. (1998). PDZ proteins organize synaptic signaling pathways. *Cell* 93:495–498.

Das, K., Bohl, J., and Vande Pol, S.B. (2000). Identification of a second transforming function in bovine papillomavirus type 1 E6 and the role of E6 interactions with paxillin, E6BP, and E6AP. *J. Virol.* 74:812–816.

Davy, C.E., Jackson, D.J., Wang, Q., Raj, K., Masterson, P.J., Fenner, N.F., Southern, S., Cuthill, S., Millar, J.B., and Doorbar, J. (2002). Identification of a G(2) arrest domain in the E1 wedge E4 protein of human papillomavirus type 16. *J. Virol.* 76:9806–9818.

Davy, C.E., Jackson, D.J., Raj, K., Peh, W.L., Southern, S.A., Das, P., Sorathia, R., Laskey, P., Middleton, K., Nakahara, T., Wang, Q., Masterson, P.J., Lambert, P.F., Cuthill, S., Millar, J.B., Doorbar, J. (2005). Human papillomavirus type 16 E1 E4-induced G2 arrest is associated with cytoplasmic retention of active Cdk1/cyclin B1 complexes. *J. Virol.*, 79(7): 3998–4011.

Defeo-Jones, D., Vuocolo, G.A., Haskell, K.M., Hanobik, M.G., Kiefer, D.M., McAvoy, E.M., Ivey-Hoyle, M., Brandsma, J.L., Oliff, A., and Jones, R.E. (1993). Papillomavirus E7 protein binding to the retinoblastoma protein is not required for viral induction of warts. *J. Virol.* 67:716–725.

Degenhardt, Y.Y., and Silverstein, S. (2001a). Interaction of zyxin, a focal adhesion protein, with the e6 protein from human papillomavirus type 6 results in its nuclear translocation. *J. Virol.* 75:11791–11802.

Degenhardt, Y.Y., and Silverstein, S.J. (2001b). Gps2, a protein partner for human papillomavirus E6 proteins. *J. Virol.* 75:151–160.

del Mar Pena, L.M., and Laimins, L.A. (2001). Differentiation-dependent chromatin rearrangement coincides with activation of human papillomavirus type 31 late gene expression. *J. Virol.* 75, 10005–10013.

DiLorenzo, T.P., and Steinberg, B.M. (1995). Differential regulation of human papillomavirus type 6 and 11 early promoters in cultured cells derived from laryngeal papillomas. *J. Virol.* 69:6865–6872.

Doorbar, J., Ely, S., Sterling, J., McLean, C., and Crawford, L. (1991). Specific interaction between HPV-16 E1–E4 and cytokeratins results in collapse of the epithelial cell intermediate filament network. *Nature* 352:824–827.

Doorbar, J., Campbell, D., Grand, R.J., Gallimore, P.H. (1986). Identification of the human papilloma virus-1a E4 gene products. EMBO J. 5(2):355–362.

Doorbar, J., Medcalf, E., and Napthine, S. (1996) Analysis of HPV1 E4 complexes and their association with keratins *in vivo*. *Virology* 218(1):114–126.

Doorbar, J., Foo, C., Coleman, N., Medcalf, L., Hartley, O., Prospero, T., Napthine, S., Sterling, J., Winter, G., Griffin, H. (1997). Characterization of events during the late stages of HPV16 infection *in vivo* using high-affinity synthetic Fabs to E4. *Virology* 238(1):40–52.

Doorbar, J., Elston, R.C., Napthine, S., Raj, K., Medcalf, E., Jackson, D., Coleman, N., Griffin, H.M., Masterson, P., Stacey, S., Mengistu, Y., Dunlop, J. (2000). E1E4 protein of human papillomavirus type 16 associates with a putative RNA helicase through sequences in its C terminus. *J. Virol.* 74(Nov.;21):10081–10095.

Dyson, N., Howley, P.M., Munger, K., and Harlow, E. (1989). The human papilloma virus-16 E7 oncoprotein is able to bind to the retinoblastoma gene product. *Science* 243:934–937.

Edmonds, C., and Vousden, K.H. (1989). A point mutational analysis of human papillomavirus type 16 E7 protein. *J. Virol.* 63:2650–2656.

Fanning, A.S., and Anderson, J.M. (1999). PDZ domains: fundamental building blocks in the organization of protein complexes at the plasma membrane. *J. Clin. Invest.* 103:767–772.

Fehrmann, F., Klumpp, D.J., and Laimins, L.A. (2003). Human papillomavirus type 31 E5 protein supports cell cycle progression and activates late viral functions upon epithelial differentiation. *J. Virol.* 77:2819–2831.

Filippova, M., Song, H., Connolly, J.L., Dermody, T.S., and Duerksen-Hughes, P.J. (2002). The human papillomavirus 16 E6 protein binds to tumor necrosis factor (TNF) R1 and protects cells from TNF-induced apoptosis. *J. Biol. Chem.* 277:21730–21739.

Flores, E.R., Allen-Hoffmann, B.L., Lee, D., and Lambert, P.F. (2000). The human papillomavirus type 16 E7 oncogene is required for the productive stage of the viral life cycle. *J. Virol.* 74:6622–6631.

Flores, E.R., Allen-Hoffmann, B.L., Lee, D., Sattler, C.A., and Lambert, P.F. (1999). Establishment of the human papillomavirus type 16 (HPV-16) life cycle in an immortalized human foreskin keratinocyte cell line. *Virology* 262:344–354.

Frattini, M.G., and Laimins, L.A. (1994). Binding of the human papillomavirus E1 origin-recognition protein is regulated through complex formation with the E2 enhancer-binding protein. *Proc. Natl. Acad. Sci. U S A* 91:12398–12402.

Frattini, M.G., Lim, H.B., Doorbar, J., and Laimins, L.A. (1997). Induction of human papillomavirus type 18 late gene expression and genomic amplification in organotypic cultures from transfected DNA templates. *J. Virol.* 71:7068–7072.

Frattini, M.G., Lim, H.B., and Laimins, L.A. (1996). *In vitro* synthesis of oncogenic human papillomaviruses requires episomal genomes for differentiation-dependent late expression. *Proc. Natl. Acad. Sci. U S A* 93:3062–3067.

Gao, Q., Kumar, A., Srinivasan, S., Singh, L., Mukai, H., Ono, Y., Wazer, D.E., and Band, V. (2000). PKN binds and phosphorylates human papillomavirus E6 oncoprotein. *J. Biol. Chem.* 275:14824–14830.

Gao, Q., Srinivasan, S., Boyer, S.N., Wazer, D.E., and Band, V. (1999). The E6 oncoproteins of high-risk papillomaviruses bind to a novel putative GAP protein, E6TP1, and target it for degradation. *Mol. Cell. Biol.* 19:733–744.

Gardiol, D., Kuhne, C., Glaunsinger, B., Lee, S.S., Javier, R., and Banks, L. (1999). Oncogenic human papillomavirus E6 proteins target the discs large tumour suppressor for proteasome-mediated degradation. *Oncogene* 18:5487–5496.

Genther, S.M., Sterling, S., Duensing, S., Munger, K., Sattler, C., and Lambert, P.F. (2003). Quantitative role of the human papillomavirus type 16 E5 gene during the productive stage of the viral life cycle. *J. Virol.* 77:2832–2842.

Giroglou, T., Florin, L., Schafer, F., Streeck, R.E., and Sapp, M. (2001) Human papillomavirus infection requires cell surface heparan sulfate. *J. Virol.* 75:1565–1570.

Glaunsinger, B.A., Lee, S.S., Thomas, M., Banks, L., and Javier, R. (2000). Interactions of the PDZ-protein MAGI-1 with adenovirus E4-ORF1 and high-risk papillomavirus E6 oncoproteins. *Oncogene* 19:5270–5280.

Grassmann, K., Rapp, B., Maschek, H., Petry, K.U., and Iftner, T. (1996). Identification of a differentiation-inducible promoter in the E7 open reading frame of human papillomavirus type 16 (HPV-16) in raft cultures of a new cell line containing high copy numbers of episomal HPV-16 DNA. *J. Virol.* 70:2339–2349.

Halbert, C.L., Demers, G.W., and Galloway, D.A. (1992). The E6 and E7 genes of human papillomavirus type 6 have weak immortalizing activity in human epithelial cells. *J. Virol.* 66:2125–2134.

Heck, D.V., Yee, C.L., Howley, P.M., and Munger, K. (1992). Efficiency of binding the retinoblastoma protein correlates with the transforming capacity of the E7 oncoproteins of the human papillomaviruses. *Proc. Natl. Acad. Sci. U S A* 89:4442–4446.

Howley, P.M., Lowy, D.R. (2001). Papillomaviruses and their replication. In P.M. Howley (ed.): *Virology*, vol. 2. Philadelphia, PA: Lippincott/The Williams & Wilkins Co, pp. 2197–2229.

Huibregtse, J.M., and Beaudenon, S.L. (1996). Mechanism of HPV E6 proteins in cellular transformation. *Semin. Cancer Biol.* 7:317–326.

Huibregtse, J.M., Scheffner, M., and Howley, P.M. (1991). A cellular protein mediates association of p53 with the E6 oncoprotein of human papillomavirus types 16 or 18. *EMBO J.* 10:4129–4135.

Hummel, M., Hudson, J.B., and Laimins, L.A. (1992). Differentiation-induced and constitutive transcription of human papillomavirus type 31b in cell lines containing viral episomes. *J. Virol.* 66:6070–6080.

Iftner, T., Elbel, M., Schopp, B., Hiller, T., Loizou, J.I., Caldecott, K.W., and Stubenrauch, F.(2002). Interference of papillomavirus E6 protein with single-strand break repair by interaction with XRCC1. *EMBO J.* 21:4741–4748.

Ishiji, T., Lace, M.J., Parkkinen, S., Anderson, R.D., Haugen, T.H., Cripe, T.P., Xiao, J.H., Davidson, I., Chambon, P., and Turek, L.P. (1992). Transcriptional enhancer factor (TEF)-1 and its cell-specific co-activator activate human papillomavirus-16 E6 and E7 oncogene transcription in keratinocytes and cervical carcinoma cells. *EMBO J.* 11:2271–2281.

Joyce, J.G., Tung, J.S., Przysiecki, C.T., Cook, J.C., Lehman, E.D., Sands, J.A., Jansen, K.U., and Keller, P.M. (1999). The L1 major capsid protein of human papillomavirus type 11 recombinant virus-like particles interacts with heparin and cell-surface glycosaminoglycans on human keratinocytes. *J. Biol. Chem.* 274:5810–5822.

Klingelhutz, A.J., Foster, S.A., and McDougall, J.K. (1996). Telomerase activation by the E6 gene product of human papillomavirus type 16. *Nature* 380:79–82.

Klumpp, D.J., and Laimins, L.A. (1999). Differentiation-induced changes in promoter usage for transcripts encoding the human papillomavirus type 31 replication protein E1. *Virology* 257:239–246.

Knight, G.L., Grainger, J.R., Gallimore, P.H., and Roberts, S. (2004). Cooperation between different forms of the human papillomavirus type 1 E4 protein to block cell cycle progression and cellular DNA synthesis. *J. Virol.* 78:13920–13933.

Kubbutat, M.H., V.K.H. (1996). Role of E6 and E7 oncoproteins in HPV-induced anogenital malignancies. *Semin. Virol.* 7:295–304.

Kuhne, C., and Banks, L. (1998). E3-ubiquitin ligase/E6-AP links multicopy maintenance protein 7 to the ubiquitination pathway by a novel motif, the L2G box. *J. Biol. Chem.* 273:34302–34309.

Kukimoto, I., Aihara, S., Yoshiike, K., and Kanda, T. (1998). Human papillomavirus oncoprotein E6 binds to the C-terminal region of human minichromosome maintenance 7 protein. *Biochem. Biophys. Res. Commun.* 249:258–262.

Kumar, A., Zhao, Y., Meng, G., Zeng, M., Srinivasan, S., Delmolino, L.M., Gao, Q., Dimri, G., Weber, G.F., Wazer, D.E., *et al.* (2002). Human papillomavirus oncoprotein E6 inactivates the transcriptional coactivator human ADA3. *Mol. Cell. Biol.* 22: 5801–5812.

Kyo, S., Klumpp, D.J., Inoue, M., Kanaya, T., and Laimins, L.A. (1997). Expression of AP1 during cellular differentiation determines human papillomavirus E6/E7 expression in stratified epithelial cells. *J. Gen. Virol.* 78:401–411.

Kyo, S., Tam, A., and Laimins, L.A. (1995). Transcriptional activity of human papillomavirus type 31b enhancer is regulated through synergistic interaction of AP1 with two novel cellular factors. *Virology* 211:184–197.

Laimins, L.A. (1993). The biology of human papillomaviruses: from warts to cancer. *Infect. Agent Dis.* 2:74–86.

Laimins, L.A. (1998). Regulation of transcription and replication by human papillomaviruses. In D.J. McCance (ed.): *Human tumor viruses.* Washington, D.C.:American Society for Microbiology, pp. 201–223.

Lambert, P.F. (1991). Papillomavirus DNA replication. *J. Virol.* 65:3417–3420.

Lee, C., and Laimins, L.A. (2004). Role of the PDZ domain-binding motif of the oncoprotein E6 in the pathogenesis of human papillomavirus type 31. *J. Virol.* 78:12366–12377.

Lee, S.S., Glaunsinger, B., Mantovani, F., Banks, L., and Javier, R.T. (2000). Multi-PDZ domain protein MUPP1 is a cellular target for both adenovirus E4-ORF1 and high-risk papillomavirus type 18 E6 oncoproteins. *J. Virol.* 74:9680–9693.

Longworth, M.S., and Laimins, L.A. (2004). The binding of histone deacetylases and the integrity of zinc finger-like motifs of the E7 protein are essential for the life cycle of human papillomavirus type 31. *J. Virol.* 78:3533–3541.

Lowy, D.R., Kirnbauer, R., and Schiller, J.T. (1994). Genital human papillomavirus infection. *Proc. Natl. Acad. Sci. U S A* 91:2436–2440.

Mack, D.H., and Laimins, L.A. (1991). A keratinocyte-specific transcription factor, KRF-1, interacts with AP-1 to activate expression of human papillomavirus type 18 in squamous epithelial cells. *Proc. Natl. Acad. Sci. U S A* 88:9102–9106.

Martin, L.G., Demers, G.W., and Galloway, D.A. (1998). Disruption of the G1/S transition in human papillomavirus type 16 E7-expressing human cells is associated with altered regulation of cyclin E. *J. Virol.* 72:975–985.

McCance, D.J., Kopan, R., Fuchs, E., and Laimins, L.A. (1988). Human papillomavirus type 16 alters human epithelial cell differentiation *in vitro*. *Proc. Natl. Acad. Sci. U S A* 85:7169–7173.

Meyers, C., Frattini, M.G., Hudson, J.B., and Laimins, L.A. (1992). Biosynthesis of human papillomavirus from a continuous cell line upon epithelial differentiation. *Science* 257:971–973.

Meyers, C., and Laimins, L.A. (1994). *In vitro* systems for the study and propagation of human papillomaviruses. *Curr. Top. Microbiol. Immunol.* 186:199–215.

Meyers, C., Mayer, T.J., and Ozbun, M.A. (1997). Synthesis of infectious human papillomavirus type 18 in differentiating epithelium transfected with viral DNA. *J. Virol.* 71:7381–7386.

Middleton, K., Peh, W., Southern, S., Griffin, H., Sotlar, K., Nakahara, T., El-Sherif, A., Morris, L., Seth, R., Hibma, M., Jenkins, D., Lambert, P., Coleman, N., Doorbar, J. (2003). Organization of human papillomavirus productive cycle during neoplastic progression provides a basis for selection of diagnostic markers. *J. Virol.* 77(19): 10186–10201.

Mittal, R., Pater, A., and Pater, M.M. (1993). Multiple human papillomavirus type 16 glucocorticoid response elements functional for transformation, transient expression, and DNA–protein interactions. *J. Virol.* 67:5656–5659.

Mohr, I.J., Clark, R., Sun, S., Androphy, E.J., MacPherson, P., and Botchan, M.R. (1990). Targeting the E1 replication protein to the papillomavirus origin of replication by complex formation with the E2 transactivator. *Science* 250:1694–1699.

Munger, K., Werness, B.A., Dyson, N., Phelps, W.C., Harlow, E., and Howley, P.M. (1989). Complex formation of human papillomavirus E7 proteins with the retinoblastoma tumor suppressor gene product. *EMBO J.* 8:4099–4105.

Nakagawa, S., and Huibregtse, J.M. (2000). Human scribble (Vartul) is targeted for ubiquitin-mediated degradation by the high-risk papillomavirus E6 proteins and the E6AP ubiquitin–protein ligase. *Mol. Cell. Biol.* 20:8244–8253.

Nakahara, T., Nishimura, A., Tanaka, M., Ueno, T., Ishimoto, A., and Sakai, H. (2002). Modulation of the cell division cycle by human papillomavirus type 18 E4. *J. Virol.* 76:10914–10920.

Nasseri, M., Hirochika, R., Broker, T.R., Chow, L.T. (1987) A human papilloma virus type 11 transcript encoding an E1–E4 protein. *Virology* 159(2):433–439.

Oberg, D., Collier, B., Zhao, X., and Schwartz, S. (2003). Mutational inactivation of two distinct negative RNA elements in the human papillomavirus type 16 L2 coding region induces production of high levels of L2 in human cells. *J. Virol.* 77:11674–11684.

Oh, S.T., Kyo, S., and Laimins, L.A. (2001). Telomerase activation by human papillomavirus type 16 E6 protein: induction of human telomerase reverse transcriptase expression through Myc and GC-rich Sp1 binding sites. *J. Virol.* 75:5559–5566.

Oh, S.T., Longworth, M.S., and Laimins, L.A. (2004). Roles of the E6 and E7 proteins in the life cycle of low-risk human papillomavirus type 11. *J. Virol.* 78:2620–2626.

Ozbun, M.A. (2002). Human papillomavirus type 31b infection of human keratinocytes and the onset of early transcription. *J. Virol.* 76:11291–11300.

Ozbun, M.A., and Meyers, C. (1998a). Human papillomavirus type 31b E1 and E2 transcript expression correlates with vegetative viral genome amplification. *Virology* 248:218–230.

Ozbun, M.A., and Meyers, C. (1998b). Temporal usage of multiple promoters during the life cycle of human papillomavirus type 31b. *J. Virol.* 72:2715–2722.

Ozbun, M.A., and Meyers, C. (1999). Two novel promoters in the upstream regulatory region of human papillomavirus type 31b are negatively regulated by epithelial differentiation. *J. Virol.* 73:3505–3510.

Palefsky, J.M., Winkler, B., Rabanus, J.P., Clark, C., Chan, S., Nizet, V., Schoolnik, G.K. (1991). Characterization of *in vivo* expression of the human papillomavirus type 16 E4 protein in cervical biopsy tissues. *J. Clin. Invest.* 87(6): 2132–2141.

Park, R.B., and Androphy, E.J. (2002). Genetic analysis of high-risk e6 in episomal maintenance of human papillomavirus genomes in primary human keratinocytes. *J. Virol.* 76:11359–11364.

Patel, D., Huang, S.M., Baglia, L.A., and McCance, D.J. (1999). The E6 protein of human papillomavirus type 16 binds to and inhibits co-activation by CBP and p300. *EMBO J.* 18:5061–5072.

Pattison, S., Skalnik, D.G., and Roman, A. (1997). CCAAT displacement protein, a regulator of differentiation-specific gene expression, binds a negative regulatory element within the 5' end of the human papillomavirus type 6 long control region. *J. Virol.* 71:2013–2022.

Peh, W.L., Middleton, K., Christensen, N., Nicholls, P., Egawa, K., Sotlar, K., Brandsma, J., Percival, A., Lewis, J., Liu, W.J., Doorbar, J. (2002). Life cycle heterogeneity in animal models of human papillomavirus-associated disease. *J. Virol.*, 76(20): 10401–10416.

Peh, W.L., Brandsma, J.L., Christensen, N.D., Cladel, N.M., Wu, X., and Doorbar, J. (2004). The viral E4 protein is required for the completion of the cottontail rabbit papillomavirus productive cycle *in vivo. J. Virol.* 78:2142–2151.

Pim, D., Thomas, M., Javier, R., Gardiol, D., and Banks, L. (2000). HPV E6 targeted degradation of the discs large protein: evidence for the involvement of a novel ubiquitin ligase. *Oncogene* 19:719–725.

Pray, T.R. and Laimins, L.A. (1995). Differentiation-dependent expression of E1–E4 proteins in cell lines maintaining episomes of human papillomavirus type 31b. *Virology* 206(1):679–685.

Remm, M., Remm, A., and Ustav, M. (1999). Human papillomavirus type 18 E1 protein is translated from polycistronic mRNA by a discontinuous scanning mechanism. *J. Virol.* 73:3062–3070.

Roberts, S., Ashmole, I., Sheehan, T., Davies, A. and Galimore,P.H. (1993). Cutaneous and mucosal human papillomavirus E4 proteins form intermediate filament-like structures in epithelial cells. *Virology* 197(1):176–187.

Roberts, S. Ashmole, I., Gibson, L.J., Rookes, S.M., Barton, G.J., and Gallimore, P.H. (1994). Mutational analysis of human papillomavirus E4 proteins: identification of structural features important in the formation of cytoplasmic E4/cytokeratin networks in epithelial cells. *J. Virol.* 68(10):6432–6445.

Roberts, S., Ashmole, I., Rookes, S.M., and Gallimore, P.H. (1997). Mutational analysis of the human papillomavirus type 16 E1–E4 protein shows that the C terminus is dispensable for keratin cytoskeleton association but is involved in inducing disruption of the keratin filaments. *J. Virol.* 71(5):3554–3562.

Rohlfs, M., Winkenbach, S., Meyer, S., Rupp, T., and Durst, M. (1991). Viral transcription in human keratinocyte cell lines immortalized by human papillomavirus type-16. *Virology* 183:331–342.

Ronco, L.V., Karpova, A.Y., Vidal, M., and Howley, P.M. (1998). Human papillomavirus 16 E6 oncoprotein binds to interferon regulatory factor-3 and inhibits its transcriptional activity. *Gene Dev.* 12:2061–2072.

Ruesch, M.N., and Laimins, L.A. (1998). Human papillomavirus oncoproteins alter differentiation-dependent cell cycle exit on suspension in semisolid medium. *Virology* 250:19–29.

Ruesch, M.N., Stubenrauch, F., and Laimins, L.A. (1998). Activation of papillomavirus late gene transcription and genome amplification upon differentiation in semisolid medium is coincident with expression of involucrin and transglutaminase but not keratin-10. *J. Virol.* 72:5016–5024.

Sanders, C.M., and Stenlund, A. (1998). Recruitment and loading of the E1 initiator protein: an ATP-dependent process catalysed by a transcription factor. *EMBO J.* 17:7044–7055.

Scheffner, M., Huibregtse, J.M., Vierstra, R.D., and Howley, P.M. (1993). The HPV-16 E6 and E6-AP complex functions as a ubiquitin-protein ligase in the ubiquitination of p53. *Cell* 75:495–505.

Scheffner, M., Werness, B.A., Huibregtse, J.M., Levine, A.J., and Howley, P.M. (1990). The E6 oncoprotein encoded by human papillomavirus types 16 and 18 promotes the degradation of p53. *Cell* 63:1129–1136.

Sedman, J., and Stenlund, A. (1998). The papillomavirus E1 protein forms a DNA-dependent hexameric complex with ATPase and DNA helicase activities. *J Virol* 72:6893–6897.

Sen, E., Alam, S., and Meyers, C. (2004). Genetic and biochemical analysis of cis regulatory elements within the keratinocyte enhancer region of the human papillomavirus type 31 upstream regulatory region during different stages of the viral life cycle. *J. Virol.* 78:612–629.

Sen, E., Bromberg-White, J.L., and Meyers, C. (2002). Genetic analysis of cis regulatory elements within the 5' region of the human papillomavirus type 31 upstream regulatory region during different stages of the viral life cycle. *J. Virol.* 76:4798–4809.

Seo, Y.S., Muller, F., Lusky, M., and Hurwitz, J. (1993). Bovine papilloma virus (BPV)-encoded E1 protein contains multiple activities required for BPV DNA replication. *Proc. Natl. Acad. Sci. U S A* 90:702–706.

Smotkin, D., Prokoph, H., and Wettstein, F.O. (1989). Oncogenic and nononcogenic human genital papillomaviruses generate the E7 mRNA by different mechanisms. *J. Virol.* 63:1441–1447.

Spink, K.M. and Laimins, L.A. (2005) Induction of the human papillomavirus type 31 late promoter requires differentiation but not DNA amplification. *J. Virol.* 79:4918–4926.

Srivenugopal, K.S., and Ali-Osman, F. (2002). The DNA repair protein, O(6)-methylguanine–DNA methyltransferase is a proteolytic target for the E6 human papillomavirus oncoprotein. *Oncogene* 21:5940–5945.

Stanley, M.A., Browne, H.M., Appleby, M., and Minson, A.C. (1989). Properties of a non-tumorigenic human cervical keratinocyte cell line. *Int. J. Cancer* 43:672–676.

Steger, G., and Corbach, S. (1997). Dose-dependent regulation of the early promoter of human papillomavirus type 18 by the viral E2 protein. *J. Virol.* 71:50–58.

Sterling, J.C., Skepper, J.N., and Stanley, M.A. (1993) Immunoelectron microscopical localization of human papillomavirus type 16 L1 and E4 proteins in cervical keratinocytes cultured *in vivo*. *J. Invest. Dermatol.* 100(2):154–158.

Straight, S.W., Herman, B., and McCance, D.J. (1995). The E5 oncoprotein of human papillomavirus type 16 inhibits the acidification of endosomes in human keratinocytes. *J. Virol.* 69:3185–3192.

Straight, S.W., Hinkle, P.M., Jewers, R.J., and McCance, D.J. (1993). The E5 oncoprotein of human papillomavirus type 16 transforms fibroblasts and effects the downregulation of the epidermal growth factor receptor in keratinocytes. *J. Virol.* 67:4521–4532.

Stubenrauch, F., Colbert, A.M., and Laimins, L.A. (1998a). Transactivation by the E2 protein of oncogenic human papillomavirus type 31 is not essential for early and late viral functions. *J. Virol.* 72:8115–8123.

Stubenrauch, F., Hummel, M., Iftner, T., and Laimins, L.A. (2000). The E8E2C protein, a negative regulator of viral transcription and replication, is required for extrachromosomal maintenance of human papillomavirus type 31 in keratinocytes. *J. Virol.* 74:1178–1186.

Stubenrauch, F., and Laimins, L.A. (1999). Human papillomavirus life cycle: active and latent phases. *Semin. Cancer Biol.* 9:379–386.

Stubenrauch, F., Lim, H.B., and Laimins, L.A. (1998b). Differential requirements for conserved E2 binding sites in the life cycle of oncogenic human papillomavirus type 31. *J. Virol.* 72:1071–1077.

Tan, S.H., Leong, L.E., Walker, P.A., and Bernard, H.U. (1994). The human papillomavirus type 16 E2 transcription factor binds with low cooperativity to two flanking sites and represses the E6 promoter through displacement of Sp1 and TFIID. *J. Virol.* 68:6411–6420.

Terhune, S.S., Hubert, W.G., Thomas, J.T., and Laimins, L.A. (2001). Early polyadenylation signals of human papillomavirus type 31 negatively regulate capsid gene expression. *J. Virol.* 75:8147–8157.

Terhune, S.S., Milcarek, C., and Laimins, L.A. (1999). Regulation of human papillomavirus type 31 polyadenylation during the differentiation-dependent life cycle. *J. Virol.* 73:7185–7192.

Thomas, J.T., Hubert, W.G., Ruesch, M.N., and Laimins, L.A. (1999). Human papillomavirus type 31 oncoproteins E6 and E7 are required for the maintenance of episomes during the viral life cycle in normal human keratinocytes. *Proc. Natl. Acad. Sci. U S A* 96:8449–8454.

Thomas, M., and Banks, L. (1998). Inhibition of Bak-induced apoptosis by HPV-18 E6. *Oncogene* 17:2943–2954.

Thomas, M., Laura, R., Hepner, K., Guccione, E., Sawyers, C., Lasky, L., and Banks, L. (2002). Oncogenic human papillomavirus E6 proteins target the MAGI-2 and MAGI-3 proteins for degradation. *Oncogene* 21:5088–5096.

Tong, X., and Howley, P.M. (1997). The bovine papillomavirus E6 oncoprotein interacts with paxillin and disrupts the actin cytoskeleton. *Proc. Natl. Acad. Sci. U S A* 94:4412–4417.

Ustav, M., and Stenlund, A. (1991). Transient replication of BPV-1 requires two viral polypeptides encoded by the E1 and E2 open reading frames. *EMBO J.* 10:449–457.

Vande Pol, S.B., Brown, M.C., and Turner, C.E. (1998). Association of Bovine Papillomavirus Type 1 E6 oncoprotein with the focal adhesion protein paxillin through a conserved protein interaction motif. *Oncogene* 16:43–52.

Wang, Q., Griffin, H., Southern, S., Jackson, D., Martin, A., McIntosh, P., Davy, C., Masterson, P.J., Walker, P.A., Laskey, P., Omary, M.B., Doorbar, J. (2004). Functional analysis of the human papillomavirus type 16 E1–E4 protein provides a mechanism for *in vivo* and *in vitro* keratin filament reorganization. *J. Virol.* 78(2):821–833.

Weintraub, S.J., Chow, K.N., Luo, R.X., Zhang, S.H., He, S., and Dean, D.C. (1995). Mechanism of active transcriptional repression by the retinoblastoma protein. *Nature* 375:812–815.

Werness, B.A., Levine, A.J., and Howley, P.M. (1990). Association of human papillomavirus types 16 and 18 E6 proteins with p53. *Science* 248:76–79.

Wilson, R., Fehrmann, F., and Laimins, L.A. (2005). Role of the E1–E4 protein in the differentiation-dependent life cycle of human papillomavirus type 31. *J. Virol.* 79(11):6732–6740.

Yang, L., Mohr, I., Fouts, E., Lim, D.A., Nohaile, M., and Botchan, M. (1993). The E1 protein of bovine papilloma virus 1 is an ATP-dependent DNA helicase. *Proc. Natl. Acad. Sci. U S A* 90:5086–5090.

Zegers, M.M., O'Brien, L.E., Yu, W., Datta, A., and Mostov, K.E. (2003). Epithelial polarity and tubulogenesis *in vitro*. *Trends Cell Biol.* 13:169–176.

zur Hausen, H. (2002). Papillomaviruses and cancer: from basic studies to clinical application. *Nat. Rev. Cancer* 2:342–350.

5
Papillomavirus Structure and Assembly

Robert L. Garcea[1] and Xiaojiang Chen[2]

[1] Section of Pediatric Oncology, University of Colorado School of Medicine, PO Box 6511, MS 8302, Aurora, CO 80045
[2] Department of Molecular and Computational Biology, University of Southern California, Los Angeles, CA 90089-1340

5.1.   Introduction

Papillomaviruses are spherical, nonenveloped DNA viruses 55–60 nm in diameter. Their coat or capsid is comprised of 72 pentamers (capsomeres) of their major capsid protein (L1) arranged on a T = 7 icosahedral lattice. An internal minor (in amount) capsid protein (L2) is associated with a subset of the L1 pentamers that forms the outer shell. Enclosed within the capsid shell is the viral genomic DNA packaged as a minichromosome by cellular nucleosomes. The overall structure of the papillomavirus virion is strikingly similar to that of the polyomaviruses, where VP1 is the major capsid protein comprising the pentameric capsomeres, and the VP2 and VP3 proteins are analogous to L2.

The structures of papilloma and polyoma viruses have been a longstanding object of investigation, and have served to model the study of other icosahedral viruses. Their symmetry provided support for the Crick-Watson postulate (Crick and Watson, 1956) that most viruses likely would have structures constructed of similar repeating subunits because of genomic coding economy, and that this constraint could be most easily accommodated with either icosahedral or helical symmetry. In particular, icosahedral geometry provides a near spherical container that most efficiently optimizes the ratio of surface area (and thus the minimal subunit number) to volume ratio. The first surprise was that the capsid consists of only 360 subunits of 72 pentamers. This all pentamer construction violated Caspar-Klug "quasi-equivalence" assembly rules, and raised the question of how identical pentamers could "bond-switch" between positions on the capsid surface having five (pentavalent) or six (hexavalent) neighbors. The atomic structures of recombinant papillomavirus virus-like particles and of SV40/polyoma virions now provide interesting explanations for these structural puzzles as well as for other biological functions.

The morphologic similarity between papillomaviruses and polyomaviruses (e.g., SV40) initially resulted in their co-classification in the papovavirus family (papilloma, polyoma, vaculoating agent [SV40]). As a consequence of

a taxonomic re-classification based upon their distinctive genomic organizations and biology, these are now two separate families, polyoma and papilloma.
Nonetheless, the morphological relationship between the two families, 72
pentameric capsomeres arranged on a T = 7 icosahedral lattice, provides
yet another puzzle—how do different proteins, L1 and VP1, without any
primary sequence homology, yield the same final assembled virion structure?
Furthermore, crystal structures of the two proteins showed the same fold in the
core domain (see below). These findings raise a very intriguing question whether
the same particle morphology is the result of convergent or divergent evolution.
Progress in the structural analysis of papillomavirus structure has been led by
findings with polyomaviruses, and comparison between structures illustrates how
common problems in assembly and biology have been solved using different
strategies. Our discussion will follow both virus families.

5.2.   Structure Determination

Electron microscopy and image analysis of negatively stained SV40, polyoma,
and papillomaviruses (Klug, 1965; Anderer et al., 1967; Finch and Klug, 1965)
established that the virion capsids were comprised of subunits (capsomeres)
arranged in an icosahedral lattice (Fig. 5.1). The T, or triangulation, number
nomenclature was derived by Caspar and Klug to explain the possible icosahedral symmetries of capsids of various sizes (Caspar and Klug, 1962). Strict
icosahedral symmetry implies $60 \times T$ identical copies of the capsid subunits.
In a T = 7 capsid, 60×7 or 420 identical subunits should be present, with 12
pentamers of these subunits at the pentavalent positions and 60 hexamers at the
hexavalent positions (T = 1 capsids have 60 subunits, or 12 pentamers, all located
at pentavalent positions). Unexpectedly, when Rayment et al. (1982) analyzed
low-resolution (25 Å) images of polyoma virions, only pentamers (72 pentamer
or 360 subunits) were identified (Fig. 5.2). This architecture was subsequently
confirmed for SV40 and papilloma virions using both cryo-electron microscopic
image reconstruction techniques as well as crystallography (Baker et al., 1989,
1991; Belnap et al., 1996; Liddington et al., 1991).
    The atomic structures of SV40 and polyomavirus preceded that of a papil-
lomavirus primarily because of the lack of an appropriate cell culture system
in which to grow and purify papillomaviruses. Initial attempts were made to
obtain the structure of bovine papilloma virions (BPV), purified from cow
warts, but the crystal quality was inadequate. Thus, a recombinant protein
approach was undertaken. After expression in *E. coli*, HPV16 L1 was purified
as pentameric capsomeres, and subjected to crystallization conditions. The
capsomeres assembled into small, uniform particles (T = 1) that were sufficiently
uniform for crystals of high resolution analysis (Chen et al., 2000).This T = 1
structure provides atomic details of the L1 monomer and capsomere, but not
inter-pentamer contacts that may be present in a larger T = 7 capsid. However,
the combination of the atomic structures of the recombinant L1 pentamers with

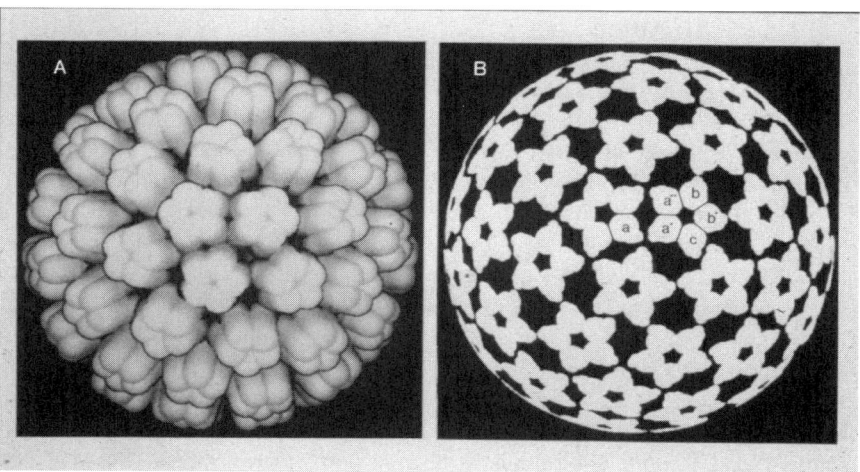

FIGURE 5.1. Schematic diagram of the papilloma (or polyoma) capsid. 72 pentamers (termed capsomeres) of the major capsid protein L1 are arranged in a T = 7 icosahedral lattice. In contrast to the rules of "quasi-equivalence" described by Caspar and Klug, the capsomeres are only pentamers and not pentamers and hexamers. Thus, pentamers must arrange themselves in two distinct positions: hexavalent (six neighboring pentamers) and pentavalent (five neighboring pentamers). The intrinsic bond switching capabilities of the L1 protein make this arrangement possible through changing conformations in the invading C-terminal arms of the protein as they bond with the neighboring pentamer (reproduced from Salunke et al., 1986).

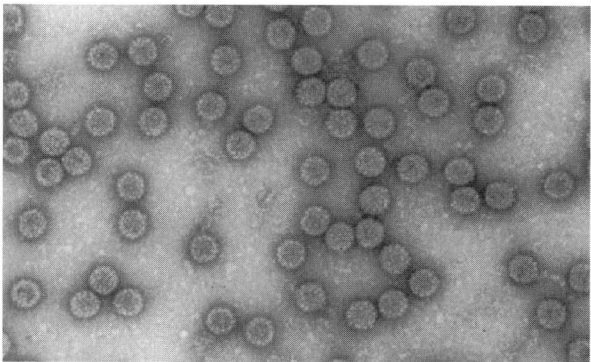

FIGURE 5.2. Electron micrograph of bovine papillomavirus. (See Plate 1).

data from the cryo-EM reconstruction of BPV at 9Å (Trus et al., 1997) have suggested a possible atomic model for contacts between L1 pentamers in the T = 7 virion (see below).

5.3.   L1 Monomer

The 3.5Å structure of T = 1 assemblies of recombinant HPV16 L1 capsomeres defines the majority of the bonding relationships within L1 monomers and pentamers (Chen et al., 2000). The core of the L1 monomer is composed of residues 20–382 (out of a total length of 504 amino acids for HPV16 L1) and appears as an eight-stranded anti-parallel β-barrel, in two distinct sheets composed of the CHEF and BIDG strands (Fig. 5.3, L1 monomer structure compared with VP1 structure). This structure is a classical "jelly roll" β-sandwich, similar to the one found in the polyomavirus VP1 structures (Liddington et al., 1991; Stehle et al., 1996), and reminiscent of the folding for picornaviruses such as the rhino and poliovirus capsid proteins (Hogle et al., 1985; Rossmann et al., 1985). Within the core domain, three elaborate loop domains (HI, DE, and FG) are located on the exterior surface in the assembled pentamer and particle. C-terminal to the core domain are residues 383–475 that fold into α-helical and long coiled structures that form the lateral projections from the core comprising the center of a pentamer. The remaining 30 residues of the C-terminus (476–505) are disordered and likely extend into the interior space of the pentamer or the particle where the basic residues in this region may

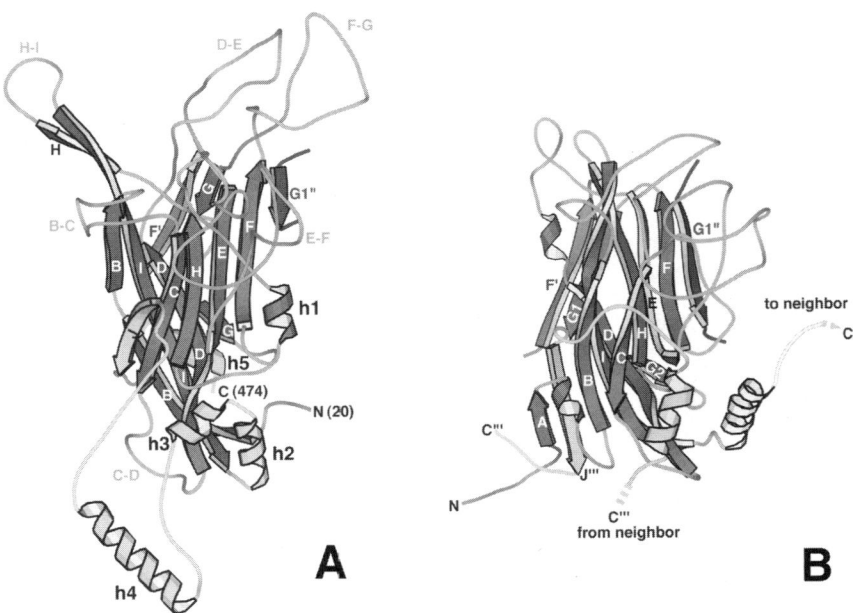

FIGURE 5.3. The monomer structures of L1 (**A**) and polyoma VP1 (**B**) (14). Both have a typical β-jelly roll structure comprised of an eight-stranded antiparallel β-barrel. The overall structural similarity is remarkable considering that L1 and VP1 have no primary amino acid homology.

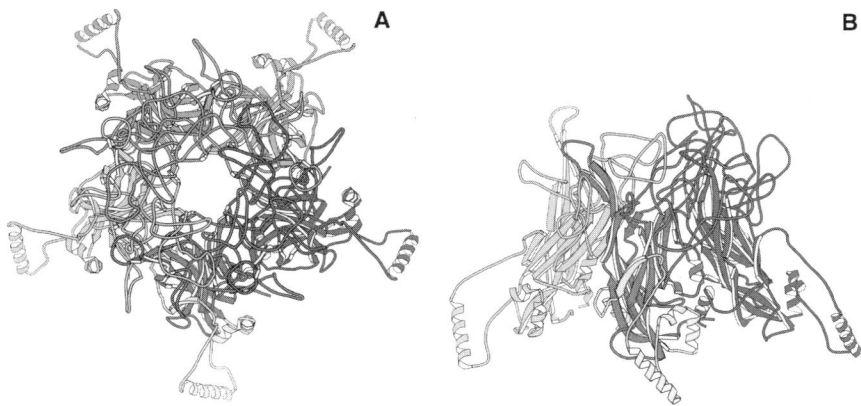

FIGURE 5.4. An HPV L1 pentamer viewing from the top (**A**) and from the side (**B**). The side view in panel B only show the three monomers in the front, with the two monomers in the back taken away to give a clearer view (Chen et al., 2000). (See Plate 2).

interact with the viral DNA. Deletions at the C-terminus of up to 30 residues, as well as fusion of additional sequences to the C-terminus (Chen et al., 2001; Muller et al., 1997; Paintsil et al., 1996), still permit pentamer formation and thus support the model that this stretch of residues at the very C-terminus is independently oriented and folded.

The contacts between L1 monomers within the pentamer are extensive. For example, the G strand of the BIDG sheet from one monomer interacts with the CHEF sheet from its neighbor. The surface loops also intertwine, with the HI loop of one monomer inserting between the FG and EF loop of the anti-clockwise neighbor. Fusion of GST to the amino terminus leaves the structure unaffected, again emphasizing the robust folding of the L1 monomer and the oligomerization into the pentamer structure (Chen et al., 2001).

5.4. Disulfide Bonds

The formation of disulfide bonds is critical for stable papilloma capsid assembly. L1 from HPV virions isolated from skin lesions is cross-linked by disulfide bonds (Doorbar and Gallimore, 1987). Sapp (Sapp et al., 1995) first demon-strated that dithiothreitol (DTT) treatment caused disassembly of L1 VLPs into capsomeres. When purified intact virons (BPV) are treated with DTT, there is no disassembly but rather a conformational change resulting in expansion of the capsids by approximately 10 percent in diameter (Li et al., 1998). This expansion allows penetration of proteases and nucleases to the interior, which can then result in virion disruption. This structural change may correspond to the "open" capsids seen by cryoEM (Belnap et al., 1996). In vitro assembly of capsids from recombinant pentamers (HPV11) is promoted by oxidation of cysteine residues, initiated by dialysis from the DTT-containing buffer that stabilized pentamers (Li et al., 1998). There are also data showing that certain constructs of L1 assemble into T = 7 particles in high salt and low pH in the presence of DTT (Chen et al., 2001), suggesting that the assembly for the empty shell of these L1s does not depend on disulfide bonds, as in the case of virion assembly. The specific cysteine residues contributing to the critical disulfide bonds have been mapped to Cys 424(HPV11)/Cys 427 (HPV33) and Cys 176 (HPV33) (Li et al., 1998; Sapp et al., 1998). In VLPs, about 50 percent of the L1 molecules are disulfide bonded trimers (Sapp et al., 1998) in contrast to almost complete disulfide bonding in virions isolated from warts. Slight differences may exist between serotypes, as HPV 16 L1 has been observed to dimerize and trimerize (Ishii et al., 2003), and BPV may have a more extensive crosslinking (Buck, personal comm.). Indeed, inclusion of cellular DNA into recombinant VLPs (HPV33) increases the L1 disulfide cross-linking to 100 percent, indicating that nucleic acid in the virion likely induces a capsid conformation that is structurally distinct from that of the VLP (Fligge et al., 2001).

5.5.   Pentamer–Pentamer Contacts

The T = 1 particle structure reveals the pentamer-pentamer bonding in this assembly (Chen et al., 2000). Because all 12 pentamers are pentavalent, all contacts between L1 monomers are identical. This bonding is accomplished by helix–helix interactions between the C-terminal lateral projections (Fig. 5.4 and Fig. 5.5). Helix 4 from each pentamer projects outward to interact with helices 2 and 3 of a neighbor, using strong hydrophobic interactions. The remainder of the C-terminal chain after residue 474 then returns to the L1 from where it originated. The structural contacts overall thus appear as three projecting helices from each L1 in a pentamer that abut neighboring helical projections around the threefold symmetric axes (Fig. 5.5).

The identification of specific residues involved in disulfide bonding allowed derivation of a T = 7 capsid model by Modis et al. (2002) (Fig. 5.6). In the T = 1 structure the key cysteine residue, cys428 (HPV16), is located at the end of helix 4, well away from any other cysteine with which it might disulfide bond. Moreover, the bonding between pentamers is distinct from that in the polyomaviruses where the C-terminal assembly domain (approximately the last 60 residues) intimately "invades" the neighboring pentamer, and does not return to the VP1 of origin. Thus, Modis et al. (2002) postulated that a T = 7 capsid may have a folding pattern distinct from that seen in the T = 1 structure. The atomic structure of the L1 pentamer (3.5Å) was therefore "fitted" into image reconstructions of BPV1 from cryoelectron microscopy data (9Å) (Trus et al., 1997) to model C-terminal structures in the T = 7 virion. The crystallographic density of residues 402–445 were found not to fit the EM map, and were therefore rebuilt using the additional constraint that cysteines 428 and 175 should disulfide bond. In the resulting new model, residues 403–413 act as a hinge, bridging

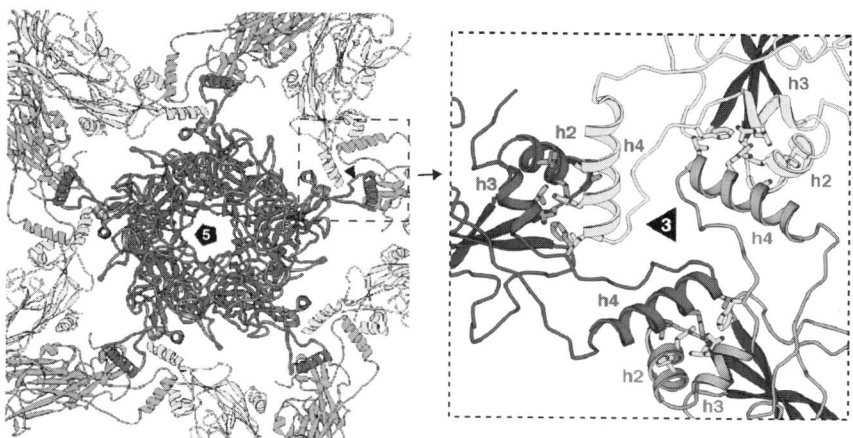

FIGURE 5.5. The threefold interactions in a T1 particle assembled with HPV16 L1 pentamers (Modis et al., 2002). In a T = 1 particle all pentamers of L1 have five neighboring pentamers, so that the inter-pentamer bonds are all equivalent. These bonds may be distinct from those seen in the complete T = 7 virion. (See Plate 3).

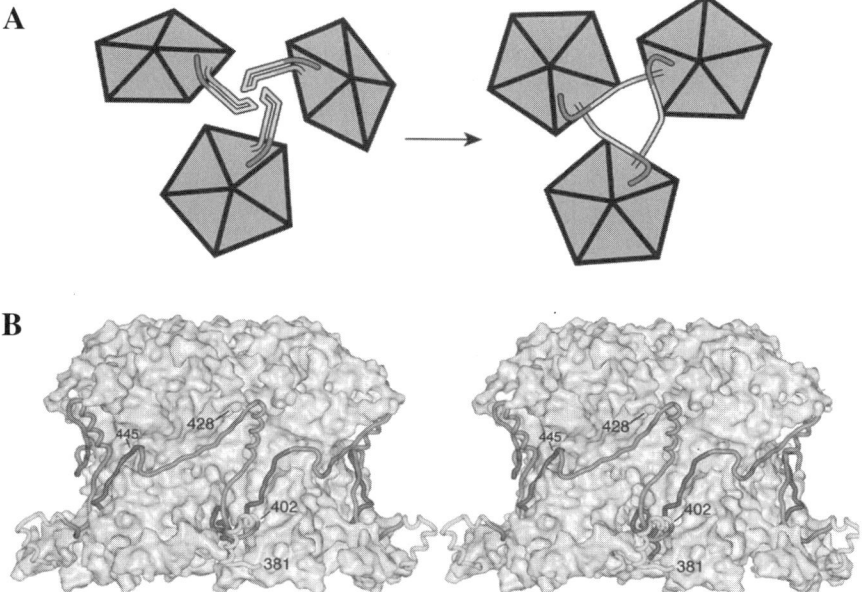

FIGURE 5.6. Schematic showing the bonding interactions modeled by Modis et al. (2002) in the complete T = 7 papilloma virion. **A**: The threefold pentamer–pentamer interactions observed in the T1 structure on the left and the possible C-terminal arm conformation at the same threefold location in a T7 particle. **B**: The invading C-terminal arms of adjacent pentamers wrapping around their neighboring pentamers allowing important juxtaposition of cysteine residues (e.g., cys428 HPV16L1) for disulfide bond formation. (See Plate 4).

donor and acceptor pentamers, and helix 4 (414–429) lies between the BC and EF loops of a neighboring L1 positioning cys428 next to the neighbor's cys175. Residues 430–446 lie around the circumference of the invaded pentamer; residues 447–474 have the same position in the invaded pentamer in the $T = 1$ structure where they resided in the pentamer of origin. In this model, the hinge region adopts different conformations for bonding between hexavalent and pentavalent pentamers, and serves as the "bond switch" (Fig. 5.6). The final structure appears as an "invading arm" model reminiscent of the carboxy-terminal arm of polyoma VP1 bonding to neighboring pentamers.

Disulfides are also important in the polyomavirus capsid (Christiansen et al., 1977; Walter and Deppert, 1974). In SV40 the invading C-terminal arm is anchored to the invaded pentamer by an interpentamer disulfide bond, and in mouse polyomavirus the invading arm is locked in place by an intrapentamer disulfide bond (Stehle et al., 1996). Unlike the polyomaviruses, however, calcium is not used to stabilize the papilloma virion. The disulfide bonds may not be completely oxidized until cell lysis (which may explain why in vitro assembly can occur in the presence of DTT). Upon cell entry, the disulfides would be reduced, opening the virion for proteolytic attack or other mechanisms that would facilitate disassembly.

Why the difference between $T = 1$ and $T = 7$ structures? The L1 protein used in the $T = 1$ structure determination was deleted for 10 residues at the amino terminus (Chen et al., 2000). These residues may normally prevent the C-terminal arms from returning to the pentamer of origin so that they can be extended into a neighboring pentamer for bonding. Alternatively, in the highly reducing buffer used for crystal formation, which would prevent disulfide bond formation, the hydrophobic contacts between C-terminal helices may be energetically more favorable and stable. Nonetheless, the Modis et al. (2002) model supports the idea that polyomaviruses and papillomaviruses share the same basic design architecture of "invading" C-terminal domains linking together capsomeric subunits.

5.6.   Surface Loops and Epitopes

Not surprisingly, alignment of L1 sequences from 49 different HPV serotypes reveals that all of the hypervariable regions lie on the external face of the pentamer (Chen et al., 2000). This hypervariability most likely represents immunologically driven evolution of serotypes, and perhaps some component of tissue specificity adaptation. Two distinct mechanisms of antibody-mediated neutralization have been described for papillomaviruses: steric inhibition of cell binding and inhibition of uncoating. CryoEM structural analysis of two different neutralizing monoclonal antibodies to BPV1 show that the Mab blocking attachment binds to the tips of capsomeres, whereas the Mab not affecting attachment binds to the sides of capsomeres, well away from the exterior surface (Booy et al., 1998). Epitopes identified for neutralizing monoclonal antibodies for HPV16 (V5, E70) and HPV11 (B2, F1, G5) can be mapped directly to surface loop domains on the capsomere (Chen et al., 2000). Serotype variants, e.g., minimal sequence changes within a serotype not yet resulting in a distinct new serotype, can also be mapped to the surface loop domains (Chen et al., 2000). Thus, most of the antibody recognition towards papillomaviruses (except a distinct class directed against the L2 protein, see below) can be rationalized based upon L1 capsomere loop structures and sequences. Having epitopes for neutralization contained within the pentamer structures is also consistent with the observation that unassembled capsomeres can induce neutralizing antibodies and are protective against papillomavirus infection in animal models (Rose et al., 1998; Yuan et al., 2001).

The surface loops also generate a variety of pockets or canyons on the capsomere surface that might be utilized as receptor binding sites. Similar canyon features are used in such a manner by polyomaviruses (Stehle and Harrison, 1996; Stehle et al., 1994). However, the receptors, either primary nor secondary, are still to be identified for papillomaviruses, but their future analysis certainly might involve co-crystallization with L1 pentamers to determine their precise contact points.

5.7. L1 Self-Assembly

Both VP1 and L1 have the amazing property of in vitro self-assembly (Chen et al., 2000; Li et al., 1997; Salunke et al., 1986). When recombinantly expressed in bacteria, both L1 and VP1 are purified as pentameric capsomeres. These purified capsomere preparations can be induced to assemble in vitro either by addition of calcium or high ionic strength (VP1) or oxidation of disulfide bonds (L1). The structure of the final assemblies are dependent upon other buffer conditions such as pH, but in "physiologic" ionic and pH buffers, structures approximating $T = 7$ capsids can be obtained (Salunke et al., 1989). Other assemblies are possible however, and for example under the conditions used for L1 crystallization (pH 5, high ionic strength) $T = 1$ (12 pentamer) assemblies are produced at sufficient fidelity and efficiency to form crystals (Chen et al., 2000). The property of in vitro self-assembly has been used as an assay for assessing the contributions of sequences and residues to capsid formation. In the first example, deletion of the carboxy-terminal domain (63 residues from VP1) led to the formation of capsomeres unable to assemble (Garcea et al., 1987), thus demonstrating the essential function of this region in inter-capsomere bonding. For L1, mutation of specific cysteine residues abrogates in vitro assembly, thus demonstrating their importance (Li et al., 1997).

The intrinsic self-assembly properties of the L1 and VP1 pentamers also suggest a common, robust nucleating event for assembly. The pentavalent capsomere has the most extensive bonding contacts with its neighbors, suggesting that it may be the favored nucleation point for further capsomere addition, e.g., the "five-around-one" nucleus (Stehle et al., 1996). Subsequently added pentamers could then be placed into hexavalent or pentavalent positions using local bonding requirements, e.g., local rules theory (Berger et al., 1994), as required for energy minimization. In order to test this hypothesis, assembly of recombinant L1 pentamers was initiated in vitro by disulfide oxidation, and followed by light scattering to determine the size and kinetics of product formation (Casini et al., 2004). As expected, pentamers assembled into capsid-like structures as a function of protein concentration. However, the kinetics indicated that assembly was second order with a nucleation size of two pentamers. Although well removed from in vivo conditions, where chaperones and viral nucleic acid encapsidation may greatly influence capsid assembly, it seems that in vitro self-assembly of capsomeres may not be explained by a five-around-one nucleus.

Recombinant expression of L1 and VP1 in eukaryotic cells (e.g., yeast, Sf9 insect cells) leads to capsid self-assembly within the cells (Hagensee et al., 1993; Kirnbauer et al., 1992; Montross et al., 1991; Neeper et al., 1996; Rose et al., 1993; Zhou et al., 1991, 1993). Remarkably, the assembled structures or virus-like particles (VLPs) are formed only in the nuclear compartment, suggesting a fundamental regulation of the site of capsid assembly (Montross et al., 1991). (DNA virus assembly occurs in the nucleus where the genome replicates, and thus capsid assembly should only occur in this compartment as well. The additional problem of assembly only around viral nucleic acid is another regulatory phenomenon that must be overcome, see below). In contrast to in vitro assemblies formed from purified proteins, these VLPs are very uniform in size, and by cryo-EM appear authentic $T = 7$ structures to at least 15Å resolution. VLPs can be purified intact from these expression systems for use as vaccine reagents (see Chapter 14).

The regulation of assembly in vivo remains largely unknown. For polyomavirus infection of mouse cells in tissue culture, the VP1 protein associates with the cellular chaperone protein hsc70 for 30–60 min post-translation, about the time required for nuclear import (Cripe et al., 1995). In recombinant expression systems (E. coli, baculovirus) VP1 is purified associated with DnaK or another hsc70 homologue (Cripe et al., 1995). Energy-dependent, chaperone-mediated assembly of VP1 has been recapitulated in vitro under otherwise non-assembly conditions, using purified protein components (Chromy et al., 2003). This reaction yields more uniform $T = 7$ assemblies than achieved by calcium addition alone. However, because of the lack of a suitable in vitro cell culture system the chaperone partners of L1 are unknown. Presumably an analogous reaction can be postulated for L1 as for VP1, and L1 can be assembled in vitro by hsp70 chaperones (Chromy and Garcea, unpublished). However, L2 appears directly associated with hsc70 in both keratinocytes and cultured cells (Florin et al., 2004). This association causes a relocalization of hsc70 to the POD nuclear domains where assembly of virions may occur. Virion assembly displaces hsc70 from L2. Thus, L2 rather than L1 may function as the chaperone co-factor in papilloma assembly in contrast to VP1 for the polyomaviruses.

5.8.   Pseudovirion Synthesis

The assembly of pseudovirions (i.e., a VLP coat with included DNA) has been instructive in identifying components essential for capsid formation and infectivity. Several strategies have been used, initially with somewhat low efficiencies and mixed results. Roden (Roden et al., 1996) expressed L1 and L2 from Semliki-Forest virus vectors that were used to infect hamster (BPHE-1) cells carrying autonomously replicating BPV1 genomes. They found that L2 was required for genome encapsidation and that there was L1-L2 specificity at least between proteins for HPV16 and BPV1. This system was further used to identify L1-L2 interactions domains (Okun et al., 2001), showing that deletion of the C-terminal 9 residues of L2 affected infectivity but not encapsidation. Unckell et al. (1997) expressed HPV 33 L1 and L2 from recombinant vaccinia virus vectors that were used to infect COS7 cells harboring a marker plasmid (SVβgal). They found that L2 was not required for encapsidation but necessary for efficient pseudoinfection. The efficiency of this system was low, with only 1 in 25,000 VLPs containing a marker plasmid (pseudovirion), and a pseudovirion to infectious units ratio of 2,000:1. Stauffer et al. (1998) also expressed HPV18 L1 and L2 from recombinant vaccinia vectors and infected 293T cells carrying plasmids replicating using an SV40 origin. They found that L2 was required for genome encapsidation and that HPV DNA sequences were not necessary for plasmid packaging. This latter finding was in contrast to Zhao et al. (1998), who used L1/L2 (BPV) expression vectors transfected into COS1 cells. Their efficacy was again low however (1 genome in 10,000 VLPs), and they found an enhancement of BPV E2 sequences for packaging. The most recent method is described by Buck et al. (2004) who codon optimized the BPV L1 and L2 genes for mammalian expression and transfected expression plasmids into human 293 TT cells (optimally expressing SV40 large T-ag). They achieved a 5 percent packaging efficiency, and a particle to infectious units ratio of approximately 8. Further optimizing the purification protocol (Buck, personal comm.) yielded milligram quantities of pseudovirions. Again, L2 was found essential for encapsidation, no E2 or viral sequences were needed, and the size preference for packaging was about 6 kb (in contrast to the native genome size of 7.9 kb). They concluded that pseudovirion assembly is sequence promiscuous but size-dependent.

Most recently, HPV16 and BPV1 have been found to replicate as episomes in yeast (*S. cervisiae*) using only a selectable marker (Angeletti et al., 2002; Zhao and Frazer, 2002). Concurrent L1 expression leads to encapsidation of both viral and cellular DNA. This system shows promise in identifying cellular factors required for episomal replication and encapsidation.

VLPs are capable of "encapsidating" foreign DNA in vitro, either by osmotic shock of VLPs or assembly of capsomeres onto the nucleic acid. Osmotic shock inclusion of nucleic acid was first described for empty polyoma particles isolated from infected mouse cells (Barr et al., 1979) and has subsequently been applied to a number of polyomavirus VLPs with variable success (Petry et al., 2003). Uptake of 3–4 kb of naked dsDNA is possible, indicating that the interstices between pentamers in VLPs are capable of "opening" sufficiently for the nucleic acid to thread into the positively charged center of the capsid. HPV VLPs have also been reassembled around reporter plasmids (or drugs [Bergsdorf et al., 2003]) using various strategies (Kawana et al., 1998; Touze and Coursaget, 1998), but the packaging efficiency, capacity, and in vivo biodistribution of the products have limited practical use.

5.9.   Role of L2

The stoichiometry of L2 within the virion is not precisely known. Densitometry of Coomassie-blue gels of BPV virions purified from warts have suggested approximately 12 copies, but the error in this determination could extend that value to 72 copies (Garcea, Schiller, Trus, unpublished data). Literature estimates range from 12 to 36, and cryoEM suggests perhaps 12 L2 monomers, one each associated with the pentavalent pentamers (Doorbar and Gallimore, 1987; Pfister, 1987; Trus et al., 1997). Either 12, 24, 36 or 72 fits well with 72-pentamer icosahedral symmetry, but one L2 for each pentameric capsomere has some additional symmetry attraction. By analogy to the VP2 and VP3 minor proteins of the polyomaviruses, L2 might be expected to associate with the L1 pentamer through a limited number of residues that extend into the interior of the capsomeric fivefold cavity (Chen et al., 1998). Indeed, coexpression of L1 and L2 in *E. coli* has identified the L2 binding region for L1 as residues 396–439 (HPV11) near the carboxy-terminus (Finnen et al., 2003). This same analysis also showed that some L1-L2 pairs could form between some different serotypes (e.g., HPV16 L2 with HPV 11 L1) but not others (e.g., HPV1a L2 with HPV11 L1). Pseudovirion analysis (Okun et al., 2001), suggested two L1 binding domains in L2 (BPV residues 129–246 and 384–460) where the more aminoterminal domain may represent a salt-dependent interaction identified by Sapp et al. (1995).

   L2 may play an important function in localizing capsid components to POD nuclear domains, which may be the sites for virion assembly. In the model proposed by Day et al. (1998), L2 is transported to the nucleus independent of L1 pentamers (a marked difference from the cotransport of polyomavirus VP1 pentamers and VP2/VP3 that associate in the cytoplasm and are co-transported), localizing to POD domains where an interaction with E2 bound to viral genomes and L1 pentamers brings together the components for virion assembly. The L2 domain involved in ND10 (nuclear domain 10) homing has been mapped to residues 390–420 (HPV33) (Becker et al., 2003), which are adjacent to or overlap the L2 domain that interacts with L1 (406–449 for HPV33, (Finnen et al., 2003)). L2 has a DNA binding domain at its amino terminus (Zhou et al., 1994) and this region is necessary for L2 to facilitate pseudovirion genome encapsidation (Roden et al., 2001)

   The position within the capsid for the majority of the L2 protein not interacting with L1 is undetermined. However, some portion of L2 may be exposed to the exterior as suggested by studies identifying a common neutralization epitope in

L2 between residues 108 and 120 (HPV16) (Embers et al., 2002; Kawana et al., 1999; Liu et al., 1997). Moreover, this region may participate in cell-surface binding (Kawana et al., 2001). However, the lack of icosahedral symmetry of this domain on the capsid surface likely renders it invisible in image-averaged cryo-EM reconstructions.

For polyoma, the "inside" appearance of the virion has been deduced by subtracting 25Å resolution structures of complete virions against empty VP1 capsids (Griffith et al., 1992). The resulting image, which should represent all of the non-VP1 density including the genome, shows 72 prongs of electron density extending from the virus core to the axial cavities of the VP1 pentamers. These "prongs" were identified as VP2 and VP3, and were interpreted as bridging the genome to the outer capsid shell. The VP1 DNA binding domain is immediately C-terminal to its VP2/3 interaction domain, similar to the configuration found for L1 with L2 suggesting a related conformation. Therefore it might be anticipated that L2 could form similar bridges in the papilloma virion as VP2/3 for polyoma.

5.10.   Nuclear Localization and DNA Binding of L1 and L2

Like most DNA viruses, both polyoma and papilloma virions are assembled in the cell nucleus. Thus the capsid proteins translated in the cytoplasm must be nuclear imported for virion assembly. For both virus families, it appears that the karyopherin α/β importin pathway is used, and that the nuclear localization signal (NLS) sequences are composed of the classical mono- or bipartite basic amino acid domains. These basic domains also seem to serve as DNA binding motifs, with a nonspecific DNA sequence preference. DNA binding is an important feature for the capsid interaction with the viral genome, but the precise structural orientation is unknown. A major difference between the nuclear import of polyoma and papillomavirus capsid proteins appears to be whether the minor proteins (VP2/3, L2) are co- or independently transported with the major capsid proteins. L2 can be independently nuclear localized in the absence of L1 (Florin et al., 2002), whereas the polyoma proteins appear to assemble as a complex in the cytoplasm and then cotransported (Delos et al., 1993; Forstova et al., 1993; Ishii et al., 1994). This distinction leads to potentially different strategies in genome recognition and sequential order of the assembly process itself.

Zhou et al. (1991) first identified an NLS with overlapping mono- and bipartite basic amino acids for HPV16 L1 near the extreme C-terminus of the protein (residues 510–530). Other HPV L1 proteins have similar motifs, and their NLS function has been confirmed (Merle et al., 1999; Nelson et al., 2000). HPV 16, 45, and 11 L1 have been shown to interact with $\alpha2\beta1$ karyopherin heterodimers to promote nuclear import as assayed in the digitonin-permeabilized cell assay (Merle et al., 1999; Nelson et al., 2000, 2002). Interestingly, HPV11 L1 has also been shown to interact with $\beta2$ and $\beta3$ importins, inhibiting import of cellular cargoes specific for these karyopherins (Nelson et al., 2003). The DNA binding domain of L1 appears to overlap these basic NLS motifs as shown by Southwestern blot assay (Li et al., 1997; Schafer et al., 2002; Touze et al., 2000) and DNA mobility shifts (Nelson et al., 2000), and no specificity for DNA sequence has been shown for any L1.

For L2 the data concerning NLS and DNA binding sequences are less straight-forward and somewhat discordant. Zhou et al. (1994) determined by South-western assay that the aminoterminal 12 residues of HPV6 L2 comprised a nonspecific DNA binding domain. However, Fay et al. (2004) showed that the carboxyterminal basic residues of HPV16 L2 were the DNA binding domain using mobility shift assays. Sun et al. (1995) found that either amino or carboxy terminal basic domains were dispensable for nuclear import in whole cell expression assays, whereas an internal domain (residues 286–306, HPV6b) was important although not sufficient by itself. Also using expression in cells, Becker et al. (2003) found two NLS motifs in HPV33 L2, one centered around residue 300 as found by Sun et al. (1995), and another near the C-terminus (residues 450–456, HPV33). Roden et al. (2001) found that a BPV L2 protein deleted at both the C and N terminal basic motifs was nuclear localized after cell expression. Using GST-fusion proteins, Darshan et al. (2004) demonstrated that the amino- or carboxyterminal but not the internal motifs of HPV16 L2 could mediate transport in the digitonin-permeabilized assay, and that $\alpha2\beta1–3$ karyopherins mediated this import. Fay et al. (2004) found the same result for BPV L2. Furthermore, L2 import appears to require association with the cellular chaperone hsc70 that also accompanies L2 to POD nuclear domains (Florin et al., 2004). Thus, the internal sequences may function by enhancing retention of L2 in the nucleus, or by another unknown mechanism. Both the DNA binding and NLS functions of L2 may be difficult to assay because of the variable folding properties for this protein in recombinant systems and its association in vivo with other proteins such as hsc70.

Identification of L1/L2 sequence specific DNA binding would raise the possi-bility of identifiable, distinct encapsidation sequences on the viral genome. Although regions of the BPV1 genome were thought to enhance DNA packaging into L1/L2 capsids (Zhao et al., 1998), there has been no confirmation by others of specific L1 DNA binding or sequence specificity for DNA packaging in pseudovirions. For SV40, cis-acting sequences on the viral genome have been identified as important for pseudovirion assembly, but again no sequence specific DNA binding has been identified for the capsid proteins VP1/VP2,3. An alter-native scenario may utilize the specific viral genome binding properties of early viral proteins, e.g., E2 or T-antigen, that have unique high affinity binding sites on the viral genome. Binding of the capsid proteins, e.g., L2 to E2 or VP1-hsc70 to T-antigen, might therefore accomplish the task of specificity in genome recog-nition concomitant with initiation of capsid assembly. L2 also has been shown to interact with E2 in vitro using reticulocyte-lysate-translated L2 with GST-E2TA protein (Heino et al., 2000). However, most pseudovirion experiments, which may or may not reflect the natural life cycle, have not shown a dependence on E2 for encapsidation.

5.11. Summary

Atomic structure data are now available for both polyoma and papilloma capsids. Despite the lack of primary sequence homology between VP1 and L1 capsid proteins, the overall conformation of the capsomeres and capsid are remarkably similar. C-terminal domain strand invasion appears to be a general principle in linking pentamers on the T = 7 lattice, although L1 and VP1 utilize different bonding partners and connectors. Self-assembly in vitro and in recombinant systems also appears to be robust for both VP1 and L1. The biologic controls of assembly must harness the intrinsic self-assembly properties of the capsid proteins such that encapsidation occurs only around the genome. Cellular chaperones and viral-encoded proteins with specific DNA binding properties likely play a role in solving these problems.

References

Anderer, F.A., Schlumberger, H.D., Koch, M.A., Frank, H., and Eggers, H.J. (1967). Structure of simian virus 40. II. Symmetry and components of the virus particle. *Virology* 32:511–523.

Angeletti, P.C., Kim, K., Fernandes, F.J., and Lambert, P.F. (2002). Stable replication of papillomavirus genomes in *Saccharomyces cerevisiae*. *J. Virol.* 76:3350–3358.

Baker, T.S., Drak, J., and Bina, M. (1989). The capsid of small papova viruses contains 72 pentameric capsomeres: direct evidence from cryo-electron-microscopy of simian virus 40. *Biophys. J.* 55:243–253.

Baker, T.S., Newcomb, W.W., Olson, N.H., Cowsert, L.M., Olson, C., and Brown, J.C. (1991). Structures of bovine and human papillomaviruses: analysis by cryoelectron microscopy and three-dimensional image reconstruction. *Biophys. J.* 60:1445–1456.

Barr, S.M., Keck, K., and Aposhian, H.V. (1979). Cell-free assembly of a polyoma-like particle from empty capsids and DNA. *Virology* 96:656–659.

Becker, K.A., Florin, L., Sapp, C., and Sapp, M. (2003). Dissection of human papillomavirus type 33 L2 domains involved in nuclear domains (ND) 10 homing and reorganization. *Virology* 314:161–167.

Belnap, D.M., Olson, N.H., Cladel, N.M., Newcomb, W.W., Brown, J.C., Kreider, J.W., Christensen, N.D., and Baker, T.S. (1996). Conserved features in papillomavirus and polyomavirus capsids. *J. Mol. Bio.* 259:249–263.

Berger, B., Shor, P.W., Tucker-Kellogg, L., and King, J. (1994). Local rule-based theory of virus shell assembly. *Proc. Natl. Acad. Sci. U S A* 91:7732–7736.

Bergsdorf, C., Beyer, C., Umansky, V., Wer, M., and Sapp, M. (2003). Highly efficient transport of carboxyflurorescenin diacetate succinimidyl ester into COS7 cells using human papillomavirus-like particles. *FEBS Lett.* 536:120–124.

Booy, F.P., Roden, R.B., Greenstone, H.L., Schiller, J.T., and Trus, B.L. (1998). Two antibodies that neutralize papillomavirus by different mechanisms show distinct binding patterns at 13Å resolution. *J. Mol. Biol.* 281:95–106.

Buck, C.B., Pastrana, D.V., Lowy, D.R., and Schiller, J.T. (2004). Efficient intracellular assembly of papillomaviral vectors. *J. Virol.* 78:751–757.

Casini, G.L., Graham, D., Heine, D., Garcea, R.L., and Wu, D.T. (2004). *In vitro* papillomavirus capsid assembly analyzed by light scattering. *Virology* 325:320–327.

Caspar, D.L.D., and Klug, A. (1962). Physical principles in the construction of regular viruses. *Cold Spring Harbor Sym. Quant. Biol.* 27:1–24.

Chen, X.S., Casini, G., Harrison, S.C., and Garcea, R.L. (2001). Papillomavirus capsid protein expression in Escherichia coli: purification and assembly of HPV11 and HPV16 L1. *J. Mol. Biol.* 307:173–182.

Chen, X.S., Garcea, R.L., Goldberg, I., Casini, G., and Harrison, S.C. (2000). Structure of small virus-like particles assembled from the L1 protein of human papillomavirus 16. *Mol. Cell* 5:557–567.

Chen, X.S., Stehle, T., and Harriscn, S.C. (1998). Interaction of polyomvirus internal protein VP2 with the major capsid protein VP1 and implications for participation of VP2 in viral entry. *EMBO J.* 17:3233–3240.

Christiansen, Landers, G.T., Griffith, J., and Berg, P. (1977). Characterization of components released by alkali disruption of simian virus 40. *J. Virol.* 21:1079–1084.

Chromy, L.R., Pipas, J.M., and Garcea, R.L. (2003). Chaperone-mediated *in vitro* assembly of polyomavirus capsids. *Proc. Natl. Acad. Sci.* 100:10477–10482.

Crick, F.H.C., and Watson, J.D. (1956). Structure of small viruses. *Nature* 177:473–475.

Cripe, T.P., Delos, S.E., Estes, P.A., and Garcea, R.L. (1995). *In vivo* and *in vitro* association of Hsc70 with the polyomavirus capsid proteins. *J. Virol.* 69:7807–7813.

Darshan, M.S., Lucchi, J., Harding, E., and Moroianu, J. (2004). The L2 minor capsid protein of human papillomavirus type 16 interacts with a network of nuclear import receptors. *J. Virol.* 78:12179–12188.

Day, P.M., Roden, R.B.S., Lowy, D.R., and Schiller, J.T. (1998). The papillomavirus minor capsid protein, L2, induces localization of the major capsid protein, L1, and the viral transcription/replication protein, E2, to PML oncogenic domains. *J. Virol.* 72:142–150.

Delos, S.E., Montross, L., Moreland, R.B., and Garcea, R.L. (1993). Expression of the polyomavirus VP2 and VP3 proteins in insect cells: coexpression with the major capsid protein VP1 alters VP2/VP3 subcellular localization. *Virology* 194:393–398.

Doorbar, J., and Gallimore, P.H. (1987). Identification of proteins encoded by the L1 and L2 open reading frames of human papillomavirus 1a. *J. Virol.* 61:2793–2799.

Embers, M.E., Budgeon, L.R., Pickel, M., and Christensen, N.D. (2002). Protective immunity to rabbit oral and cutaneous papillomaviruses by immunization with short peptides of L2, the minor capsid protein. *J. Virol.* 76:9798–9805.

Fay, A., Yutzy IV, W.H., Roden, R.B., and Moroianu, J. (2004). The positively charged termini of L2 minor capsid protein required for bovine papillomavirus infection function separately in nuclear import and DNA binding. *J. Virol.* 78:13447–13454.

Finch, J.T., and Klug, A. (1965). Structure of viruses of the papilloma-polyoma type. III. Structure of rabbit papilloma virus. *J. Mol. Biol.* 13:1–12.

Finnen, R.L., Ericson, K.D., Chen, X.S., and Garcea, R.L. (2003). Interactions between papillomavirus L1 and L2 capsid proteins. *J. Virol.* 77:4818–4826.

Fligge, C., Schafer, F., Selinka, H.-C., Sapp, C., and Sapp, M. (2001). DNA-induced structural changes in the papillomavirus capsid. *J. Virol.* 75:7727–7731.

Florin, L., Becker, K.A., Sapp, C., Lambert, C., Sirma, H., Muller, M., Streeck, R.E., and Sapp, M. (2004). Nuclear translocation of papillomavirus minor capsid protein L2 requires hsc70. *J. Virol.* 78:5546–5553.

Florin, L., Sapp, C., Streeck, R.E., and Sapp, M. (2002). Assembly and translocation of papillomavirus capsid proteins. *J Virol.* 76:10009–10014.

Forstova, J., Krauzewicz, N., Wallace, S., Street, A.J., Dilworth, S.M., Beard, S., and Griffin, B.E. (1993). Cooperation of structural proteins during late events in the life cycle of polyomavirus. *J. Virol.* 67:1405–1413.

Garcea, R.L., Salunke, D.M., and Caspar, D.L.D. (1987). Site-directed mutation affecting polyomavirus self-assembly *in vitro*. *Nature (London)* 329:86–87.

Griffith, J.P., Griffith, D.L., Rayment, I., Murakami, W.T., and Caspar, D.L.D. (1992). Inside polyomavirus at 25-Å resolution. *Nature* 355:652–654.

Hagensee, M.E., Yaegashi, N., and Galloway, D.A. (1993). Self-assembly of human papillomavirus type 1 capsids by expression of the L1 protein alone or by coexpression of the L1 and L2 capsid proteins. *J. Virol.* 67:315–322.

Heino, P., Zhou, J., and Lambert, P.F. (2000). Interaction of the papillomavirus transcription/replication factor, E2, and the viral capsid protein, L2. *Virology* 276:304–314.

Hogle, J.M., Chow, M., and Filman, D.J. (1985). Three-dimensional structure of poliovirus at 2.9 Å resolution. *Science* 229:1358–1365.

Ishii, N., Nakanishi, A., Yamada, M., Macalalad, M.A., and Kasamatsu, H. (1994). Functional complementation of nuclear targeting-defective mutants of simian virus 40 structural proteins. *J. Virol.* 68:8209–8216.

Ishii, Y., Tanaka, K., and Kanda, T. (2003). Mutational analysis of human papillomavirus type 16 major capsid protein L1: the cysteines affecting the intermolecular bonding and structure of L1-capsids. *Virology* 308:128–136.

Kawana, K., Yoshikawa, H., Taketani, Y., Yoshiike, K., and Kanda, T. (1999). Common neutralization epitope in minor capsid protein L2 of human papillomavirus types 16 and 6. *J. Virol.* 73:6188–6190.

Kawana, K., Yoshikawa, H., Taketani, Y., Yoshiike, K., and Kanda, T. (1998). *In vitro* construction of pseudovirions of human papillomavirus type 16: incorporation of plasmid DNA into reassembled L1/L2 capsids. *J. Virol.* 72:10298–10300.

Kawana, Y., Kawana, K., Yoshikawa, H., Taketani, Y., Yoshiike, K., and Kanda, T. (2001). Human papillomavirus type 16 minor capsid protein L2 N-terminal region containing a common neutralization epitope binds to the cell surface and enters the cytoplasm. *J. Virol.* 75:2331–2336.

Kirnbauer, R., Booy, F., Cheng, N., Lowy, D.R., and Schiller, J.T. (1992). Papillomavirus L1 major capsid protein self-assembles into virus-like particles that are highly immunogenic. *Proc. Natl. Acad. Sci. U S A* 89:12180–12184.

Klug, A.J. (1965). Structure of viruses of the papilloma-polyoma type II. Comments on other work. *J. Mol. Biol.* 11:424–431.

Li, M., Beard, P., Estes, P.A., Lyon, M.K., and Garcea, R.L. (1998). Intercapsomeric disulfide bonds in papillomavirus assembly and disassembly. *J. Virol.* 72:2160–2167.

Li, M., Cripe, T.P., Estes, P.A., Lyon, M.K., Rose, R.C., and Garcea, R.L. (1997). Expression of the human papillomavirus type-11 L1 capsid protein in *Escherichia coli*: characterization of protein domains involved in DNA-binding and capsid assembly. *J. Virol.* 71:2988–2995.

Liddington, R.C., Yan, Y., Moulai, J., Sahli, R., Benjamin, T.L., and Harrison, S.C. (1991). Structure of simian virus 40 at 3.8-Å resolution. *Nature* 354:278–284.

Liu, W.-J., Gissmann, L., Sun, X.-Y., Kanjanahaluethai, A., Muller, M., Doorbar, J., and Zhou, J. (1997). Sequence close to the N terminus of L2 protein is displayed on the surface of bovine papillomavirus type 1 virions. *Virology* 227:474–483.

Merle, E., Rose, R.C., LeRoux, L., and Moroianu, J. (1999). Nuclear import of HPV 11 L1 capsid protein is mediated by karyopherin $\alpha 2\beta 1$ heterodimers. *J. Cell Biochem.* 74:628–637.

Modis, Y., Trus, B.L., and Harrison, S.C. (2002). Atomic model of the papillomavirus capsid. *EMBO J.* 21:4754–4762.

Montross, L., Watkins, S., Moreland, R.B., Mamon, H., Caspar, D.L.D., and Garcea, R.L. (1991). Nuclear assembly of polyomavirus capsids in insect cells expressing the major capsid protein VP1. *J. Virol.* 65:4991–4998.

Muller, M., Zhou, J., Reed, T.D., Rittmuller, C., Burger, A., Gabelsberger, J., Braspenning, J., and Gissmann, L. (1997). Chimeric papillomavirus-like particles. *Virology* 234:93–111.

Neeper, M.P., Hofmann, K.J., and Jansen, K.U. (1996). Expression of the major capsid protein of human papillomavirus type 11 in *Saccharomyces cerevisae*. *Gene* 180:1–6.

Nelson, L.M., Rose, R.C., LeRoux, L., Lane, C., Bruya, K., and Moroianu, J. (2000). Nuclear import and DNA binding of human papillomavirus type 45 L1 capsid protein. *J. Cell Biochem.* 79:225–238.

Nelson, L.M., Rose, R.C., and Moroianu, J. (2003). The L1 major capsid protein of human papillomavirus type 11 interacts wtih Kap β2 and kap β3 nuclear import receptors. *Virology* 306:162–169.

Nelson, L.M., Rose, R.C., and Moroianu, J. (2002). Nuclear import strategies of high risk HPV16 L1 major capsid protein. *J. Biol. Chem.* 277: 23958–23964.

Okun, M.M., Day, P.M., Greenstone, H.L., Booy, F.P., Lowy, D.R., Schiller, J.T., and Roden, R.B. (2001). L1 interaction domains of papillomavirus l2 necessary for viral genome encapsidation. *J. Virol.* 75:4332–4342.

Paintsil, J., Muller, M., Picken, M., Gissmann, L., and Zhou, J. (1996). Carboxyl terminus of bovine papillomavirus type-1 L1 protein is not required for capsid formation. *Virology* 223:238–244.

Petry, H., Goldmann, C., Ast, O., and Lüke, W. (2003). The use of virus-like particles for gene transfer. *Curr. Opin. Mol. Ther.* 5:524–528.

Pfister, H. (1987). Papillomaviruses: general description, taxonomy, and classification. In N. P., a. H., P. M. Salzman (eds): *Papovaviridiae*, vol. 2. New York: Plenum Press, pp. 1–38.

Rayment, I., Baker, T.S., Caspar, D.L.D., and Murakami, W.T. (1982). Polyoma virus capsid structure at 22.5 Å resolution. *Nature (London)* 295:110–115.

Roden, R.B., Day, P.M., Bronzo, B.K., Yutzy, W.H.I., Yang, Y., Lowy, D.R., and Schiller, J.T. (2001). Positively charged termini of the L2 minor capsid protein are necessary for papillomavirus infection. *J. Virol.* 75:10493–10497.

Roden, R.B.S., Greenstone, H.L., Kirnbauer, R., Booy, F.P., Jessie, J., Lowy, D.R., and Schiller, J.T. (1996). In vitro generation and type-specific neutralization of a human papillomavirus type 16 virion pseudotype. *J. Virol.* 70:5875–5883.

Rose, R.C., Bonnez, W., Reichman, R.C., and Garcea, R.L. (1993). Expression of human papillomavirus type 11 L1 protein in insect cells: *in vivo* and *in vitro* assembly of viruslike particles. *J. Virol.* 67:1936–1944.

Rose, R.C., White, W., Li, M., Suzich, J., Lane, C., and Garcea, R.L. (1998). Human papillomavirus type 11 recombinant L1 capsomeres induce virus-neutralizing antibodies. *J Virol.* Jul;72(7):6151-6154.

Rossmann, M.G., Arnold, E., Erickson, J.W., Hecht, H.-J., Johnson, J.E., Kamer, G., Luo, M., Mosser, A.G., Rueckert, R.R., Sherry, B., and Vriend, G. (1985). Structure of a human common cold virus and functional relationship to other picornaviruses. *Nature* 317:145–153.

Salunke, D.M., Caspar, D.L.D., and Garcea, R.L. (1989). Polymorphism in the assembly of polyomavirus capsid protein VP1. *Biophys. J.* 56:887–900.

Salunke, D.M., Caspar, D.L.D., and Garcea, R.L. (1986). Self-assembly of purified polyomavirus capsid protein VP1. *Cell* 46:895–904.

Sapp, M., Fligge, C., Petzak, I., Harris, J.R., and Streeck, R.E. (1998). Papillomavirus assembly requires trimerization of the major capsid protein by disulfides between two highly conserved cysteines. *J. Virol.* 72:6186–6189.

Sapp, M., Volpers, C., Muller, M., and Streeck, R.E. (1995). Organization of the major and minor capsid proteins in human papillomavirus type 33 virus-like particles. *J. Gen. Virol.* 76:2407–2412.

Schafer, F., Florin, L., and Sapp, M. (2002). DNA binding of L1 is required for human papillomavirus morphogenesis in vivo. *Virology* 295:172–181.

Stauffer, Y., Raj, K., Masternak, K., and Beard, P. (1998). Infectious human papillomavirus type 18 pseudovirions. *J. Mol. Biol.* 283:529–536.

Stehle, T., Gamblin, S.J., Yan, Y., and Harrison, S.C. (1996). The structure of simian virus 40 refined at 3.1 Å resolution. *Structure* 4:165–182.

Stehle, T., and Harrison, S.C. (1996). Crystal structures of murine polyomavirus in complex with straight-chain and branched-chain sialyloligosaccharide receptor fragments. *Structure* 4:183–194.

Stehle, T., Yan, Y., Benjamin, T.L., and Harrison, S.C. (1994). Structure of murine polyomavirus complexed with an oligosaccharide receptor fragment. *Nature* 369:160–163.

Sun, X.-Y., Frazer, I.H., Muller, M., Gissmann, L., and Zhou, J. (1995). Sequences required for the nuclear targeting and accumulation of human papillomavirus type 6B L2 protein. *Virology* 213:321–327.

Touze, A., and Coursaget, P. (1998). *In vitro* gene transfer using human papillomavirus-like particles. *Nucleic Acids Res.* 26:1317–1323.

Touze, A., Mahe, D., El Mehdaoui, S., Dupuy, C., Combita-Rojas, A.-l., Bousarghin, L., Sizaret, P.-Y., and Coursaget, P. (2000). The nine C-terminal amino acids of the major capsid protein of the human papillomavirus type 16 are essential for DNA binding and gene transfer capacity. *FEMS Microbiol. Lett.* 189:121–127.

Trus, B.L., Roden, R.B.S., Greenstone, H.L., Vrhel, M., Schiller, J.T., and Booy, F.P. (1997). Novel structural features of bovine papillomavirus capsid revealed by a three-dimensional reconstruction to 9Å resolution. *Nat. Struct. Biol.* 4:413–420.

Unckell, G., Streeck, R.E., and Sapp, M. (1997). Generation and neutralization of pseudovirions of human papillomavirus type 33. *J.Virol.* 71:2934–2939.

Volpers, C., Schirmacher, P., Streeck, R.E., and Sapp, M. (1994). Assembly of the major and the minor capsid protein of human papillomavirus type 33 into virus-like particles and tubular structures in insect cells. *Virology* 200:504–512.

Walter, G., and Deppert, W. (1974). Intermolecular disulphide bonds: an important structural feature of the polyoma virus capsid. *Cold Spring Harbor Symp. Quant. Biol.* 39:255–257.

Yuan, H., Estes, P.A., Chen, Y., Newsome, J., Olcese, V.A., Garcea, R.L., and Schlegel, R. (2001). Immunization with a pentameric L1 fusion protein protects against papillomavirus infection. *J. Virol.* 75:7848–7853.

Zhao, K.-N., and Frazer, I.H. (2002). *Saccharomyces cerevisiae* is permissive for replication of bovine papillomvirus type 1. *J. Virol.* 76:12265–12273.

Zhao, K.-N., Sun, X.-Y., Frazer, I.H., and Zhou, J. (1998). DNA packaging by L1 and L2 capsid proteins of bovine papillomavirus type 1. *Virology* 243:482–491.

Zhou, J., Doorbar, J., Sun, X.-Y., Crawford, L.V., McLean, C.S., and Frazer, I.H. (1991). Identification of the nuclear localization signal of human papillomavirus type 16 L1 protein. *Virology* 185:625–632.

Zhou, J., Stenzel, D.J., Sun, X.Y., and Frazer, I.H. (1993). Synthesis and assembly of infectious bovine papillomavirus particles *in vitro*. *J. Gen. Virol.* 74:763–768.

Zhou, J., Sun, X.Y., Louis, K., and Frazer, I.H. (1994). Interaction of human papillomavirus (HPV) type 16 capsid proteins with HPV DNA requires an intact L2 N-terminal sequence. *J. Virol.* 68:619–625.

Zhou, J., Sun, X.Y., Stenzel, D.J., and Frazer, I.H. (1991). Expression of vaccinia recombinant HPV16 L1 and L2 ORF proteins in epithelial cells is sufficient for assembly of HPV virion-like particles. *Virology* 185:251–257.

6
Viral Entry and Receptors

Rolf E. Streeck,[1] Hans-Christoph Selinka,[1] Martin Sapp[1,2]

[1] *Institute for Medical Microbiology and Hygiene, University of Mainz, D-55101 Mainz, Germany and*
[2] *Department of Microbiology and Immunology, Feist-Weiller Cancer Center, Louisiana State University Shreveport, LA 71130–3932*

6.1. Introduction

Studying viral infection in the laboratory requires viruses, permissive cells, and an assay of infection. None of these elements has been easily available for papillomaviruses. Only a few types of papillomavirus can be obtained in sufficiently large quantity, e.g., bovine papillomavirus types 1 and 4 (BPV1, 4), cottontail rabbit papillomavirus (CRPV), and human papillomavirus types 1 or 11 (HPV1, 11), prepared from cutaneous warts and squamous lesions, respectively. Therefore these viruses have been the first to be used in studies of papillomavirus infection. A few additional viruses, including HPV16, 18, and 31, have been prepared in small quantity using the xenograft and the raft culture system. A more recent system for in vivo studies has been canine oral papillomavirus (CoPV).

Because of the strict epitheliotropism of papillomaviruses, matching epithelia were used in early studies of viral infection and neutralization: fetal bovine skin for BPV1, rabbit ear for CRPV, and neonatal human foreskin for HPV11 (Christensen and Kreider, 1990). The infected epithelia were transplanted beneath the renal capsule of athymic mice, where the tissues develop into lesions similar to naturally occurring lesions in the course of three to five months. These lesions have been a valuable source of a small number of authentic virions, most notably HPV11, and this system has been used to identify neutralizing antibodies. However, xenografts are not easily amenable to genetic or biochemical approaches and therefore are of limited value for the analysis of molecular mechanisms of viral entry into cells. The organotypic (raft) epithelial culture system (Meyers et al., 1992) has been most valuable for the analysis of the papillomavirus life cycle (Laimins, this volume) and the synthesis of infectious papillomavirus (Meyers et al., 1997, 2002; Ozbun, 2002), but has not been used for the study of viral entry.

Substituting cell culture for skin grafts has been a major advancement for the analysis of papillomavirus infection. Primary keratinocyte, but also keratinocytes immortalized spontaneously or by SV40, are susceptible to infection by authentic HPV virions (Smith et al., 1995; White et al., 1998; Shafti-Keramat et al., 2003). Since late transcription is not activated in such cells, no viral particles are produced, and infection is usually assayed by quantification of early transcripts, most commonly E1^E4 mRNA, the most abundant viral transcript. Surprisingly, papillomaviruses can also transiently infect heterologous cell lines of diverse origin and species, notwithstanding their strict species and tissue specificity for productive infection (Chow et al., 1987; Christensen et al., 1995; Zhou et al., 1995; Liu et al., 2001b). Infection of mouse fibroblast C127 cells by BPV1 yields stable transformants, and the number of foci is thus a quantitative assay of infection, which can easily be scored (Dvoretzky et al., 1980). Unfortunately, HPVs cannot be used in this assay because they lack transforming capacity in this test.

In vitro generation of BPV1 virions was first achieved by using vaccinia virus to express BPV1 L1 and L2 in a cell line harboring episomal BPV1 DNA (Zhou et al., 1993). HPV16 pseudoviruses encapsidating BPV1 DNA were produced in hamster cells containing 50–200 copies of BPV1 DNA, by expression of HPV16 L1 and L2 using recombinant Semliki Forest Virus (Roden et al., 1996). Similarly, HPV18 pseudotype virus encapsidating HPV16 DNA was obtained by expression of HPV18 L1 and L2 in the W12 keratinocyte cell line, which carries a high copy number of episomal HPV16 DNA (Stauffer et al., 1998). Another major breakthrough was the generation of virus-like particles (VLPs) by expression of capsid proteins in insect and mammalian cells or yeast (Kirnbauer et al., 1992; Hagensee et al., 1993; Rose et al., 1993; Volpers et al., 1994), and the encapsidation of heterologous plasmids into VLPs yielding pseudovirions (Unckell et al., 1997; Stauffer et al., 1998; Zhao et al., 1998; Rossi et al., 2000; Buck et al., 2004). Since no packaging signal has been detected in a papillomavirus genome, viral capsids seem to package any DNA of appropriate size. Various marker genes, such as those encoding E.coli β-galactosidase, green fluorescent protein (GFP), secreted alkaline phosphatase, or puromycin resistance have been included on the encapsidated plasmids to facilitate the assay of infection. As shown recently, the yield of pseudovirions can be greatly increased by using codon-optimized L1 and L2 genes (Leder et al., 2001; Buck et al., 2004). In contrast, only low yields of pseudovirions are obtained by dissociation of VLPs into capsomeres and reassociation in the presence of DNA (Kawana et al., 1998; Touzé and Coursaget, 1998). Finally, DNA has also been bound covalently or attached non-covalently to the outside of VLPs (Müller et al., 1995; Yeager et al., 2000; Bousarghin et al., 2003a). Such particles do not, of course, correspond to virions, but have been of some use, e.g., for the identification of neutralizing antibodies.

The generation of pseudovirions in cultured cells permits both biochemical and genetic study of the capsid proteins and the encapsidated DNA. Mutant pseudovirions are helpful, if not indispensable, in unravelling the interactions of virions with the cell surface and the endocytic machinery. Virions carrying mutations in the viral genome can be obtained using raft cultures, but only mutations not interfering with the papillomavirus life cycle are tolerated in this system. This limitation is not true for pseudovirions, offering greater flexibility. However, one must be cognizant of the possibility that the structure of pseudovirions and authentic viral particles may differ. Viral particles obtained from natural lesions and virions produced in raft cultures should therefore be included in studies of viral entry, whenever possible.

6.2.   Binding to the Cell Surface

Viral entry into cells requires energy. Therefore, binding of viruses to a cell surface without passage to the inside can be monitored at low temperature. Studies of papillomavirus binding to cells demonstrated from the outset that the high specificity of papillomavirus infection cannot be attributed to selective attachment to a keratinocyte-specific receptor. BPV1 was shown to bind to a surprisingly large variety of cells (Roden et al., 1994a,b; Müller et al., 1995), and this promiscuity was confirmed for HPV11 (Culp and Christensen, 2004) as well as for pseudovirions and VLPs of many papillomavirus types. Indeed, finding cells to which papillomaviruses do not bind has been a major problem for the analysis of papillomavirus infection.

Although papillomaviruses seem to be promiscuous with respect to the type of cell, the binding specificity is nevertheless high. Raising the virus to cell ratio in the binding assay reveals that the binding capacity of a cell can be saturated. At saturation, approximately 10^5 VLPs/cell of HPV16 were found to bind to the immortalized keratinocyte cell line KH-SV (Shafti-Keramat et al., 2003), whereas 1×10^4 and 2×10^4 binding sites per cell were determined for HPV6b and HPV33 VLPs on CV-1 (Qi et al., 1996) and HeLa cells (Volpers et al., 1995), respectively. The bound viral particles thus cover only a small fraction of the cell surface, suggesting binding to a specific receptor.

VLPs of BPV1, HPV6b, HPV11, and HPV16 have been shown to compete with BPV1 virions for binding to C127 mouse cells and human keratinocytes (Roden et al., 1994; Müller et al., 1995). This competition confirms that VLPs and the capsids of different types of virus are structurally closely related and suggests that different papillomaviruses bind to the same receptor. On the other hand, the absence of competition by SV40 virions and HPV VLPs (Volpers et al., 1995) demonstrates that a specific tertiary structure and sequence is required for the binding to the cellular receptor.

The type-specific inhibition of binding by antisera and monoclonal antibodies to VLPs provides further proof of specificity for papillomavirus binding to a cell surface. Binding is not inhibited by antibodies raised against L1 or L2 fusion proteins (Volpers et al., 1995) and denatured VLPs (Roden et al., 1994). Since antisera to VLPs consisting only of L1 or of both L1 and L2, respectively, were equally effective in inhibiting the binding of virions to cells (Roden et al., 1994), L2 is unlikely to contribute to the initial binding of papillomaviruses to the cell surface. This finding is in agreement with the fact that no differences were found between the potential of L1-VLPs and L1/L2-VLPs to compete with BPV1 virions for cell binding (Roden et al., 1994; Müller et al., 1995) and the observation of almost identical kinetics for internalization of L1 and L1/L2 particles (Selinka et al., 2003)

As a first approach to determine the chemical nature of papillomavirus receptors, virus binding was assayed after enzymatic modification of the cell surface. Whereas neuraminidase and PNGase F, which cleave terminal sialic acid and N-linked carbohydrates, respectively, had no effect on the binding of papillomavirus, binding of HPV to various cell lines was strongly reduced by trypsin treatment, indicating the proteinaceous nature of the binding receptor (Volpers et al., 1995; Qi et al., 1996). High salt concentrations and high pH also prevented the binding of VLPs, suggesting that ionic interactions involving positively charged basic amino acid residues are necessary for high-affinity attachment (Volpers et al., 1995). Trypsin treatment also prevented the hemagglutination of mouse erythrocytes by papillomavirus VLPs (Roden et al., 1995).

For further analyses of virus receptors, binding studies alone are, of course, inadequate and may be misleading. Viral entry into cells and expression of genes carried inside the viral particles must be demonstrated to be strictly dependent on the presumptive receptor. Using the incubation of HPV6b VLPs with cell extracts as the binding assay, α6 integrin was proposed to be a papillomavirus receptor (Evander et al., 1997). Although binding of HPV6b VLPs to a human B lymphoma cell line was increased by expression of α6 integrin (McMillan et al., 1999), restoration of viral entry was not demonstrated. In contrast, α6 integrin is dispensable for infection of cells by BPV4 (Sibbet et al., 2000) and HPV11 virions (Shafti-Keramat et al., 2003), and by HPV33 pseudovirions (Giroglou et al., 2001).

Binding studies of HPV16 VLPs to antigen-presenting immune cells, e.g. dendritic cells, has led to the identification of CD16 as another potential receptor or co-receptor for papillomavirus infection, (Da Silva et al., 2001). However, a role for CD16 in the cellular uptake of viral particles has not been demonstrated.

Until now, the best-studied candidate receptors for attachment of papillomaviruses to cells are heparan sulfate proteoglycans (HSPGs). HSPGs are ubiquitously expressed at the surface of mammalian cells and in extracellular matrices, and they play diverse and fundamental roles in a variety of biological processes, including cell adhesion and facilitation of internalization of ligands (Lindahl et al., 1998). These complex molecules are composed of a core protein and covalently attached glycosaminoglycans with alternating disaccharide units of hexuronic acid and N-substituted glucosamine or galactosamine. Characteristic features of HSPGs are spatially discrete sulfation and acetylation of the sugar residues to various degrees. Syndecans, attached to the plasma membrane by a transmembrane domain, and glycosyl-phosphatidyl-inositol-(GPI)-anchored glypicans comprise the major families of cell surface HSPGs.

Seminal studies of Joyce et al. (1999) showed that VLPs of HPV11 interact with heparin and bind to heparan sulfate glycosaminoglycans resembling heparin on HaCaT and CHO cells. VLP binding to cells was specifically inhibited by free heparin, and removal of heparan sulfate by treatment of cells with heparinase resulted in a 90 per cent reduction of VLP binding. Reducing sulfation by growth of cells in the presence of chlorate also reduced the binding of VLPs. Binding to a CHO mutant cell line deficient in glycosaminoglycan synthesis was also severely restricted (Joyce et al., 1999). This result suggested for the first time that glycosaminoglycans are crucial for the interaction of papillomaviruses with the cell surface.

These data were confirmed and extended by infection of HeLa and COS7 cells with pseudovirions of HPV16 and HPV33 that contained a GFP gene (Giroglou et al., 2001). Synthesis of GFP in the infected cells was strictly dependent on cell-surface heparan sulfate. Additional evidence for a role of HSPG in papillomavirus binding and uptake was obtained using other forms of internalization assays (Combita et al., 2001; Drobni et al., 2003). More recently, infection of human keratinocytes (KHSV cells) by authentic HPV11 virions was also found to depend on HSPG (Shafti-Keramat et al., 2003). Cells deficient in syndecan and glypican synthesis bound VLPs poorly and were refractory to papillomavirus infection. VLP binding and susceptibility to HPV11 were restored by transfection with genes encoding syndecan-1 and, to a lesser degree, syndecan-4 and glypican, suggesting that all can function as uptake receptor (Shafti-Keramat et al., 2003). This finding also indicates that the core protein of the HSPG is of minor importance for the interaction with papillomaviruses. Since syndecan-1 is the most abundant HSPG expressed on human keratinocytes it may be most often used as primary HPV receptor in vivo.

Joyce et al. (1999) suggested that the carboxy-terminal sequence of L1 is essential for binding heparin since HPV11 VLPs consisting of C-terminally truncated L1 were unable to bind to heparin-Sepharose and HaCaT cells. In contrast, C-terminally truncated and wild-type HPV33 L1VLPs bound to heparin with similar efficiency (Giroglou et al., 2001). Carboxy-terminal peptides did not compete with binding of HPV11 (Joyce et al., 1999) or HPV33 (Giroglou et al., 2001) VLPs, suggesting that heparin binding requires a specific three-dimensional structure (Romme et al., 2005).

Despite their almost ubiquitous expression on the cell surface, HSPGs display a high degree of heterogeneity and selectivity, due to the variety in sugar units and modifications by N- and O-sulfation and N-acetylation (Fig. 6.1A) (Lindahl et al., 1998). Because of this diversity, the capacity of HSPGs to serve as primary receptor for papillomaviruses is most likely restricted to a limited number of HSPGs with specific molecular shape, internal mobility, and specific substitution patterns. When enzymatically modified heparin was tested as inhibitor of cell binding of VLPs and pseudovirus infection (Giroglou et al., 2001), heparin lacking sulfation of hydroxyl groups was completely inactive. In contrast, de-N-sulfated heparin significantly reduced cell attachment of VLPs but did not interfere with infection of HPV pseudovirions (Selinka et al., 2003). These experiments indicate that both N- and O-sulfation are required for HSPG to function as initiator of the infectious entry pathway of papillomaviruses. Although VLPs and virions compete for HSPG attachment sites, productive infection has more

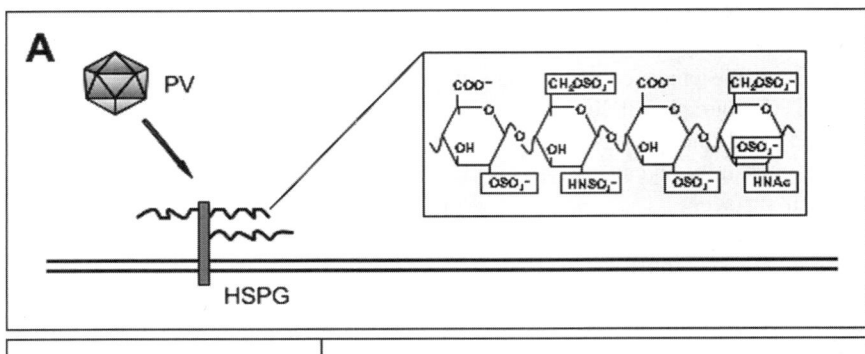

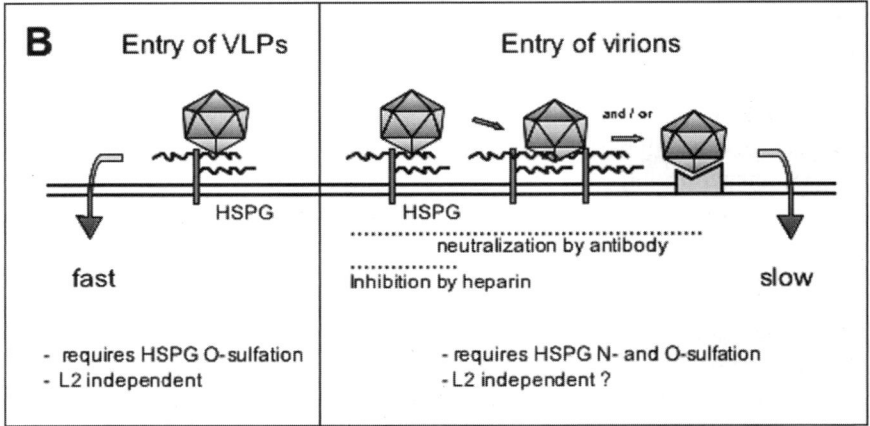

FIGURE 6.1. Interaction with the cell surface. A: Attachment to heparan sulfate proteoglycan (HSPG). The box shows potential modifications of glycans: sulfation of 2'-, 3'-, 6'-O and 2'-N, acetylation of 2'-N. B: Cell entry of virus-like particles (VLPs) and virions. Entry of VLPs is faster than entry of virions, does not require N-sulfation, and is independent of L2. Virions bound to the cell surface pass from a heparin-sensitive to a heparin-refractory binding site, suggesting the involvement of a potential secondary receptor. Viral entry is a slow process, which requires both O- and N-sulfation. A role for L2 in this process is not established.

stringent requirements than binding to the cell surface with regard to sulfation of the HSPG receptor molecule. Preliminary data regarding the interaction of HPV16 VLPs with selectively desulfated heparin preparations suggest the occurrence of two different binding sites, one dependent on N-, 2-O, and 6-O-sulfate groups, the other on 6-O-sulfates only (Knappe et al., 2006, submitted). Most likely, the more complex binding site is required for initiating viral entry, whereas the 6-O dependent site may be sufficient for binding (Fig. 6.1B).

The exact spatial arrangement of the sulfate groups in heparan sulfate that promote the productive interaction with papillomaviruses has not been determined. However, mutational analysis of HPV16 pseudovirions has shown that the surface-exposed L1 lysines 278, 351, and 361 are essential for cell binding and infection (Knappe et al., 2006; submitted). These basic residues are highly conserved, in agreement with the ability of papillomaviruses to compete for cell binding, suggesting use of a common receptor. On the other hand, neutralizing antibodies to L1 are highly type specific, and their epitopes should thus correspond to specific sequences. Conformational epitopes of neutralizing antibodies that prevent cell binding of virions have been mapped to the variable loop regions of L1 (Christensen et al., 1996; Roden et al., 1997; Ludmerer et al., 1996, 2000; White et al., 1999; Yeager et al., 2000; McClements et al., 2001; Roth et al., submitted) located on the outer surface of the capsomere structure (Chen et al., 2000). These data are compatible with a model of cell binding suggesting that the basic residues contacting sulfate groups are located at the floor of shallow grooves, the brim of which is formed by the variable loop sequences (Fig. 6.2) (Chen et al., 2000).

Recently, Culp et al. (2006a) described the preferential binding of multiple HPV types to a secreted, keratinocyte-specific extracellular matrix (ECM) protein. This ECM protein has now been identified as laminin 5 and the interaction has been shown to be transient (Culp et al., 2006b). Even though this interaction with the basal ECM protein could not be blocked by heparin, successful infection with HPV11 virions was still sensitive to the pretreatment of virions with heparin. These authors proposed a role of laminin 5 in natural infection by transferring HSPG-bound virions to mitotically active migrating keratinocytes.

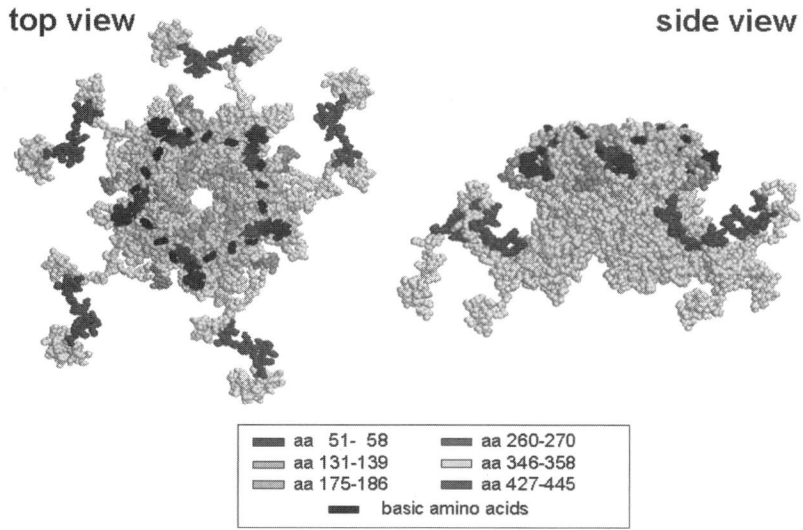

FIGURE 6.2. Location of neutralization epitopes and surface-exposed basic amino acids. The structure of the HPV16 capsomere is taken from Chen *et al.* (2000). References for the neutralization epitopes are Roden et al. (1997), McClements et al. (2001), and Carter *et al.* (2003).

6.3.   Contribution of L2 to Viral Entry

Virus-like particles consisting of L1 only or of both L1 and L2 are equally effective in competing with virions for cell binding (Roden et al., 1994; Müller et al., 1995). These findings indicate that the free energy of binding of virions to cells is mainly contributed by interactions of L1 with the cell surface. Although L2 is unlikely to contribute to the cell binding of virions, peptides derived from the N-terminal region of L2 bind to cells. For example, a peptide comprising HPV16 L2 residues 108–120 fused to GFP bound to HeLa and other cells at 4 °C and was internalized at 37 °C (Kawana et al., 2001). Yang et al. (2003a) identified a different cell binding motif in HPV16 L2, which was mapped to residues 13–31. This peptide competed with BPV1 L2 residues 1–88 in binding to the cell surface. Surprisingly, binding was unaffected by trypsin and heparinase pretreatment of cells. Mutation of residues 108–111 (Kawana et al., 2001) and 18/19 or 21/22 (Yang et al., 2003a) of HPV16 L2 prevented cell binding. However, no differences in binding between VLPs containing either wild-type L2 or the L2 mutants were observed. Although the indicated peptide sequences are thus not essential for the initial binding of virions to cells, they are nevertheless important for viral entry, since pseudovirions containing the 13–31 or 108–111 L2 mutants instead of wild-type L2 were no longer infectious. These findings indicate that L2 plays a critical role in a later step of the infectious process following binding to the cell surface.

The propensity of the L2 protein, and in particular of its amino-terminal 170 amino acids, to elicit neutralizing antibodies is a further proof of the importance of L2 for viral entry. Calves vaccinated with three peptides from the BPV4 L2 amino acid sequence 101–170 were completely protected against viral challenge (Campo et al., 1997). A common neutralization epitope for HPV6 and HPV16 was mapped to L2 amino acids 108–120 (Kawana et al., 1999). Immunization of rabbits with peptides derived from amino acids 94–122 of CRPV and ROPV L2 yielded serum that neutralized the homologous virus in vitro and protected rabbits against viral challenge (Embers et al., 2002). Antisera raised in sheep against L2 of HPV6, 16, and 18 were neutralizing against the homologous and, to a lesser degree, the heterologous HPV types (Roden et al., 2000). Antibodies against amino acids 1–88 of BPV-1 L2 were recently shown to be even more cross-reactive, neutralizing all pseudoviruses tested including HPV6, 16, 18, and 31 as well CRPV and HPV11 virions (Pastrana et al., 2005). Presumably the neutralizing antibodies inhibit some essential functions of L2 required for viral entry. These functions will be discussed further below.

6.4.   A Passage to Cytoplasm and Nucleus

Viruses exploit normal cellular structures and mechanisms that have evolved for the internalization of substances, including caveolae and clathrin-coated pits. One characteristic feature of papillomavirus infection is the slow internalization that occurs. Authentic BPV1, CRPV, and HPV11 virions can be identified on the cell surface as late as 8 h after addition to cells, and complete internalization may take up to 48 h (Christensen et al., 1995). Half-times of 10 to 14 h and 7 to 8 h were reported for internalization of HPV33 pseudovirions, when measured by neutralization and FACS analysis, respectively (Giroglou et al., 2001; Selinka et al., 2003). Whereas FACS analysis measures both productive and non-productive uptake, only productive infection is measured by the neutralization assay. This endpoint may explain the difference between the binding measurements obtained using the two assays. Nonproductive uptake is considerably faster, indeed, uptake of VLPs was reported to occur at a half-time of about 3 h (Fig. 6.1B) (Selinka et al., 2003). Slow uptake kinetics are unusual for any of the classical cell entry pathways. Receptor-mediated endocytosis via clathrin-coated pits is usually a particularly fast process, with half-times of less than 15 min. Nevertheless, evidence is accumulating that most papillomaviruses follow this pathway. Inhibitors of clathrin-coated pit formation and receptor-mediated endocytosis interfere with infection by BPV1 virions (Day et al., 2003) and HPV33 pseudovirions (Selinka et al., 2002) (Fig. 6.3). In contrast, inhibitors of lipid raft-dependent internalization via caveolae displayed only marginal effects in these studies. However, it cannot be ruled out that some types of papillomaviruses follow different uptake pathways. Internalization of HPV31 was reported to require caveolae, in contrast to HPV16 and 58

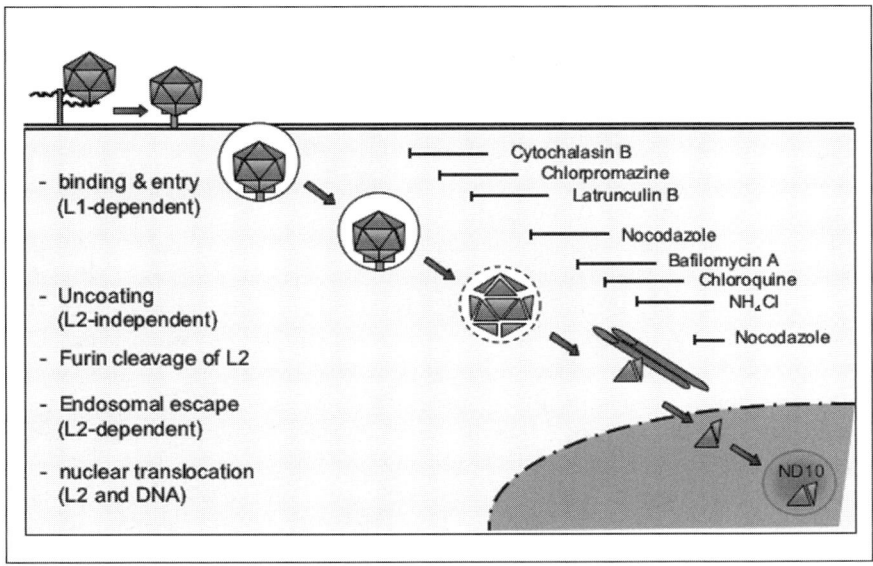

FIGURE 6.3. Translocation from the cell surface to the nucleus. Inhibitors for individual steps of vesicular transport are indicated. L2 is not required for cell binding, but is essential for endosomal escape and translocation of the viral genome to the ND10 (POD) nuclear substructures.

(Bousarghin et al., 2003b). Since the particles used in this study were obtained by simply mixing VLPs and DNA, these observations may not apply to HPV31 pseudovirions or natural virions.

Viral entry has been assayed by post-attachment neutralization using antibodies and, more recently, using competitors of virus/cell binding, like heparin. The rationale for this approach is that when viruses are inside cells they can no longer be neutralized by addition of antibodies or inhibitors. Using HPV33 pseudovirions the mode of cell binding was found to change with time (Giroglou et al., 2001). Up to 90% of pseudovirions bound to COS-7 cells were neutralized by antibodies 6 h after attachment. Cell-attached viruses were initially inhibited by excess heparin. However, with time virions became refractory to heparin, although they were still neutralized by antibodies. The shift in the mode of binding was a slow process with a half-time of approximately 6 to 10 h (Giroglou et al., 2001). The underlying biochemical mechanism is not clear yet, but the recruitment of additional HSPG molecules or the attachment to additional sulfate groups of a given heparan sulfate chain may be involved. Slow transfer of virions to a secondary, non-HSPG receptor is an alternative explanation (Fig. 6.1). In the infection assays presently used, HSPG is indispensable for productive infection. These kinetics indicate that the affinity for the putative secondary receptor is either too low to allow cell binding in the absence of HSPG, or that a conformational change of the virion, induced by HSPG, is required for binding to the secondary uptake receptor. The specific nature of the putative keratinocyte-specific secondary receptor (Patterson et al., 2005; Culp et al. 2006a,b) is still unclear. It remains to be determined if keratinocytes express higher levels of this secondary receptor, obviating the need of a primary attachment receptor. In vivo, papillomavirus infection may occur by initial adsorption to HS throughout the wounded epithelium, followed by a slow transfer of virions, via a transient association with the secreted ECM receptor, mitotically to active migrating keratinocytes which invade the wound to reestablish the basal epithelial layer and the basal membrane (Culp et al., 2006a, b).

Evidence for a conformational change of cell-bound virions comes from pre- and post-attachment neutralization studies in vitro, using monoclonal antibodies. The HPV33 specific neutralizing antibody H33.J3 readily binds to both free and cell-bound virions but efficiently neutralizes only cell-bound virus. Cell attachment changes the mode of H33.J3 binding, which may be induced by a conformational change in the viral particle. The neutralization epitope was mapped to the BC loop of L1 (amino acid residues 51 to 58), located at the vertices of capsomeres (Roth et al., submitted).

In contrast to biochemical measurements, a faster uptake of papillomaviruses has been reported using microscopy (Liu et al., 2001a) or detection of early transcripts (Ozbun, 2002; Sibbet et al., 2000; Day et al., 2003). These differences do not essentially contradict the observations outlined above, since fluorescence and electron microscopy may not adequately distinguish an infectious and noninfectious pathway of entry. The RT-PCR techniques used to monitor infections are highly sensitive and will detect a small number of productive infections. Careful analysis of the kinetics of viral entry (Culp and Christensen, 2004) confirmed that a few virions can enter cells and initiate transcription within a few hours after infection, however, the bulk of applied viruses are taken up at much slower kinetics. Furthermore, differences in uptake kinetics may depend not only on the assays but also on the particular cell lines which may have different densities of primary and secondary receptors on their cell surface.

The importance of L2 for the infection process was first demonstrated by the use of HPV33 pseudovirions lacking L2 (Unckell et al., 1997). The important role played by L2 in the infection process after binding of virions to cells was also confirmed by the finding that BPV1 virions containing L2 mutants truncated either at the N- or the C-terminus by eight or nine amino acids were not infectious in the focal transformation assay using C127 cells (Roden et al., 2001). Infectivity was restored when the positively charged C-terminus, but not the N-terminus, was substituted by a scrambled version of the original sequence, suggesting sequence-independent binding of the L2 C-terminus to DNA, possibly for delivery of the viral genome to the nucleus.

Passage through the cell membrane is followed by transport across the cytoplasm, further uncoating, and nuclear entry of the genome. Two studies have analyzed the function of actin and microtubules for these processes. Actin plays a critical role in endocytosis, being essential for cross-linking of receptor-ligand complexes as well as budding and release of clathrin-coated pits from the plasma membrane. Microtubules, in contrast, are involved in later steps including vesicular transport from early to late endosomes. Using HPV33 pseudovirions and various inhibitors, it was observed that actin is involved early in infection, whereas microtubules act at a later stage (Selinka et al., 2002). Internalization of virus particles was completely blocked by cytochalasin D, a drug inducing depolymerization of actin filaments. Several hours after attachment the inhibitory effect of cytochalasin D was less pronounced, in contrast to nocodazole, which was more effective later in infection (Selinka et al., 2002) (Fig. 6.3). Nocodazole is a drug interfering with the polymerization of tubulin. These results were essentially confirmed and extended using BPV1 virions and an RT-PCR to assay infection (Day et al., 2003).

Viral entry has also been analyzed by immunofluorescence and electron microscopy. Liu et al. (2001a) added BPV1 to CV-1 cells at 37 °C, and observed virions in close association with microtubules immediately after infection of the cells. When the cells were kept at 37 °C, L1 protein accumulated in a perinuclear region. When the temperature was shifted from 37 °C to 4 °C shortly after viral entry, leading to depolymerization of microtubules, BPV1 remained in numerous small vesicle-like structures scattered throughout the cytoplasm. Pretreatment with nocodazole similarly prevented nuclear trafficking of BPV1, although binding and viral entry were perfectly normal (Liu et al., 2001a).

In an earlier study of BPV1 entry into CV-1 and C127 cells, viral particles were detected in large cytoplasmic vesicles with diameters of up to 800 nm, and shortly afterwards in the nucleus or the perinuclear region (Zhou et al., 1995). Most likely, this distribution is not characteristic of papillomavirus infection but rather a consequence of the extremely high m.o.i. of 10^6 particles per cell used. In another early study of viral entry using electron microscopy, VLPs of HPV33 were detected in HeLa cells in small endocytic vesicles, but not in the nucleus (Volpers et al., 1995).

The maturation of early to late endosomes seems to be an essential step in papillomavirus infection. A wide range of inhibitors interfering with acidification, a characteristic feature of this process, completely abrogates infection by BPV1 virions (Day et al., 2003) and HPV33 pseudovirions (Selinka et al., 2002). The conversion of early into late endosomes most likely also triggers unpackaging of the viral genome. When pseudovirions of BPV1 containing BrdU-labeled DNA and epitope HA-tagged L2 were used to infect 293 cells, neither the DNA nor L2 could be detected intracellularly until about 6 hrs after infection, and then were found in small endocytic vesicles colocalizing with lamp-2, a marker of late endosomes (Day et al., 2004). Since DNA and the L2 tag at the C-terminus are inaccessible to antibody inside viral capsids, their simultaneous detection indicates that disassembly of viral particles starts inside cytoplasmic vesicles approximately 6 hrs after infection. The most likely mechanism of uncoating is reduction of intercapsomeric disulfide bonds in late endosomes, followed by cleavage of the C-terminal arm of L1 and further proteolysis, as previously suggested by Li et al. (1998). This mechanism is in agreement with the observation of Day et al. (2004) that the L1 protein of infecting particles, either virions or pseudovirions, could not be detected in the nucleus, using a variety of antibodies and fixation conditions. Disassembly of viral particles prior to nuclear translocation has also been proposed by Merle et al. (1999), since papillomavirus capsids are too large to pass through nuclear pores.

It has been shown recently that papillomavirus infection requires furin cleavage of L2 (Richards et al., 2006). Viruses mutated in the conserved furin cleavage site near the L2 N-terminus are not infectious, and CHO cells lacking furin cannot be infected. Furin cleavage is neither necessary for virus morphogenesis nor for virus entry into cells but seems to be essential for trafficking of L2 and the viral genome out of the endocytic compartment. Since furin, a ubiquitous proprotein convertase, cycles between the trans-Golgi network/endosomal system and the cell surface, L2 cleavage could occur at either site.

A different entry pathway was postulated by Bossis et al. (2005). These authors observed that the BPV1 L2 protein interacts with syntaxin 18, an ER-resident SNAP receptor protein, and concluded that BPV1 virions traffic through the ER. Pseudovirions generated with L2 mutants deleted for the potential syntaxin 18 interaction site (KILK, amino acids 40–44) were noninfectious. However, since the FLAG-labeled syntaxin 18 used in this study causes aggregation of the ER, the importance of syntaxin 18 for intracellular trafficking of papillomavirus needs to be confirmed under more physiological conditions.

A 23-amino-acid sequence located near the C-terminus of L2 was identified by Kämper et al. (2006) to harbor strong microbicidal, fungicidal, and cytotoxic activity due to membrane disrupting activity. A characteristic feature of this sequence is a hydrophobic cluster of amino acids adjacent to a stretch of basic amino acid residues, which is conserved among papillomaviruses. The peptide mediates membrane integration when fused to GFP. In the absence of the chaperone Hsc 70, it also induces insertion of wild-type L2 into cellular membranes. Mutations within this sequence completely abrogate L2-enhanced infection, and neither L2 nor DNA can be detected in nuclei of infected cells. The L2 protein thus most likely mediates endosomal escape of the viral genome.

L2 was also found to bind dynein via the C-terminal 70 to 80 amino acids, and it has therefore been proposed to mediate binding to microtubules, facilitating the intracytoplasmic transport of the viral genome (Florin et al., 2006). Mutations within L2 affecting dynein binding reduce the infectivity of pseudovirions. Colocalization of wt but not mutant L2 with microtubules during the infectious process has been demonstrated (Florin et al., 2006). L2 together with the encapsidated DNA reaches the nucleus about 20 hrs after infection, where both colocalize with PML at ND10 (PML bodies, PODs) nuclear substructures. PML enhances expression of the viral genes and is therefore important for establishment of a papillomavirus infection (Day et al., 2004).

BPV1 and HPV16 L2 were also shown to bind actin via the amino-terminal residues 25–45 (Yang et al., 2003a,b). It was suggested that intact viral particles interact with actin early in infection. However, this association is unlikely since it would require release of complete virions from the endosomes. This sequence of events is also inconsistent with the observation that uncoating occurs in late endosomes prior to release into the cytoplasm (Day et al., 2004). In contrast to L1 (Nelson et al., 2002), at least three import pathways mediated by Kap $\alpha2\beta1$, Kap$\beta2$, and Kap$\beta3$, respectively, have been reported for nuclear entry of HPV16 L2 (Darshan et al., 2004). This redundancy is consistent with the identification of several nuclear localization signals (NLSs) in the L2 protein of HPV6b (Sun et al., 1995) and HPV33 (Becker et al., 2003). L2 may use a different NLS and import pathway in the initial phase of infection and the productive phase, which take place in undifferentiated and differentiated epithelial cells, respectively (Darshan et al., 2004).

6.5. Conclusion

The development of systems for generation of (pseudo)virions has considerably increased our knowledge of papillomavirus interaction with cells. Based on these advances, the most likely scenario for viral entry is as follows: Cell-surface heparan sulfate proteoglycans with specific modifications at the amino-, 2-hydroxyl, and 6-hydroxyl groups mediate binding of virions to cells. This binding is dependent only on conformationally correct L1 protein and does not require L2. A long delay in internalization is accompanied by changes in the mode of binding and possible transfer to a secondary receptor. There is at yet no evidence that L2 is involved in this early process. The virions are internalized via a clathrin-dependent endocytic pathway, which initially is dependent on actin for budding of vesicles and later on requires microtubules for vesicular transport.

Reduction of intercapsomeric disulfide bonds and acidification of endosomes probably trigger uncoating of the viral genome. L2 protein mediates endosomal escape and perhaps chaperones the viral genome to the nucleus. L2-mediated homing of the genome to PML bodies promotes its transcription and consequently establishment of a papillomavirus infection.

Despite these significant advances and the emergence of a general picture of the infectious entry pathway, many details remain to be clarified. These questions include the nature of the putative secondary receptor, the cellular machinery involved in vesicular transport and sorting, the uncoating mechanism, and the viral determinants and cellular factors involved in nuclear translocation. It is also unlikely that all of the numerous types of papillomavirus use a common entry pathway. Since the experimental approach has been established, these unsolved issues hopefully will be soon addressed.

References

Becker, K.A., Florin, L., Sapp, C., and Sapp, M. (2003). Dissection of human papillomavirus type 33 L2 domains involved in nuclear domains (ND) 10 homing and reorganization. *Virology* 314:161–167.

Bossis, I., Roden, R.B.S., Gambhira, R., Yang, R., Tagaya, M., Howley, P.M., and Meneses, P.I. (2005). Interaction of tSNARE syntaxin 18 with the papillomavirus minor capsid protein mediates infection. *J. Virol.* 79:6723–6731.

Bousarghin, L., Touzé, A., Combita-Rojas, A.L., and Coursaget, P. (2003a). Positively charged sequences of human papillomavirus type 16 capsid proteins are sufficient to mediate gene transfer into target cells via the heparan sulfate receptor. *J. Gen. Virol.* 84:157–164.

Bousarghin, L., Touzé, A., Sizaret, P.Y., and Coursaget, P. (2003b). Human papillomavirus types 16, 31, and 58 use different endocytosis pathways to enter cells. *J. Virol.* 77:3846–3850.

Buck, C.B., Pastrana, D.V., Lowy, D.R., and Schiller, J.T. (2004). Efficient intracellular assembly of papillomaviral vectors. *J. Virol.* 78:751–757.

Campo, M.S., O'Neil, B.W., Grindlay, G.J., Curtis, F., Knowles, G., and Chandrachud, L. (1997). A peptide encoding a B-cell epitope from the N-terminus of the capsid protein L2 of bovine papillomavirus-4 prevents disease. *Virology* 234:261–266.

Carter, J.J., Wipf, G.C., Benki, S.F., Christensen, N.D., and Galloway, D.A. (2003). Identification of a human papillomavirus type 16-specific epitope on the C-terminal arm of the major capsid protein L1. *J. Virol.* 77:11625–11632.

Chen, X.S., Garcea, R.L., Goldberg, I., Casini, G., and Harrison, S.C. (2000). Structure of small virus-like particles assembled from the L1 protein of human papillomavirus 16. *Mol. Cell* 5:557–567.

Chow, L.T., Reilly, S.S., Broker, T.R., and Taichman, L.B. (1987). Identification and mapping of human papillomavirus type 1 RNA transcripts recovered from plantar warts and infected epithelial cell cultures. *J. Virol.* 61:1913–1918.

Christensen, N.D., Cladel, N.M., and Reed, C.A. (1995). Postattachment neutralization of papillomaviruses by monoclonal and polyclonal antibodies. *Virology* 207:136–142.

Christensen, N.D., and Kreider, J.W. (1990). Antibody-mediated neutralization *in vivo* of infectious papillomaviruses. *J. Virol.* 64:3151–3156.

Christensen, N.D., Reed, C.A., Cladel, N.M., Hall, K., and Leiserowitz, G.S. (1996). Monoclonal antibodies to HPV-6 L1 virus-like particles identify conformational and linear neutralizing epitopes on HPV-11 in addition to type-specific epitopes on HPV-6. *Virology* 224:477–486.

Combita, A.L., Touzé, A., Bousarghin, L., Sizaret, P.Y., Munoz, N., and Coursaget, P. (2001). Gene transfer using human papillomavirus pseudovirions varies according to virus genotype and requires cell surface heparan sulfate. *FEMS Microbiol. Lett.* 204:183–188.

Culp, T.D., and Christensen, N.D. (2004). Kinetics of *in vitro* adsorption and entry of papillomavirus virions. *Virology* 319:152–161.

Culp, T.D., Budgeon, L.R., and Christensen, N.D. (2006a). Human papilloma-viruses bind a basal extracellular matrix component secreted by keratinocytes which is distinct from a membrane-assocuated receptor. *Virology* 347:147–159.

Culp, T.D., Budgeon, L.R., Marinkovich, M.P., Meneguzzi, G., and Christensen, N.D. (2006b). Keratinocyte-secreted laminin 5 can function as a transient receptor for human papillomaviruses by binding virions and transferring them to adjacent cells. *J. Virol.* 80:8940–8950.

Da Silva, D., Velders, M.P., Nieland, J.D., Schiller, J.T., Nickoloff, B.J., Kast, W.M. (2001). Physical interaction of human papillomavirus virus-like particles with immune cells. *Internat. Immunol.* 13:633–641.

Darshan, M.S. Lucchi, J., Harding, E., Moroianu, J. (2004). The L2 minor capsid protein of human papillomavirus type 16 Interacts with a network of nuclear import receptors. *J. Virol.* 78:12179–12188.

Day, P.M., Baker, C.C., Lowy, D.R., and Schiller, J.T. (2004). Establishment of papillo-mavirus infection is enhanced by promyelocytic leukemia protein (PML) expression. *Proc. Natl. Acad. Sci. U S A* 101:14252–14257.

Day, P.M., Lowy, D.R., and Schiller, J.T. (2003). Papillomaviruses infect cells via a clathrin-dependent pathway. *Virology* 307:1–11.

Drobni, M., Mistry, N., McMillan, N.A., Evander, M. (2003). Carboxy-fluorescein diacetate, succinimidyl ester labeled papillomavirus virus-like particles fluoresce after internalization and interact with heparan sulfate for binding and entry. *Virology* 310:163–172.

Dvoretzky, I., Shober, R., Chattopadhyay, S.K., and Lowy, D.R. (1980). A quantitative *in vitro* focus assay for bovine papilloma virus. *Virology* 103:369–375.

Embers, M.E., Budgeon, L.R., Pickel, M., and Christensen, N.D. (2002). Protective immunity to rabbit oral and cutaneous papillomaviruses by immunization with short peptides of L2, the minor capsid protein. *J. Virol.* 76:9798–9805.

Evander, M., Frazer, I.H., Payne, E., Qi, Y.M., Hengst, K., and McMillan, N.A. (1997). Identification of the alpha-6 integrin as a candidate receptor for papillomaviruses. *J. Virol.* 71:2449–2456.

Florin, L., Becker, K.A., Lambert, C., Nowak, T., Sapp, C., Strand, D., Streeck, R.E., and Sapp, M. (2006). Identification of a dynein interacting domain in the papillomavirus minor capsid protein L2. *J. Virol.* 80:6691–6696.

Giroglou, T., Florin, L., Schäfer, F., Streeck, R.E., and Sapp, M. (2001). Human papillo-mavirus infection requires cell surface heparan sulfate. *J. Virol.* 75:1565–1570.

Hagensee, M.E., Yeagashi, N., and Galloway, D.A. (1993). Self-assembly of human papillomavirus type 1 capsids by expression of the L1 protein alone or by coexpression of the L1 and L2 capsid proteins. *J. Virol.* 67:315–322.

Joyce, J.G., Tung, J.S., Przysiecki, C.T., Cook, J.C., Lehman, E.D., Sands, J.A., Jansen, K. U., and Keller, P.M. (1999). The L1 major capsid protein of human papillomavirus type 11 recombinant virus-like particles interacts with heparin and cell-surface glycosaminoglycans on human keratinocytes. *J. Biol. Chem.* 274:5810–5822.

Kämper, N., Day, P.M., Nowak, T., Selinka, H.-C., Florin, L., Bolscher, J., Hilbig, L., Schiller, J.T., and Sapp, M. (2006). A membrane-destabilizing peptide in capsid protein L2 is required for egress of papillomavirus genomes from endosomes. *J. Virol.* 80:759–768.

Kawana, K., Yoshikawa, H., Taketani, Y., Yoshiike, K., and Kanda, T. (1998). *In vitro* construction of pseudovirions of human papillomavirus type 16: Incorporation of plasmid DNA into reassembled L1/L2 capsids. *J. Virol.* 72:10298–10300.

Kawana, K., Yoshikawa, H., Taketani, Y., Yoshiike, K., and Kanda, T. (1999). Common neutralization epitope in minor capsid protein L2 of human papillomavirus types 16 and 6. *J. Virol.* 73:6188–6190.

Kawana, Y., Kawana, K., Yoshikawa, H., Taketani, Y., Yoshiike, K., and Kanda, T. (2001). Human papillomavirus type 16 minor capsid protein L2 N-terminal region containing a common neutralization epitope binds to the cell surface and enters the cytoplasm. *J. Virol.* 75:2331–2336.

Kirnbauer, R., Booy, F., Cheng, N., Lowy, D.R., and Schiller, J.T. (1992). Papillomavirus L1 major capsid protein self-assembles into virus-like particles that are highly immunogenic. *Proc. Natl. Acad. Sci. U S A* 89:12180–12184.

Knappe, M., Selinka, H.-C., Bodevin, S., Spillman, D., Streeck, R.E., Chen, X.S., Lindahl, U., and Sapp, M. (2006). Surface-exposed amino acids of HPV16 L1 protein mediating interaction with cell surface heparan sulfate. (submitted for publication).

Leder, C., Kleinschmidt, J.A., Wiethe, C., and Müller, M. (2001). Enhancement of capsid gene expression: preparing the human papillomavirus type 16 major structural gene L1 for DNA vaccination purposes. *J. Virol.* 75:9201–9209.

Li, M., Beard, P., Estes, P.A., Lyon, M.K., and Garcea, R.L. (1998). Intercapsomeric disulfide bonds in papillomavirus assembly and disassembly. *J. Virol.* 72:2160–2167.

Lindahl, U., Kusche-Gullberg, M., and Kjellen, L. (1998). Regulated diversity of heparan sulfate. *J. Biol. Chem.* 273:24979–24982.

Liu, W.J., Qi, Y.M., Zhao, K.N., Liu, Y.H., Liu, X.S., and Frazer, I.H. (2001a). Association of bovine papillomavirus type 1 with microtubules. *Virology* 282:237–244.

Liu, Y., You, H., Chiriva-Internati, M., Korourian, S., Lowery, C.L., Carey, M.J., Smith, C.V., and Hermonat, P.L. (2001b). Display of complete life cycle of human papillomavirus type 16 in cultured placental trophoblasts. *Virology* 290:99–105.

Ludmerer, S.W., Benincasa, D., and Mark, G. E., 3rd (1996). Two amino acid residues confer type specificity to a neutralizing, conformationally dependent epitope on human papillomavirus type 11. *J. Virol.* 70:4791–4794.

Ludmerer, S.W., McClements, W.L., Wang, X.M., Ling, J.C., Jansen, K.U., and Christensen, N.D. (2000). HPV11 mutant virus-like particles elicit immune responses that neutralize virus and delineate a novel neutralizing domain. *Virology* 266:237–245.

McClements, W.L., Wang, X.M., Ling, J.C., Skulsky, D.M., Christensen, N.D., Jansen, K.U., and Ludmerer, S.W. (2001). A novel human papillomavirus type 6 neutralizing domain comprising two discrete regions of the major capsid protein L1. *Virology* 289:262–268.

McMillan, N.A., Payne, E., Frazer, I.H., and Evander, M. (1999). Expression of the alpha-6 integrin confers papillomavirus binding upon receptor-negative B-cells. *Virology* 261:271–279.

Merle, E., Rose, R.C., LeRoux, L., and Moroianu, J. (1999). Nuclear import of HPV11 L1 capsid protein is mediated by karyopherin alpha2beta1 heterodimers. *J. Cell. Biochem.* 74:628–637.

Meyers, C., Frattini, M.G., Hudson, J.B., and Laimins, L.A. (1992). Biosynthesis of human papillomavirus from a continuous cell line upon epithelial differentiation. *Science* 257:971–973.

Meyers, C., Mayer, T.J., and Ozbun, M.A. (1997). Synthesis of infectious human papillomavirus type 18 in differentiating epithelium transfected with viral DNA. *J. Virol.* 71:7381–7386.

Meyers, C., Bromberg-White, J.L., Zhang, J., Kaupas, M.E., Bryan, J.T., Lowe, R.S., and • Jansen, K.U. (2002). Infectious virions produced from a human papillomavirus type 18/16 genomic DNA chimera. *J. Virol.* 76:4723–4733.

Müller, M., Gissmann, L., Cristiano, R.J., Sun, X.Y., Frazer, I.H., Jenson, A.B., Alonso, A., Zentgraf, H., and Zhou, J. (1995). Papillomavirus capsid binding and uptake by cells from different tissues and species. *J. Virol.* 69:948–954.

Nelson, L.M., Rose, R.C., and Moroianu, J. (2002). Nuclear import strategies of high risk HPV16 L1 major capsid protein. *J. Biol. Chem.* 277:23958–23964.

Ozbun, M. A. (2002). Infectious human papillomavirus type 31b: Purification and infection of an immortalized human keratinocyte cell line. *J. Gen. Virol.* 83:2753–2763.

Pastrana, D.V., Gambhira, R., Buck, C.B., Pang, Y.-Y.S, Thompsom, C.D., Culp, T.D., Christensen, N.D., Lowy, D.R., Schiller, J.T., and Roden, R.B.S. (2005). Cross-neutralization of cutaneous and mucosal papillomavirus types with anti-sera to the amino terminus of L2. *Virology* 337:365–371.

Patterson, N.A., Smith, J.L., and Ozbun, M.A. (2005). Human papillomavirus type 31b infection of human keratinocytes does not require heparan sulfate. *J. Virol.* 79:6838–6847.

Qi, Y.M., Peng, S.W., Hengst, K., Evander, M., Park, D.S., Zhou, J., and Frazer, I.H. (1996). Epithelial cells display separate receptors for papillomavirus VLPs and for soluble L1 capsid protein. *Virology* 216:35–45.

Richards, R.M., Lowy, D.R., Schiller, J.T., and Day, P.M. (2006). Cleavage of the papillomavirus minor capsid protein, L2, at a furin consensus site is necessary for infection. *Proc. Natl. Acad. Sci. U S A* 103:1522–1527.

Roden, R.B., Kirnbauer, R., Jenson, A.B., Lowy, D.R., and Schiller, J.T. (1994a). Interaction of papillomaviruses with the cell surface. *J. Virol.* 68:7260–7266.

Roden, R.B., Weissinger, E.M., Henderson, D.W., Booy, F., Kirnbauer, R., Mushinski, J.F., Lowy, D.R., and Schiller, J.T. (1994b). Neutralization of bovine papillomavirus by antibodies to L1 and L2 capsid proteins. *J. Virol.* 68:7570–7574.

Roden, R.B., Hubbert, N.L., Kirnbauer, R., Breitburd, F., Lowy, D.R., and Schiller, J.T. (1995). Papillomavirus L1 capsids agglutinate mouse erythrocytes through a proteinaceous receptor. *J. Virol.* 69:5147–5151.

Roden, R.B., Greenstone, H.L., Kirnbauer, R., Booy, F.P., Jessie, J., Lowy, D.R., and Schiller, J.T. (1996). *In vitro* generation and type-specific neutralization of a human papillomavirus type 16 virion pseudotype. *J. Virol.* 70:5875–5883.

Roden, R.B., Armstrong, A., Haderer, P., Christensen, N.D., Hubbert, N.L., Lowy, D.R., Schiller, J.T., and Kirnbauer, R. (1997). Characterization of a human papillomavirus type 16 variant-dependent neutralizing epitope. *J. Virol.* 71:6247–6252.

Roden, R.B., Yutzy, W.H. IV, Fallon, R., Inglis, S., Lowy, D.R., and Schiller, J.T. (2000). Minor capsid protein of human genital papillomaviruses contains subdominant, cross-neutralizing epitopes. *Virology* 270:254–257.

Roden, R.B., Day, P.M., Bronzo, B.K., Yutzy, W.H. IV, Yang, Y., Lowy, D.R., and Schiller, J.T. (2001). Positively charged termini of the L2 minor capsid protein are necessary for papillomavirus infection. *J. Virol.* 75:10493–10497.

Rommel, O., Dillner, J., Fligge, C., Bersdorf, C., Wang, X., Selinka, H.-C., Sapp, M. (2004). Heparan sulfate proteoglycans interact exclusively with conformationally intact HPV L1 assemblies: Basis for a virus-like particle ELISA. *J. Med. Virol.*:114–121.

Rose, R.C., Bonnez, W., Reichman, R.C., and Garcea, R.L. (1993). Expression of human papillomavirus type 11 L1 protein in insect cells: *In vivo* and *in vitro* assembly of viruslike particles. *J. Virol.* 67:1936–1944.

Rossi, J.L., Gissmann, L., Jansen, K., and Müller, M. (2000). Assembly of human papillomavirus type 16 pseudovirions in *Saccharomyces cerevisiae*. *Hum. Gene Ther.* 11:1165–1176.

Roth, S., Sapp, M., Streeck, R.E., and Selinka, H.-C. (2006). Characterization of neutralizing epitopes within the major capsid protein of human papilloma-virus type 33. Manuscript submitted for publication.

Selinka, H.C., Giroglou, T., and Sapp, M. (2002). Analysis of the infectious entry pathway of human papillomavirus type 33 pseudovirions. *Virology* 299:279–287.

Selinka, H.C., Giroglou, T., Nowak, T., Christensen, N.D., and Sapp, M. (2003). Further evidence that papillomavirus capsids exist in two distinct conformations. *J. Virol.* 77:12961–12967.

Shafti-Keramat, S., Handisurya, A., Kriehuber, E., Meneguzzi, G., Slupetzky, K., and Kirnbauer, R. (2003). Different heparan sulfate proteoglycans serve as cellular receptors for human papillomaviruses. *J. Virol.* 77:13125–13135.

Sibbet, G., Romero-Graillet, C., Meneguzzi, G., and Saveria Campo, M. (2000). Alpha-6 integrin is not the obligatory cell receptor for bovine papillomavirus type 4. *J. Gen. Virol.* 81:327–334.

Smith, L.H., Foster, C., Hitchcock, M.E., Leiserowitz, G.S., Hall, K., Isseroff, R., Christensen, N.D., and Kreider, J.W. (1995). Titration of HPV-11 infectivity and antibody neutralization can be measured *in vitro*. *J. Invest. Dermatol.* 105:438–444.

Stauffer, Y., Raj, K., Masternak, K., and Beard, P. (1998). Infectious human papillomavirus type 18 pseudovirions. *J. Mol. Biol.* 283:529–536.

Sun, X.Y., Frazer, I., Müller, M., Gissmann, L., and Zhou, J. (1995). Sequences required for the nuclear targeting and accumulation of human papillomavirus type 6B L2 protein. *Virology* 213:321–327.

Touzé, A., and Coursaget, P. (1998). *In vitro* gene transfer using human papillomavirus-like particles. *Nucleic Acids Res.* 26:1317–1323.

Unckell, F., Streeck, R.E., and Sapp, M. (1997). Generation and neutralization of pseudovirions of human papillomavirus type 33. *J. Virol.* 71:2934–2939.

Volpers, C., Schirmacher, P., Streeck, R.E., and Sapp, M. (1994). Assembly of the major and the minor capsid protein of human papillomavirus type 33 into virus-like particles and tubular structures in insect cells. *Virology* 200:504–512.

Volpers, C., Unckell, F., Schirmacher, P., Streeck, R.E., and Sapp, M. (1995). Binding and internalization of human papillomavirus type 33 virus-like particles by eukaryotic cells. *J. Virol.* 69:3258–3264.

White, W.I., Wilson, S.D., Bonnez, W., Rose, R.C., Koenig, S., and Suzich, J.A. (1998). *In vitro* infection and type-restricted antibody-mediated neutralization of authentic human papillomavirus type 16. *J. Virol.* 72:959–964.

White, W.I., Wilson, S.D., Palmer-Hill, F.J., Woods, R.M., Ghim, S.J., Hewitt, L.A., Goldman, D.M., Burke, S.J., Jenson, A.B., Koenig, S., and Suzich, J.A. (1999). Characterization of a major neutralizing epitope on human papillomavirus type 16 L1. *J. Virol.* 73:4882–4889.

Yang, R., Day, P.M., Yutzy, W. H. IV, Lin, K.Y., Hung, C.F., and Roden, R.B. (2003a). Cell surface-binding motifs of L2 that facilitate papillomavirus infection. *J. Virol.* 77:3531–3541.

Yang, R., Yutzy, W.H. IV, Viscidi, R.P., and Roden, R.B. (2003b). Interaction of L2 with beta-actin directs intracellular transport of papillomavirus and infection. *J. Biol. Chem.* 278:12546–12553.

Yeager, M.D., Aste-Amezaga, M., Brown, D.R., Martin, M.M., Shah, M.J., Cook, J.C., Christensen, N.D., Ackerson, C., Lowe, R.S., Smith, J.F., Keller, P., and Jansen, K.U. (2000). Neutralization of human papilloma-virus (HPV) pseudovirions: A novel and efficient approach to detect and characterize HPV neutralizing antibodies. *Virology* 278:570–577.

Zhao, K.N., Sun, X.Y., Frazer, I.H., and Zhou, J. (1998). DNA packaging by L1 and L2 capsid proteins of bovine papillomavirus type 1. *Virology* 243:482–491.

Zhou, J., Stenzel, D.J., Sun, X.Y., and Frazer, I.H. (1993). Synthesis and assembly of infectious bovine papillomavirus particles *in vitro*. *J. Gen. Virol.* 74:763–768.

Zhou, J., Gissmann, L., Zentgraf, H., Müller, H., Picken, M., and Müller, M. (1995). Early phase in the infection of cultured cells with papillomavirus virions. *Virology* 214:167–176.

7
Human Papillomavirus Transcription

Louise T. Chow and Thomas R. Broker

University of Alabama at Birmingham, Department of Biochemistry and Molecular Genetics, Birmingham, AL 35294-0005

7.1. Introduction

The human papillomaviruses (HPVs) comprise a large family of pathogens of the cutaneous or mucosal squamous epithelia. Primary infection requires wounding to expose the basal and parabasal keratinocytes. Following establishment, the viral DNA is maintained as low-copy-number nuclear plasmids in these relatively undifferentiated layers, and viral transcripts are generally below the sensitivity of detection. Productive infections result in benign, hyperproliferative lesions variously called warts, papillomas, or condylomas, in which high levels of viral transcription, genome amplification, and virion assembly take place in the middle to upper layers of cells that are undergoing terminal differentiation. Although these lesions often spontaneously regress, persistent infections by the high-risk mucosotropic HPVs may convert, over a period of months or years, from benign lesions or low-grade squamous intraepithelial dysplasias (SIL) to high-grade SIL and, occasionally, progress to carcinomas in situ and invasive cancers. Associated with the transition from high-grade SIL to carcinoma, the viral DNA typically becomes integrated, the viral oncogenes become overexpressed, while other viral genes are silenced, and progeny viruses are no longer produced. This chapter provides an overview of papillomavirus transcription regulation, with an emphasis on the normal programs of viral persistence and reproduction, as well as on the changes that occur during neoplastic progression to high grade SIL and carcinomas. We also propose new interpretations of the observed transcription patterns based upon considerations of the encoded protein functions.

7.1.1. Papillomaviridae

The papillomaviruses were initially classified together with simian virus 40 (SV40) and mouse polyomavirus in the family *Papovaviridae*, based on their small double-stranded circular DNA genomes, nonenveloped protein capsid with icosahedral symmetry, general mode of replication, and oncogenic potential.

Sequencing of the first papillomavirus genomes in the mid-1980s revealed that all the open reading frames were encoded by a single DNA strand. This arrangement is in sharp contrast to the SV40/polyoma viruses, which have divergent transcription from early and late promoters. Ultimately, the substantial genomic differences and the identification of extraordinarily large numbers of distinct human and animal papillomavirus types led to the reclassification of these small DNA tumor viruses into separate families of *Papillomaviridae* and *Polyomaviridae* (de Villiers et al., 2004).

7.1.2.   Papillomavirus Infections

The human papillomaviruses (HPV) are highly prevalent pathogens that prefer-entially target different body sites and epithelial tissues types (de Villiers et al., 2004). Distinct members of the large family of HPVs infect either cutaneous skin or mucosal epithelia. It was discovered in the early nineteen eighties that certain HPV genotypes are closely associated with cervical and other ano-genital cancers (Dürst et al., 1983; reviewed by zur Hausen, 2002), and cervical cancers and cervical carcinoma cell lines contain integrated HPV DNA which continues to transcribe the viral E6 and E7 genes (Schwarz et al., 1985; Yee et al., 1985; Schneider-Gädicke and Schwarz, 1986; Smotkin and Wettstein, 1986; Baker et al., 1987; Inagaki et al., 1988; Pater and Pater, 1988; Shirasawa et al., 1988).

Primary infection by papillomaviruses is thought to occur through cuts, abrasions, and other wounding of the stratified epithelium and is established in the long living basal and parabasal keratinocytes. Infections are often asymp-tomatic and can persist in a latent state for months or years. Infections by the low-risk genotypes such as HPV-6 and HPV-11 generally remain benign (though clinically serious) causing papillomas, condylomata, and low-grade SIL. These lesions often regress spontaneously to a subclinical latent state, typically after 9–12 months of activity. HPV-6 and HPV-11 also cause recurrent laryngeal papil-lomas, benign growths that nonetheless can result in potentially life-threatening obstruction of the airway. HPV infections of the cervix, penis and anus by the so-called high-risk or oncogenic genotypes, including HPV-16, -31, -33, -35, -52, and -58, and HPV-18, -39, -45, -59, -68, and related types, can occasionally progress from the benign stages to high-grade SIL and cancer. Epidemiological studies worldwide indicate that HPVs are necessary agents for the emergence of cervical cancers (Walboomers et al., 1999). In addition, some laryngeal, pharyngeal and tonsillar cancers are also caused by HPV-16 (reviewed by zur Hausen, 2002). As will be discussed, profound alterations in viral transcription accompany such neoplastic changes.

7.1.3.   HPV Genome Organization

The genome organization is highly conserved among human and animal papil-lomaviruses (Fig. 7.1). Each has a double-stranded circular DNA of 7,400 to 8,200 bp, which replicates as an extrachromosomal plasmid in the nucleus of

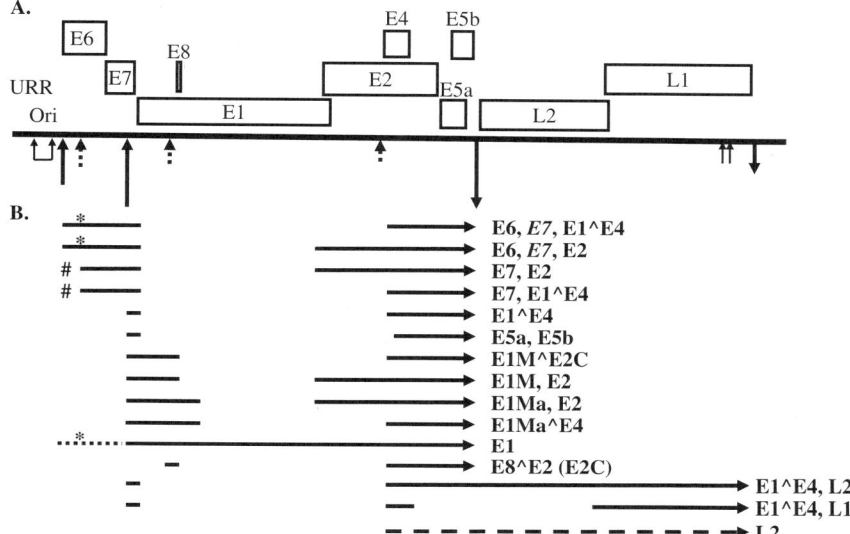

FIGURE 7.1. General schematic diagram of HPV genome organization. **A**: Genome organization. The circular genome is linearized for convenience. The URR contains promoters, transcription factor binding sites and an origin of replication (see Fig. 7.3). Open boxes represent ORFs in the E and L regions, each followed by a polyadenylation site (downward arrows). All the ORFs are encoded by the same DNA strand, but the reading frame of a given gene relative to other genes may be different among HPV genotypes. The various virus types may have 0, 1, or 2 E5 ORFs. Upward solid, thick arrows denote promoters that are common to both high-risk and low-risk HPV types. The dotted arrows represent promoters observed in the low-risk HPV-6 or HPV-11 (within E6 and E1 genes) or for HPV-31b (immediately upstream of the E4 ORF). The lengths of the arrows reflect the relative strength of the promoters or polyA sites. Thin upward arrows or linked arrows represent clusters of minor and less well characterized promoters reported for HPV-16 or HPV-31b. **B**: Transcripts and proven or deduced protein coding capacity. The transcription map is a composite of species defined in both high- and low-risk mucosotropic HPV types. Many of the species are also found for HPV-1. Each mRNA is depicted by an arrow. Gaps represent introns. *, E6 intragenic splices (E6*I, II) may exist in the high-risk HPVs. The donor site can also be spliced to acceptors in downstream ORFs (E6*III, IV). Unspliced species also encodes E7, but only in the high-risk HPV types. #, unique promoter for the low-risk HPV-6 and HPV-11 to generate mRNA capable of encoding the E7 protein. ...., extended 5'sequence or RNA species in the high-risk HPV types. - - -, mRNA from the P_{E4} promoter of HPV31b. All E region mRNAs and two of the three L region mRNA have the potential to encode E5 proteins. Not illustrated are additional spliced species for L1 transcripts reported for HPV-16 or HPV-31b because the translation of the upstream ORF would have terminated downstream of the L1 ORF, making L1 protein synthesis inefficient or unlikely. For splice sites of HPV-1, -6, -11, -16, -18, and -31, see Table 7.1.

infected keratinocytes. All open reading frames (ORFs) are encoded along the same DNA strand, and the early (E) and late (L) regions are each followed by a polyadenylation (polyA) site. The E4 ORF is embedded within the longer E2 ORF in a different translational reading frame, and the tiny E8 ORF is located within the E1 ORF. Between the end of the L-region ORFs and the beginning of the E-region ORFs is a segment of 450–700 bp which harbors no ORF and is designated as the upstream regulatory region (URR), or alternatively, the long control region (LCR). The URR contains numerous regulatory sequence elements, including transcription factor binding sites, promoters, and the origin of viral DNA replication.

The properties of most viral proteins have been investigated in isolation, but the functions of some of the proteins are not completely understood in the context of the intact viral genome during productive infections. In broad terms, the E-region proteins promote the initial expansion of the newly infected cell population (E6, E7, and E5); support the maintenance of the viral DNA in undifferentiated cells at low copy number (E1 DNA helicase, E2 origin-binding protein, E6 and E7); reestablish a permissive S-phase milieu in postmitotic, differentiated squamous keratinocytes (E7 and possibly E6) to support productive amplification of the viral DNA (continuing to require E1 and E2); and attenuate immune surveillance of viral infection (E5, E6, and E7). The L region encodes the L1 and L2 capsid proteins which self-assemble and package the progeny viral DNA into virions.

7.2.   Overview of Viral Transcription in Productive Infections

Early efforts to propagate papillomavirus in conventional cell culture systems comparable to those used for many other mammalian DNA viruses proved unsuccessful (Butel, 1972). Simply put, HPVs are not lytic viruses and no plaque assay exists. To understand this difficulty, two issues were addressed in the 1980s: to appreciate the architecture of the host tissues in which the papillomaviruses reproduce and to develop tools to investigate the viral activities in naturally infected human lesions.

The squamous epithelium is comprised of distinct cell strata: the basal, the parabasal, the lower and upper spinous layers and, additionally in the cutaneous skin, the stratum granulosum. The parabasal cells divide approximately daily. One daughter cell remains parabasal and continues to divide for two to three months. The other daughter withdraws from the cycle and begins the process of terminal differentiation as it is pushed upward. In contrast, the basal cells are normally quiescent, effectively serving as a long-term reservoir for replacement of the cycling parabasal cells as they exhaust their proliferative potential and undergo terminal differentiation (reviewed by Fuchs, 1993). The superficial cells undergo programmed cell death during which the cellular DNA and RNA are degraded and the envelope proteins become cross-linked, forming a tough but flexible physical barrier. The superficial dead cell layers (the cornified envelopes) are then sloughed off, allowing for the ongoing, steady-state renewal of the epithelium. Depending upon anatomic site, the transit time of a keratinocyte from dividing parabasal layer to desquamation takes one to several weeks.

In situ hybridization assays in thin sections of productively infected patient specimens reveal that the papillomaviral DNA and mRNA are below detection in the basal and parabasal cells but both dramatically increase in abundance in a subset of the spinous cells. The L1 major capsid protein is expressed and virions are assembled in a small number of terminally differentiated superficial cells (Stoler and Broker, 1986; Crum et al., 1988; Stoler et al., 1989, 1992; Higgins et al., 1992; reviewed by Chow and Broker, 1997a). The cornified envelopes, the integrity of which is compromised as a consequence of HPV infection (Lehr et al., 2002), are then sloughed off, carrying with them high concentrations of virions ready to infect a new host (Bryan and Brown, 2001). Thus, one can conclude that viral DNA is present as extrachromosomal plasmids in low copy numbers in the basal cells and that the productive phase of viral DNA replication takes place upon squamous differentiation. However, some cutaneous human and animal papilomaviruses are precocious in that viral DNA amplification and capsid protein synthesis can be detected starting from parabasal cells (Egawa et al., 2000; Nicholls et al., 2001; Peh et al., 2002).

7.3.   Viral RNA Mapping in Warty Lesions

The mRNAs of the closely related genotypes HPV-11 and HPV-6 recovered from benign genital condylomata have been investigated by a variety of RNA mapping techniques (Chow et al., 1987a; Nasseri et al., 1987; Ward and Mounts, 1989; Smotkin et al., 1989; DiLorenzo and Steinberg, 1995). Three promoters were inferred on the basis of the locations of the 5′ ends of the RNA: P1 immediately upstream of the E6 ORF, P2 within the E6 ORF, and P3 within the E7 ORF. In addition, putative minor promoters appear to exist just in front of the E8 ORF. All the transcripts are polyadenylated at the end of the E region or the L region, and virtually all are spliced (Fig. 7.1). The primary transcripts are alternatively spliced between various combinations of donor and acceptor sites (Table 7.1) (Rotenberg et al., 1989; Chiang et al., 1991; Sherman et al., 1992b). The mRNAs are inferred to encode proteins from individual ORFs as well as proteins derived from different ORFs brought together by RNA splicing (Table 7.1). The relative abundance of the different RNA species varies a great deal.

RNA mapping data and the results of in situ hybridization demonstrated that promoter activities and the choice of polyA sites are highly regulated during differentiation. Although the overlapping exon sequences in different mRNAs has made it very difficult to distinguish precisely one RNA species from another, the P3 promoter generates a predominant E1^E4 mRNA which is greatly up-regulated upon squamous differentiation (Chow et al., 1987a; Stoler et al., 1989). This mRNA was previously designated E1i^E4, to denote that the E1 exon contributes only the initiation plus four more codons. P3 is sometimes referred

TABLE 7.1. Coding Capacities of HPV mRNAs*

HPV type	Nucleotide position of splice sites [Location]						poly A#	Protein(s) Encoded%
	D	A	D	A	D	A		
1			827[E1]	3200[E4]			E	E6, E7, E1^E4
			827[E1]	2545[E1]			E	E2
			1231[E1,E8]	3200[E2]			E	E1M, E8^E2C (i.e. E2C)
			827[E1]	3200[E4]	3592[E2]	5431[L1]	L	E1^E4, L1
			827[E1]	3200[E4]		5431[L1]	L	E1^E4, L2, L1
	7710[URR]						L	L1
	7710[URR]			3200[E4]			L	E5, L2, L1
6, 11*			847[E1]	2622[E1]			E	E6, E7, E2
			847[E1]	3325[E4]			E	E6, E7 E1^E4
			847[E1]	3377[E2,E4]			E	E1*; E5a, E5b
			1272[E1]	2622[E1]			E	E2
			1272[E1, E8]	3325[E2]			E	E1M, E8^E2C (i.e. E2C)
			1272[E1]	3377[E2]			E	E1M^E2C
			1459[E1]	2622[E1]			E	E1Ma, E2
			1459[E1]	3325[E4]			E	E1Ma^E4
			847[E1]	3325[E4]	3596[E2]	5701[L1]	L	E1^E4, L1
			847[E1]	3325[E4]			L	E1^E4, L2, L1
16~	(226	409)[E6]	880[E1]	3357[E4]			E	E6, E6*1, E7, E1^E4
	(226	526)[E6]	880[E1]	3357[E4]			E	E6, E6*II, E7, E1^E4
	226			3357[E4]			E	E6*III,
	226			2708[E1]			E	E6*IV, E2
			880[E1]	2708[E1]			E	E2
			1301[E1,E8]	2708[E1]			E	E2
			1301[E8]	3357[E2]			E	E8^E2C
	226	409	880[E1]		3632[E2]	5639[L1]	L	E6*1, E7, E1*, L1
			1301[E8]			5639[L1]	L	E8*, L1
			1301[E8]	3357[E2]		5639[L1]	L	E8^E2C*, L1

Type								Products
18	(223,	416)[E6]	929[E1]		3434[E4]		E	E6, E6*1, E7, E1^E4
	(223,	546)[E6]	929[E1]		3434[E4]		E	E6*II, E7, E1^E4
	223				3434[E4]		E	E6*III,
	223				2779[E1]		E	E6*IV, E2
31	(210,	413)[E6]	877[E1]		3295[E4]		E	E6, E6*1, E7, E1^E4
	(210,	413)[E6]	877[E1]		2646[E1]		E	E6, E6*1, E7, E2
	(210,	413)[E6]	877[E1]		2510[E1]		E	E6, E6*1, E7
	210				3295[E4]		E	E6*III,
			1296[E8]		3295[E2]		E	E8^E2C
			877[E1]	3590[E2]	3295[E4]	5552[L1]	L	E1^E4, L1
			877[E1]			5552[L1]	L	E1*, L1
	210	413[E6]	877[E1]			5552[L1]	L	E6*1, E7, E1*, L1
	210	413[E6]	877[E1]		3295[E4]		L	E6*1, E7, E1^E4, L2, L1

*This table lists only mRNAs that encode proteins known to be expressed or consistently observed in multiple virus types. Rare or novel spliced transcripts are not presented. The splice sites were taken from numerous papers cited in this review. For simplicity, the promoters discussed in the text are not indicated. Virtually all the mRNA splice donor (D) and acceptor (A) sites are conserved among different HPV genotypes. [], the gene in which the splice site is located. The sites are often used in multiple message species derived from different promoters. %, the ability to encode E6 and E7 depends on the promoter used. &, nt positions are based on HPV-11. Equivalent nt positions in the highly homologous HPV-6 could be off by a few nts but can easily be deduced from sequence alignment. *, translation is terminated shortly after the splice and thus can only encode a short peptide. (), species without the splice also exists. #, all transcripts that end at the E region polyA site have the potential to encode the E5 proteins (but whether they do or not is largely unknown). The only exception is HPV-1, where the putative E5 termination codon is downstream of the polyA site. It can only be translated from species that are polyadenylated at the end of the L region. E1^E4 was previously designated as E1i^E4. E8^E2C was previously called E2C. Transcripts capable of encoding the E1 protein are very rare, and transcripts encoding E1 and E2 proteins may possibly initiate from the promoter preceding the E6 gene, a promoter within the E6 gene (for the low-risk HPVs), or a promoter within the E7 gene (see Fig. 7.1). ~, the nucleotide positions for HPV-16 are based on sequences corrected for the E1 (on the basis of sequence homology to other HPVs) and E5 genes (Halbert and Galloway, 1988) from the original publication (Seedorf et al., 1985), making the total length of 7,906 base pairs.

to as a late promoter because it is highly active in the differentiated upper strata, but its activity and the encoded protein functions clearly begin at the earliest times in the basal and parabasal keratinocytes (Stoler et al., 1990). The HPV-11 E1^E4 mRNA derived from the P3 promoter comprises over 90 percent of all viral transcripts. The 5' ends are heterogenously located between nts 685 and 716, and the mRNA is polyadenylated at the distal end of the E region (~nt 4400) (Nasseri et al., 1987; Ward and Mounts, 1989). Accordingly, E1^E4 is the most abundant viral protein in the condyloma (Brown et al., 1988). A minority of the transcripts derived from the P3 promoter is spliced to a position about 100 bases upstream of the E2 initiation codon and encodes the viral origin binding E2 protein. Additional transcripts from the P3 promoter contain a splice from the dominant donor site at nt 847 or alternatively from the site at 1272 to an acceptor at nt 3377. The former has the potential to encode the E5a and E5b peptides after the termination of a short E1* peptide, whereas the latter could encode a fusion protein between the amino terminal portion of E1 to the carboxyl terminal half of the E2 protein (E1M^E2C) (Chiang et al., 1991). An unspliced mRNA capable of encoding the viral replicative helicase E1 was rarely detected, if at all, but clearly the E1 protein provides an essential function for plasmid maintenance and amplification in undifferentiated and differentiated keratinocytes (Chow and Broker, 2006a). We suggest that, in the basal cells, the viral promoter activity must be minimal and periodic to produce the E1 and E2 proteins necessary to maintain the DNA copy number when these cells reenter the S-phase and divide.

Only a very low percentage of the viral transcripts are derived from the P1 promoter, located upstream of the E6 ORF. The P1 RNA initiation site utilized during in vitro transcription of HPV-11 has been mapped to nucleotide 94 (Hou et al., 2000). P1 transcripts in infected cells span the E6 and E7 ORFs and are polyadenylated at the end of the E region. They each contain a splice from the dominant E1 donor site to the alternative downstream acceptor sites that are also utilized by the P3 transcripts. Transcripts from the P2 promoter have 5' ends near nt 260 within the E6 ORF. Similarly, the transcripts are alternatively spliced using the same donor and acceptor sites described above, and are polyadenylated at the end of the E region. Only mRNA derived from the P2 promoter is efficiently translated into the E7 protein (Smotkin et al., 1989). An additional, infrequently observed RNA 5' end is located upstream of the overlapping E1/E8 ORFs, and this small E8 exon is spliced to the dominant acceptor site in the E2/E4 ORFs, generating E8^E2C (i.e., E2C) (Table 7.1) (Chow et al., 1987a; Rotenberg et al., 1989; Chiang et al., 1991).

In the differentiated strata, the P3 promoter is also responsible for the L-region transcripts. A minority of the transcripts from this promoter extends continuously to the L-region polyA site, whereas others contain a second splice, after the translation termination of the E1 ^ E4 directly to the initiation codon of the L1 ORF, skipping the E5 and L2 ORFs. Thus for HPV-6 and HPV-11, the "early-to-late" stage transition is not achieved by turning on a new late promoter, but rather by the up-regulation of the P3 promoter, increased read-through past the early polyadenylation site, elongation of primary transcripts to the late poly-A site for processing and maturation, and stabilization of the L1 and L2 mRNAs. In vitro, the expression of the L-region transcripts of the high-risk HPVs and bovine papillomavirus type 1 (BPV-1) has indeed been demonstrated experimentally to be controlled at the levels of transcription, splicing, polyadenylation, nuclear retention, and RNA stability mediated by cellular proteins that bind to untranslated as well as to coding sequences (Dietrich-Goetz et al., 1997; Terhune et al., 1999, 2001; Koffa et al., 2000; Cumming et al., 2002; Collier et al., 2002; Wiklund et al., 2002; Oberg et al., 2003; Zhao et al., 2004, 2005; Zheng, 2004).

In summary, all the E-region transcripts contain a downstream 3′ exon which includes the E2C/E4/E5a/E5b ORFs, whereas all the L region transcripts contain a 3′ exon which encodes the L1 major capsid protein (Table 7.1). Consequently, most mRNAs contain multiple ORFs and some indeed are polycistronic (Brown et al., 1996, 1998). For example, the E1^E4 mRNA has the potential to reinitiate translation of E5a, based upon experiments with a downstream surrogate green fluorescent protein reporter in place of E5a, and the E1^E4/L1 transcripts can be translated in vitro into both the E1^E4 and L1 proteins.

Similar HPV-1 transcripts are derived from P1 and P3 in RNA recovered from plantar warts (Chow et al., 1987b). Some L1 region transcripts have a short 5′ exon located entirely within the URR, and this noncoding exon is spliced directly to the L1 AUG or to the dominant acceptor at nt 3200 used for the E1^E4 protein (Palermo-Dilts et al., 1990) (Table 7.1). However, this latter species does not encode a variant E4 protein because it lacks an AUG. The termination codon of the putative HPV-1 E5 gene is located downstream of the E region polyA site. Thus, if an HPV-1 E5 protein actually exists, it can be translated only from transcripts that are polyadenylated at the end of the L region.

Similar mRNA splice combinations were detected by RT-PCR of RNA from precancerous lesions and from cell lines that harbor HPV-16 DNA (Doorbar et al., 1990; Dilts et al., 1990; Rohlfs et al., 1991; Sherman et al., 1992a). Additional HPV-16 promoter activities were also detected within the 5′ region of the URR, in the L1 region, and within E6 ORF (Glahder et al., 2003; Tan et al., 2003). One splice donor and two alternative splice acceptor two alternative splice acceptor sites within the E6 transcript are unique to the high-risk HPV types (Schwarz et al., 1985; Schneider-Gädicke and Schwarz, 1986; Smotkin and Wettstein, 1986). The donor site is also used in combination with splice acceptor sites for the E1^E4 or the E2 mRNAs (Table 7.1). Thus, it is possible to encode four related and truncated E6 proteins (E6*I, II, III, IV), each with a common amino terminus and terminating within a few amino acids after the splice. The E6*I is the most abundant E6-containing transcript. The species in which the donor in E6 is spliced to the acceptor in E4 does not encode an E4 variant protein because it lacks an AUG initiation codon after the termination of E6 translation.

A cell line called W12, which harbors extrachromosomal HPV-16 DNA, contains three spliced transcripts involving the L1 region. In one transcript, the dominant splice donor at nt 880 is directly spliced to the AUG codon of the the L1 exon at nt 5639 (Tan et al., 2003). Another report described two alternative species, with the donor site at nt 1301 spliced to either nt 3357 or nt 5639 (Table 7.1) (Doorbar et al., 1990). The former species has a second splice between nt 3632 and nt 5639. These latter species were also observed in a precancerous lesion (Sherman et al., 1992a). All three L1 transcripts are predicted to encode a short peptide which initiates in the E1 or E8 ORF and terminates downstream of the L1 AUG (designated E1* and or E8* or E8^E2C*), placing L1 in a very unfavorable sequence context for translation reinitiation. This RNA sequence organization is very different from the bicistronic E1^E4/L1 mRNAs observed in other HPVs in which the two ORFs do not overlap. Thus, these transcripts might not be functional L1 mRNAs, at least not efficient L1 mRNAs.

7.4. Organotypic Raft Cultures of Primary Human Keratinocytes as a Model System to Study HPVs

Organotypic or raft cultures of primary human keratinocytes (PHKs) grown at the medium:air interface achieve stratification and squamous differentiation in vitro (Asselineux and Prunieras, 1984), suggesting that they may support the productive phase of papillomavirus infection (Broker and Botchan, 1986). Initial efforts with this culture system recapitulated high-grade squamous intra-epithelial neoplasia with keratinocytes transfected with recombinant HPV-16 plasmid (McCance et al., 1988). The culture system was then optimized to generate from native primary human keratinocytes a squamous epithelium that closely resembles the native tissues (Wilson et al., 1992), with the exception that the basal cells are actively cycling rather than being quiescent as in native tissues. This culture condition allowed, for the first time, the productive phase of the HPV infection in the outgrowth of explanted condylomatous tissues (Dollard et al., 1992) or, alternatively, in the cell line CIN-612 9E, which was established from a dysplasia and harbors HPV-31b plasmids, upon treatment with an agent which promoted differentiation (Meyers et al., 1992).

In conjunction with acute retrovirus-mediated gene transfer, which enables highly efficient introduction of viral sequences into PHKs, the raft cultures were then used for genetic analysis of HPV regulatory regions and encoded proteins. The transduced cells are used to develop a fully stratified and differentiated epithelium without extensive population expansion or bias in growth selection in submerged cultures (reviewed in Chow and Broker, 1997b). The constitutive expression of the HPV E6 and E7 oncogenes from the retrovirus LTR promoter established the first experimental model for dysplasias (Halbert et al., 1992; Blanton et al., 1992), while the differentiation-dependent expression of E7 from the URR demonstrated its role in reactivating S-phase in post-mitotic spinous cells (Cheng et al., 1995). This system allowed the genetic dissection of the E7 domains critical in this function and for investigating the novel host responses to the unusual S phase state in the differentiated keratinocytes. These virus-host interactions lead to two cell populations. In one, cells unable to reenter S-phase accumulate high levels of cyclin E and cyclin E/cdk2 inhibitors p21cip1 and p27kip1. In the other population, cells that re-enter S phase and do not accumulate detectable levels of these host proteins (Jian et al., 1998, 1999; Noya et al., 2001; Chien et al., 2000, 2002). Since viral DNA amplification requires cyclin E/cdk2 (Ma, 1999; Lin et al., 2000; Deng et al., 2004), the inactivation of cdk2 by the inhibitors provides an explanation for the long-standing puzzle that only a fraction of the differentiated cells have abundant viral DNA or RNA (Stoler et al., 1989, 1992; Jian et al., 1999; reviewed by Chow and Broker, 2006b). Using a reporter gene, the raft system was also used to validate the differentiation-dependent activation of the E6 (P1) promoter and details of its regulation (Parker et al., 1997; Zhao et al., 1997, 1999a, b).

Further advances were made when the viral productive phase was reproduced in raft cultures of PHKs or HPV-negative, spontaneously immortalized human keratinocyte cell lines. In these experiments, HPV DNA excised from recombinant plasmids is transfected into PHKs, and colonies containing extra-chromosomal HPV DNA are expanded and put into raft cultures that generate infectious virions in a small fraction of the cells. This approach has been success-fully applied to several high-risk mucosotropic HPV types (Frattini et al., 1996, 1997b; Meyers et al., 1997; Flores et al., 1999; Ozbun, 2002b; Lehr et al., 2003; McLaughlin-Drubin et al., 2003, 2004) and to the low-risk HPV-11 (Fang et al., 2006a, b). The HPV-16 productive program has also been reproduced in raft cultures of an HPV-16 containing cell line established from a vulvar intraep-ithelial neoplasm (Grassmann et al., 1996), as well as in PHKs into which the HPV-16 genome has been introduced via an adenovirus, followed by Cre-LoxP excision to generate an autonomous HPV-16 replicon (Lee et al., 2004).

7.4.1.   Viral RNAs in Organotypic Model Systems

By using these experimental systems, transcription programs of several HPV types have been investigated in detail. Furthermore, viral RNA species recovered from submerged PHKs and raft cultures have been compared to identify differentiation-dependent activation of promoters and the selective use of the polyadenylation sites. However, it should be noted that a submerged PHK culture is akin to the wounding-healing state, rather than being comparable to the basal cells in an established epithelium, as judged from URR-reporter assays (Parker et al., 1997; Zhao et al., 1997).

In general, the locations of the several major and minor promoters as well as splice donor and acceptor sites identified in raft culture systems of the high-risk HPV are equivalent or identical to those identified in patient specimens or in the cell lines just described. The P1 promoter immediately upstream of the E6 gene can generate mRNAs that encode each of the E region proteins. Differentiation-dependent up-regulation of P670 in HPV-16 (Grassmann et al., 1996) and p742 in HPV-31b (Hummel et al., 1992; Klumpp and Laimins, 1999, also see below), equivalent to the P3 of the low-risk HPVs, was observed, as found in tissue sections of patient specimens (Crum et al., 1988; Higgins et al., 1992; Stoler et al., 1992). The transcripts are predicted to encode the E1^E4 and E5 proteins and perhaps E1 and E2 proteins as well. However, there are several differences. a) A P2 promoter within the E6 gene for the expression of E7 was not observed. Rather, the mechanism for E7 protein translation (to be discussed later) is distinctly different from that of the low-risk HPVs. b) The minor promoter immediately upstream of the E8 ORF has not been observed for any of the high-risk HPVs. However, a spliced transcript with the potential to encode the E8^E2 protein has been detected and the function of the protein demonstrated (Stubenrauch et al., 2000; Zobel et al., 2003). Since no transcript with a more remote 5′ end has been reported, it most likely comes from a promoter near the E8 ORF. c) An HPV-31b transcript from the P_{E4} promoter (immediately upstream of the E4 ORF) has the potential to encode E5a, E5b, L2 and L1 proteins (Ozbun and Meyer, 1998). This species is distinct from the putative E5 transcript of HPV-11 described above. d) There are many more RNA start sites, primarily in the URR, than observed in patient specimens or HPV-containing cell lines, possibly because the quality of viral RNAs recovered from raft cultures is superior, because patient specimens might not have been collected and preserved under optimal conditions. It is also likely the 2-week-old raft cultures represent a more dynamically regulated state than is characteristic of long-established patient lesions. Further, submerged cultures of stable cell lines of course lack temporal regulation of transcription.

By far the most comprehensive investigations into transcription regulation were conducted in the cell line CIN612 9E, which contains HPV-31b DNA, and in immortalized human epithelial HaCaT cells infected with recovered HPV-31b virions (Hummel et al., 1992; Ozbun and Meyer, 1997, 1998, 1999; Klumpp and Laimins, 1999; Terhune et al., 1999, 2001; Ozbun, 2002a, b). Even though the frequently used splice donor and acceptor sites were analogous to those for HPV-6, -11, and -16 just described, a larger number of RNA species was detected, primarily because of the presence of the splice donor site in the E6 ORF and of additional 5′ RNA termini. There are four significant promoters, and the transcripts contain virtually all possible combinations of exons. P99 is the major promoter and is analogous to P1 of the low-risk HPV which immediately precedes the E6 ORF. Alternatively spliced transcripts from this promoter can encode all the E region proteins. A second, much weaker promoter P77 (or the P_L promoter) generates primarily L region transcripts, but some transcripts are polyadenylated at the end of the E region as well. They have the capacity to encode various combinations of E6, E6*s, E7, E1^E4, E5a, E5b, L2, and L1. Both P77 and P99 are active in submerged cultures and in raft cultures. P742 (equivalent to P670 of HPV-16 or P3 of the low-risk HPV types) is embedded within the E7 ORF, and the transcripts may encode E1^E4, L2 and L1 and possibly E1 and E2 as well. Its activation is dependent on keratinocyte differentiation, but, surprisingly, viral DNA amplification is not required for its up-regulation (Spink and Laimins, 2005; Bodily and Meyers, 2005). The constitutive P_{E4} promoter was described above. Additional minor promoters were mapped to the URR, mainly in submerged cultures, two of which are down-regulated upon differentiation. L-region transcripts were also observed in submerged cultures. However such transcripts must be unstable as they were not detected in the mid-spinous cells by in situ hybridization of patient specimens (Stoler et al., 1989, 1992). It is conceivable that the intrinsic instability of L region transcripts in all but the most differentiated superficial cells could have affected their recovery from the patient specimens. Alternatively, their detection in the model system could reflect a difference between conditions in vitro and in vivo.

7.5. Where in the Stratified Squamous Epithelium Are E6 and E7 Proteins Expressed in Productively Infected Lesions?

High-risk HPV mRNAs capable of encoding the full-length E6 and E7 proteins are extremely rare and have not been detected by in situ methods in productively infected patient specimens (Bohm et al., 1993), whereas those of the low-risk types were reported to be restricted to the lower cell strata (Iftner et al., 1992). This pattern of distribution does not explain the S-phase reentry by the spinous cell when E7 is expressed from its native promoter, nor the need for E7 to enable viral DNA amplification in raft cultures (Cheng et al., 1995; Flores et al., 2000; Garner-Hamrick et al., 2004; McLaughlin-Drubin et al., 2005).

Nevertheless, one should be able to infer the expression patterns during different stages of the virus infection from the known or proposed protein functions. Both the E6 and the E7 proteins are required to maintain the viral plasmids in submerged cultures of primary keratinocytes (Thomas et al., 1999; Park and Androphy, 2002; Oh et al., 2004). In addition, the E6 promoter is also thought to give rise to polycistronic mRNAs that encode the E1 and E2 replication proteins (Ozbun and Meyer, 1998; Stacey et al., 1995, 2000; Remm et al., 1999), and the E6 intragenic splice has been implicated in regulating E1 and E2 expression (Hubert and Laimins, 2002). We reason that most if not all of the E region proteins are expressed for a short duration when viruses first gain entry into the basal cells. This timing would extend the wound healing state and period of rapid cell cycling to establish the population of infected cells, each with a minimal number of copies of the extrachromosomal viral DNA, perhaps 20 or more copies per cell. In conjunction with the host replication enzymes and proteins, the E1 and E2 proteins can replicate and maintain the viral plasmids in the cycling cells. The E7 protein targets the pRB/E2F pathway, driving the basal cells from the G0 state into the cell cycle (Fig. 7.2A). The best known activity of the high-risk HPV E6 protein is its collaboration with the host E6-associated protein (E6AP) to function as a ubiquitin E3 ligase, marking many host proteins for degradation (Scheffner et al., 1993; reviewed by Banks et al., 2003). Among them, the p53 protein is likely the primary target as far as viral DNA replication is concerned. The expression the E7 protein stabilizes the p53 protein in submerged PHK cultures (Demers et al., 1994; Eichten et al., 2002). Since a high level of p53 protein can inhibit HPV DNA replication in transfected cells (Lepik et al., 1998), E6 protein would then be needed to inactivate p53.

A. Extrachromosomal HPV DNA Replication in Productive Infections

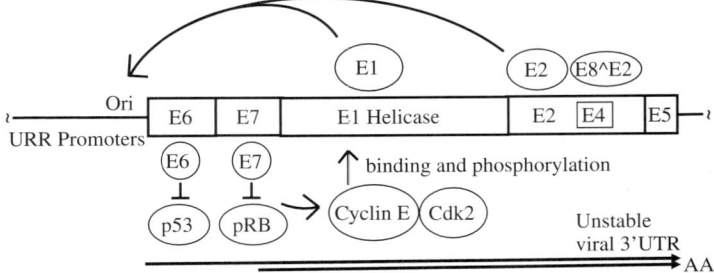

B. Clonal Selection for a Cancer Cell with a Single Transcribed HPV Copy
at the Downstream Integration Junction

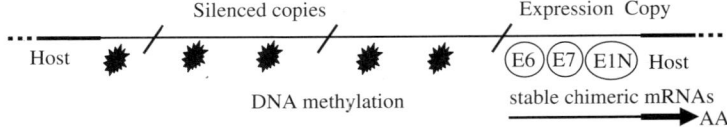

FIGURE 7.2. Changing transcription patterns in productive infections relative to carcinomas. **A**: Productive infections. The viral DNA exists as extrachromosomal plasmids. Only the E region is illustrated. For simplicity, the transcripts are represented as two horizontal arrows from the two major promoters. All use the E region polyA site and have a relatively unstable 3' untranslated region (UTR). The E6 protein inactivates p53 and E7 inactivates pRB to establish a permissive cell environment for viral DNA amplification in differentiated cells. The Cdk2/cyclin E or Cdk/cyclin A complex phosphorylates the E1 protein and allows its nuclear retention and initiation of replication. The E1, E2, and E8^E2 proteins control the levels of viral DNA replication. **B**: Clonal selection for an HPV-induced cancer cell. Within any cancer, only one viral copy at the downstream integration junction is transcriptionally active, regardless of the viral copy number or integration loci. Illustrated is a locus with tandemly integrated viral DNA with a disrupted E1 or E2 gene. The chimeric virus-host mRNAs have a stable 3' UTR and polyA site captured from the downstream host sequence. Thus, the cell expresses elevated but optimal levels of viral oncoproteins to achieve continuous long-term proliferation. It also expresses at least the aminoterminal portion of the E1 protein (E1N). All upstream viral DNA copies and viral DNAs at other integration sites are silenced by DNA methylation. Expression of at least the aminoterminal portion of the E1 protein and the inability to express an intact E2 protein may be required for long-term survival of the cells.

Indeed, p53 is also inactivated by other small DNA tumor viruses such as SV40, polyomaviruses and adenoviruses.

Once the establishment stage is over, the viral promoter(s) would be downregulated and the basal cells would return to the G0 state. Thereafter, in the state of persistent maintenance, the HPV plasmid would replicate in pace with host chromosomes only when an occasional G0 basal cell reenters the cell cycle to replenish a parabasal cell lost to terminal differentiation. The virus should require no more than the viral E1 and E2 replication proteins. Were the viral promoter(s) to express the viral oncogenes continuously in basal cells, these cells would proliferate continuously. Such continuous cell cycling clearly does not occur in normal, productively infected condylomas or papillomas, as only very few basal cells are positive for the proliferating cell nuclear antigen necessary for engaging in DNA replication (Demeter et al., 1994). Thus, it is reasonable to conclude that the E6 and E7 genes are rarely or lowly expressed in basal cells, unless the cells are subjected to repeated wounding–healing.

The mRNAs expressing E6, E7, E1, and E2 proteins should be up-regulated in the differentiated cells during the productive phase of the virus life cycle (Fig. 7.2A): E1 and E2 proteins to amplify the viral DNA (Chow and Broker, 2006), the E7 protein to reestablish an S-phase milieu necessary to support viral DNA amplification (Cheng et al., 1995; Flores et al., 2000; Garner-Hamrick et al., 2004; McLaughlin-Drubin et al., 2005), and the E6 protein to down-modulate p53 (Jian et al., 1998).

7.6. Mechanisms That Control the Expression of the E6 and E7 Proteins

7.6.1. Transcriptional Regulation

The enhancer-E6 promoter elements in the URR are particularly conducive to genetic dissection because they do not overlap any protein-coding regions. Experimental analyses have been conducted in the natural host, the primary keratinocytes, grown in submerged monolayer cell cultures and in organotypic rafts (Thierry et al., 1992; Dollard et al., 1993; Farr et al., 1995; O'Connor et al., 1996; Parker et al., 1997; Kyo et al., 1997; Kanaya et al., 1997; Zhao et al., 1997, 1999a, b; Hubert et al., 1999; Ai et al., 1999, 2000; O'Connor et al., 2000; del Mar Pena and Laimins, 2001; Hubert and Laimins, 2002; Sen et al., 2002, 2004; Bouallaga et al., 2003; Tan et al., 1994; Bromberg-White et al., 2003). To summarize, the E6 promoter overlaps the origin and is regulated by numerous host proteins. The sequence elements in the URR bind both positive and negative cellular transcription factors (Fig. 7.3). The most important include recognition sites for positive factors, such as AP1, Sp1, Oct1, CBP/p300, TEF1, and sites for negative factors such as YY1, c/EBP, and CDP. The URR also contains binding sites for progesterone receptor. The regulation by progesterone is likely an important factor in the higher pathogenicity of mucosotropic HPV infections in women than in men. In addition, nuclear matrix attachment sites that modulate HPV transcription have been mapped in HPV-16 DNA (Tan et al., 1998; Stunkel et al., 2000). Many of these topics have been expertly reviewed by Bernard (2002) and will not be repeated here.

However, several interesting observations are worth noting. First, factors important for high promoter activity in cell lines and in proliferating submerged keratinocyte cultures are also critical for high activity in the differentiated keratinocytes and many of the key factors function synergistically (see Parker

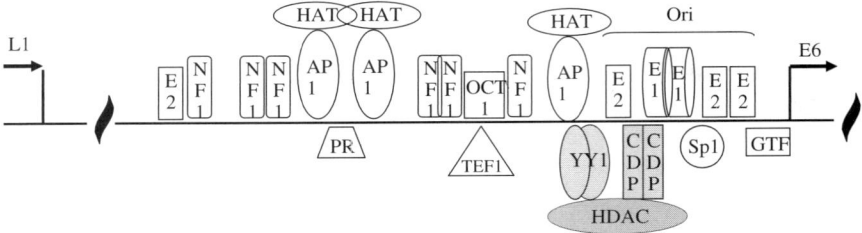

FIGURE 7.3. The E6 promoter-proximal transcription and replication regulatory region of mucosotropic HPVs. The great majority of the regulatory elements are located in this region. The origin is conserved among all mucosotropic HPVs and consists of an imperfect palindrome where E1 assembles into a dihexamer and three binding sites where three E2 dimers bind. There are numerous transcription factor-binding sites in this region, which is illustrated here for HPV-16 (adapted from Bernard, 2002). Most HPVs have an Oct1 binding site, two or more sites for AP1, one Sp1 site, and several NF1-, YY1-, and CDP-binding sites. AP1 recruits histone acetyltransferase (HAT), whereas YY1 and CDP recruit histone deacetylases (HDAC). Positive factors are shown in white background and negative factors in gray background. The role of NF1 is not clear as mutations of all four sites in HPV-11 had only minimal effects on promoter activity in raft cultures (Zhao et al., 1999a). GTF, general transcription factor; PR, progesterone receptor; TEF, transcription enhancer factor. Additional factors, such as KRF, PEF, cEBP, have been reported for other HPV types.

et al., 1997; Zhao et al., 1997). Second, transcription regulation in basal proliferating cells in raft cultures is distinct from that in submerged proliferating cells. For example, the E6 promoter, which is active in submerged cultures, is not detectably active in the basal cells once they have overlying strata. Thus, submerged, subconfluent cells may simulate a wound-and-healing state distinct from the maintenance proliferation found in a stratified epithelium. Third, the E6 promoter is regulated by chromatin remodeling via histone acetylation/deacetylation. Therefore, trichostatin A, an inhibitor of histone deacetylase, up-regulates this promoter in the basal cells (Zhao et al., 1999b; del Mar Pena and Laimins, 2001). Fourth, viral promoters are profoundly influenced by DNA methylation, which recruits histone deacetylases, leading to condensed chromatin. Methylation of HPV DNA has been described in cell lines and in patient specimens (Rösl et al., 1993; Badal et al., 2003, 2004; Van Tine et al., 2004b; Kalantari et al., 2004). Moreover, in a cell line containing HPV plasmids, the regulatory region is hypermethylated in the undifferentiated state but is hypomethylated in the differentiated state, implicating DNA methylation as an important mechanism to regulate viral gene expression (Kim et al., 2003).

7.6.2. Post-Transcriptional Regulation

The 5′ end of the mRNA from the E6 promoter is located within a few bases of the initiation codon of the E6 ORF. This situation would certainly make the translation of the E6 protein extremely inefficient or even unlikely. For HPV-31b in raft cultures, the presence of the additional minor 5′ RNA ends farther upstream in the URR could potentially improve the efficiency of translation initiation. Similar minor 5′ ends have been detected for HPV-18 in the HeLa cervical carcinoma cell line (Thierry and Yaniv, 1987). The regulation of these presumptive minor promoters has not yet been investigated.

The E6 and E7 ORFs often overlap or are separated by only a few nucleotides, making the translation of E7 protein less than optimal after the termination of E6 translation. The oncogenic HPV types have E6 intragenic splices to generate mRNAs that encode truncated E6*(I, II) peptides. These splices were initially thought to facilitate the initiation of translation of the downstream E7 protein by lengthening the untranslated sequence following the translation termination of E6*I or E6*II peptides (Smotkin et al., 1989). However, in vitro studies suggest that E7 protein can be translated efficiently from unspliced mRNA and that access to the E7 ORF is facilitated by an inefficient ribosome scanning mechanism which often skips the upstream E6 ORF (Stacey et al., 1995, 2000; Remm et al., 1999). Furthermore, in raft cultures, an E6-E7 operon incapable of splicing due to a mutation in the splice donor sequence is as effective in promoting S-phase entry as the wild-type operon (Cheng et al., 1995). Nevertheless, one cannot rule out the possibility that splicing within E6 has a minor effect on the translation of E7. E6 intragenic and intergenic splices of course prevent expression of the full-length E6 protein and result in the synthesis of E6* peptides that inhibit the E6 activity (Shally et al., 1996; Pim et al., 1997; Pim and Banks, 1999; Guccione et al., 2004). Thus the viruses appear to limit the activities of the full-length E6 protein, at least under certain conditions, although it is not known whether this occurs in a manner dependent on the stage of infection. In contrast, the low-risk viruses do not have E6 intragenic splice, and the E7 mRNA is derived from the P2 promoter (P260) to ensure ample supplies of the E6 and E7 proteins, both of which are far less potent than the high-risk HPV oncoproteins in inactivating their host targets.

7.7. What Is the Role of E2?

The understanding of the roles of E2 proteins has been evolving over the years. The full-length E2 protein and alternative forms (E2C or E8^E2 as well as E1M^E2C) were originally identified as transcription regulatory factors in reporter assays using a surrogate or native viral promoter in transfected cells (Spalholz et al., 1985; Phelps and Howley, 1987; Hirochika et al., 1987, 1988; Hawley-Nelson et al., 1988; Chiang et al., 1991). In transient assays, low amounts of HPV E2 protein can stimulate the activity of the native P1 (E6) promoter slightly, but high amounts of all forms of E2 can have a significant repressive effect on the P1 promoter. There are three E2 protein binding sites within the origin of HPV DNA replication, which overlaps the P1 promoter and its regulatory elements. Binding of E2 or E2-related proteins to the tandem E2 protein binding sites in the origin occludes the host Sp1 and TATA binding proteins from binding the promoter (Fig. 7.3) (Chin et al., 1988, 1989; Bernard et al., 1989; Romanczuk et al., 1990; Thierry and Howley, 1991; Hou et al., 2000, 2002; Dostatni et al., 1991; Dong et al., 1994a; Tan et al., 1994; Rapp et al., 1997; Alloul and Sherman, 1999; Stubenrauch et al., 2000).

However, it should be noted that there is no evidence that the endogenous E2 protein in productively infected cells ever achieves levels as high as ectopic E2, which is invariably expressed from a strong surrogate promoter. In fact, high levels of the ectopic E2 protein can cause chromosome polyploidy or even cell death in cells with or without HPV oncogenes (Frattini et al., 1997a; Webster et al., 2000; Demeret et al., 2003, and see below). Moreover, some reports also cast doubt on a major role of E2 as a transcription factor during the normal virus life cycle. In one, a high level of HPV-16 E2 expressed from an adenovirus did not affect transcription from the endogenous E6 promoter in a cell line containing high copies of HPV-16 DNA plasmids, but it did severely inhibit the E6 promoter in an isogenic cell line which harbors integrated viral DNA (Bechtold et al., 2003). In another, a normal productive phase in the raft cultures is displayed by an HPV-31b mutant in which the E2 protein is inactive in its transcription regulatory function but retains its replication activity. The only defects detected were a slight reduction DNA copy number and L region mRNAs (Staubenrauch et al., 1998). Thus, the primary functions of E2 during the normal viral life cycle are likely to be in the initiation of viral DNA replication (for reviews, see Stenlund, 2003; Chow and Broker, 2006a) and in viral plasmid persistence in dividing cells (Lehman and Botchan, 1998; Skiadapolous and McBride, 1998; Ilves et al., 1999; Van Tine et al., 2004a; You et al., 2004; Baxter et al., 2005).

In contrast to HPV E2 proteins, the BPV-1 E2 protein has a very strong repressing effect on the HPV-URR-E6 promoter (Thierry and Yaniv, 1987; Chin et al., 1989; Thierry and Howley, 1991; Dostatni et al., 1991; Tan et al., 1994) and on the endogenous E6 promoter in cervical carcinoma cells, leading to senescence or apoptosis (Hwang et al., 1993; Dowhanick et al., 1995; Desaintes et al., 1997). This property of BPV-1 E2 protein has been exploited to examine the roles of E6 and E7 in regulating cell growth, apoptosis, or senescence in these cervical carcinoma cell lines (Goodwin and DiMaio, 2000; Francis et al., 2000; Wells et al., 2000; Lee et al., 2002; DeFilippis et al., 2003; Horner et al., 2004; Psyrri et al., 2004). Interestingly, only the full-length BPV-1 E2 with the aminoterminal *trans*-acting domain can repress the viral URR promoter from an integrated state, whereas BPV1 E2 proteins lacking transactivating activity have little or no repressing effect (Goodwin et al., 1998; Nishimura et al., 2000).

7.8.   Why Are There So Many Promoters and Spliced Transcripts?

Despite the large number of transcripts resulting from alternative promoters and splice sites in the high-risk HPV genotypes (Table 7.1), the proteins inferred are not particularly different from those encoded by the much less complex mRNAs of the low-risk HPVs. The only notable exceptions are the truncated E6* peptides that are unique to the high-risk HPVs. The use of different promoters may allow the virus to fine tune the timing and levels of various viral proteins in different epithelial strata for the establishment, maintenance and amplification phases of the infection. Selective promoter usage coupled with alternative mRNA splicing obviously allows the ribosomes to access different ORFs, thereby producing different proteins in a coordinated manner from a single promoter. On the other hand, such a large number of alternatively spliced transcripts may also simply reflect the absence of a mechanism to prevent promiscuous assortment of exons. Because these events occur at a low frequency, they do not disrupt the production of the intended proteins or interfere with their functions.

Certain major alternative splices undoubtedly serve important purposes, as may be the case with the E2 protein and its alternative forms. The E1 and E2 proteins presumably are expressed from the same promoter. Different levels of the full-length E2 protein are produced by mRNAs with different E1 intragenic splices (Alloul and Sherman, 1999; Deng et al., 2003). Thus, RNA splicing regulates the amounts of E1 and E2 proteins, depending upon whether the primary transcript is spliced and how it is spliced. Because the amounts of E1 and E2 proteins control the extent of viral DNA replication in transiently transfected cells (Deng et al., 2003), we infer that alternative mRNA splicing may be a means to regulate viral DNA replication in the different strata of the squamous epithelium.

Moreover, alternative mRNA splicing, coupled with alternative promoter use in some instances, creates proteins that are antagonistic to the full-length protein, such as E8^E2 (i.e., E2C) and E1M^E2C described above. These proteins may compete with E2 by binding to the E2 binding sites, thus regulating viral DNA replication (Chiang et al., 1992b; Liu et al., 1995; Lim et al., 1998). In particular, E8^E2 protein of HPV-31b is a broad range, strong repressor of extrachromosomal plasmid DNA replication, independent of E2 binding sites (Zobel et al., 2003). HPV-31 mutants unable to produce E8^E2 replicate to high copy number in transient assays but are unable to maintain extrachromosomal plasmids in long term cultures. Thus, the E8^E2 protein plays a role in maintaining an appropriate plasmid copy number (Stubenrauch et al., 2000).

Another interesting issue concerns the expression of the E6 protein. If p53 inhibits viral DNA replication, why would the virus want to limit the expression or activity of the full-length E6 protein by unfavorable translation sequence context and by producing the antagonistic E6* peptides? We hypothesize that too much E6 protein or, conversely, too little p53 might not be beneficial for the long-term persistent survival of monomeric, extrachromosomal viral DNA, by virtue of the important role of p53 in preventing excessive DNA recombination (reviewed by Sengupta and Harris, 2005). Total elimination of p53 might lead not only to host chromosomal instability but also to the generation of multimeric HPV genomes that cannot be packaged into virions. Thus, the amount of E6 must be carefully controlled throughout all phases of the virus infection. This hypothesis can also account for the high levels of E6*I transcript in cancers and in the highly aneuploid cervical cancer cell lines that, remarkably, have not continued to degenerate cytogenetically since their isolation many decades ago.

Alternative splicing might also be a safeguard to ensure the synthesis of viral proteins. For example, the E1^E4 protein is critical for the productive phase of cottontail rabbit papillomavirus and HPV-31b (Peh et al., 2004; Wilson et al., 2005). When the dominant splice receptor site normally used for generating the E1^E4 mRNA is mutated, a cryptic acceptor site one codon away is utilized (Klumpp et al., 1997). An analogous alternative E4 acceptor site previously predicted for HPV-16 was recently confirmed in the CaSki cervical carcinoma cell line (Chow et al., 1987a; Van Tine et al., 2004b). Finally, there are a very large number of spliced transcriptions containing portions of the L1 ORF. They may represent spurious mRNA splicing or fragments resulting from read-through of the E region RNAs (Ozbun and Meyer, 1997).

7.9.   Basis For HPV Oncogenesis

Papillomaviruses encode proteins that enable them to reproduce in postmitotic, differentiated keratinocytes in the spinous strata. However, expressing such powerful regulatory proteins is a double-edged sword for the high-risk viruses, as viral-mediated neoplastic progression is an occasional sequella. High-grade SIL and cancer can occur under conditions in which the quasi-equilibrium between the virus and host is lost when the viral oncogenes are repeatedly up-regulated in the normally quiescent basal cells. This condition could occur during multiple rounds of wounding and healing of the squamous epithelium. The inactivation of the p53 and pRB proteins by the viral oncoproteins drives excessive cell proliferation and host genome instability resulting in progression to high-grade dysplasias and carcinomas (for a review, see Münger et al., 2004).

7.9.1.   Transcription in Carcinomas and Cervical Carcinoma Cell Lines

Carcinomas are accidental dead ends for the virus as far as production of progeny virus is concerned. HPV-associated cancers typically contain integrated DNA, sometimes in tandem arrays, or harbor a mixture of extrachromosomal genomes and integrated viral DNA (reviewed by Wentzensen et al., 2004). In such tissues, viral oncogene expression is highly up-regulated and viral mRNA is easily detected by in situ hybridization (Higgins et al., 1991; Stoler et al., 1992). Clearly, an elevated level of viral oncogene expression in basal or basal-like cells is obligatory in HPV-induced cancers, as only such cells can divide and have the opportunity to pass on to daughter cells cellular mutations and genomic instability associated with neoplastic progression.

HPV DNA integration is random with regard to host chromosomes, except for a predilection for fragile sites (Wentzensen et al., 2004). In contrast, the virus–host junction is usually located within the E1 or E2 gene, severing viral promoters from the viral polyA sites and disrupting the expression of an intact E1, E2 and downstream genes (Ziegert et al., 2003). In situ hybridization reveals that, regardless of copy number of HPV genomes or number of sites of integration, there is usually only one chromosomal locus that actively transcribes the viral DNA. This active viral genome always lies at the downstream integration junction, whereas all upstream copies with intact E1 and E2 genes are silenced (Fig. 7.2B). Direct cloning and sequencing of the virus-host chimeric RNAs from many tumors indeed shows a predominant RNA species with an interrupted or completely deleted E2 gene (Ziegert et al., 2003). The viral transcription center in cancer cells appears to be selected to achieve an optimal level of oncogene expression appropriate for the particular growth environment (Van Tine et al., 2004b). The silenced copies can be experimentally reactivated upon growth in the presence of 5-azacytidine, an inhibitor of DNA methylation. Thus, all but one or at most a few copies of the integrated HPV DNA are silenced by DNA methylation (Van Tine et al., 2004b).

7.9.2. Are There Additional Viral Proteins Contributing to Viral Carcinogenesis?

What are the reasons for such a pattern of viral DNA integration and gene expression? Traditionally, two factors were thought to contribute to the growth advantage of the transformed cancer cells. First, integration provides an opportunity to capture a host 3′ untranslated region (UTR) and polyadenylation site which confers a higher stability of the viral E6 and E7 mRNAs, increased rounds of translation, and growth advantage of the cells (Jeon et al., 1995; Jeon and Lambert, 1995). Second, the inability to encode the E2 family of proteins would disrupt the negative feedback regulation of the viral oncogenes, increasing their transcription. In tumors where integrated HPV cannot be demonstrated, mutations in viral or host DNA may have up-regulated the viral E6 promoter (May et al., 1994; Dong et al., 1994b). However, mutations in the E2 binding sites proximal to the E6 promoter have not been observed in these viral genomes. This result suggests that loss of E2 repression may not be a major factor in the up-regulation of the E6 promoter in these cells and implies that there may be some other reasons for the cessation of E2 protein expression in cancers containing integrated HPV genomes.

In view of recent findings concerning viral protein functions, we propose two additional potential factors to consider with regard to the molecular basis for neoplastic progression. Continued expression of E1 and E2 proteins from intact internal copies of the viral genome in tandem arrays could cause repeated viral *ori*-dependent replication in situ (endo-reduplication in an onion-skin fashion), leading to chromosomal branching which could be fragile and subject to breakage. Moreover, the HPV E2 protein associates with mitotic spindles during mitosis. In the presence of the E2 protein, the cluster of E2 binding sites in the replication origin of the integrated HPV DNA might function as a viro-centromere. Segregation of the viral sequences to the opposite pole from that selected by the host chromosome centromere may cause the chromosome to break and initiate a breakage-fusion-bridge cycle (Van Tine et al., 2004a). If such E2-promoted cycle remains unchecked, it would lead to mitotic catastrophe. Thus, transcription from the internal viral DNA copies is silenced by DNA methylation to interrupt the cycle of breakage and nonhomologous end joining. In the cancer cell which eventually emerges, only one integration junction copy, which cannot express an intact E2 protein, is transcriptionally active. This hypothesis remains to be tested. Nevertheless, progression from high-grade SIL to invasive cancer is accompanied by viral DNA integration and increasing host chromosome instability in the form of polysomy and aneusomy. In contrast, low-grade SIL is associated with extrachromosomal viral DNA and with disomy and some tetrasomy (Klaes et al., 1999; Hopman et al., 2004; Melsheimer et al., 2004), consistent with the ability of the E7 protein to promote endo-reduplication in differentiated cells (Chien et al., 2002).

We propose that there is another important contributor to viral oncogenesis. As already discussed, the transcriptionally active integration junction copy in cancers and cervical carcinoma cell lines contains either the entire E1 region or the amino terminal region of the E1 gene. We suggest that there are a number of reasons for the preservation of at least a portion of the E1 protein. First, the predominant splice donor site near the 5' end of the E1 ORF is important for efficient immortalization of primary human keratinocytes (Belaguli et al., 1992). Splicing from this site would allow the efficient capture of a somewhat remote 3' UTR and polyA addition site in the downstream host sequence without the retention of intervening viral and host RNA, which could be long and unstable. In the cancers surveyed, the predominant virus-host chimeric mRNAs indeed use this splice donor site (Ziegert et al., 2003). In addition, mRNA splicing itself may also increase message transport into the cytoplasm.

The mechanisms that regulate viral DNA replication suggest another reason for retention of at least a portion of the E1 sequences in cervical cancers. The aminoterminal region of the E1 protein binds with high affinity to multiple classes of cyclins (Ma, 1999), and cyclin E/cdk2 is essential for nuclear retention of the E1 protein as well as for initiation of viral DNA replication (Lin et al., 2000; Deng et al., 2004). Cyclin E is up-regulated by the E7 protein and is necessary to initiate cellular DNA synthesis. However, high levels of ectopic cyclin E protein shorten the G1 phase and, beyond some threshold number of cell cycles, may not allow sufficient time for the cell to replenish its cytoplasm and nucleoplasm (Ohtsubo and Roberts, 1993). Such accelerated cell cycling would be detrimental to the long-term cell survival and the cell lineage would "burn out." We hypothesize that the amino terminal fragment of the E1 protein (E1N in Fig. 7.2B) in the cervical carcinomas or cell lines might sequester the excess cyclin E to avoid shrinking cell sizes and the ultimate loss of cell viability.

In summary, a number of factors collectively lead to emergence of a rare cancer cell in which the E6 and E7 oncogenes of integrated, high-risk HPV genotypes are expressed at optimal levels to sustain continued proliferation of the tumor cells, with E6 apparently modulated so that it does not totally eliminate the p53 protein. At an early stage after integration, the E2 protein most likely contributes to viral carcinogenesis by inducing host chromosome breakage, but its expression is eventually extinguished to prevent mitotic catastrophe. This is achieved by a clonal selection for a cell in which integration disrupts the expression of E2 but has captured a stable host polyA site to confer elevated viral oncogene expression, and by transcriptional silencing of intact upstream copies mediated by DNA methylation. Furthermore, excess cyclin E is titrated out by an aminoterminal segment of the E1 helicase to maintain an optimal duration of the G1 phase for long-term cell proliferation. All these features are molecular attributes of HPV-induced cancer cells (Fig. 7.2B).

7.10. Concluding Remarks

In summary, HPVs use alternative promoters, alternative splice sites and two polyadenylation sites to regulate the production of their messenger RNAs and proteins and to modulate the protein ratios and activities, enabling the virus to exist in perpetuity and to amplify as a transmissible agent. Integration of the viral genome can lead to significant alteration of transcription controls and drive the host cells toward neoplasia.

Acknowledgments. The authors are supported by USPHS grants CA36200, CA83679, and CA107338. We gratefully acknowledge the invaluable contributions of past and present lab members and the many collaborators who have contributed to our investigations into the pathobiology of the human papillomaviruses.

References

Ai, W., Narahari, J., and Roman, A. (2000). Yin yang 1 negatively regulates the differentiation-specific E1 promoter of human papillomavirus type 6. *J. Virol.* 74:5198–5205.

Ai, W., Toussaint, E., and Roman, A. (1999). CCAAT displacement protein binds to and negatively regulates human papillomavirus type 6 E6, E7, and E1 promoters. *J. Virol.* 73:4220–4229.

Alloul, N., and Sherman, L. (1999). The E2 protein of human papillomavirus type 16 is translated from a variety of differentially spliced polycistronic mRNAs. *J. Gen. Virol.* 80:29–37.

Asselineau, D., and Prunieras, M. (1984). Reconstruction of 'simplified' skin: control of fabrication. *Br. J. Dermatol.* 111(27 Suppl.):219–222.

Badal, S., Badal, V., Calleja-Macias, I.E., Kalantari, M., Chuang, L.S., Li, B.F., and Bernard, H.-U. (2004). The human papillomavirus-18 genome is efficiently targeted by cellular DNA methylation. *Virology* 324:483–492.

Badal, V., Chuang, L.S., Tan, E.H., Badal, S., Villa, L.L., Wheeler, C.M., Li, B.F., and Bernard, H.-U. (2003). CpG methylation of human papillomavirus type 16 DNA in cervical cancer cell lines and in clinical specimens: genomic hypomethylation correlates with carcinogenic progression. *J. Virol.* 77:6227–6234.

Baker, C.C., Phelps, W.C., Lindgren, V., Braun, M.J., Gond, M.A., and Howley, P.M. (1987). Structural and transcriptional analysis of human papillomavirus type 16 sequences in cervical carcinoma cell lines. *J. Virol.* 61:962–971.

Banks, L., Pim, D., and Thomas, M. (2003). Viruses and the 26S proteasome: hacking into destruction. *Trends Biochem. Sci.* 28:452–459.

Baxter, M.K., McPhillips, M.G., Ozato, K., and McBride, A.A. (2005). The mitotic chromosome binding activity of the papillomavirus E2 protein correlates with interaction with the cellular chromosomal protein, Brd4. *J. Virol.* 79:4806–4818.

Bechtold, V., Beard, P., and Raj, K. (2003). Human papillomavirus type 16 E2 protein has no effect on transcription from episomal viral DNA. *J. Virol.* 77:2021–2128.

Belaguli, N.S., Pater, M.M., and Pater, A. (1992). Nucleotide 880 splice donor site required for efficient transformation and RNA accumulation by human papillomavirus type 16 E7 gene. *J. Virol.* 66:2724–2730.

Bernard, B.A., Bailly, C., Lenoir, M.C., Darmo, M., Thierry, F., and Yaniv, M. (1989). The human papillomavirus type 18 (HPV18) E2 gene product is a repressor of the HPV18 regulatory region in human keratinocytes. *J. Virol.* 63:4317–4324.

Bernard, H.-U. (2002). Gene expression of genital human papillomaviruses and considerations on potential antiviral approaches. *Antivir. Ther.* 7:219–237.

Blanton, R.A., Coltrera, M.D., Gown, A.M., Halbert, C.L., and McDougall, J.K. (1992). Expression of the HPV16 E7 gene generates proliferation in stratified squamous cell cultures which is independent of endogenous p53 levels. *Cell Growth Differ.* 3:791–802.

Bodily, J.M., and Meyers, C. (2005). Genetic analysis of the human papillomavirus type 31 differentiation-dependent late promoter. *J. Virol.* 79:3309–3321.

Bohm, S., Wilczynski, S.P., Pfister, H., and Iftner, T. (1993). The predominant mRNA class in HPV16-infected genital neoplasias does not encode the E6 or the E7 protein. *Int. J. Cancer* 55:791–798.

Bouallaga, I., Teissier, S., Yaniv, M., and Thierry, F. (2003). HMG-I(Y) and the CBP/p300 coactivator are essential for human papillomavirus type 18 enhanceosome transcriptional activity. *Mol. Cell. Biol.* 23:2329–2340.

Broker, T.R., and Botchan, M. (1986). Papillomaviruses: Retrospectives and prospectives. *Cancer Cells* 4:17–36.

Bromberg-White, J.L., Sen, E., Alam, S., Bodily, J.M., and Meyers, C. (2003). Induction of the upstream regulatory region of human papillomavirus type 31 by dexamethasone is differentiation dependent. *J. Virol.* 77:10975–10983.

Brown, D.R., Chin, M.T., and Strike, D.G. (1988). Identification of human papillomavirus type 11 E4 gene products in human tissue implants from athymic mice. *Virology* 165:262–267.

Brown, D.R., McClowry, T.L., Sidner, R.A., Fife, K.H., and Bryan, J.T. (1998). Expression of the human papillomavirus type 11 E5A protein from the E1E4,E5 transcript. *Intervirology* 41:47–54.

Brown, D.R., Pratt, L., Bryan, J.T., Fife, K.H., and Jansen, K. (1996). Viruslike particles and E1E4 protein expressed from the human papillomavirus type 11 bicistronic E1E4L1 transcript. *Virology* 222:43–50.

Bryan, J.T., and Brown, D.R. (2001). Transmission of human papillomavirus type 11 infection by desquamated cornified cells. *Virology* 281:35–42.

Butel, J.S. (1972). Studies with human papilloma virus modeled after known papovavirus systems. *J. Natl. Cancer Inst.* 48:285–299.

Cheng, S., Schmidt-Grimminger, D.-C., Murant, T., Broker, T.R., and Chow, L.T. (1995). Differentiation-dependent up-regulation of the human papillomavirus E7 gene reactivates cellular DNA replication in suprabasal differentiated keratinocytes. *Genes Dev.* 9:2335–2349.

Chiang, C.-M., Broker, T.R., and Chow, L.T. (1991). An E1M^E2C fusion protein encoded by human papillomavirus type 11 is a sequence specific transcription repressor. *J. Virol.* 65:3317–3329.

Chiang, C.-M., Ustav, M., Stenlund, A., Ho, T.F., Broker, T.R., and Chow, L.T. (1992b). Viral E1 and E2 proteins support replication of homologous and heterologous papillomaviral origins. *Proc. Natl. Acad. Sci. USA* 89:5799–5803.

Chien, W.-M., Noya, F., Benedict-Hamilton, H.M., Broker, T.R., and Chow, L.T. (2002). Alternative fates of keratinocytes transduced by human papillomavirus type 18 E7 during squamous differentiation. *J. Virol.* 76:2964–2972.

Chien, W.-M., Parker, J.N., Schmidt-Grimminger, D.-C., Broker, T.R., and Chow, L.T. (2000). Casein kinase II phosphorylation of the human papillomavirus-18 E7 protein is critical for promoting S phase entry. *Cell Growth Differ.* 11:425–435.

Chin, M.T., Hirochika, R., Hirochika, H., Broker, T.R., and Chow, L.T. (1988). Regulation of the human papillomavirus type 11 enhancer and E6 promoter by activating and repressing proteins from the E2 open reading frame: functional and biochemical studies. *J. Virol.* 62:2994–3002.

Chin, M.T., Broker, T.R., and Chow, L.T. (1989). Identification of a novel constitutive enhancer element and its associated binding protein: Implications for human papillomavirus type 11 enhancer regulation. *J. Virol.* 63:2967–2976.

Chow, L.T., and Broker, T.R. (1997a). Small DNA tumor viruses. In N. Nathanson (ed.): *Viral Pathogenesis.* Philadelphia, PA: Lippincott-Raven Publishers, pp. 267–302.

Chow, L.T., and Broker, T.R. (1997b). In vitro experimental systems for HPV: Epithelial raft cultures for viral reproduction and pathogenesis and for genetic analyses of viral proteins and regulatory sequences. *Clin. Dermatol.* 15:217–227.

Chow, L.T., and Broker, T.R. (2006a). Mechanisms and regulation of papillomavirus DNA replication. In M. S. Campo (ed.): *Recent Advances in Papillomavirus Research.* Caister Academic Press. pp.53–71.

Chow, L.T., and Broker, T.R. (2006b). Human papillomavirus infections: warts or cancer? In M. DePamphilis (ed.): *DNA Replication and Human Disease.* pp.609–625, chapter 30. Cold Spring Harbor Laboratory Press.

Chow, L.T., Nasseri, M., Wolinsky, S.M., and Broker, T.R. (1987a). Human papilloma virus type 6 and 11 mRNAs from genital condylomata acuminata. *J. Virol.* 61:2581–2588.

Chow, L.T., Reilly, S., Broker, T.R., and Taichman, L. (1987b). Identification and mapping of human papillomavirus type 1 RNA transcripts recovered from plantar warts and infected epithelial cell cultures. *J. Virol.* 61:1913–1918.

Collier, B., Oberg, D., Zhao, X., and Schwartz, S. (2002). Specific inactivation of inhibitory sequences in the 5′ end of the human papillomavirus type 16 L1 open reading frame results in production of high levels of L1 protein in human epithelial cells. *J. Virol.* 76:2739–2752.

Crum, C.P, Nuovo, G., Friedman, D., and Silverstein, S.J. (1988). Accumulation of RNA homologous to human papillomavirus type 16 open reading frames in genital precancers. *J. Virol.* 62:84–90.

Cumming, S.A., Repellin, C.E., McPhillips, M., Radford, J.C., Clements, J.B., and Graham, S.V. (2002). The human papillomavirus type 31 late 3′ untranslated region contains a complex regulatory element. *J. Virol.* 76:5993–6003.

DeFilippis, R.A., Goodwin, E.C., Wu, L., and DiMaio, D. (2003). Endogenous human papillomavirus E6 and E7 proteins differentially regulate proliferation, senescence, and apoptosis in HeLa cervical carcinoma cells. *J. Virol.* 77:1551–1563.

del Mar Pena, L.M., and Laimins, L.A. (2001). Differentiation-dependent chromatin rearrangement coincides with activation of human papillomavirus type 31 late gene expression. *J. Virol.* 75:10005–10013.

Demeret, C., Garcia-Carranca, A., and Thierry, F. (2003). Transcription-independent triggering of the extrinsic pathway of apoptosis by human papillomavirus 18 E2 protein. *Oncogene* 16:168–175.

Demers, G.W., Halbert, C.L., and Galloway, D.A. (1994). Elevated wild-type p53 protein levels in human epithelial cell lines immortalized by the human papillomavirus type 16 E7 gene. *Virology* 198:169–174.

Demeter, L.M., Stoler, M.H., Broker T.R., and Chow, L.T. (1994). Induction of proliferating cell nuclear antigen (PCNA) in differentiated epithelium in productively infected human papillomavirus-infected lesions. *Hum. Pathol.* 25:343–348.

Deng, W., Jin, G., Lin, B.Y., Van Tine, B.A., Broker, T.R., and Chow, L.T. (2003). mRNA splicing regulates human papillomavirus type 11 E1 protein production and DNA replication. *J. Virol.* 77:10213–10226.

Deng, W., Lin, B.Y., Jin, G., Wheeler, C., Ma, T., Harper, J.W., Broker, T.R., and Chow, L.T. (2004). Cyclin/CDK regulates the nucleo-cytoplasmic localization of the human papillomavirus E1 DNA helicase. *J. Virol.* 78:13954–13965.

Desaintes, C., Demeret, C., Goyat, S., Yaniv, M., and Thierry, F. (1997). Expression of the papillomavirus E2 protein in HeLa cells leads to apoptosis. *EMBO J.* 16:504–514.

de Villiers, E.-M., Fauquet, C., Broker, T.R., Bernard, H.-U., and zur Hausen, H. (2004). Classification of papillomaviruses. *Virology* 324:17–27.

DiLorenzo, T.P., and Steinberg, B.M. (1995). Differential regulation of human papillomavirus type 6 and 11 early promoters in cultured cells derived from laryngeal papillomas. *J. Virol.* 69:6865–6872.

Dietrich-Goetz, W., Kennedy, I.M., Levins, B., Stanley, M.A., and Clements, J.B. (1997). A cellular 65-kDa protein recognizes the negative regulatory element of human papillomavirus late mRNA. *Proc. Natl. Acad. Sci. USA* 94:163–168.

Dilts, D.P., Broker, T.R., and Chow, L.T. (1990). The structures of human papillomavirus type 1 and type 16 messenger RNAs determined by polymerase chain reaction. In P. Howley, and T. Broker (eds): *Papillomaviruses. UCLA Symp. Mol. Cell. Biol.* 124:533–540. New York, NY: Alan R. Liss.

Dollard, S.C., Broker, T.R., and Chow L.T. (1993). Regulation of the human papillomavirus type-11 E6 promoter by viral and host transcription factors in primary human keratinocytes. *J. Virol.* 67:1721–1726.

Dollard, S.C., Wilson, J.L., Demeter, L.M., Bonnez, W., Reichman, R.C., Broker, T.R., and Chow, L.T. (1992). Production of human papillomavirus and modulation of the infectious program in epithelial raft cultures. *Genes Dev.* 6:1131–1142.

Dong, G., Broker, T.R., and Chow, L.T. (1994a). Human papillomavirus type-11 E2 proteins repress the homologous E6 promoter by interfering with the binding of host transcription factors to adjacent elements. *J. Virol.* 68:1115–1127.

Dong, X.P., Stubenrauch, F., Beyer-Finkler, E., and Pfister, H. (1994b). Prevalence of deletions of YY1-binding sites in episomal HPV 16 DNA from cervical cancers. *Int. J. Cancer* 58:803–808.

Doorbar, J., Parton, A., Hartley, K., Banks, L., Crook, T., Stanley, M., and Crawford, L. (1990). Detection of novel splicing patterns in a HPV16-containing keratinocyte cell line. *Virology* 178:254–262.

Dostatni, N., Lambert, P.F., Sousa, R., Ham, J., Howley, P.M, and Yaniv, M. (1991). The functional BPV-1 E2 trans-activating protein can act as a repressor by preventing formation of the initiation complex. *Genes Dev.* 5:1657–1671.

Dowhanick, J.J., McBride, A.A., and Howley, P.M. (1995). Suppression of cellular proliferation by the papillomavirus E2 protein. *J. Virol.* 69:7791–7799.

Dürst, M., Gissmann, L., Ikenberg, H., and zur Hausen, H. (1983). A papillomavirus DNA from a cervical carcinoma and its prevalence in cancer biopsy samples from different geographic regions. *Proc. Natl. Acad. Sci. USA* 80:3812–3815.

Egawa, K., Iftner, A., Doorbar, J., Honda Y., and Iftner, T. (2000). Synthesis of viral DNA and late capsid protein L1 in parabasal spinous cell layers of naturally occurring benign warts infected with human papillomavirus type 1. *Virology* 268:281–293.

Eichten, A., Westfall, M., Pietenpol, J.A., and Münger, K. (2002). Stabilization and functional impairment of the tumor suppressor p53 by the human papillomavirus type 16 E7 oncoprotein. *Virology* 295:74–85.

Fang, L., Budgeon, L.R., Doorbar, J., Briggs, E.R., and Howett, M.K. (2006a). The human papillomavirus type 11 E1 ˆ E4 protein is not essential for viral genome amplification. *Virology* 351:271–279.

Fang, L., Meyers, C., Budgeon, and Howett, M.K. (2006b). Induction of productive human papillomavirus type 11 life cycle in epithelial cells grown in organotypic raft cultures. *Virology* 347:28–35.

Farr, A., Pattison, S., Youn, B.S., and Roman, A. (1995). Detection of silencer activity in the long control regions of human papillomavirus type 6 isolated from both benign and malignant lesions. *J. Gen. Virol.* 76:827–835.

Flores, E.R., Allen-Hoffmann, B.L., Lee, D., and Lambert, P.F. (2000). The human papillomavirus type 16 E7 oncogene is required for the productive stage of the viral life cycle. *J. Virol.* 74:6622–6631.

Flores, E.R., Allen-Hoffmann, B.L., Lee, D., Sattler, C.A., and Lambert, P.F. (1999). Establishment of the human papillomavirus type 16 (HPV-16) life cycle in an immortalized human foreskin keratinocyte cell line. *Virology* 262:344–354.

Francis, D.A., Schmid, S.I., and Howley, P.M. (2000). Repression of the integrated papillomavirus E6/E7 promoter is required for growth suppression of cervical cancer cells. *J. Virol.* 74:2679–2686.

Frattini, M.G., Hurst, S.D., Lim, H.B., Swaminathan, S., and Laimins, L.A. (1997a). Abrogation of a mitotic checkpoint by E2 proteins from oncogenic human papillomaviruses correlates with increased turnover of the p53 tumor suppressor protein. *EMBO J.* 16:318–331.

Frattini, M.G., Lim, H.B., Doorbar, J., and Laimins, L.A. (1997b). Induction of human papillomavirus type 18 late gene expression and genomic amplification in organotypic cultures from transfected DNA templates. *J. Virol.* 71:7068–7072.

Frattini, M.G., Lim, H.B., and Laimins, L.A. (1996). In vitro synthesis of oncogenic human papillomaviruses requires episomal genomes for differentiation-dependent late expression. *Proc. Natl. Acad. Sci. USA* 93:3062–3067.

Fuchs, E. (1993). Epidermal differentiation and keratin gene expression. *J. Cell. Sci. Suppl.* 17:197–208.

Garner-Hamrick, P.A., Fostel, J.M., Chien, W.-M., Banerjee, N.S., Chow, L.T., Broker, T.R., and Fisher, C. (2004). Global effects of human papillomavirus 18 (HPV-18) E6/E7 in an organotypic culture system. *J. Virol.* 78:9041–9050.

Glahder, J.A., Hansen, C.N., Vinther, J., Madsen, B.S., and Norrild, B. (2003). A promoter within the E6 ORF of human papillomavirus type 16 contributes to the expression of the E7 oncoprotein from a monocistronic mRNA. *J. Gen. Virol.* 84: 3429–3441.

Goodwin, E.C., and DiMaio, D. (2000). Repression of human papillomavirus oncogenes in HeLa cervical carcinoma cells causes the orderly reactivation of dormant tumor suppressor pathways. *Proc. Natl. Acad. Sci. USA* 97:12513–12518.

Goodwin, E.C, Naeger, L.K., Breiding, D.E., Androphy, E.J., and DiMaio, D. (1998). Transactivation-competent bovine papillomavirus E2 protein is specifically required for efficient repression of human papillomavirus oncogene expression and for acute growth inhibition of cervical carcinoma cell lines. *J. Virol.* 72:3925–3934.

Grassmann, K., Rapp, B., Maschek, H., Petry, K.U., and Iftner, T. (1996). Identification of a differentiation-inducible promoter in the E7 open reading frame of human papillomavirus type 16 (HPV-16) in raft cultures of a new cell line containing high copy numbers of episomal HPV-16 DNA. *J. Virol.* 70:2339–2349.

Guccione, E., Pim, D., and Banks, L. (2004). HPV-18 E6*I modulates HPV-18 full-length E6 functions in a cell cycle dependent manner. *Int. J. Cancer* 110:928–933.

Halbert, C.L., Demers, G.W., and Galloway, D.A. (1992). The E6 and E7 genes of human papillomavirus type 6 have weak immortalizing activity in human epithelial cells. *J. Virol.* 66:2125–2134.

Halbert, C.L., and Galloway, D.A. (1988). Identification of the E5 open reading frame of human papillomavirus type 16. *J. Virol.* 62:1071–1075.

Hawley-Nelson, P., Androphy, E.J., Lowy, D.R., and Schiller, J.T. (1988). The specific DNA recognition sequence of the bovine papillomavirus E2 protein is an E2-dependent enhancer. *EMBO J.* 7:525–531.

Higgins, G.D., Uzelin, D.M., Phillips, G.E., and Burrell, C.J. (1991). Presence and distribution of human papillomavirus sense and antisense RNA transcripts in genital cancers. *J. Gen. Virol.* 72:885–895.

Higgins, G.D., Uzelin, D.M., Phillips, G.E., McEvoy, P., Marin, R., and Burrell, C.J. (1992). Transcription patterns of human papillomavirus type 16 in genital intraepithelial neoplasia: evidence for promoter usage within the E7 open reading frame during epithelial differentiation. *J. Gen. Virol.* 73:2047–2057.

Hirochika, H., Broker, T.R., and Chow, L.T. (1987). Enhancers and trans-acting E2 transcriptional factors of papilloma viruses. *J. Virol.* 61:2599–2606.

Hirochika, H., Hirochika, R., Broker, T.R., and Chow, L.T. (1988). Functional mapping of the human papillomavirus transcriptional enhancer and its complexes with the trans-acting E2 protein. *Genes Dev.* 2:54–67.

Hopman, A.H.N., Smedts, F., Dignef, W., Ummelen, M., Sonke, G., Mravunac, M., Vooijs, G.P., Speel, E.J., and Ramaekers, F.C. (2004). Transition of high-grade cervical intraepithelial neoplasia to micro-invasive carcinoma is characterized by integration of HPV 16/18 and numerical chromosome abnormalities. *J. Pathol.* 202:23–33.

Horner, S.M., DeFilippis, R.A., Manuelidis, L., and DiMaio, D. (2004). Repression of the human papillomavirus E6 gene initiates p53-dependent, telomerase-independent senescence and apoptosis in HeLa cervical carcinoma cells. *J. Virol.* 78:4063–4073.

Hou, S.Y., Wu, S.-Y., and Chiang, C.-M. (2002). Transcriptional activity among high and low risk human papillomavirus E2 proteins correlates with E2 DNA binding. *J. Biol. Chem.* 277:45619–45629.

Hou, S.Y., Wu, S.-Y., Zhou, T., Thomas, M.C., and Chiang, C.-M. (2000). Alleviation of human papillomavirus E2-mediated transcriptional repression via formation of a TATA binding protein (or TFIID)-TFIIB-RNA polymerase II-TFIIF preinitiation complex. *Mol. Cell Biol.* 20:113–125.

Hubert, W.G., Kanaya, T., and Laimins, L.A. (1999). DNA replication of human papillomavirus type 31 is modulated by elements of the upstream regulatory region that lie 5′ of the minimal origin. *J. Virol.* 73:1835–1845.

Hubert, W.G., and Laimins, L.A. (2002). Human papillomavirus type 31 replication modes during the early phases of the viral life cycle depend on transcriptional and posttranscriptional regulation of E1 and E2 expression. *J. Virol.* 76:2263–2273.

Hummel, M., Hudson, J.B., and Laimins, L.A. (1992). Differentiation-induced and constitutive transcription of human papillomavirus type 31b in cell lines containing viral episomes. *J. Virol.* 66:6070–6080.

Hwang, E.S., Riese, D.J. II, Settleman, J., Nilson, L.A., Honig, J., Flynn, S., and DiMaio, D. (1993). Inhibition of cervical carcinoma cell line proliferation by the introduction of a bovine papillomavirus regulatory gene. *J. Virol.* 67:3720–3729.

Iftner, T., Oft, M., Bohm, S., Wilczynski, S.P., and Pfister, H. (1992). Transcription of the E6 and E7 genes of human papillomavirus type 6 in anogenital condylomata is restricted to undifferentiated cell layers of the epithelium. *J. Virol.* 66:4639–4646.

Ilves, I., Kivi, S., and Ustav, M. (1999). Long-term episomal maintenance of bovine papillomavirus type 1 plasmids is determined by attachment to host chromosomes, which is mediated by the viral E2 protein and its binding sites. *J. Virol.* 73:4404–4412.

Inagaki, Y., Tsunokawa, Y., Takebe, N., Nawa, H., Nakanishi, S., Terada, M., and Sugimura, T. (1988). Nucleotide sequences of cDNAs for human papillomavirus type 18 transcripts in HeLa cells. *J. Virol.* 62:1640–1646.

Jeon, S., Allen-Hoffmann, B.L., and Lambert, P.F. (1995). Integration of human papillomavirus type 16 into the human genome correlates with a selective growth advantage of cells. *J. Virol.* 69:2989–2997.

Jeon, S., and Lambert, P.F. (1995). Integration of human papillomavirus type 16 DNA into the human genome leads to increased stability of E6 and E7 mRNAs: implications for cervical carcinogenesis. *Proc. Natl. Acad. Sci. USA* 92:1654–1658.

Jian, Y., Schmidt-Grimminger, D.-C., Wu, X., Broker, T.R., and Chow, L.T. (1998). Post-transcriptional induction of p21cip1 protein by HPV E7 in differentiated epithelial cells inhibits reactivated unscheduled DNA synthesis. *Oncogene* 17:2027–2038.

Jian, Y., Van Tine, B.A., Chien, W.-M., Shaw, G.M., Broker, T.R., and Chow, L.T. (1999). Concordant induction of cyclin E and p21cip1 in differentiated keratinocytes by the HPV E7 protein inhibits cellular and viral DNA synthesis. *Cell Growth Differ.* 10:101–111.

Kalantari, M., Calleja-Macias, I.E., Tewari, D., Hagmar, B., Lie, K., Barrera-Saldana, H.A., Wiley, D.J., and Bernard, H.-U. (2004). Conserved methylation patterns of human papillomavirus type 16 DNA in asymptomatic infection and cervical neoplasia. *J. Virol.* 78:12762–12772.

Kanaya, T., Kyo, S., and Laimins, L.A. (1997). The 5′ region of the human papillomavirus type 31 upstream regulatory region acts as an enhancer which augments viral early expression through the action of YY1. *Virology* 237:159–169.

Kim, K., Garner-Hamrick, P.A., Fisher, C., Lee, D., and Lambert, P.F. (2003). Methylation patterns of papillomavirus DNA, its influence on E2 function, and implications in viral infection. *J. Virol.* 77:12450–12459.

Klaes, R., Woerner, S.M., Ridder, R., Wentzensen, N., Dürst, M., Schneider, A., Lotz, B., Melsheimer, P., and von Knebel Doeberitz, M. (1999). Detection of high-risk cervical intraepithelial neoplasia and cervical cancer by amplification of transcripts derived from integrated papillomavirus oncogenes. *Cancer Res.* 59:6132–6136.

Klumpp, D.J., and Laimins, L.A. (1999). Differentiation-induced changes in promoter usage for transcripts encoding the human papillomavirus type 31 replication protein E1. *Virology* 257:239–246.

Klumpp, D.J., Stubenrauch, F., and Laimins, L.A. (1997). Differential effects of the splice acceptor at nucleotide 3295 of human papillomavirus type 31 on stable and transient viral replication. *J. Virol.* 71:8186–8194.

Koffa, M.D., Graham, S.V., Takagaki, Y., Manley, J.L., and Clements, J.B. (2000). The human papillomavirus type 16 negative regulatory RNA element interacts with three proteins that act at different posttranscriptional levels. *Proc. Natl. Acad. Sci. USA* 97:4677–4682.

Kyo, S., Klumpp, D.J., Inoue, M., Kanaya, T., and Laimins, L.A. (1997). Expression of AP1 during cellular differentiation determines human papillomavirus E6/E7 expression in stratified epithelial cells. *J. Gen. Virol.* 78:401–411.

Lee, J.H., Yi, S.M., Anderson, M.E., Berger, K.L, Welsh, M.J., Klingelhutz, A.J., and Ozbun, M.A. (2004). Propagation of infectious human papillomavirus type 16 by using an adenovirus and Cre/LoxP mechanism. *Proc. Natl. Acad. Sci. USA* 101:2094–2099.

Lee, C.J., Suh, E.J., Kang, H.T., Im, J.S., Um, S.J., Park, J.S., and Hwang, E.S. (2002). Induction of senescence-like state and suppression of telomerase activity through inhibition of HPV E6/E7 gene expression in cells immortalized by HPV16 DNA. *Exp. Cell Res.* 277:173–182.

Lehman, C.W., and Botchan, M.R. (1998). Segregation of viral plasmids depends on tethering to chromosomes and is regulated by phosphorylation. *Proc. Natl. Acad. Sci. USA* 95:4338–4343.

Lehr, E., Jarnik, M., and Brown, D.R. (2002). Human papillomavirus type 11 alters the transcription and expression of loricrin, the major cell envelope protein. *Virology* 298:240–247.

Lehr, E.E, Qadadri, B, Brown, C.R., and Brown, D.R. (2003). Human papillomavirus type 59 immortalized keratinocytes express late viral proteins and infectious virus after calcium stimulation. *Virology* 314:562–571.

Lepik, D., Ilves, I., Kristjuhan, A., Maimets, T., and Ustav, M. (1998). p53 protein is a suppressor of papillomavirus DNA amplificational replication. *J. Virol.* 72:6822–6831.

Lim, D.A., Gossen, M., Lehman, C.W., and Botchan, M.R. (1998). Competition for DNA binding sites between the short and long forms of E2 dimers underlies repression in bovine papillomavirus type 1 DNA replication control. *J. Virol.* 72:1931–1940.

Lin, B.Y., Ma, T., Liu, J.-S., Kuo, S.-R., Jin, G., Broker, T.R., Harper, J.W., and Chow, L.T. (2000). HeLa cells are phenotypically limiting in cyclin E/CDK2 for efficient human papillomavirus DNA replication. *J. Biol. Chem.* 275:6167–6174.

Liu, J.-S., Kuo, S.-R., Broker, T.R., and Chow, L.T. (1995). The functions of human papillomavirus type 11 E1, E2 and E2C proteins in cell-free DNA replication. *J. Biol. Chem.* 270:27283–27291.

Ma, T., Zou, N., Lin, B.Y., Chow, L.T., and Harper, J.W. (1999). Interaction between cyclin-dependent kinases and human papillomavirus replication initiation protein E1 is required for efficient viral replication. *Proc. Natl. Acad. Sci. USA* 96:382–387.

May, M., Dong, X.P., Beyer-Finkler, E., Stubenrauch, F., Fuchs, P.G., and Pfister, H. (1994). The E6/E7 promoter of extrachromosomal HPV16 DNA in cervical cancers escapes from cellular repression by mutation of target sequences for YY1. *EMBO J.* 13:1460–1466.

McCance, D.J., Kopan, R., Fuchs, E., and Laimins, L.A. (1988). Human papillomavirus type 16 alters human epithelial cell differentiation in vitro. *Proc. Natl. Acad. Sci. USA* 85:7169–7173.

McLaughlin-Drubin, M.E., Bromberg-White, J.L., and Meyers, C. (2005). The role of the human papillomavirus type 18 E7 oncoprotein during the complete viral life cycle. *Virology* 338:61–68.

McLaughlin-Drubin, M.E., Christensen, N.D., and Meyers, C. (2004). Propagation, infection, and neutralization of authentic HPV16 virus. *Virology* 322: 213–219.

McLaughlin-Drubin, M.E., Wilson, S. Mullikin, B., Suzich, J., and Meyers, C. (2003). Human papillomavirus type 45 propagation, infection, and neutralization. *Virology* 312:1–7.

Melsheimer, P., Vinokurova, S., Wentzensen, N., Bastert, G., and von Knebel Doeberitz, M. (2004). DNA aneuploidy and integration of human papillomavirus type 16 E6/E7 oncogenes in intraepithelial neoplasia and invasive squamous cell carcinoma of the cervix uteri. *Clin. Cancer Res.* 10:3059–3063.

Meyers, C., Frattini, M.G., Hudson, J.B., and Laimins, L.A. (1992). Biosynthesis of human papillomavirus from a continuous cell line upon epithelial differentiation. *Science* 257:971–973.

Meyers, C., Mayer, T.J., and Ozbun, M.A. (1997). Synthesis of infectious human papillomavirus type 18 in differentiating epithelium transfected with viral DNA. *J. Virol.* 71:7381–7386.

Münger, K., Baldwin, A., Edwards, K.M., Hayakawa, H., Nguyen, C.L., Owens, M., Grace, M., and Huh, K. (2004). Mechanisms of human papillomavirus-induced oncogenesis. *J. Virol.* 78:11451–11460.

Nasseri, M., Hirochika, R., Broker, T.R., and Chow, L.T. (1987). A human papillomavirus type 11 transcript encoding an E1 E4 protein. *Virology* 159:433–439.

Nicholls, P.K., Doorbar, J., Moore, R.A., Peh, W., Anderson, D.M., and Stanley, M.A. (2001). Detection of viral DNA and E4 protein in basal keratinocytes of experimental canine oral papillomavirus lesions. *Virology* 284:82–98.

Nishimura, A., Ono, T., Ishimoto, A., Dowhanick, J.J., Frizzell, M.A., Howley, P.M., and Sakai, H. (2000). Mechanisms of human papillomavirus E2-mediated repression of viral oncogene expression and cervical cancer cell growth inhibition. *J. Virol.* 74:3752–3760.

Noya, F., Chien, W.-M., Broker, T.R., and Chow, L.T. (2001). p21cip1 degradation in differentiated keratinocytes is abrogated by co-stabilization with cyclin E induced by HPV E7. *J. Virol.* 75:6121–6134.

Oberg, D., Collier, B., Zhao, X., and Schwartz, S. (2003). Mutational inactivation of two distinct negative RNA elements in the human papillomavirus type 16 L2 coding region induces production of high levels of L2 in human cells. *J. Virol.* 77:11674–11684.

O'Connor, M.J., Stunkel, W., Koh, C.H., Zimmermann, H., and Bernard, H.-U. (2000). The differentiation-specific factor CDP/Cut represses transcription and replication of human papillomaviruses through a conserved silencing element. *J. Virol.* 74:401–410.

O'Connor, M.J., Tan, S.H., Tan, V.H. and Bernard, H.-U. (1996). Bernard.YY1 represses human papillomavirus type 16 transcription by quenching AP-1 activity. *J. Virol.* 70:6529–6539.

Oh, S.T., Longworth, M.S., and Laimins, L.A. (2004). Roles of the E6 and E7 proteins in the life cycle of low-risk human papillomavirus type 11. *J. Virol.* 78:2620–2626.

Ohtsubo, M., and Roberts, J.M. (1993). Cyclin-dependent regulation of G1 in mammalian fibroblasts. *Science* 259:908–1912.

Ozbun, M.A. (2002a). Human papillomavirus type 31b infection of human keratinocytes and the onset of early transcription. *J. Virol.* 76:11291–11300.

Ozbun, M.A. (2002b). Infectious human papillomavirus type 31b: purification and infection of an immortalized human keratinocyte cell line. *J. Gen. Virol.* 83:2753–2763.

Ozbun, M.A., and Meyers, C. (1997). Characterization of late gene transcripts expressed during vegetative replication of human papillomavirus type 31b. *J. Virol.* 71:5161–5172.

Ozbun, M.A., and Meyers, C. (1998). Temporal usage of multiple promoters during the life cycle of human papillomavirus type 31b. *J. Virol.* 72:2715–2722.

Ozbun, M.A., and Meyers, C. (1999). Two novel promoters in the upstream regulatory region of human papillomavirus type 31b are negatively regulated by epithelial differentiation. *J. Virol.* 73:3505–3510.

Palermo-Dilts, D.A., Broker, T.R., and Chow, L.T. (1990). Human papillomavirus type 1 produces redundant as well as polycistronic messenger RNAs in plantar warts. *J. Virol.* 64:3144–3149.

Park, R.B., and Androphy, E.J. (2002). Genetic analysis of high-risk E6 in episomal maintenance of human papillomavirus genomes in primary human keratinocytes. *J. Virol.* 76:11359–11364.

Parker, J.N., Zhao, W., Askins, K.J., Broker, T.R., and Chow, L.T. (1997). Mutational analyses of differentiation-dependent human papillomavirus type-18 enhancer elements in epithelial raft cultures of neonatal foreskin keratinocytes. *Cell Growth Differ.* 8:751–762.

Pater, M.M., and Pater, A. (1988). Expression of human papillomavirus types 16 and 18 DNA sequences in cervical carcinoma cell lines. *J. Med. Virol.* 26:185–195.

Peh, W.L., Brandsma, J.L., Christensen, N.D., Cladel, N.M., Wu, X., and Doorbar, J. (2004). The viral E4 protein is required for the completion of the cottontail rabbit papillomavirus productive cycle in vivo. *J. Virol.* 78:2142–2151.

Peh, W.L., Middleton, K., Christensen, N., Nicholls, P., Egawa, K., Sotlar, K., Brandsma, J., Percival, A., Lewis, J., Liu, W.J., and Doorbar, J. (2002). Life cycle heterogeneity in animal models of human papillomavirus-associated disease. *J. Virol.* 76:10401–10416.

Phelps, W.C., and Howley, P.M. (1987). Transcriptional trans-activation by the human papillomavirus type 16 E2 gene product. *J. Virol.* 61:1630–1638.

Pim, D., and Banks, L. (1999). HPV-18 E6*I protein modulates the E6-directed degradation of p53 by binding to full-length HPV-18 E6. *Oncogene* 18:7403–7408.

Pim, D., Massimi, P., and Banks, L. (1997). Alternatively spliced HPV-18 E6* protein inhibits E6 mediated degradation of p53 and suppresses transformed cell growth. *Oncogene* 15:257–264.

Psyrri, A., DeFilippis, R.A., Edwards, A.P., Yates, K.E., Manuelidis, L., and DiMaio, D. (2004). Role of the retinoblastoma pathway in senescence triggered by repression of the human papillomavirus E7 protein in cervical carcinoma cells. *Cancer Res.* 64:3079–3086.

Rapp, B., Pawellek, A., Kraetzer, F., Schaefer, M., May, C., Purdie, K., Grassmann, K., and Iftner, T. (1997). Cell-type-specific separate regulation of the E6 and E7 promoters of human papillomavirus type 6a by the viral transcription factor E2. *J. Virol.* 71:6956–6966.

Remm, M., Remm, A., and Ustav, M. (1999). Human papillomavirus type 18 E1 protein is translated from polycistronic mRNA by a discontinuous scanning mechanism. *J. Virol.* 73:3062–3070.

Rohlfs, M., Winkenbach, S., Meyer, S., Rupp, T., and Dürst, M. (1991). Viral transcription in human keratinocyte cell lines immortalized by human papillomavirus type-16. *Virology* 183:331–342.

Romanczuk, H., Thierry, F., and Howley, P.M. (1990). Mutational analysis of cis elements involved in E2 modulation of human papillomavirus type 16 P97 and type 18 P105 promoters. *J. Virol.* 64:2849–2859.

Rösl, F., Arab, A., Klevenz, B., and zur Hausen, H. (1993). The effect of DNA methylation on gene regulation of human papillomaviruses. *J. Gen. Virol.* 74:791–801.

Rotenberg, M O., Chow, L.T., and Broker, T.R. (1989). Characterization of rare human papillomavirus type 11 mRNAs coding for regulatory and structural proteins using the polymerase chain reaction. *Virology* 172:489–497.

Scheffner, M., Huibregtse, J.M., Vierstra, R.D., and Howley, P.M. (1993). The HPV-16 E6 and E6-AP complex functions as a ubiquitin-protein ligase in the ubiquitination of p53. *Cell* 75:495–505.

Schneider-Gädicke, A., and Schwarz, E. (1986). Different human cervical carcinoma cell lines show similar transcription patterns of human papillomavirus type 18 early genes. *EMBO J.* 5:2285–2292.

Schwarz, E., Freese, U.K., Gissmann, L., Mayer, W., Roggenbuck, B., Stremlau, A., and zur Hausen, H. (1985). Structure and transcription of human papillomavirus sequences in cervical carcinoma cells. *Nature* 314:111–114.

Seedorf, K., Krammer, G., Dürst, M., Suhai, S., and Rowekamp, W.G. (1985). Human papillomavirus type 16 DNA sequence. *Virology* 145:181–185.

Sen, E., Alam, S., and Meyers, C. (2004). Genetic and biochemical analysis of cis regulatory elements within the keratinocyte enhancer region of the human papillomavirus type 31 upstream regulatory region during different stages of the viral life cycle. *J. Virol.* 78:612–629.

Sen, E., Bromberg-White, J.L., and Meyers, C. (2002). Genetic analysis of cis regulatory elements within the 5ʹ region of the human papillomavirus type 31 upstream regulatory region during different stages of the viral life cycle. *J. Virol.* 76:4798–4809.

Sengupta, S., and Harris, C.C. (2005). p53: traffic cop at the crossroads of DNA repair and recombination. *Nat. Rev. Mol. Cell. Biol.* 6:44–55.

Shally, M., Alloul, N., Jackman, A., Müller, M., Gissmann, L. and Sherman, L. (1996). The E6 variant proteins E6I-E6IV of human papillomavirus 16: expression in cell free systems and bacteria and study of their interaction with p53. *Virus Res.* 42:81–96.

Sherman, L., Alloul, H., Golan, I., Dürst, M., and Baram, A. (1992a). Expression and splicing patterns of human papillomavirus type-16 mRNAs in pre cancerous lesions and carcinomas of the cervix, in human keratinocytes immortalized by HPV 16, and in cell lines established from cervical cancers. *Int. J. Cancer* 50:356–364.

Sherman, L., Golan, Y., Mitrani-Rosenbaum, S., and Baram, A. (1992b). Differential expression of HPV types 6 and 11 in condylomas and cervical preneoplastic lesions. *Virus Res.* 25:23–36.

Shirasawa, H., Tomita, Y., Kubota, K., Kasai, T., Sekiya, S., Takamizawa, H., and Simizu, B. (1988). Transcriptional differences of the human papillomavirus type 16 genome between precancerous lesions and invasive carcinomas. *J. Virol.* 62: 1022–1027.

Skiadopoulos, M.H., and McBride, A.A. (1998). Bovine papillomavirus type 1 genomes and the E2 transactivator protein are closely associated with mitotic chromatin. *J. Virol.* 72:2079–2088.

Smotkin, D., Prokoph, H., and Wettstein, F.O. (1989). Oncogenic and nononcogenic human genital papillomaviruses generate the E7 mRNA by different mechanisms. *J. Virol.* 63:1441–1447.

Smotkin, D., and Wettstein, F.O. (1986). Transcription of human papillomavirus type 16 early genes in a cervical cancer and a cancer-derived cell line and identification of the E7 protein. *Proc. Natl. Acad. Sci. USA* 83:4680–4684.

Spalholz, B.A., Yang, Y.C., and Howley, P.M. (1985). Transactivation of a bovine papilloma virus transcriptional regulatory element by the E2 gene product. *Cell* 42:183–191.

Spink, K.M., and Laimins, L.A. (2005). Induction of the human papillomavirus type 31 late promoter requires differentiation but not DNA amplification. *J. Virol.* 79:4918–4926.

Stacey, S.N., Jordan, D., Snijders, P.J., Mackett, M., Walboomers, J.M., and Arrand, J.R. (1995). Translation of the human papillomavirus type 16 E7 oncoprotein from bicistronic mRNA is independent of splicing events within the E6 open reading frame. *J. Virol.* 69:7023–7031.

Stacey, S.N., Jordan, D., Williamson, A.J., Brown, M., Coote, J.H., and Arrand, J.R. (2000). Leaky scanning is the predominant mechanism for translation of human papillomavirus type 16 E7 oncoprotein from E6/E7 bicistronic mRNA. *J. Virol.* 74:7284–7297.

Stenlund, A. (2003). Initiation of DNA replication: lessons from viral initiator proteins. *Nat. Rev. Mol. Cell. Biol.* 4:777–785.

Stoler, M.H., and Broker, T.R. (1986). In situ hybridization detection of human papilloma virus DNAs and messenger RNAs in genital condylomas and a cervical carcinoma. *Hum. Pathol.* 17:1250–1257.

Stoler, M.H., Rhodes, C.R., Whitbeck, A., Wolinsky, S., Chow, L.T., and Broker, T.R. (1992). Gene expression of human papillomavirus type 16 and 18 in cervical neoplasias. *Hum. Pathol.* 23:117–128.

Stoler, M.H., Whitbeck, A., Wolinsky, S.M., Broker, T.R., Chow, L.T., Howett, M.K., and Kreider, J.W. (1990). Infectious cycle of human papillomavirus type 11 in human foreskin xenografts in nude mice. *J. Virol.* 64:3310–3318.

Stoler, M.H., Wolinsky, S.M., Whitbeck, A., Broker, T.R., and Chow, L.T. (1989). Differentiation-linked human papillomavirus types 6 and 11 transcription in genital condylomata revealed by in situ hybridization with message-specific RNA probes. *Virology* 172:331–340.

Stubenrauch, F., Colber, A.M., and Laimins, L.A. (1998). Transactivation by the E2 protein of oncogenic human papillomavirus type 31 is not essential for early and late viral functions. *J. Virol.* 72:8115–8123.

Stubenrauch, F., Hummel, M., Iftner, T., and Laimins, L.A. (2000). The E8E2C protein, a negative regulator of viral transcription and replication, is required for extrachromosomal maintenance of human papillomavirus type 31 in keratinocytes. *J. Virol.* 74:1178–1186.

Stunkel, W., Huang, Z., Tan, S.H., O'Connor, M.J., and Bernard, H.-U. (2000). Nuclear matrix attachment regions of human papillomavirus type 16 repress or activate the E6 promoter, depending on the physical state of the viral DNA. *J. Virol.* 74:2489–2501.

Tan, S.H., Baker, C.C., Stunkel, W., and Bernard, H.-U. (2003). A transcriptional initiator overlaps with a conserved YY1 binding site in the long control region of human papillomavirus type 16. *Virology* 305:486–501.

Tan, S.H., Bartsch, D., Schwarz, E., and Bernard, H.-U. (1998). Nuclear matrix attachment regions of human papillomavirus type 16 point toward conservation of these genomic elements in all genital papillomaviruses. *J. Virol.* 72:3610–3622.

Tan, S.H., Leong, L.E., Walker, P.A., and Bernard, H.-U. (1994). The human papillomavirus type 16 E2 transcription factor binds with low cooperativity to two flanking
sites and represses the E6 promoter through displacement of Sp1 and TFIID. *J. Virol.*
68:6411–6420.

Terhune, S.S., Hubert, W.G., Thomas, J.T., and Laimins, L.A. (2001). Early polyadenylation signals of human papillomavirus type 31 negatively regulate capsid gene
expression. *J. Virol.* 75:8147–8157.

Terhune, S.S., Milcarek, C., and Laimins, L.A. (1999). Regulation of human papillomavirus type 31 polyadenylation during the differentiation-dependent life cycle.
J. Virol. 73:7185–7192.

Thierry, F. and Yaniv, M. (1987). The BPV1-E2 trans-acting protein can be either an
activator or a repressor of the HPV18 regulatory region. *EMBO J.* 6:3391–3397.

Thierry, F., Heard, J.M., Dartmann, K., and Yaniv, M. (1987). Characterization of a
transcriptional promoter of human papillomavirus 18 and modulation of its expression
by simian virus 40 and adenovirus early antigens. *J. Virol.* 61:134–142.

Thierry, F., and Howley, P.M. (1991). Functional analysis of E2-mediated repression of
the HPV18 P105 promoter. *New Biol.* 3:90–100.

Thierry, F., Spyrou, G., Yaniv, M., and Howley, P. (1992). Two AP1 sites binding JunB
are essential for human papillomavirus type 18 transcription in keratinocytes. *J. Virol.*
66:3740–3748.

Thomas, J.T., Hubert, W.G., Ruesch, M.N., and Laimins, L.A. (1999). Human papillomavirus type 31 oncoproteins E6 and E7 are required for the maintenance of episomes
during the viral life cycle in normal human keratinocytes. *Proc. Natl. Acad. Sci. USA*
96:8449–8454.

Thomas, J.T., Oh, S.T., Terhune, S.S., and Laimins, L.A. (2001). Cellular changes induced
by low-risk human papillomavirus type 11 in keratinocytes that stably maintain viral
episomes. *J. Virol.* 75:7564–7571.

Van Tine, B.A., Dao, L.D., Wu, S.-Y., Sonbuchner, T.M., Lin, B.Y., Zou, N.,
Chiang, C.-M., Broker, T.R., and Chow, L.T. (2004a). HPV origin binding protein
associates with mitotic spindles to enable viral DNA partitioning. *Proc. Natl. Acad.
Sci. USA* 101:4030–4035.

Van Tine, B.A, Kappes, J.C., Banerjee, N.S., Knops, J., Lai, L., Steenbergen, R.D.M.,
Meijer, C.L.J.M., Snijders, P.J.F., Chatis, P., Broker, T.R., Moen, Jr., P.T., and Chow, L.T.
(2004b). Clonal selection for transcriptionally active viral oncogenes during progression
to cancer by DNA methylation-mediated silencing. *J. Virol.* 78:11172–11186.

Walboomers, J.M., Jacobs, M.V., Manos, M.M., Bosch, F.X., Kummer, J.A., Shah, K.V.,
Snijders, P.J., Peto, J., Meijer, C.J., and Muñoz, N. (1999). Human papillomavirus is
a necessary cause of invasive cervical cancer worldwide. *Pathology* 189:12–19.

Ward, P., and Mounts, P. (1989). Heterogeneity in mRNA of human papillomavirus
type-6 subtypes in respiratory tract lesions. *Virology* 68:1–12.

Webster, K., Parish, J., Pandya, M., Stern, P.L., Clarke, A.R., and Gaston, K. (2000). The
human papillomavirus (HPV) 16 E2 protein induces apoptosis in the absence of other
HPV proteins and via a p53-dependent pathway. *J. Biol. Chem.* 275:87–94.

Wells, S.I., Francis, D.A., Karpova, A.Y., Dowhanick, J.J., Benson, J.D., and
Howley, P.M. (2000). Papillomavirus E2 induces senescence in HPV-positive cells via
pRB- and p21(CIP)-dependent pathways. *EMBO J.* 19:5762–5771.

Wentzensen, N., Vinokurova, S., and von Knebel Doeberitz, M. (2004). Systematic review
of genomic integration sites of human papillomavirus genomes in epithelial dysplasia
and invasive cancer of the female lower genital tract. *Cancer Res.* 64:3878–3884.

Wiklund, L., Sokolowski, M., Carlsson, A., Rush, M., and Schwartz, S. (2002). Inhibition of translation by UAUUUAU and UAUUUUUAU motifs of the AU-rich RNA instability element in the HPV-1 late 3′ untranslated region. *J. Biol. Chem.* 277:40462–40471.

Wilson, J.L., Dollard, S.C., Chow, L.T., and Broker, T.R. (1992). Epithelial-specific gene expression during differentiation of stratified primary human keratinocyte cultures. *Cell Growth Differ.* 3:471–483.

Wilson, R., Fehrmann, F., and Laimins, L.A. (2005). Role of the E1^E4 protein in the differentiation-dependent life cycle of human papillomavirus type 31. *J. Virol.* 79:6732–6740.

Yee, C., Krishnan-Hewlett, I., Baker, C.C., Schlegel, R., and Howley, P.M. (1985). Presence and expression of human papillomavirus sequences in human cervical carcinoma cell lines. *Am. J. Pathol.* 119:361–366.

You, J., Croyle, J.L., Nishimura, A., Ozato, K., and Howley, P.M. (2004). Interaction of the bovine papillomavirus E2 protein with Brd4 tethers the viral DNA to host mitotic chromosomes. *Cell* 117:349–360.

Zhao, W., Broker, T.R., and Chow, L.T. (1997). Transcriptional activities of human papillomavirus type-11 promoter-proximal elements in raft and submerged cultures of foreskin keratinocytes. *J. Virol.* 71:8832–8840.

Zhao, W., Chow, L.T., and Broker, T.R. (1999a). A distal element in the HPV-11 upstream regulatory region contributes to promoter repression in basal keratinocytes of squamous epithelium. *Virology* 253:219–229.

Zhao, W., Noya, F., Chen, W.Y., Townes, T.M., Chow, L.T., and Broker, T.R. (1999b). Trichostatin A up-regulates HPV-11 URR-E6 promoter activity in undifferentiated primary human keratinocytes. *J. Virol.* 73:5026–5033.

Zhao, X., Oberg, D., Rush, M., Fay, J., Lambkin, H., and Schwartz, S. (2005). A 57-nucleotide upstream early polyadenylation element in human papillomavirus type 16 interacts with hFip1, CstF-64, hnRNP C1/C2, and polypyrimidine tract binding protein. *J. Virol.* 79:4270–4288.

Zhao, X., Rush, M., and Schwartz, S. (2004). Identification of an hnRNP A1-dependent splicing silencer in the human papillomavirus type 16 L1 coding region that prevents premature expression of the late L1 gene. *J. Virol.* 78:10888–10905.

Zheng, Z.M. (2004). Regulation of alternative RNA splicing by exon definition and exon sequences in viral and mammalian gene expression. *J. Biomed. Sci.* 113:278–294.

Ziegert, C., Wentzensen, N., Vinokurova, S., Kisseljov, F., Einenkel, J., Hoeckel, M., and von Knebel Doeberitz, M. (2003). A comprehensive analysis of HPV integration loci in anogenital lesions combining transcript and genome-based amplification techniques. *Oncogene* 22:3977–3984.

Zobel, T., Iftner, T., and Stubenrauch, F. (2003). The papillomavirus E8-E2C protein represses DNA replication from extrachromosomal origins. *Mol. Cell. Biol.* 23:8352–8362.

zur Hausen, H. (2002). Papillomaviruses and cancer: from basic studies to clinical application. *Nat. Rev. Cancer* 2:342–350.

8
DNA Replication of Papillomaviruses

Arne Stenlund

Cold Spring Harbor Laboratory, Cold Spring Harbor, NY 11724

8.1. Introduction

The molecular biology of the papillomaviruses has a relatively short history, and consequently virtually all that is known about viral DNA replication has come to light during the last 15–20 years. The study of replication of the viral DNA has centered on a small number of papillomavirus types. In particular, the studies of the bovine papillomavirus type 1 (BPV) and the human papillomavirus (HPV-11) have dominated the field. These are both representative members of the papillomavirus family: BPV is an animal papillomavirus, which gives rise to fibropapillomas, and HPV-11 is a member of the large group of human papillomaviruses that cause papillomas of the genital epithelium. Overall, the replication properties of these viruses are very similar and it is likely that the differences that exist represent variations on a theme, rather than fundamental differences. For this reason the description here will focus on a consensus PV model for replication rather than a description of the individual viruses. The consensus model is to a significant extent based on the bovine virus, which for many aspects of replication is the best understood. However, there clearly are some differences and distinctions between different virus types. Some of these known differences are reflected in differences in ori structure, as will be discussed below. I will point out perceived differences between the different viruses in the cases where the differences are understood. As we have come to understand more about the replication properties of the papillomaviruses, the similarities to other virus groups have become more and more apparent. It may not be surprising to find similarities to the polyomaviruses. However, the similarities unexpectedly extend much further to distantly related viruses such as those from the parvovirus family, such as adeno-associated viruses (AAV), and even plant viruses such as the Gemini virus tomato leaf yellow curl virus (TLYCV).

The earliest studies of replication of the viral DNA in an experimental system was the important observations made in the early 1980s (Dvoretzky et al., 1980; Lancaster, 1981; Law et al., 1981), that virus particles, or viral DNA, when introduced into the mouse cell line C127 cells could produce transformed foci from which transformed cell lines could be grown out. At the time, one surprise was that although virus particles were not produced, the viral DNA could be recovered as a plasmid from the nucleus of the cells. Furthermore, cells from the transformed foci could be cultured for long periods of time, and the viral DNA persisted in the nucleus as a replicating plasmid. Obviously, this fact raised the possibility that this viral system could be used to answer a number of interesting questions both about the DNA replication properties of papillomaviruses as well as questions about how the viral DNA is stably inherited. It also demonstrated that the block to productive viral infection did not result from a failure to replicate the viral DNA.

8.2.   Assays for Replication of Viral DNA

After this very promising start, further details of viral DNA replication proved difficult to obtain. Although the transformation assay could be used to identify which viral open reading frames (ORFs) were required for the appearance of replicated viral DNA, identification of viral proteins and cis-acting elements that were directly required for DNA replication remained problematic, since indirect effects, such as effects on viral gene expression also would affect DNA replication (Berg et al., 1986; Lusky and Botchan, 1985; Sarver et al., 1984). For a number of years only modest progress was made and viral DNA replication therefore remained very poorly understood. Two new approaches were instrumental in resolving this deadlock, short-term in vivo replication assays (transient assays) and a cell-free DNA replication system (in vitro replication assays).

8.2.1.   Short-Term Replication Assays

Although many attempts were made over the years to develop a short-term replication assay where DNA replication could be measured directly after introduction of the viral DNA into the cells, a robust method failed to emerge for many years. In retrospect, it is likely that these difficulties resulted from two major causes: Firstly, since this procedure requires a relatively high transfection efficiency, some transfection protocols were just too inefficient. Secondly, and maybe most importantly, transfection procedures that resulted in high-level transfection also frequently resulted in temporary cell cycle arrest, which obviously is not conducive to viral DNA replication, since viral DNA replication occurs only in S-phase (Ravnan et al., 1992). When these considerations were taken into account, the transient replication assays were surprisingly easy to establish and rapidly led to new insights (Ustav and Stenlund, 1991; Ustav et al., 1991). Once the required BPV E1 and E2 polypeptides and the ori sequences were identified, a simple replication system was established. Up until this point all that was known about viral replication resulted from the study of BPV-1; however, the definition of the minimal viral requirements for DNA replication now allowed establishment of transient replication systems for virtually any papillomavirus by expression of the cognate E1 and E2 proteins in the presence of the cognate cis-acting sequences (Chiang et al., 1992b; Del Vecchio et al., 1992; Remm et al., 1992). This modification also allowed many different cell lines to be tested for the ability to replicate viral DNA, since expression of the viral genes could be uncoupled from viral control. It rapidly became clear that the papillomaviruses show little cell type or species specificity at the level of DNA replication, and that the cell type specificity observed for the virus is related to limitations on expression of viral genes.

8.2.2.   Viral DNA Replication In Vitro

A cell free PV replication system was first developed by Botchan and co-workers (Yang et al., 1991), followed by several similar systems using the same approach (Kuo et al., 1994; Muller et al., 1994; Sedman and Stenlund, 1995). These in vitro systems were modeled directly on the in vitro DNA replication system that had been developed in the early 1980's for simian virus 40 (SV40). In this system, recombinant purified E1 and E2 proteins from BPV-1 or HPV were added to a crude extract from mammalian cells. In the presence of plasmids containing viral ori sequences, DNA synthesis could be detected by incorporation of labeled dNTPs. Both of these types of assays allowed the identification of the viral gene products and cis-acting sequences required for viral DNA replication and demonstrated bidirectional replication. Results from both in vivo and in vitro DNA replication demonstrated an absolute dependence on E1, and on ori sequences. However, replication in vitro showed only a modest, conditional, dependence on E2.

8.2.3.   Is There More Than One Way to Replicate Viral DNA?

The DNA replication measured in transformed tissue culture cells, such as mouse C127 or NIH 3T3 cells, has the hallmarks of the latent or early stage of the viral life cycle, where early genes are expressed, capsid proteins are not expressed, and the viral DNA is replicated and maintained at a relatively low level (100s of copies). An unanswered question is whether DNA replication required for the much higher copy numbers that are associated with productive infection and generation of new virus particles, proceeds by the same or a different mechanism as DNA replication during latent infection. An early-to-late switch exists for viral gene expression since viral transcripts encoding capsid proteins, absent during latent infection, appear in more superficial layers of the papilloma. The nature of this switch is not understood. Similarly, there are hints of a switch in the mode of DNA replication and it has been proposed that a switch to rolling circle replication may occur at late times, although conclusive evidence for such a switch is lacking (Flores and Lambert, 1997).

8.3. Plasmid Maintenance

Plasmid maintenance is a term that has been used for the ability of papillomavirus genomes to persist as stably replicating plasmids in infected or transfected cells. Long-term maintenance of a viral plasmid obviously requires replication to maintain the viral DNA in an expanding cell population; however, DNA synthesis does not appear to be sufficient for plasmid maintenance since viral ori plasmids are not maintained even in cells where E1 and E2 are supplied continuously and transient replication occurs (Piirsoo et al., 1996). The missing factor was found to be other cis-acting sequences in addition to the ori. In the presence of the URR/LCR, ori plasmids are stably maintained. Multimerized binding sites for the E2 protein could functionally replace the LCR/URR indicating that E2 in some way was involved in stable plasmid maintenance (Piirsoo et al., 1996). An obvious possibility was that the activity, which could be provided by the URR/LCR, but is lacking in the ori, was an activity related to plasmid segregation. It is well established that extrachromosomal elements from a variety of organisms, including other viruses (e.g., Epstein–Barr virus), require segregation or partitioning mechanisms for faithful inheritance. Several lines of evidence indicated that this was the case. Both the E2 protein and the viral DNA were detected in close association with the host chromosomes (Bastien and McBride, 2000; Ilves et al., 1999; Lehman and Botchan, 1998; Skiadopoulos and McBride, 1998). More recently, a cellular interaction partner for E2 has been characterized, and a tentative mechanism for papillomvirus segregation provided (You et al., 2004). The process appears to rely on the interaction between E2 and a cellular chromatin bound factor, Brd4. Association of E2 with Brd4 via the activation domain, and with the viral DNA via the E2 binding sites (BS) present in the viral genome, result in tethering of the viral DNA to the host chromosomes, which can then be transported into the daughter cells.

8.4. The Relationship Between DNA Replication and Morphological Transformation

Early observations indicated that a relationship existed between the ability of BPV to transform mouse cells in culture and the capacity of the viral DNA to replicate, such that viral plasmids defective for DNA replication were also defective for morphological transformation and produced foci at a much lower rate than plasmids that were replication competent. Similar observations from another virus indicated that this phenomenon was related to the ability of the viral DNA to persist, rather than direct effects on transformation. It has been well established that in plasmids replicated from the EBV oriP, which are stably maintained, marker genes such as drug resistance markers are expressed with greatly increased frequency compared to plasmids lacking oriP. This effect is similar to high frequency transformation (HFT) observed with plasmids containing ARS elements in yeast and likely has the same cause (Stinchcomb et al., 1979). In plasmids that are incapable of replication and maintenance, the low frequency at which DNA is integrated into the host chromosome by illegitimate recombination limits the frequency at which cells can become transformed. DNA that is replication competent and can be maintained as a plasmid, transforms with much higher frequency precisely because the DNA is maintained in the cell without integration.

8.5. The Basic Requirements for DNA Replication: E1, E2, and Ori

Replication of the viral DNA is dependent on two viral polypeptides, which play a direct role in replication (Fig. 8.1). These are the full-length products from the E1 and E2 open reading frames (ORFs). In addition, the presence of the viral replicator sequence (ori), is required. These three components are necessary and sufficient for viral DNA replication in mammalian cells. Other virus-encoded factors and cis-acting sequences likely have effects on viral DNA replication and may be regulatory: for example, variants of both the E1 and E2 proteins, generated by alternative splicing, are encoded by many papillomaviruses. The N-terminus of E1 (the M-protein) exists as a separate polypeptide due to alternative splicing and has been shown to be present in BPV transformed cells, but is of unknown significance (Hubert and Lambert, 1993; Thorner et al., 1988). At least two forms of E2, in addition to the full-length E2, are encoded by BPV (Choe et al., 1989; Lambert et al., 1989; Lambert et al., 1987) and other papillomaviruses also have additional

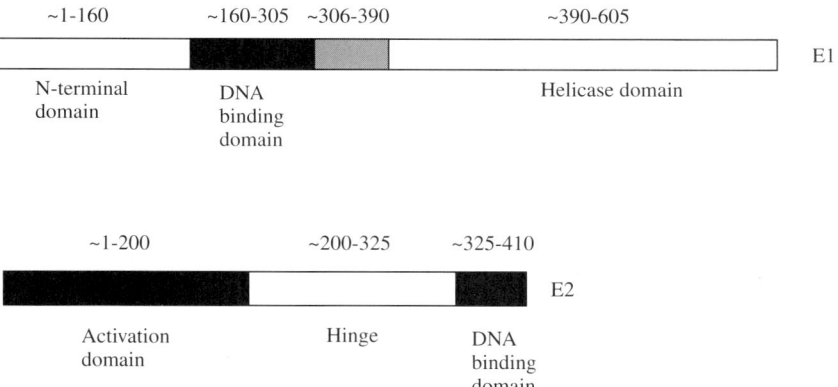

FIGURE 8.1. Domain structure of the E1 and E2 polypeptides. Schematic representation of the organization of the E1 and E2 proteins from BPV-1. The E1 and E2 proteins from all papillomviruses are highly conserved and have similar domain structures. The boxed regions (domains) represent regions with known function. The amino acid numbering, delineating regions (domains) with known function is approximate and refers to the BPV E1 and E2 proteins.

forms of the E2 protein (Chiang et al., 1991; Stubenrauch et al., 2000; Zobel et al., 2003). These lack the N-terminus (the transactivation domain) of E2 but contain parts of the hinge and the C-terminal DNA binding domain (DBD). Whether or not these polypeptides have direct effects on DNA replication in a physiological setting is presently unclear. However, the effect of over expression of the truncated forms of E2, as expected, is reduced DNA replication, likely due to competition for binding to the E2 BS present in the ori (Chiang et al., 1992a; Lim et al., 1998; Liu et al., 1995). Similarly, it is clear that the URR/LCR region, which flanks the minimal ori, has effects on the efficiency by which the ori is utilized and likely affects the regulation of DNA replication.

DNA replication in short-term in vivo replication assays, requires two viral proteins, E1 and E2, for virtually all papillomaviruses tested to date (Chiang et al., 1992b; Ustav and Stenlund, 1991). In general, replication requires the full-length proteins, although some residual activity has been detected for truncations lacking the N-terminus of E1 (Ferran and McBride, 1998). Similarly, in some settings, high levels of expression of the E1 protein results in replication even in the absence of E2 for some papillomavirus types (Gopalakrishnan and Khan, 1994). Also, replication can be achieved using cis-acting sequences consisting of multimerized E2 BS only (Sverdrup and Khan, 1995). However in no case is replication observed in the absence of E1. These observations together indicate that the E2 protein serves an auxiliary function. Also consistent with an auxiliary role for E2 are observations from several laboratories that DNA replication in vitro shows a modest dependence on E2 (Muller et al., 1994; Sedman and Stenlund, 1995; Yang et al., 1991). As will be discussed at greater length below, the critical step in initiation of papillomavirus replication appears to be the recruitment of E1 to DNA. Due to a strong physical interaction between the E1 and E2 proteins, recruitment of E1 can be accomplished in several different ways, providing an explanation for the ability of E2 BS to under certain circumstances function as an ori.

8.6.   The Requirement for Cellular Factors in Viral DNA Replication

DNA replication of papillomaviruses displays little species and cell-type specificity. In transient replication assays where replication does not depend on viral gene expression, the replication machinery present in many types of cells can support papillomavirus DNA replication. This is well illustrated by E1, E2, and ori from BPV, which has been shown to function in many different cell lines from a variety of species, including human, monkey, bovine, mouse, and hamster (Ustav et al., 1993; Ustav and Stenlund, 1991). It therefore seems clear that the interaction between viral proteins and the cellular replication machinery is highly conserved. An unexplained observation is the replication of viral DNA from BPV and HPV in the yeast S. cerevisiae, which appears independent of E1 and the ori sequences (Angeletti et al., 2002; Kim et al., 2005; Zhao and Frazer, 2002).

A benefit of a cell-free DNA replication system is that it can be used to determine which cellular factors are required for viral DNA replication. Due to the similarities of papillomavirus to SV40, the requirement for cellular factors for papillomavirus DNA replication has not been pursued in depth, and the general assumption, which is likely correct, is that a similar set of factors as those that have been defined for DNA replication of SV40, are also required for papillomavirus replication (Waga and Stillman, 1998). There are however some interesting distinctions. The original SV40 replication system was based on an S-100 extract, essentially a cytosolic extract, where replication factors from the nucleus leak out. In such an extract, which replicates SV40 efficiently, the BPV replicon is not functional. However, addition of small quantities of nuclear extract restores replication, indicating that a factor present in the nuclear extract is required for BPV replication, but not for SV40 replication (Melendy et al., 1995; Sedman and Stenlund, 1995). Further fractionation localized this difference to the Fraction IIA, which contains the activity together with the DNA polymerases α and δ. It is currently unclear what this activity is, but one interesting possibility is that the nuclear activity may be a kinase that phosphorylates E1. It has been demonstrated that SV40 T-antigen expressed in *E. coli*, which consequently is unphosphorylated, is inactive in S100 extracts, due to a lack of phosphorylation and can be activated by cyclin dependent kinases. Although this would be an appealing resolution to these observations, experiments from the Hurwitz laboratory which compared the *E. coli* expressed E1 with E1 purified from baculovirus-infected insect cells, which is phosphorylated, detected no significant difference in DNA replication for the two E1 preparations, and in addition showed that the E1 protein expressed in *E. coli* was functional for DNA synthesis with purified factors, in the absence of a kinase (Muller et al., 1994). The exact difference between SV40 replication and papillomavirus replication in vitro, therefore remains unresolved.

8.7.   Description of the Parts

8.7.1.   The Viral Origin of DNA Replication

The cis-acting sequences required for papillomavirus DNA replication have been defined for a number of papillomaviruses using transient replication assays with E1 and E2 expressed from heterologous promoters (Chiang et al., 1992b; Del Vecchio et al., 1992; Lu et al., 1993; Remm et al., 1992; Sverdrup and Khan, 1995; Ustav et al., 1991). In all cases, a short sequence (60–80 bp) located between the URR/LCR and the early open reading frames (ORFs), has been identified as a minimal origin of replication (Fig. 8.2). This position appears to be completely conserved in all papillomaviruses examined. The sequence is modestly conserved, but contains conserved features. These features are binding sites for the E2 protein (E2BS, $ACCN_6GGT$), which due to a high-level sequence conservation are easily identified, and binding sites for E1, (E1BS), which are less well conserved. The E2 BS are present in different arrangements in different types of viruses, but as will be discussed below, the position and number of

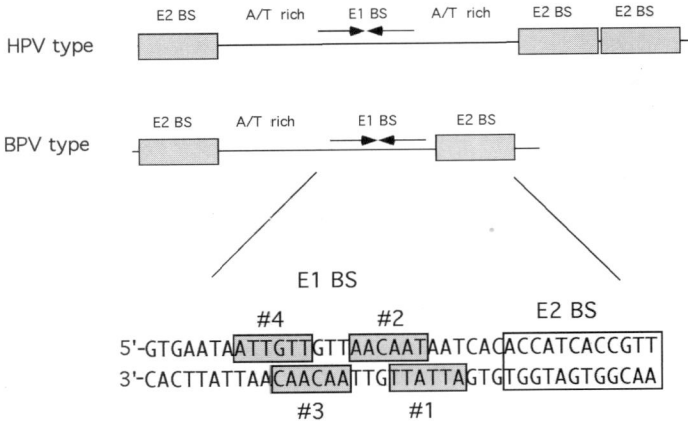

FIGURE 8.2. Schematic representation of papillomavirus ori structures. The HPV type ori represents the ori structure found in the majority of human papillomaviruses. The BPV type ori is found in a small number of animal papillomaviruses. The E2 BS are indicated by boxes and the E1 BS are indicated by an inverted pair of arrows. "A/T-rich" refers to the A + T rich region of the ori. In the bottom part of the figure, the arrangement of the four overlapping E1BS (E1 BS 1–4, shaded boxes) is indicated. The boxes enclose the complementary strand of sites #2 and #3.

sites does not appear to be critical for function in DNA replication, although the presence of at least one E2 BS is required. E1 BS are well defined only for BPV-1 and HPV-11 but the majority of papillomaviruses have sequence motifs that resemble these. However, whether all papillomavirus E1 proteins bind to exactly the same recognition sequence is currently unknown. The third ori component, the A/T rich region is poorly defined and poorly conserved in most viruses, but in BPV and in HPV-11 clearly has a function in DNA replication based on mutational analysis. In summary, a generic papillomavirus ori contains binding sites for E1, for E2, and short stretches of A/T rich sequences.

8.7.1.1.   The Binding Sites for E1

The sequence that E1 recognizes was originally defined in BPV as an imperfect palindromic sequence 18 bp in length with the sequence 5'-ATTGTTGTTAACAATAAT-3' (Chiang et al., 1992b; Holt et al., 1994; Holt and Wilson, 1995; Ustav et al., 1991; Wilson and Ludes-Meyers, 1991; Yang et al., 1991). This sequence upon closer examination was revealed to contain four individual E1 binding sites, which have been extensively characterized. (Chen and Stenlund, 1998, 2001, 2002; Enemark et al., 2002; Sedman et al., 1997). The recognition sequence for the E1 protein, i.e., the sequences recognized by the E1 DNA binding domain (DBD) has been defined as the hexanucleotide 5'-ATTGTT-3'. Extensive mutational analysis of this sequence has been performed and significant variation is allowed, especially in positions 5 and 6 where most substitutions have only modest effects on binding. Four such recognition sequences are present in the BPV ori, two have the consensus sequence ATTGTT, one has the sequence ATTATT, and one the sequence GTTGTT. All four sites have been shown to be bona fide binding sites that can bind E1 DBD (Chen and Stenlund, 2002). The E1 BS are present as pairs of sites, where each site is separated from its partner by 3 bp.

The E1 DBD binds to these pairs of sites as a dimer, although the protein is a monomer in solution. A truly remarkable feature of the binding sites is that the two pairs of E1 BS sites are overlapping but staggered by 3 bp (see Fig. 8.2). Due to the overlap of the recognition sequences, the constraint on the sequences is greater than that observed for binding to individual pairs of E1 BS since some position serve dual function. For example, the basepairs in positions 4–6 in site 2 are also used as positions 1–3 in site 1. This provides an explanation for the greater than expected conservation at these positions, which contribute only modestly to DNA binding of site 2 but are essential for binding in site 1. The dimer interaction appears to be required for high-affinity site-specific DNA binding since studies using either a mutation in the dimer interface, or probes lacking paired E1 binding sites reduced the affinity of binding ~20-fold (Titolo et al., 2003). Similar studies using HPV-11 have given rise to virtually identical results (Titolo et al., 2003). The arrangement of the E1 BS and binding by the E1 DBD as two dimers to the two pairs of sites results in a complex that partially surrounds the double helix (Chen and Stenlund, 2002; Enemark et al., 2002) This arrangement of binding sites is similar to the arrangement of binding sites for large T-ag in SV40 indicating that T-ag and E1 most likely form similar protein complexes at the ori.

8.7.1.2.   The Binding Sites for E2

Binding sites for E2 have the sequence ($ACCN_6GGT$). These sites show variability both in the number of sites and in the position of the sites. Most HPVs have three high affinity E2 BS in the ori region. Deletion or mutation of any of these sites results in a small reduction in replication (Chiang et al., 1992a; Lu et al., 1993; Remm et al., 1992; Russell and Botchan, 1995; Sverdrup and Khan, 1994, Sverdrup and Khan, 1995). Removal of any two sites results in a severe drop in replication activity. It is likely that some of these sites are required for transcriptional control. BPV contains 2 E2 BS in the ori region, a distal, upstream site, E2 BS 11, and a proximal downstream site, E2 BS 12 (Li et al., 1989). E2 BS 11 is of intermediate affinity, while E2 BS 12 has very low affinity and binding of E2 to this site is most readily detected in the presence of E1 when the two proteins bind cooperatively. Either of these two E2 BS can function for DNA replication; however, the proximal site has been studied in greater detail. For example, a clear relationship exists between the affinity of the site and the ability to function at a distance from the E1 BS (Sedman and Stenlund, 1995; Ustav et al., 1993). The functional importance of the E2 BS and their positions will be further discussed below.

8.7.1.3.   The A/T-Rich Sequence

A/T-rich sequences are important features of both eukaryotic and prokaryotic oris, and various functions have been assigned to such sequences ranging from sites for DNA bending to preferred duplex melting. The A/T-rich sequences in papillomaviruses are poorly characterized, and neither the exact sequence requirement, nor the function of the A/T rich region has been defined. Deletion and linker substitution analysis in HPVs have failed to show strong effects of mutations in these elements (Lu et al., 1993; Russell and Botchan, 1995). However, in BPV the A/T-rich sequence in the context of a minimal ori is clearly important since deletions affect ori function significantly (Ustav et al., 1991). A possible function for the A/T-rich region in BPV has been suggested to be the induction of structural changes in this sequence in response to E1 binding (Gillette et al., 1994; Sanders and Stenlund, 2001). As will be discussed below, it is likely that the A/T-rich region constitutes a binding site for the helicase domain of E1.

8.8.   Understanding the E1 Protein

The E1 protein is a polypeptide of 605–650 aa, depending on virus type. The polypeptide consists of several functionally defined regions or domains. Sequence alignments of E1 proteins reveal a poorly conserved N-terminal portion encompassing the first 140–150 aa. This N-terminal portion has few distinguishing features, and its specific function in DNA replication is not known, but this region contains the nuclear localization signal (NLS), which controls the cellular localization of the protein. This region contains recognition sequences for cyclin-dependent kinases and is a substrate for cyclin E/cdk2 (Deng et al., 2004b; Lentz et al., 1993). It is also the most protease sensitive part of the E1 protein, indicating that it may not be tightly folded.

8.8.1.   The E1 DBD

C-terminal to the N-terminal domain is a well-conserved region (aa $\sim$ 150–300), which corresponds to the very well characterized E1 DNA-binding domain (DBD). This domain provides sequence-specific binding of E1 to the E1 BS present in the ori. The structures of the E1 DBD from BPV-1 and HPV-18 have been solved by X-ray crystallography (Auster and Joshua-Tor, 2004; Enemark et al., 2000, 2002). Remarkably, the structure of the BPV E1 DBD is very similar to the NMR structure of the DBD from SV40 T-ag (Luo et al., 1996). This similarity was completely unexpected since the two DBDs share no sequence homology ($\sim$ 6% identity). Subsequently, structures of DBDs from two other viral initiator proteins, Rep proteins from adeno-associated virus (AAV-5) from the parvovirus family and the tomato yellow leaf curl virus (TYLCV) from Gemini viruses have been solved and been show to share significant structural homology with the E1 and T-ag DBDs (Campos-Olivas et al., 2002; Hickman et al., 2002). These two virus groups, the parvo and Gemini viruses clearly belong to different virus groups. The high degree of structural similarity indicates that these domains are indeed highly related, and perform related functions. At the sequence level however, little or no homology with other viral initiator proteins can be detected.

The structure of the E1 DBD is very compact, providing an explanation for the high level of protease resistance of this region of E1. Interestingly, although the E1 DBD in solution is a monomer even at very high concentrations, the crystal structure contained a DBD dimer, held together by an interaction between two α3 helices. This dimer interaction has been shown by biochemical means to be significant, and mutations that disrupt this interaction are defective for DNA binding in vitro and for DNA replication in vivo. (Enemark et al., 2000; Schuck and Stenlund, 2005). A similar interaction is likely to occur in HPV-6 E1 since mutations on the surface of the HPV-11 DBD, corresponding to the mutations that disrupt dimerization of BPV E1 DBD, have defects in DNA binding and DNA replication in vivo (Titolo et al., 2003).

8.8.1.1.   DNA Binding by E1 DBD

DNA binding by the E1 DBD has been investigated by biochemical, biophysical, and structural approaches (Chen and Stenlund, 1998, 2001, 2002; Enemark et al., 2002; Titolo et al., 2003). The E1 DBD binds DNA as a dimer, making contacts with the phosphate backbone of the DNA as well as with particular bases (Fig. 8.3). These DNA contacts are contributed by two separate structural elements in the DBD, a DNA-binding loop and a DNA-binding helix where the DNA binding loop contacts one strand exclusively and the other strand is contacted exclusively by the DNA-binding helix. Due to the arrangement of the E1 BS, two E1 DBD dimers can bind simultaneously to overlapping binding sites, generating a complex that partially encircles the DNA. The two dimers are held together by a dimer interface consisting of two α3 helices. The biochemical and mutational data indicate that these features are likely to be general features of papillomavirus E1 proteins, and certainly the DNA binding properties of HPV-11 E1 DBD are virtually identical to those of BPV E1 DBD, including a requirement for dimerization (Titolo et al., 2003). It is however difficult to generalize. The structure of the HPV-18 DBD deviates somewhat from the BPV DBD structure and the orientation of the α3 helices in HPV-18 DBD appear to preclude dimer formation and simultaneous DNA binding, indicating that variations on this theme are likely to exist (Auster and Joshua-Tor, 2004).

8.8.1.2.   The E1 Helicase Domain

C-terminal to the E1 DBD, which ends around amino acid 300 in BPV E1, is a region with modest conservation, followed by a very highly conserved region of approximately 200 amino acids. This 200 aa region shows a high degree of sequence similarity within the papillomavirus family, and in contrast to the DBD, also shows significant similarity to initiator proteins from other viruses,

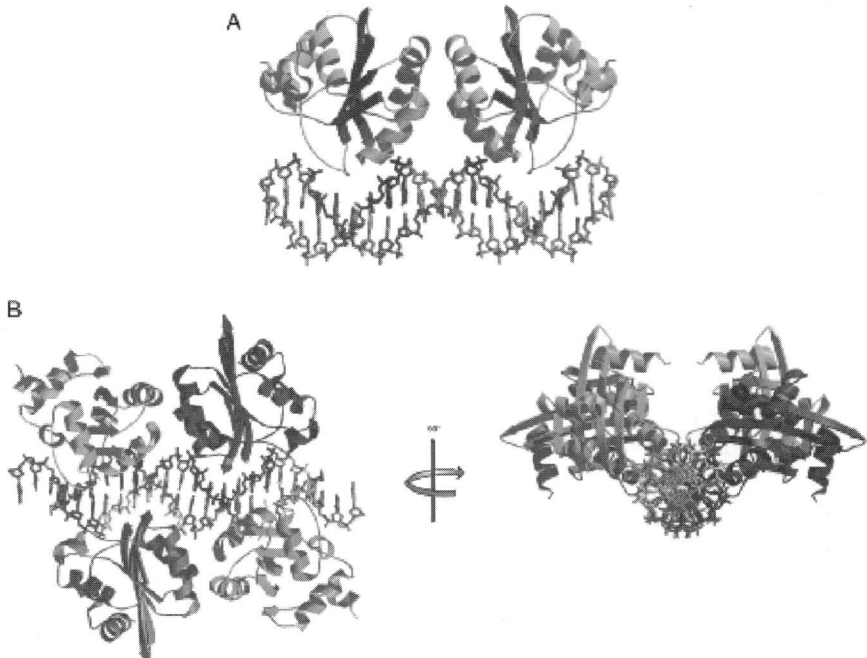

FIGURE 8.3. **A**: Structure of the E1 DBD dimer bound to E1 BS 2 and 4. The binding site nucleotides identified by mutagenesis are colored in blue and phosphate contacts identified by ethylation interference are colored red. **B**: Structure of the E1 DBD tetramer bound to E1 binding sites 1–4 viewed perpendicular and parallel to the DNA helical axis. The four monomers are individually colored: one dimer is colored cyan and purple; the other, dimer pink and green; and the two sets of binding sites are colored blue (E1 BS 2 and 4) and yellow (E1 BS 1 and 3). Adapted from Enemark et al. (2002). Crystal structures of two intermediates in the assembly of the papillomavirus replication initiation complex. *EMBO J.* 21:1487–1496. (See Plate 6).

including other papovaviruses and parvoviruses. Furthermore this region has features common to other ATP binding proteins. This region constitutes the helicase domain, which is involved in ATP binding and hydrolysis by E1, and is required for oligomerization and DNA helicase activity.

8.8.1.3. Structure of the E1 Helicase Domain

The helicase domain of the E1 protein has been difficult to analyze both biochemically and genetically due to its complexities and apart from the conserved residues required for ATP binding and hydrolysis little information has been available about this domain (Sun et al., 1990; Titolo et al., 1999). However, recent advances provide excellent information about the overall structure of the E1 helicase domain. A structure of the helicase domain (amino acids

428–631) from HPV-18 E1 was solved by X-ray crystallography (Abbate et al., 2004). In this structure the E1 helicase domain was crystallized as a complex with the activation domain from the E2 protein. The E1 helicase core essentially consists of a largely globular structure with extended loops emanating in two opposite directions. The five β-strands form a sheet which is sandwiched between several α-helices, giving rise to a topology characteristic of P-loop ATPases and confirming the classification of the E1 helicase domain as a AAA+ protein belonging to the super family 3 (SF 3) group of helicases (Abbate et al., 2004).

Although this structure provides information about the structure of this domain, it does not provide information about the functional form, which likely is a hexamer. However, a recent structure of the domain of SV40 T-antigen solved as a hexameric complex provides excellent information (Li et al., 2003). Due to the high degree of structural similarity, the E1 helicase domain can be modeled on the hexameric SV40 T-ag structure (Fig. 8.4). Thus, the E1 helicase domain likely forms a hexameric ring structure, where DNA passes through the center of the ring. Furthermore, several additional T-ag structures generated in the presence of ADP and ATP provide excellent information about nucleotide binding and hydrolysis, which due to the high degree of sequence similarity can be applied also to the E1 proteins (Gai et al., 2004).

8.8.2. Activities Associated with the E1 Protein

8.8.2.1. DNA Binding by Full-Length E1

The full-length E1 protein binds to the E1 BS in ori via its DBD, as discussed above. However, other parts of the E1 protein also contribute to the DNA binding. Specifically, the E1 helicase domain can be UV cross-linked to the sequences flanking the E1 BS (e.g., the A/T-rich sequence), indicating that the helicase domain contacts the sequences flanking the E1 BS directly (Stenlund, 2003a). Such an interaction is also indicated by footprinting studies, which clearly demonstrate that while the DBD footprint encompasses only the E1 BS, full-length E1 generates a much larger footprint of approximately 80 bp (Sanders and Stenlund, 2000). As will be discussed at greater length below, the DNA binding properties of the full-length E1 protein are significantly different than the isolated E1 DBD, and full-length E1 binds DNA with low sequence specificity.

8.8.2.2. ATP Binding and Hydrolysis by E1

The E1 protein can bind and hydrolyze ATP (Sedman and Stenlund, 1998; Seo et al., 1993; Sun et al., 1990; Titolo et al., 1999, 2000; White et al., 2001). Based on sequence conservation, the E1 helicase domain belongs to the family of SF3 helicases, which include a large number of viral helicases including SV40 large T-ag. Recent results from SV40 large T-ag indicate that, as in many other hexameric helicases, ATP is bound in the interface between two subunits and is contacted by residues from both subunits (Gai et al., 2004). This indicates

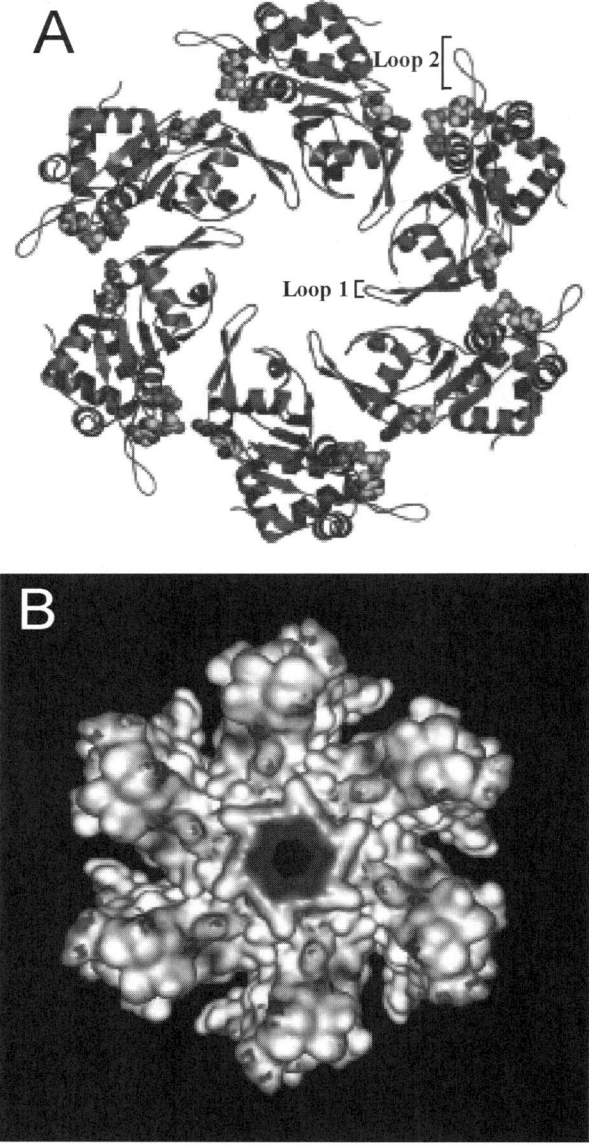

FIGURE 8.4. Structure of the helicase domain from HPV-18 E1. **A**: The X-ray crystal structure of a fragment (aa 428–631) corresponding to the helicase domain of HPV-18 E1 was solved, and the resulting structure was modeled on the hexameric structure of the SV40 T-ag helicase domain. **B**: Electrostatic potential surface representation (GRASP) of the E1 hexamer. Areas of positive potential are blue and negative potential red. Adapted from Abbate et al. (2004). The X-ray structure of the papillomavirus helicase in complex with its molecular matchmaker E2. *Genes Dev.* 18:1981–1996. (See Plate 7).

that some level of oligomerization is required for ATP binding and hydrolysis. Most of the residues that are involved in binding and hydrolyzing ATP in SV40 T-ag are conserved in the E1 proteins, indicating that ATP likely binds and is hydrolyzed in a similar manner. Because the E1 helicase domain clearly is involved in multiple functions in addition to the helicase activity, such as DNA binding and melting, it is likely that ATP binding and hydrolysis also plays a role in these processes.

8.8.2.3. Hexamer Formation by E1

The E1 protein from BPV can be induced to oligomerize in the presence of short single-stranded oligonucleotides and ATP, generating complexes consistent with an E1 hexamer (Fouts et al., 1999; Sedman and Stenlund, 1998). The stoichiometry of binding is close to 6:1 (E1:DNA) in this complex, indicating that one DNA molecule is contained in the E1 hexamer. Interestingly, the isolated hexamer has helicase activity, indicating that a preformed hexamer can function as a helicase. Furthermore, the ATPase activity, which in monomeric form is strongly stimulated by the presence of DNA, in the isolated hexamer is not stimulated by the addition of DNA, indicating that the ATPase stimulation by DNA may be indirect and likely is due to the requirement for DNA for hexamerization of E1.

8.8.2.4. Melting of the ori by E1

To initiate DNA replication on double-stranded DNA, the two DNA strands have to be separated to expose the bases. This process is referred to as melting. Such an activity is required because DNA helicases, which carry out the processive unwinding of the dsDNA in front of the replication fork, require a single stranded region to initiate the unwinding reaction. DNA helicases therefore are incapable of unwinding completely double stranded DNA. Melting of double stranded DNA has traditionally been detected by the chemical oxidation of T-residues that are not base paired, for example by treating the DNA with a strong oxidizing agent such as KMnO4. This modification results in sensitivity of the modified base to cleavage with acid. The permanganate reactivity that results from binding of the E1 protein has been extensively studied (Gillette et al., 1994; Sanders and Stenlund, 2000, Sanders and Stenlund, 2001). Permanganate reactivity is observed on both flanks of the E1 BS, particularly in the A/T-rich region, demonstrating that these sequences are melted by E1 binding. Apart from the positions of melting, remarkably little information is available about the melting process and the specific form of the E1 protein that is responsible for melting is not known.

8.8.2.5. The DNA Helicase Activity of the E1 Protein

The E1 protein has DNA helicase activity in assays using artificial substrates, such as fragment displacement assays (Sedman and Stenlund, 1998; Seo et al.,

1993; Titolo et al., 1999, 2000; Yang et al., 1993). E1 helicase has a 3' to 5' polarity, similar to that of T-antigen, and displaces oligonucleotides in an ATP-dependent manner. It was originally demonstrated for SV40 T-ag that an initiator protein, in the presence of a topoisomerase and a single-stranded DNA binding protein, can generate extensive unwinding of hundreds to thousands of basepairs of a plasmid containing an origin of replication. This assay (a form-U assay) measures a combination of initiator activities including DNA binding, melting, and helicase activities. Similar to SV40 T-ag, HPV-11 E1 can also unwind a double stranded plasmid; surprisingly however, this reaction does not appear to be origin dependent (Lin et al., 2002).

8.9. The Interaction Between E1 and Cellular and Viral Factors Involved in DNA Replication

The E1 protein, as expected for an initiator protein, interacts with several proteins of the cellular replication machinery. E1 from both BPV and HPV-11 have been shown to interact with DNA polymerase α (Conger et al., 1999; Park et al., 1994; Masterson et al., 1998) and subsequently BPV E1 has also been shown to interact with the single stranded DNA binding protein RPA (Loo and Melendy, 2004). For SV40 T-ag additional interactions with topoisomerase I have been demonstrated and it is likely that these are shared with E1.

8.9.1. Modifications of E1

E1 can be phosphorylated, ubiquitinylated, and sumoylated. These modifications are likely to regulate the activity, localization, and levels of the E1 protein and therefore are potential regulators of DNA replication. A very interesting inter-action, with clear potential consequences for DNA replication, is an interaction between E1 and cellular Cyclin/Cdk complexes (Cueille et al., 1998; Lentz et al., 1993; Ma et al., 1999). This interaction is specific for a certain subset of Cyclin/Cdk complexes, and as a result, E1 is phosphorylated by both A-type and E-type Cdk-complexes. A well-conserved cyclin binding motif is present in the N-terminal portion of E1 and mutations in this sequence were found to down-regulate phosphorylation of E1. These same mutations in E1 also reduced DNA replication in vivo, suggesting that the E1 phosphorylation affects DNA replication (Ma et al., 1999). Furthermore, Cyclin E/Cdk stimulates replication in vitro indicating that phosphorylation affects the activity of E1 directly (Cueille et al., 1998; Ma et al., 1999). Exactly how phosphorylation affects the activity of E1 is not clear. However, the phosphorylation and interaction with the cylin/cdk complexes might function at several different levels. One recent report demon-strates that phosphorylation of E1 also affects its cellular localization (Deng et al., 2004b), and another indicates that interaction with cyclin/cdk complexes inhibits ubiquitination and degradation of E1 (Malcles et al., 2002).

It has long been understood that for some replicons, molecular chaperones play an important role in replication. A classical example is initiation of DNA replication in phage λ which requires the combined activities of the three *E. coli* heat-shock proteins DnaK, DnaJ, and GrpE, that together serve to release the lambda P protein from the oriλ-O-P-DnaB complex and thereby allow the DnaB helicase to initiate unwinding. Recent data indicates that molecular chaperones may play a role also in replication of papillomaviruses (Liu et al., 1998, 2002). When added to in vitro DNA replication assays, the two related heat shock proteins Hsp70 and Hsp40 stimulate DNA replication significantly and also stimulate formation of large oligomeric E1 complexes on ori. Exactly what step in initiation is affected by the presence of heat shock proteins is not known, but one interesting possibility is that the displacement of E2 from its interaction with E1 (see below) may require the assistance of a chaperone (Lin et al., 2002).

8.9.2. The Role of E2 in DNA Replication

The biochemical properties of E1, including the ability to bind to the ori, to bind and hydrolyze ATP, as well as its DNA helicase activity, are completely consistent with a role in DNA replication. The function of E2 in DNA replication has remained more difficult to identify. The first indication of a cooperative role of E1 and E2 was the discovery of a physical interaction between E1 and E2 and it became clear that the E1 and E2 proteins form a stable complex (Blitz and Laimins, 1991; Mohr et al., 1990). This physical interaction results in cooperative DNA binding, and E1 and E2 together form a specific complex (E1-E2) on the ori. The evidence from in vivo DNA replication experiments indicated that E2 is essential for DNA replication. However, experiments using cell free replication, demonstrated that E2 was likely to play an auxiliary role in DNA replication since DNA synthesis in vitro clearly occurs in the absence of E2. It was also demonstrated that at very low levels of E1, or in the presence of nonspecific competitor DNA, replication could be made completely dependent on E2 (Sedman and Stenlund, 1995; Yang et al., 1991). The DNA binding properties of the E1 protein provide an explanation for these phenomena. When E1 was tested for site-specific DNA binding by challenge with nonspecific DNA, the protein had a surprisingly poor ability to distinguish between specific and nonspecific DNA, in other words, E1 alone binds DNA with low sequence specificity (Sedman and Stenlund, 1995). The cooperative E1-E2 complex formed on the ori however, showed > 100-fold higher specificity, indicating that the E2 protein in some way could confer specificity to E1 through the formation of the $E1_2E2_2$ complex. There are clear indications that this interaction is directly required for DNA replication. Mutations in the E1 binding site which resulted in loss of DNA replication activity, also resulted in loss of the cooperative $E1_2 E2_2$ complex. However, both replication activity and $E1_2 E2_2$ complex formation could be restored by mutations that increase the affinity of the E2 BS, indicating that the formation of the cooperative complex is essential for DNA replication (Sedman and Stenlund, 1995).

8.9.2.1.   The Interaction Between the E1 and E2 Proteins

Both the interaction between E1 and E2, and the interaction of the two proteins with ori, are conserved features in papillomviruses. In transient DNA replication assays, a considerable degree of mixing and matching of E1, E2, and ori from different viruses is allowed, indicating that not only are the general interactions conserved, but that the interaction surfaces in these proteins are sufficiently conserved between viruses of different types to allow a productive interaction (Chiang et al., 1992b). There are however, some exceptions. For example, it was observed that for replication of the BPV ori, only the BPV E2 protein was functional, although for replication of the HPV-11 ori, BPV E1 could cooperate with HPV-11 E2 (Chiang et al., 1992b). Importantly, this dependence on the identity of the E2 protein was observed only when the E2 binding site in the BPV ori was in its proximal position, and indeed the HPV-11 E2 protein was fully functional for replication when the E2 binding site was moved 22 bp away from the E1 BS. By using chimeric proteins between HPV-11 and BPV E2 it was established that both the hinge and the activation domain from HPV-11 E2 could functionally substitute for the BPV E2 hinge and activation domain, and that the specificity resided in the E2 DBD (Berg and Stenlund, 1997). Although the interaction with the activation domain showed no specificity, the presence of the activation domain was still required for strong cooperative DNA binding, indicating that the interaction between E1 and E2 takes place primarily via the activation domain of E2 and the helicase domain of E1.

The explanation for this behavior was that the DBD from BPV E2 interacts physically with the E1 DBD and that specific mutations on the surface of the E2 DBD can abolish this interaction (Berg and Stenlund, 1997; Chen and Stenlund, 2000). This particular interaction does not appear to be present between the HPV proteins. As a result of the interaction between the two DBDs bound to adjacent sites, a sharp bend is induced in the ori DNA between the binding sites (Gillitzer et al., 2000). This bend appears to be required for the interaction between the E2 activation domain and the E1 helicase domain, which clearly is the interaction that is required for initiation of DNA replication in vivo. In conclusion, replication of both BPV and HPV rely on a physical interaction between the E1 helicase domain and the E2 activation domain when the proteins are bound to the ori. The distinction between BPV and HPV-11 is that a different route is used to accomplish this interaction (Fig. 8.5). This is necessary due to the difference in ori structure, i.e. the proximal position of the E2-binding site in the BPV type ori. The requirement for the interaction between the two DBDs can be alleviated by increasing the distance between the E1 and E2 binding sites to two turns of the helix (21 bp).

8.10.   A Conflict Between Specific and Nonspecific DNA-Binding Activities in E1

It is clear that the interaction between the E1 helicase domain and the E2 activation domain contributes something very important to viral DNA replication

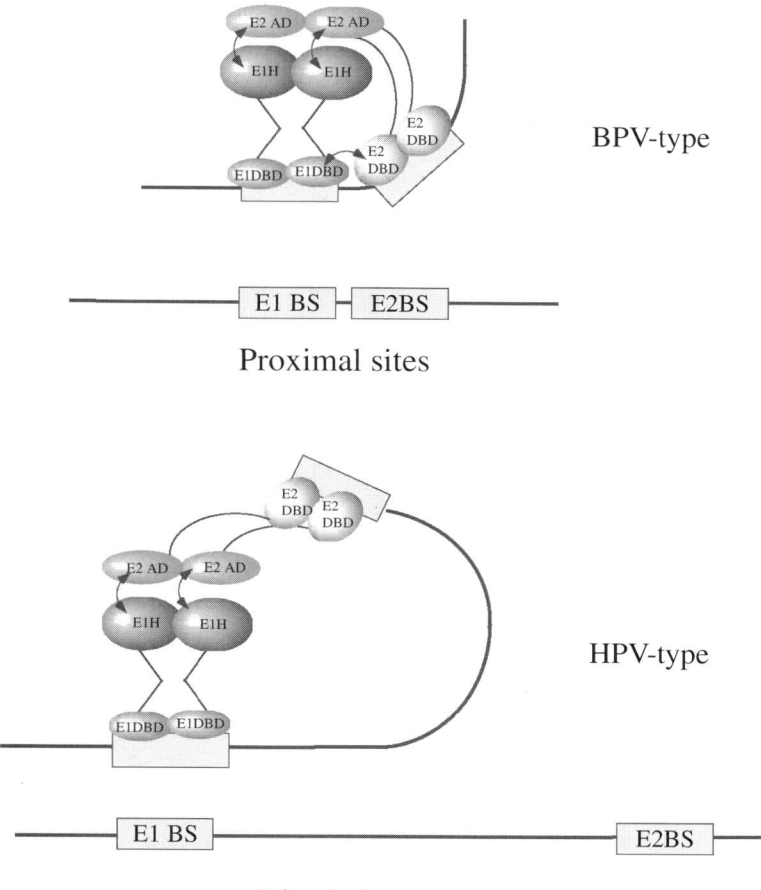

BPV-type

Proximal sites

HPV-type

Distal sites

FIGURE 8.5. Interactions between the E1 and E2 proteins in the context of oris with proximal and distal binding sites for E1 and E2. The E1 and E2 proteins can interact with each other in two different ways, depending on the ori structure. On an ori with proximal E1 and E2 BS (BPV type) the interaction between the two proteins depends on two physical interactions, one between the E1 and E2 DBDs, which generates a sharp bend in the DNA and a second interaction between the E1 helicase domain (E1 H) and the E2 activation domain (E2 AD). On an ori with distal E1 and E2 BS (HPV type), E1 and E2 interact via the E2 AD and E1 H only, likely through looping of the intervening DNA. AD, activation domain; H, helicase domain; DBD, DNA-binding domain.

likely related to the specificity of E1 DNA binding. This raises interesting questions about why such an elaborate mechanism is required to achieve high specificity DNA binding.

A hallmark of viral initiator proteins is that they are multifunctional proteins. Some of these multiple activities seem to be incompatible. For example, in order to recognize the ori, an initiator has to be able to perform highly sequence specific DNA binding. However, to function as a DNA helicase, a general non-sequence-specific DNA binding activity is required. E1 is expected to be capable of both kinds of DNA binding, creating an apparent conflict. A possible resolution of this conflict could be that more than one DNA-binding activity is present in E1. This turns out to be the case, when DNA binding activities of the isolated DBD and full-length E1 were compared, the DBD bound with high specificity while the full-length protein bound with low sequence specificity (Stenlund, 2003a). This demonstrates that the DNA-binding properties of the DBD are not representative of the full-length protein and vice versa. The low-specificity DNA-binding activity is present in the E1 helicase domain, which contacts the DNA flanking the E1 binding sites in the course of E1 DNA binding (Stenlund, 2003a). The presence of this additional nonspecific DNA binding activity results in an overall low binding specificity of the E1 protein, since the nonspecific activity masks the specificity of the DBD.

8.11.   A Mechanism of Action for E2

With these results in mind, high-sequence-specificity DNA binding by E1 could be generated in two different ways, either by stimulating specific DNA binding or by inhibiting nonspecific DNA binding. The DNA-binding properties of E1 and the nature of the interaction with E2 immediately suggested a possible mechanism for how E2 could provide specificity for E1 (Fig. 8.6). As mentioned earlier, the E2 activation domain interacts directly with the E1 helicase domain. If this interaction masks or inhibits the nonspecific DNA binding activity of the helicase domain, high-specificity DNA binding would result since only the DBD can contribute to binding, and the DBD binds with high specificity. In BPV, because the E1 BS proximal E2 BS (E2 BS12) is a very weak site ($\sim$ 100-fold lower than a consensus E2 BS [Li et al., 1989]), the direct contribution from E2 DNA binding to the combined E1-E2 complex likely is modest. However, in HPVs, where the E1 BS proximal E2 BS generally are high-affinity sites, E2 DNA binding is likely to contribute also directly to the specificity of the E1-E2 complex.

8.12.   The E1-E2 Complex is a Precursor for Formation of Larger E1 Complexes

It has long been hypothesized that the manner in which the multifunctional viral initiator proteins can provide all the activities that are required for initiation of DNA replication is that different oligomeric forms of the protein have different activities. To understand initiation, therefore, it is of importance to be able to

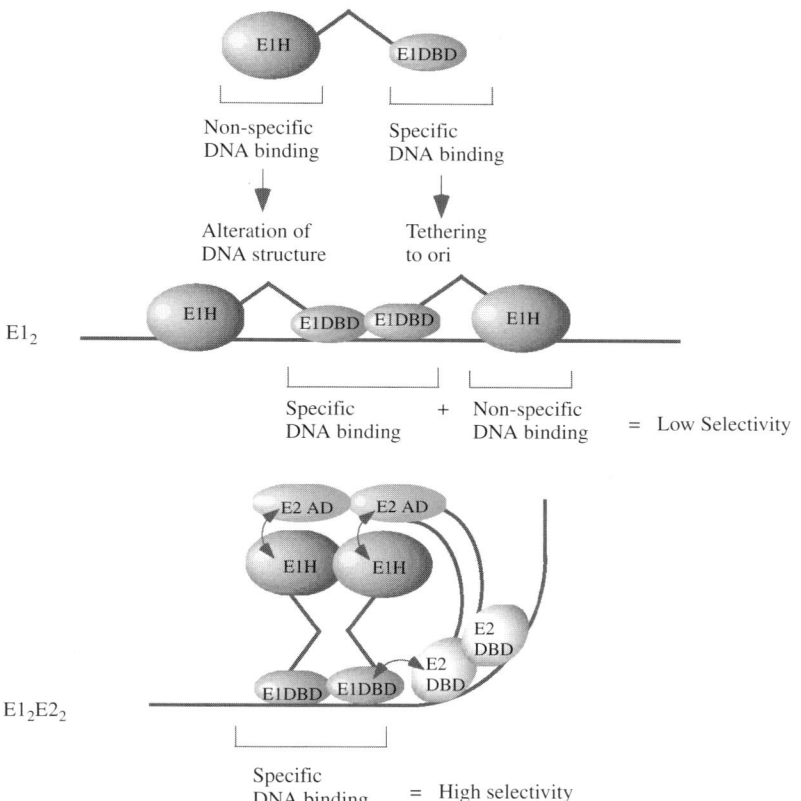

FIGURE 8.6. Summary of the dual DNA-binding activities of E1 and the consequences for DNA binding specificity. E1 contains two distinct DNA-binding activities. A highly sequence-specific DNA-binding activity is present in the E1 DBD. This activity is responsible for the tethering of E1 to the ori. The second DNA-binding activity is present in the E1 helicase domain, and binds DNA with little sequence specificity and is responsible for melting of the sequences flanking the E1 BS. In the absence of E2, the nonspecific DNA binding activity masks the specificity of the E1 DBD, resulting in a net low specificity for E1 DNA binding. In the presence of E2, however, the interaction between the E2 AD and the E1 helicase domain blocks the nonspecific DNA binding activity of the helicase domain, resulting in sequence-specific DNA binding via the E1 DBD. Adapted from Stenlund (2003a). E1 initiator DNA binding specificity is unmasked by selective inhibition of non-specific DNA binding. *EMBO J.* 22:954–963.

describe these forms and their activity. The cooperative E1-E2 complex clearly fits the bill for a highly sequence-specific complex required to recognize the viral ori initially in the infected cell. In the presence of ATP, in a process that is poorly understood, this complex can serve as a precursor for formation of larger E1 complexes, and E2 is displaced from the resulting complex (Sanders and Stenlund, 1998). This process appears to be a mechanism by which a complex

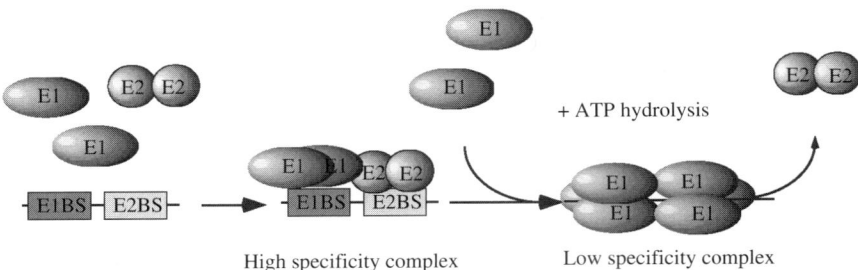

FIGURE 8.7. ATP-dependent loading of E1. The E1 and E2 proteins bind cooperatively to the ori, which contains binding sites for both proteins (E1 BS and E2 BS, respectively). The resulting $E1_2$-$E2_2$ complex binds ori with a high degree of sequence specificity. In a process that requires ATP hydrolysis, additional E1 molecules are added to the complex and E2 is displaced from the DNA. The net result of this process is to deposit the low-specificity E1-complex onto a specific site. Adapted from Stenlund (2003b). Initiation of DNA replication: Lessons from viral initiator proteins. *Natl. Rev. Mol. Cell. Biol.* 4:777–785.

with intrinsically low sequence specificity loads onto a specific site (Fig. 8.7). An interesting possibility is that the displacement of E2—which is not well understood and is a very slow step—in vivo is dependent on chaperones such as the Hsp proteins for the release of the of the E2 protein from its interaction with the E1 helicase domain, (Lin et al., 2002).

What happens next has until recently been shrouded in mystery, although a double hexameric helicase is likely the end product. In the past year, however, it has become clear that the larger form of E1, which results from the displacement of E2, is a completely novel form of E1, a double trimer (Fig. 8.8). Formation of this complex requires ATP binding but not ATP hydrolysis and represents yet another way that E1 can bind to DNA utilizing both the E1 DBD and the helicase domain. The double trimer complex is responsible for melting of the ori DNA and also serves as a direct, and required, precursor for the formation of the double hexameric (DH) helicase (Schuck and Stenlund, 2005). Together, these data now provide a clear overall description of how E1 progresses from an ori specific DNA binder together with E2, to a complex that melts the ori, and finally forms the double hexamer that generates the processive DNA helicase that is required for DNA replication.

8.13.   The Viral DNA Replication Machinery as a Target for Small Molecule Intervention

In spite of significant effort to discover suitable targets for small molecule inhibitors of papillomavirus infection over the years, no drug is as yet in use. One reason is the paucity of traditional antiviral targets. Papillomaviruses encode only

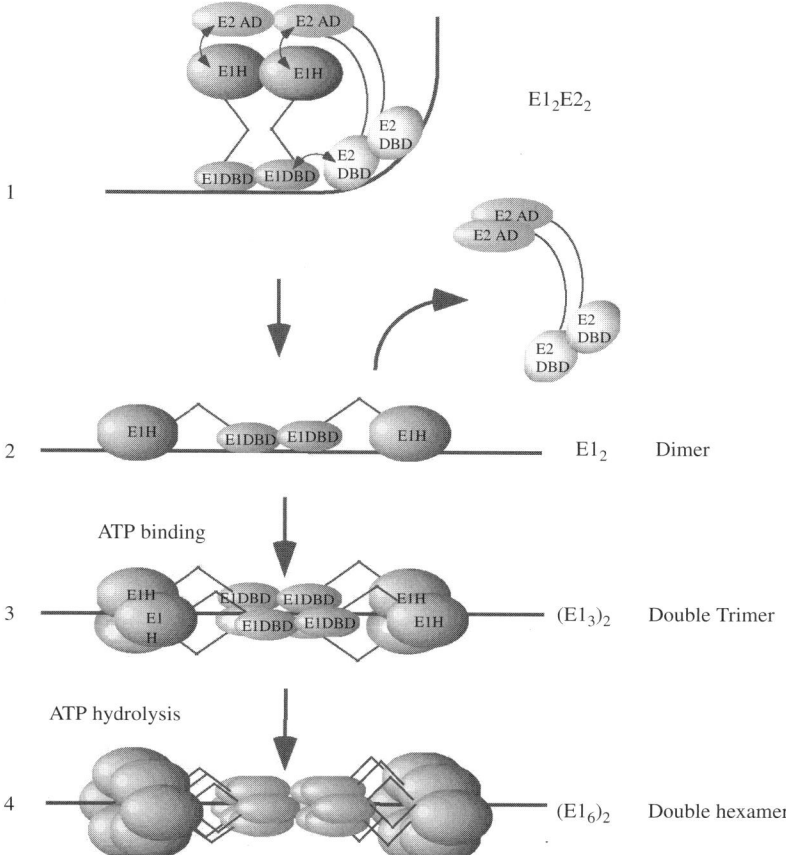

FIGURE 8.8. A pathway for the assembly of E1 into a double hexameric helicase. **1**: E1 binds cooperatively with the E2 protein forming an E1-E2 complex that due to the interaction between the E1 helicase domain and the E2 activation domain binds DNA with a high degree of sequence specificity. **2**: E2 is displaced in a process that is poorly understood, resulting in release of the E1 helicase domain, allowing contact with flanking DNA sequences. **3**: In a process that requires ATP binding but not ATP hydrolysis, a double trimer of E1 is formed which melts the ori DNA. **4**: As a result of melting of the ori DNA, and in a process requiring ATP hydrolysis, the double trimer is converted to a double hexameric helicase. Adapted from Stenlund (2003a). E1 initiator DNA binding specificity is unmasked by selective inhibition of nonspecific DNA binding. *EMBO J.* 22:954–963.

one known enzyme, the E1 ATPase/helicase, and sufficient information about how E1 functions has, until very recently, not been available. Interestingly, as we have acquired a better understanding of the viral replication machinery and its high degree of sophistication the number of potential targets have increased, and it is likely that many more potential targets will come to light as we acquire a true mechanistic understanding of the roles the E1 and E2 proteins play in initiation. For example, the protein–protein interaction between the E1 helicase domain and the E2 activation domain, which as discussed, is essential for DNA replication, has been targeted and has yielded small molecule inhibitors that currently are being evaluated (Davidson et al., 2004; Deng et al., 2004a; Wang et al., 2004; Yoakim et al., 2003). Whether this target will ultimately result in a small molecule inhibitor that can be used broadly as a drug remains to be seen. However, it is clear is that in the near future, many new targets in the viral DNA replication machinery are likely to be characterized.

References

Abbate, E.A., Berger, J.M., and Botchan, M.R. (2004). The X-ray structure of the papillomavirus helicase in complex with its molecular matchmaker E2. *Genes Dev.* 18:1981–1996.

Angeletti, P.C., Kim, K., Fernandes, F.J., and Lambert, P.F. (2002). Stable replication of papillomavirus genomes in *Saccharomyces cerevisiae. J. Virol.* 76:3350–3358.

Auster, A.S., and Joshua-Tor, L. (2004). The DNA-binding domain of human papillomavirus type 18 E1. Crystal structure, dimerization, and DNA binding. *J. Biol. Chem.* 279:3733–3742.

Bastien, N., and McBride, A.A. (2000). Interaction of the papillomavirus E2 protein with mitotic chromosomes. *Virology* 270:124–134.

Berg, L., Lusky, M., Stenlund, A., and Botchan, M.R. (1986). Repression of bovine papilloma virus replication is mediated by a virally encoded trans-acting factor. *Cell* 46:753–762.

Berg, M., and Stenlund, A. (1997). Functional interactions between papillomavirus E1 and E2 proteins. *J. Virol.* 71:3853–3863.

Blitz, I.L., and Laimins, L.A. (1991). The 68-kilodalton E1 protein of bovine papillomavirus is a DNA binding phosphoprotein which associates with the E2 transcriptional activator in vitro. *J. Virol.* 65:649–656.

Campos-Olivas, R., Louis, J.M., Clerot, D., Gronenborn, B., and Gronenborn, A.M. (2002). The structure of a replication initiator unites diverse aspects of nucleic acid metabolism. *Proc. Natl. Acad. Sci. U S A* 99:10310–10315.

Chen, G., and Stenlund, A. (1998). Characterization of the DNA-binding domain of the bovine papillomavirus replication initiator E1. *J. Virol.* 72:2567–2576.

Chen, G., and Stenlund, A. (2000). Two patches of amino acids on the E2 DNA binding domain define the surface for interaction with E1. *J. Virol.* 74:1506–1512.

Chen, G., and Stenlund, A. (2001). The E1 initiator recognizes multiple overlapping sites in the papillomavirus origin of DNA replication. *J. Virol.* 75:292–302.

Chen, G., and Stenlund, A. (2002). Sequential and ordered assembly of E1 initiator complexes on the papillomavirus origin of DNA replication generates progressive structural changes related to melting. *Mol. Cell Biol.* 22:7712–7720.

Chiang, C.M., Broker, T.R., and Chow, L.T. (1991). An E1M–E2C fusion protein encoded by human papillomavirus type 11 is asequence-specific transcription repressor. *J. Virol.* 65:3317–3329.

Chiang, C.M., Dong, G., Broker, T.R., and Chow, L.T. (1992a). Control of human papillomavirus type 11 origin of replication by the E2 family of transcription regulatory proteins. *J. Virol.* 66:5224–5231.

Chiang, C.M., Ustav, M., Stenlund, A., Ho, T.F., Broker, T.R., and Chow, L.T. (1992b). Viral E1 and E2 proteins support replication of homologous and heterologous papillomaviral origins. *Proc. Natl. Acad. Sci. U S A* 89:5799–5803.

Choe, J., Vaillancourt, P., Stenlund, A., and Botchan, M. (1989). Bovine papillomavirus type 1 encodes two forms of a transcriptional repressor: structural and functional analysis of new viral cDNAs. *J. Virol.* 63:1743–1755.

Conger, K.L., Liu, J.S., Kuo, S.R., Chow, L.T., and Wang, T.S. (1999). Human papillomavirus DNA replication. Interactions between the viral E1 protein and two subunits of human DNA polymerase alpha/primase. *J. Biol. Chem.* 274:2696–2705.

Cueille, N., Nougarede, R., Mechali, F., Philippe, M., and Bonne-Andrea, C. (1998). Functional interaction between the bovine papillomavirus virus type 1 replicative helicase E1 and cyclin E-Cdk2. *J. Virol.* 72:7255–7262.

Davidson, W., McGibbon, G.A., White, P.W., Yoakim, C., Hopkins, J.L., Guse, I., Hambly, D.M., Frego, L., Ogilvie, W.W., Lavallee, P., and Archambault, J. (2004). Characterization of the binding site for inhibitors of the HPV11 E1-E2 protein interaction on the E2 transactivation domain by photoaffinity labeling and mass spectrometry. *Anal. Chem.* 76:2095–2102.

Del Vecchio, A.M., Romanczuk, H., Howley, P.M., and Baker, C.C. (1992). Transient replication of human papillomavirus DNAs. *J. Virol.* 66:5949–5958.

Deng, S.J., Pearce, K.H., Dixon, E.P., Hartley, K.A., Stanley, T.B., Lobe, D.C., Garvey, E.P., Kost, T.A., Petty, R.L., Rocque, W.J., et al. (2004a). Identification of peptides that inhibit the DNA binding, trans-activator, and DNA replication functions of the human papillomavirus type 11 E2 protein. *J. Virol.* 78:2637–2641.

Deng, W., Lin, B.Y., Jin, G., Wheeler, C.G., Ma, T., Harper, J.W., Broker, T.R., and Chow, L.T. (2004b). Cyclin/CDK regulates the nucleocytoplasmic localization of the human papillomavirus E1 DNA helicase. *J. Virol.* 78:13954–13965.

Dvoretzky, I., Shober, R., Chattopadhyay, S.K., and Lowy, D.R. (1980). A quantitative in vitro focus assay for bovine papilloma virus. *Virology* 103:369–375.

Enemark, E.J., Chen, G., Vaughn, D.E., Stenlund, A., and Joshua-Tor, L. (2000). Crystal structure of the DNA binding domain of the replication initiation protein E1 from papillomavirus. *Mol. Cell* 6:149–158.

Enemark, E.J., Stenlund, A., and Joshua-Tor, L. (2002). Crystal structures of two intermediates in the assembly of the papillomavirus replication initiation complex. *Embo J.* 21:1487–1496.

Ferran, M.C., and McBride, A.A. (1998). Transient viral DNA replication and repression of viral transcription are supported by the C-terminal domain of the bovine papillomavirus type 1 E1 protein. *J. Virol.* 72:796–801.

Flores, E.R., and Lambert, P.F. (1997). Evidence for a switch in the mode of human papillomavirus type 16 DNA replication during the viral life cycle. *J. Virol.* 71:7167–7179.

Fouts, E.T., Yu, X., Egelman, E.H., and Botchan, M.R. (1999). Biochemical and electron microscopic image analysis of the hexameric E1 helicase. *J. Biol. Chem.* 274:4447–4458.

Gai, D., Zhao, R., Li, D., Finkielstein, C.V., and Chen, X.S. (2004). Mechanisms of conformational change for a replicative hexameric helicase of SV40 large tumor antigen. *Cell* 119:47–60.

Gillette, T.G., Lusky, M., and Borowiec, J.A. (1994). Induction of structural changes in the bovine papillomavirus type 1 origin of replication by the viral E1 and E2 proteins. *Proc. Natl. Acad. Sci. U S A* 91:8846–8850.

Gillitzer, E., Chen, G., and Stenlund, A. (2000). Separate domains in E1 and E2 proteins serve architectural and productive roles for cooperative DNA binding. *Embo J.* 19:3069–3079.

Gopalakrishnan, V., and Khan, S.A. (1994). E1 protein of human papillomavirus type 1a is sufficient for initiation of viral DNA replication. *Proc. Natl. Acad. Sci. U S A* 91:9597–9601.

Hickman, A.B., Ronning, D.R., Kotin, R.M., and Dyda, F. (2002). Structural unity among viral origin binding proteins: crystal structure of the nuclease domain of adeno-associated virus *Rep. Mol. Cell* 10:327–337.

Holt, S.E., Schuller, G., and Wilson, V.G. (1994). DNA binding specificity of the bovine papillomavirus E1 protein is determined by sequences contained within an 18-base-pair inverted repeat element at the origin of replication. *J. Virol.* 68:1094–1102.

Holt, S.E., and Wilson, V.G. (1995). Mutational analysis of the 18-base-pair inverted repeat element at the bovine papillomavirus origin of replication: identification of critical sequences for E1 binding and in vivo replication. *J. Virol.* 69:6525–6532.

Hubert, W.G., and Lambert, P.F. (1993). The 23-kilodalton E1 phosphoprotein of bovine papillomavirus type 1 is nonessential for stable plasmid replication in murine C127 cells. *J. Virol.* 67:2932–2937.

Ilves, I., Kivi, S., and Ustav, M. (1999). Long-term episomal maintenance of bovine papillomavirus type 1 plasmids is determined by attachment to host chromosomes, which is mediated by the viral E2 protein and its binding sites. *J. Virol.* 73:4404–4412.

Kim, K., Angeletti, P.C., Hassebroek, E.C., and Lambert, P.F. (2005). Identification of cis-acting elements that mediate the replication and maintenance of human papillomavirus type 16 genomes in *Saccharomyces cerevisiae. J. Virol.* 79:5933–5942.

Kuo, S.R., Liu, J.S., Broker, T.R., and Chow, L.T. (1994). Cell-free replication of the human papillomavirus DNA with homologous viral E1 and E2 proteins and human cell extracts. *J. Biol. Chem.* 269:24058–24065.

Lambert, P.F., Dostatni, N., McBride, A.A., Yaniv, M., Howley, P.M., and Arcangioli, B. (1989). Functional analysis of the papilloma virus E2 trans-activator in *Saccharomyces cerevisiae. Genes Dev.* 3:38–48.

Lambert, P.F., Spalholz, B.A., and Howley, P.M. (1987). A transcriptional repressor encoded by BPV-1 shares a common carboxy-terminal domain with the E2 transactivator. Cell 50:69–78.

Lancaster, W.D. (1981). Apparent lack of integration of bovine papillomavirus DNA in virus-induced equine and bovine tumor cells and virus-transformed mouse cells. *Virology* 108:251–255.

Law, M.F., Lowy, D.R., Dvoretzky, I., and Howley, P.M. (1981). Mouse cells transformed by bovine papillomavirus contain only extrachromosomal viral DNA sequences. *Proc. Natl. Acad. Sci. U S A* 78:2727–2731.

Lehman, C.W., and Botchan, M.R. (1998). Segregation of viral plasmids depends on tethering to chromosomes and is regulated by phosphorylation. *Proc. Natl. Acad. Sci. U S A* 95:4338–4343.

Lentz, M.R., Pak, D., Mohr, I., and Botchan, M.R. (1993). The E1 replication protein of bovine papillomavirus type 1 contains an extended nuclear localization signal that includes a p34cdc2 phosphorylation site. *J. Virol.* 67:1414–1423.

Li, D., Zhao, R., Lilyestrom, W., Gai, D., Zhang, R., DeCaprio, J.A., Fanning, E., Jochimiak, A., Szakonyi, G., and Chen, X.S. (2003). Structure of the replicative helicase of the oncoprotein SV40 large tumour antigen. *Nature* 423:512–518.

Li, R., Knight, J., Bream, G., Stenlund, A., and Botchan, M. (1989). Specific recognition nucleotides and their DNA context determine the affinity of E2 protein for 17 binding sites in the BPV-1 genome. *Genes Dev.* 3:510–526.

Lim, D.A., Gossen, M., Lehman, C.W., and Botchan, M.R. (1998). Competition for DNA binding sites between the short and long forms of E2 dimers underlies repression in bovine papillomavirus type 1 DNA replication control. *J. Virol.* 72:1931–1940.

Lin, B.Y., Makhov, A.M., Griffith, J.D., Broker, T.R., and Chow, L.T. (2002). Chaperone proteins abrogate inhibition of the human papillomavirus (HPV) E1 replicative helicase by the HPV E2 protein. *Mol. Cell. Biol.* 22:6592–6604.

Liu, J.S., Kuo, S.R., Broker, T.R., and Chow, L.T. (1995). The functions of human papillomavirus type 11 E1, E2, and E2C proteins in cell-free DNA replication. *J. Biol. Chem.* 270:27283–27291.

Liu, J.S., Kuo, S.R., Makhov, A.M., Cyr, D.M., Griffith, J.D., Broker, T.R., and Chow, L.T. (1998). Human Hsp70 and Hsp40 chaperone proteins facilitate human papillomavirus-11 E1 protein binding to the origin and stimulate cell-free DNA replication. *J. Biol. Chem.* 273:30704–30712.

Loo, Y.M., and Melendy, T. (2004). Recruitment of replication protein A by the papillomavirus E1 protein and modulation by single-stranded DNA. *J. Virol.* 78:1605–1615.

Lu, J.Z., Sun, Y.N., Rose, R.C., Bonnez, W., and McCance, D.J. (1993). Two E2 binding sites (E2BS) alone or one E2BS plus an A/T-rich region are minimal requirements for the replication of the human papillomavirus type 11 origin. *J. Virol.* 67:7131–7139.

Luo, X., Sanford, D.G., Bullock, P.A., and Bachovchin, W.W. (1996). Solution structure of the origin DNA-binding domain of SV40 T-antigen. *Nat. Struct. Biol.* 3:1034–1039.

Lusky, M., and Botchan, M.R. (1985). Genetic analysis of bovine papillomavirus type 1 trans-acting replication factors. *J. Virol.* 53:955–965.

Ma, T., Zou, N., Lin, B.Y., Chow, L.T., and Harper, J.W. (1999). Interaction between cyclin-dependent kinases and human papillomavirus replication-initiation protein E1 is required for efficient viral replication. *Proc. Natl. Acad. Sci. U S A* 96:382–387.

Malcles, M.H., Cueille, N., Mechali, F., Coux, O., and Bonne-Andrea, C. (2002). Regulation of bovine papillomavirus replicative helicase e1 by the ubiquitin–proteasome pathway. *J. Virol.* 76:11350–11358.

Masterson, P.J., Stanley, M.A., Lewis, A.P., and Romanos, M.A. (1998). A C-terminal helicase domain of the human papillomavirus E1 protein binds E2 and the DNA polymerase alpha-primase p68 subunit. *J. Virol.* 72:7407–7419.

Melendy, T., Sedman, J., and Stenlund, A. (1995). Cellular factors required for papillomavirus DNA replication. *J. Virol.* 69:7857–7867.

Mohr, I.J., Clark, R., Sun, S., Androphy, E.J., MacPherson, P., and Botchan, M.R. (1990). Targeting the E1 replication protein to the papillomavirus origin of replication by complex formation with the E2 transactivator. *Science* 250:1694–1699.

Muller, F., Seo, Y.S., and Hurwitz, J. (1994). Replication of bovine papillomavirus type 1 origin-containing DNA in crude extracts and with purified proteins. *J Biol Chem* 269:17086–17094.

Park, P., Copeland, W., Yang, L., Wang, T., Botchan, M.R., and Mohr, I.J. (1994). The cellular DNA polymerase alpha-primase is required for papillomavirus DNA replication and associates with the viral E1 helicase. *Proc. Natl. Acad. Sci. U S A* 91:8700–8704.

Piirsoo, M., Ustav, E., Mandel, T., Stenlund, A., and Ustav, M. (1996). *Cis* and *trans* requirements for stable episomal maintenance of the BPV-1 replicator. *Embo. J.* 15:1–11.

Ravnan, J.B., Gilbert, D.M., Ten Hagen, K.G., and Cohen, S.N. (1992). Random-choice replication of extrachromosomal bovine papillomavirus (BPV) molecules in heterogeneous, clonally derived BPV-infected cell lines. *J. Virol.* 66:6946–6952.

Remm, M., Brain, R., and Jenkins, J.R. (1992). The E2 binding sites determine the efficiency of replication for the origin of human papillomavirus type 18. *Nucleic Acids Res.* 20:6015–6021.

Russell, J., and Botchan, M.R. (1995). *cis*-Acting components of human papillomavirus (HPV) DNA replication: linker substitution analysis of the HPV type 11 origin. *J. Virol.* 69:651–660.

Sanders, C.M., and Stenlund, A. (1998). Recruitment and loading of the E1 initiator protein: an ATP-dependent process catalysed by a transcription factor. *Embo. J.* 17:7044–7055.

Sanders, C.M., and Stenlund, A. (2000). Transcription factor-dependent loading of the E1 initiator reveals modular assembly of the papillomavirus origin melting complex. *J. Biol. Chem.* 275:3522–3534.

Sanders, C.M., and Stenlund, A. (2001). Mechanism and requirements for bovine papillomavirus, type 1, E1 initiator complex assembly promoted by the E2 transcription factor bound to distal sites. *J. Biol. Chem.* 276:23689–23699.

Sarver, N., Rabson, M.S., Yang, Y.C., Byrne, J.C., and Howley, P.M. (1984). Localization and analysis of bovine papillomavirus type 1 transforming functions. *J. Virol.* 52:377–388.

Schuck, S., and Stenlund, A. (2005). Assembly of a double hexameric helicase. *Mol. Cell* 20:377–389.

Sedman, J., and Stenlund, A. (1995). Co-operative interaction between the initiator E1 and the transcriptional activator E2 is required for replicator specific DNA replication of bovine papillomavirus in vivo and in vitro. *Embo. J.* 14:6218–6228.

Sedman, J., and Stenlund, A. (1998). The papillomavirus E1 protein forms a DNA-dependent hexameric complex with ATPase and DNA helicase activities. *J. Virol.* 72:6893–6897.

Sedman, T., Sedman, J., and Stenlund, A. (1997). Binding of the E1 and E2 proteins to the origin of replication of bovine papillomavirus. *J. Virol.* 71:2887–2896.

Seo, Y.S., Muller, F., Lusky, M., and Hurwitz, J. (1993). Bovine papilloma virus (BPV)-encoded E1 protein contains multiple activities required for BPV DNA replication. *Proc. Natl. Acad. Sci. U S A* 90:702–706.

Skiadopoulos, M.H., and McBride, A.A. (1998). Bovine papillomavirus type 1 genomes and the E2 transactivator protein are closely associated with mitotic chromatin. *J. Virol.* 72:2079–2088.

Stenlund, A. (2003a). E1 initiator DNA binding specificity is unmasked by selective inhibition of non-specific DNA binding. *Embo. J.* 22:954–963.

Stenlund, A. (2003b). Initiation of DNA replication: lessons from viral initiator proteins. *Nat. Rev. Mol. Cell Biol.* 4:777–785.

Stinchcomb, D.T., Struhl, K., and Davis, R.W. (1979). Isolation and characterisation of a yeast chromosomal replicator. *Nature* 282:39–43.

Stubenrauch, F., Hummel, M., Iftner, T., and Laimins, L.A. (2000). The E8E2C protein, a negative regulator of viral transcription and replication, is required for extrachromosomal maintenance of human papillomavirus type 31 in keratinocytes. *J. Virol.* 74:1178–1186.

Sun, S., Thorner, L., Lentz, M., MacPherson, P., and Botchan, M. (1990). Identification of a 68-kilodalton nuclear ATP-binding phosphoprotein encoded by bovine papillomavirus type 1. *J. Virol.* 64:5093–5105.

Sverdrup, F., and Khan, S.A. (1994). Replication of human papillomavirus (HPV) DNAs supported by the HPV type 18 E1 and E2 proteins. *J. Virol.* 68:505–509.

Sverdrup, F., and Khan, S.A. (1995). Two E2 binding sites alone are sufficient to function as the minimal origin of replication of human papillomavirus type 18 DNA. *J. Virol.* 69:1319–1323.

Thorner, L., Bucay, N., Choe, J., and Botchan, M. (1988). The product of the bovine papillomavirus type 1 modulator gene (M) is a phosphoprotein. *J. Virol.* 62:2474–2482.

Titolo, S., Brault, K., Majewski, J., White, P.W., and Archambault, J. (2003). Characterization of the minimal DNA binding domain of the human papillomavirus e1 helicase: fluorescence anisotropy studies and characterization of a dimerization-defective mutant protein. *J. Virol.* 77:5178–5191.

Titolo, S., Pelletier, A., Pulichino, A.M., Brault, K., Wardrop, E., White, P.W., Cordingley, M.G., and Archambault, J. (2000). Identification of domains of the human papillomavirus type 11 E1 helicase involved in oligomerization and binding to the viral origin. *J. Virol.* 74:7349–7361.

Titolo, S., Pelletier, A., Sauve, F., Brault, K., Wardrop, E., White, P.W., Amin, A., Cordingley, M.G., and Archambault, J. (1999). Role of the ATP-binding domain of the human papillomavirus type 11 E1 helicase in E2-dependent binding to the origin. *J. Virol.* 73:5282–5293.

Ustav, E., Ustav, M., Szymanski, P., and Stenlund, A. (1993). The bovine papillomavirus origin of replication requires a binding site for the E2 transcriptional activator. *Proc. Natl. Acad. Sci. U S A* 90:898–902.

Ustav, M., and Stenlund, A. (1991). Transient replication of BPV-1 requires two viral polypeptides encoded by the E1 and E2 open reading frames. *Embo. J.* 10:449–457.

Ustav, M., Ustav, E., Szymanski, P., and Stenlund, A. (1991). Identification of the origin of replication of bovine papillomavirus and characterization of the viral origin recognition factor E1. *Embo. J.* 10:4321–4329.

Waga, S., and Stillman, B. (1998). The DNA replication fork in eukaryotic cells. *Annu. Rev. Biochem.* 67:721–751.

Wang, Y., Coulombe, R., Cameron, D.R., Thauvette, L., Massariol, M.J., Amon, L.M., Fink, D., Titolo, S., Welchner, E., Yoakim, C., et al. (2004). Crystal structure of the E2 transactivation domain of human papillomavirus type 11 bound to a protein interaction inhibitor. *J. Biol. Chem.* 279:6976–6985.

White, P.W., Pelletier, A., Brault, K., Titolo, S., Welchner, E., Thauvette, L., Fazekas, M., Cordingley, M.G., and Archambault, J. (2001). Characterization of recombinant HPV6 and 11 E1 helicases: effect of ATP on the interaction of E1 with E2 and mapping of a minimal helicase domain. *J. Biol. Chem.* 276:22426–22438.

Wilson, V.G., and Ludes-Meyers, J. (1991). A bovine papillomavirus E1-related protein binds specifically to bovine papillomavirus DNA. *J. Virol.* 65:5314–5322.

Yang, L., Li, R., Mohr, I.J., Clark, R., and Botchan, M.R. (1991). Activation of BPV-1 replication in vitro by the transcription factor E2. *Nature* 353:628–632.

Yang, L., Mohr, I., Fouts, E., Lim, D.A., Nohaile, M., and Botchan, M. (1993). The E1 protein of bovine papilloma virus 1 is an ATP-dependent DNA helicase. *Proc. Natl. Acad. Sci. U S A* 90:5086–5090.

Yoakim, C., Ogilvie, W.W., Goudreau, N., Naud, J., Hache, B., O'Meara, J.A., Cordingley, M.G., Archambault, J., and White, P.W. (2003). Discovery of the first series of inhibitors of human papillomavirus type 11: inhibition of the assembly of the E1–E2-Origin DNA complex. *Bioorg. Med. Chem. Lett.* 13:2539–2541.

You, J., Croyle, J.L., Nishimura, A., Ozato, K., and Howley, P.M. (2004). Interaction of the bovine papillomavirus E2 protein with Brd4 tethers the viral DNA to host mitotic chromosomes. *Cell* 117:349–360.

Zhao, K.N., and Frazer, I.H. (2002). *Saccharomyces cerevisiae* is permissive for replication of bovine papillomavirus type 1. *J. Virol.* 76:12265–12273.

Zobel, T., Iftner, T., and Stubenrauch, F. (2003). The papillomavirus E8–E2C protein represses DNA replication from extrachromosomal origins. *Mol. Cell Biol.* 23:8352–8362.

9
Papillomavirus E5 Proteins

Daniel DiMaio

Departments of Genetics and Therapeutic Radiology, Yale University School of Medicine, New Haven, CT 06520-8005

9.1.   Introduction

The E5 proteins are small, hydrophobic transforming proteins encoded by the 3′ end of the early region of many but not all papillomavirus genomes. The best understood E5 protein is encoded by bovine papillomavirus type 1 (BPV1), the prototype of the fibropapillomaviruses, a distinct group of animal papillomaviruses that induce papillomas that contain proliferating dermal fibroblasts as well as epithelial cells. The most distinctive genetic feature of the fibropapillomaviruses is the presence of an E5 gene that encodes a highly conserved 43- or 44-amino-acid hydrophobic protein capable of efficiently transforming cultured fibroblasts. An open reading frame at the 5′ end of the early region of BPV type 4 encodes a 42-amino-acid hydrophobic protein that was originally designated E8. Although this protein does not show a clear evolutionary relationship to other E5 proteins (Bravo and Alonso, 2004), it shares a number of biological activities with the canonical E5 proteins and has been renamed the E5 protein by some workers (O'Brien et al., 2001). A number of human papillomaviruses also encode E5 proteins, but a comprehensive phylogenetic analysis of putative E5 proteins suggests that many so-called HPV E5 proteins are spurious (Bravo and Alonso, 2004). The most studied HPV E5 protein is the 83-amino acid E5 protein encoded by the high-risk genital HPV type 16 (HPV16). The HPV16 E5 protein shares extensive sequence similarity to the E5 proteins from related HPV types, but other than their overall hydrophobicity, there is no significant sequence similarity between the BPV1 and the HPV E5 proteins. Consistent with their small size and hydrophobic composition, the E5 proteins are not enzymes but rather appear to exert their effects by modulating the activities of cellular membrane proteins.

9.2.   The BPV1 E5 Protein

The 44-amino-acid BPV1 E5 protein has a very hydrophobic central segment and a hydrophilic carboxy-terminus, which contains two conserved cysteine residues that mediate the formation of a disulfide-linked dimer of two identical

N-MET-PRO-ASN-LEU-TRP-PHE-LEU-LEU-PHE-LEU-GLY-
LEU-VAL-ALA-ALA-MET-GLN-LEU-LEU-LEU-LEU-LEU-
PHE-LEU-LEU-LEU-PHE-PHE-LEU-VAL-TYR-TRP-ASP-
HIS-PHE-GLU-CYS-SER-CYS-THR-GLY-LEU-PRO-PHE-C

FIGURE 9.1. Amino acid sequence of the BPV1 E5 protein.

E5 monomers (Fig. 9.1) (Horwitz et al., 1988; Schlegel et al., 1986). The E5 dimer is a transmembrane protein localized largely to the endoplasmic reticulum and Golgi apparatus in a type II orientation, i.e., with the carboxy-terminus of the protein extending into the lumen of these organelles (Burkhardt et al., 1989; Schlegel et al., 1986; Surti et al., 1998). The BPV1 E5 protein induces focus formation, anchorage independence, and tumorigenic transformation of immortalized rodent fibroblasts; morphologic transformation of primary bovine and human fibroblasts; and tumorigenic transformation of mouse keratinocytes (Bergman et al., 1988; Burkhardt et al., 1987; DiMaio et al., 1986; Leptak et al., 1991; Petti and Ray, 2000; Schiller et al., 1986). Acute expression of the E5 protein also induces cellular DNA synthesis in quiescent fibroblasts (Settleman et al., 1989).

9.2.1.   The Cellular Platelet-Derived Growth Factor β-Receptor as a Target of the BPV1 E5 Protein

Initial studies on the mechanism of BPV1 E5 transformation carried out in mouse fibroblasts overexpressing exogenous epidermal growth factor (EGF) or colony-stimulating factor-1 receptors implicated receptor tyrosine kinases as important cellular targets of the viral protein (Martin et al., 1989). It is now accepted that the BPV1 E5 protein induces cell transformation primarily by activating the endogenous platelet-derived growth factor (PDGF) β-receptor, although it may also utilize other pathways (Lai et al., 2005; Petti et al., 1991; Sparkowski et al., 1996; Suprynowicz et al., 2000, 2002). During cell transformation, the E5 protein forms a stable complex with the PDGF β-receptor, a transmembrane receptor tyrosine kinase (Petti and DiMaio, 1992). This interaction results in the dimerization of two receptor molecules, trans-phosphorylation of the receptors on tyrosine, and association with SH2 domain-containing cellular signal transduction proteins to generate a sustained mitogenic signal (Fig. 9.2) (Drummond-Barbosa et al., 1995; Lai et al., 1998, 2000; Petti et al., 1991). In contrast, the E5 protein does not bind or activate other growth factor receptors unless they are overexpressed (Cohen et al., 1993; Goldstein et al., 1994; Petti and DiMaio, 1994; Staebler et al., 1995). Interestingly, activation of the PDGF β-receptor by the E5 protein appears to take place at intracellular membranes as well as at the cell surface (Petti et al., 1991). In addition, the BPV1 E5 protein can cause down-regulation of the endogenous PDGF β-receptor in mouse and human fibroblasts (Nilson et al., 1995; Petti and Ray, 2000).

   In several cell systems, the interaction between the PDGF β-receptor and the BPV1 E5 protein is required for transformation. In these cells, the transformed

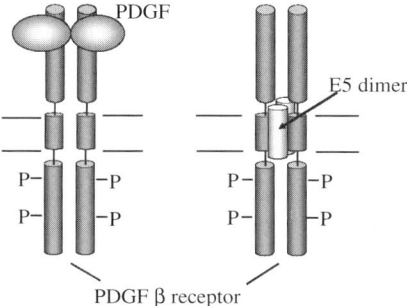

FIGURE 9.2. Schematic models of the PDGF β-receptor activated by its normal ligand, dimeric PDGF (**left**), and by a dimer of the BPV1 E5 protein (**right**). The horizontal lines represent the cellular membrane lipid bilayer. Note that PDGF binds the extracellular domain of the receptor and that the E5 protein binds the transmembrane domain, but both induce receptor dimerization and autophosphorylation. (Receptor domains not to scale).

phenotype is reverted by PDGF receptor-specific tyrosine kinase inhibitors (Klein et al., 1998; Lai et al., 2005). Furthermore, cells lacking the receptor are resistant to E5-mediated mitogenic signaling and transformation, but trans-duction of an exogenous PDGF β-receptor gene renders the cells susceptible, whereas genes encoding other growth factor receptors are not active in this assay (Drummond-Barbosa et al., 1995; Freeman-Cook et al., 2004; Goldstein et al., 1994; Nilson and DiMaio, 1993; Riese and DiMaio, 1995). Transformation and cell signaling are eliminated by mutations in the E5 protein or the PDGF β-receptor that abrogate complex formation between these two proteins Freeman-Cook et al., 2005; Klein et al., 1998; 1999; Nappi and Petti, 2002; Nappi et al., 2002; Nilson et al., 1995; Petti et al., 1997). Notably, PDGF β-receptor mutants lacking the extracellular domain and unable to bind PDGF still interact with the E5 protein and support E5-mediated transformation (Cohen et al., 1993; Drummond-Barbosa et al., 1995; Goldstein et al., 1992a; Staebler et al., 1995). Thus, the BPV1 E5 protein activates the PDGF β-receptor by a ligand-independent mechanism. These experiments define a novel mechanism of viral oncogene action and establish that receptor tyrosine kinases can be activated by viral proteins unrelated to their normal ligand. The Friend retrovirus gp55 protein uses a similar mechanism to activate the erythropoietin receptor (Constantinescu, 1999).

The interaction between the BPV1 E5 protein and the PDGF β-receptor is not restricted to cultured cell systems. The E5 protein of BPV type 2, which is identical to the BPV1 E5 protein, is expressed in naturally occurring urinary bladder carcinomas in cattle (Borzacchiello et al., 2003). Recently, it was shown that this E5 protein forms a stable complex with tyrosine-phosphorylated PDGF β-receptor in these tumors (Borzacchiello et al., 2006), suggesting that E5-induced activation of the PDGF β-receptor plays a role in carcino-genesis in vivo. In addition, the cell types susceptible to BPV1-mediated tumorigenesis in experimental animals are of mesenchymal origin, cells that are rich in PDGF β-receptor.

Several studies indicate that the interaction between the BPV1 E5 protein and the PDGF β-receptor is complex. For example, some E5 mutants induce PDGF β-receptor tyrosine phosphorylation but fail to transform cells or induce down-regulation of the receptor (Mattoon et al., 2001; Nilson et al., 1995). These mutants may be transformation defective because they do not interact productively with some other cellular protein required for cell transformation. Alternatively, these mutants may cause phosphorylation of the PDGF β-receptor at an incomplete or inappropriate set of tyrosines or in an abnormal cellular location (Sparkowski et al., 1995), so that mitogenic signaling and receptor down-regulation are not induced. Preferential phosphorylation of particular sites on the PDGF β-receptor may also explain why some highly transforming E5 mutants induce relatively low levels of receptor tyrosine phosphorylation.

9.2.2. Model for the Transmembrane Interaction Between the BPV1 E5 Protein and the PDGF β-Receptor

It was initially suggested that the BPV1 E5 protein may directly mimic PDGF by binding to the ligand binding site in the extracellular domain of the receptor (Petti et al., 1991), but subsequent studies convincingly showed that complex formation is largely driven by specific interactions involving the transmembrane domains of the two proteins (Cohen et al., 1993; Drummond-Barbosa et al., 1995; Freeman-Cook et al., 2004; Goldstein et al., 1992a; Klein et al., 1998; Meyer et al., 1994; Nappi et al., 2002; Petti et al., 1997; Staebler et al., 1995). The hydrophobic segment of the E5 protein is thought to interact directly in an antiparallel orientation with the transmembrane domain of the PDGF β-receptor (Surti et al., 1998). Furthermore, dimerization of the E5 protein is required for stable complex formation with the PDGF β-receptor and receptor activation (Horwitz et al., 1988; Mattoon et al., 2001; Meyer et al., 1994; Nilson et al., 1995). These results suggest that dimerization of the BPV1 E5 protein creates binding sites for two molecules of the PDGF β-receptor, one on each face of the E5 dimer, resulting in receptor dimerization and activation (Fig. 9.2) (Surti et al., 1998). Detailed mutational analysis and computational modeling predicted the existence of an essential hydrogen bond between Gln17 in the E5 protein and Thr513 in the PDGF β-receptor transmembrane domain and an essential salt bridge between Asp33 and Lys499 in the juxtamembrane region (Klein et al., 1998, 1999; Meyer et al., 1994; Petti et al., 1997; Surti et al., 1998), although additional contacts between the transmembrane domains of these two proteins are also likely to exist (Nappi and Petti, 2002; Nappi et al., 2002, Ely and DiMaio, in preparation). On the basis of computational modeling, it was concluded that the E5 dimer transits the membrane as a symmetric left-handed coiled-coil and that each E5 monomer contributes the same amino acids, including Gln17, to the E5 dimer interface (Surti et al., 1998). This dimer interface was confirmed in transformation experiments in which the two E5 monomers were forced to adopt different symmetric rotational orientations relative to one another in the E5 dimer (Mattoon et al., 2001). Finally, several of the residues in the PDGF β-receptor transmembrane region that are required for E5 recognition lie along a single α-helical face of the transmembrane domain of the receptor, implying that these residues directly contact the E5 dimer (Nappi and Petti, 2002; Petti et al., 1997).

9.2.3. Selection and Analysis of Small Transmembrane Transforming Proteins Modeled on the BPV1 E5 Protein

Early experiments showed that the transmembrane segment of the E5 protein could be replaced by some random hydrophobic sequences without impairing transformation (Horwitz et al., 1989; Meyer et al., 1994). More recently, small proteins with novel transmembrane sequences that activate the PDGF β-receptor and transform fibroblasts were selected from complex libraries of E5-like proteins containing random transmembrane domains (Freeman-Cook et al., 2004, 2005). Small transmembrane proteins with a variety of sequences transform cells by binding and activating the PDGF β-receptor, and computational and genetic analysis demonstrate that at least two divergent transmembrane sequence motifs can interact with the PDGF β-receptor and induce transformation (Freeman-Cook et al., 2004). When a library is used that contains a low level of hydrophilic amino acids at 16 randomized transmembrane positions, several library proteins were identified that activate not only the wild-type PDGF β-receptor but also receptor missense mutants that are not activated by the wild-type E5 protein (Freeman-Cook et al., 2005). Both the position of hydrophilic amino acids and the sequence of surrounding hydrophobic amino acids seem important in determining the specificity of transmembrane recognition. These experiments suggest that it may be possible to influence the activity of a variety of viral and cellular transmembrane proteins by using the E5 dimer as a scaffold to display diverse hydrophobic amino acid sequences and by selecting small transmembrane proteins with the desired activity (Freeman-Cook and DiMaio, 2005).

9.2.4. Alternative Models of BPV1 E5 Transformation

It has been reported that some E5 mutants transform cells and activate phosphoinositol 3' kinase and pp60[c-src] without binding or activating the PDGF β-receptor, leading to the conclusion that the E5 protein can transform fibroblasts via an alternative, PDGF β-receptor-independent mechanism, which appears to require *src* activation (Adduci and Schlegel, 1999; Sparkowski et al., 1996; Suprynowicz et al., 2000, 2002). The study of these and additional alanine substitution mutants suggest a model for E5 dimerization and receptor binding in which the E5 protein contains two distinct homo-dimerization motifs, such that each E5 monomer contributes different residues to the homodimer interface, and a third face of the E5 protein comprises the binding site for the PDGF β-receptor (Adduci and Schlegel, 1999). However, recent detailed analysis of a number of these mutants demonstrates that they can form specific stable complexes with the PDGF β-receptor, resulting in receptor activation, which in turn is required for cellular signaling and transformation (Klein et al., 1998; Lai et al., 2005). There is conflicting evidence regarding the ability of PDGF receptor kinase inhibitors to reverse various biochemical and growth properties of cells expressing these mutants. It is clear that the conclusion that some E5 mutants transform cells via a PDGF β-receptor-independent mechanism requires further scrutiny, as do models derived from the study of these mutants.

It has also been reported that the E5 protein can transform keratinocytes (Leptak et al., 1991). Because keratinocytes are nominally devoid of PDGF β-receptors, this result suggests the existence of an alternative E5 target in these cells. However, it should be noted that PDGF β-receptor is expressed in human cervical keratinocytes (Mayer et al., 2000) and in bovine bladder epithelial cells (Borzacchiello et al., 2006), so it remains possible that the PDGF β-receptor is a target of the BPV1 E5 protein in epithelial cells as well as in fibroblasts. The potential role of the vacuolar H^+-ATPase in various activities of the E5 protein, including transformation, is described in a later section.

9.3.   The BPV4 E5 Protein

In place of the E6 gene at the 5′ end of the early region, bovine papillomavirus type 4 encodes a short hydrophobic protein that has been designated the E5 protein. This 42-amino-acid protein contains a hydrophilic amino acid (asparagine) in the middle of its central hydrophobic segment and a carboxy-terminal cysteine, and it is localized primarily to the Golgi apparatus, features that resemble the BPV1 E5 protein (Marchetti et al., 2002). However, in contrast to the fibropapillomavirus BPV1, BPV4 is purely epitheliotropic. Nevertheless, the BPV4 E5 protein can transform established rodent fibroblasts and, in cooperation with the ras oncogene and HPV E7, embryonic bovine palate cells (O'Brien and Campo, 1998; Zago et al., 2004).

The BPV4 E5 protein induces expression of the cyclin A gene by influencing the activity of cellular transcription factors, including the CAAT box binding protein, p110 CBP (Grindlay et al., 2005; O'Brien et al., 1999; O'Brien and Campo, 1998). The resulting increase in cyclinA-cdk2 kinase activity is thought to be responsible for the transforming activity of the BPV4 E5 protein. Although BPV4 E5 can also increase expression of the cyclin-dependent kinase inhibitor p27^{KIP1} post-transcriptionally, this does not result in cell cycle arrest because the elevated p27^{KIP1} is sequestered into cyclinD-cdk4 complexes (O'Brien et al., 2001). There is no evidence that the BPV4 E5 protein affects the activity of receptor tyrosine kinases.

9.4.   The Human Papillomavirus E5 Proteins

The HPV E5 proteins are difficult to detect in cells and have relatively weak biological activity in cultured cells. Nevertheless, in a number of cultured cell systems the HPV E5 proteins can transform cells and affect cellular behavior. These activities have been best characterized for the HPV16 E5 protein. Like the BPV1 E5 protein, the HPV16 E5 protein is very hydrophobic and localized primarily to the membranes of the Golgi apparatus and the endoplasmic reticulum (Conrad et al., 1993). The HPV16 E5 protein can induce transformation of rodent fibroblasts and keratinocytes, stimulate DNA synthesis in primary human keratinocytes, and enhance the transforming activity of other oncogenes, including ras, high-risk HPV E7, and activated EGF receptor (Bouvard et al., 1994; Chen et al., 1996a; Leechanachai et al., 1992; Leptak et al., 1991; Pim et al., 1992; Straight et al., 1993; Vallee and Banks, 1995). In addition, E5 mutations inhibit keratinocyte immortalization by full-length HPV16 DNA, but immortalization is not restored by E5 expression *in trans*, so the basis for this defect remains unclear (Conrad-Stoppler et al., 1996). Transgenic mice expressing the HPV16 E5 gene in the epidermis display increased DNA synthesis in suprabasal epithelial cells, epidermal hyperplasia, and tumor formation (Genther-Williams et al., 2005). As described in later sections, the HPV16 E5 protein can also affect apoptosis, gap-junction activity, vacuolar-ATPase function and endosome acidification, and expression of histocompatibility antigens (Fig. 9.3).

In cultured cells, the HPV16 E5 protein activates a variety of cellular signal transduction pathways, such as the ras, protein kinase C, phospholipase Cγ, and MAP kinase pathways, and induces the expression of several nuclear oncogenes including c-jun, junB, and c-fos (Bouvard et al., 1994; Chen et al., 1996a; Crusius et al., 1997, 2000; Gu and Matlashewski, 1995; Leechanachai et al., 1992). In many cases, these activities are potentiated by treating cells with EGF but not with PDGF. Similar activities are also displayed by E5 proteins from HPV types that are not associated with cervical carcinoma (e.g. Cartin and Alonso, 2003, Chen and Mounts, 1990). A recent report demonstrated that the

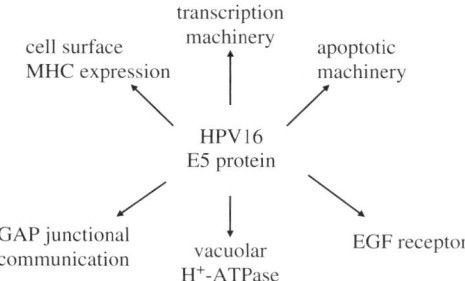

FIGURE 9.3. Schematic diagram of proposed targets of the HPV16 E5 protein. The E5/V-ATPase interaction has been postulated to mediate the effects of the HPV16 E5 protein on several of its other targets.

HPV16 E5 protein failed to activate a number of signal transduction pathways activated by the BPV1 E5 protein (Suprynowicz et al., 2005). The HPV11 and 16 E5 proteins also repress the transcription of the cyclin-dependent kinase (cdk) inhibitor, p21^{KIP1} (Tsao et al., 1996).

It has been reported that the HPV16 E5 protein affects the composition and physical characteristics of cellular membranes (Bravo and Alonso, 2004; Crusius et al., 1999). Although such global effects on membranes could obviously exert profound effects on cellular function, it is not clear how such a mechanism would generate specific effects on particular signaling pathways.

9.4.1. HPV16 E5 Protein and the EGF Receptor

The biochemical mechanisms of HPV16 E5 action are not understood in detail, but receptor tyrosine kinases have emerged as a likely target of the viral protein. It was initially reported that the HPV6 but not the HPV16 E5 protein forms stable complexes with the EGF, ErbB2, and PDGF β-receptors when they are overexpressed in COS cells (Conrad et al., 1994), but a different study in the same cell type showed that the HPV16 E5 protein stably associates with a variety of transmembrane proteins including the EGF receptor (Hwang et al., 1995). Because physical complexes between the HPV E5 proteins and growth factor receptors have been demonstrated only in overexpressing systems, the specificity of these interactions in natural settings is not known.

The HPV16 E5 protein perturbs EGF receptor function. Although the basal levels and tyrosine phosphorylation of the EGF receptor are not increased in cells expressing the HPV16 E5 protein, EGF treatment results in increased tyrosine phosphorylation of the EGF receptor in these cells compared to cells devoid of the HPV16 E5 protein (Crusius et al., 1998; Straight et al., 1993). In addition, E5-expressing cells are more sensitive to EGF stimulation, suggesting that the HPV16 E5 protein enhances ligand-dependent signaling from the EGF receptor (Crusius et al., 1997; Gu and Matlashewski, 1995; Leechanachai et al., 1992; Pim et al., 1992; Rodriguez et al., 2000; Zhang et al., 2002). In normal cells, cell surface EGF receptor is rapidly internalized in response to EGF treatment and then degraded in lysosomes. In contrast, although EGF treatment causes internalization of the EGF receptor in monolayer cultures of human keratinocytes expressing the HPV16 E5 protein, the receptor is not degraded but rather recycled back to the cell surface where it accumulates and displays increased tyrosine-phosphorylation (Straight et al., 1993), which is thought to result in prolonged mitogenic signaling. Similarly, EGF-treated cultures of stratified epithelial cells expressing the HPV16 E5 protein display reduced down-regulation of EGF receptor and increased receptor tyrosine phosphorylation (Tomakidi et al., 2000).

Impaired degradation of the EGF receptor may be due to the ability of the HPV16 E5 protein to disrupt the association of the receptor with c-Cbl, an E3 ubiquitin ligase that normally targets activated EGF receptor for proteosomal degradation (Zhang et al., 2005). In support of this model, ubiquitination of the ligand-stimulated EGF receptor is decreased in cells expressing the E5 protein.

Alternatively, the association of the HPV16 E5 protein with a subunit of the vacuolar H^+-ATPase may affect the activity and metabolism of the EGF receptor (see next section). However, increased EGF receptor signaling is not due solely to increased receptor recycling in E5-expressing cells, because elevated EGF receptor tyrosine phosphorylation is evident five minutes after EGF treatment, long before recycled receptor reappears on the cell surface (Crusius et al., 1998; Straight et al., 1993). Moreover, not all HPV16 E5 activities involve the EGF receptor. For example, the E5 protein also enhances signaling by the G-protein-coupled endothelin 1 receptor, and elevated cellular signaling can occur in an EGF receptor-independent fashion in E5-expressing cells (Crusius et al., 1997, 1999, 2000; Venuti et al., 1998).

The best evidence for the role of the EGF receptor in HPV E5 function was obtained in transgenic mice co-expressing the HPV16 E5 protein and a dominant negative mutant EGF receptor (Genther-Williams et al., 2005). E5-induced epidermal hyperplasia and DNA synthesis was inhibited by this mutant, demonstrating that EGF receptor function was required for these E5 activities *in vivo*. Similarly, HPV16 E5-induced upregulation of VEGF in cultured cells requires EGF receptor function (Kim et al., 2006). It is interesting that the EGF receptor is implicated in HPV16 E5 action, whereas the PDGF β-receptor is involved in BPV1 E5 function. This difference parallels the tissue tropism of these viruses in nature. HPV16 exclusively infects epithelial cells, which typically express abundant EGF receptor, whereas BPV1 induces tumors in fibroblasts, which express PDGF β receptor.

9.5. Interaction of E5 Proteins with the Vacuolar H^+-ATPase

The E5 proteins of all tested papillomaviruses form a stable complex with the 16kDa transmembrane pore-forming subunit c of the vacuolar H^+-ATPase (V-ATPase) (Fig. 9.3) (Adam et al., 2000; Ashby et al., 2001; Conrad et al., 1993; Faccini et al., 1996; Gieswein et al., 2003; Goldstein et al., 1991; Rodriguez et al., 2000). This subunit, referred to here as the 16K protein, is also known as ductin or proteolipid. Most binding studies were carried out in COS cells over-expressing the interacting proteins, in in vitro translation reactions, or in yeast, but the interaction also appears to take place in mouse fibroblasts. Although initial work suggested that glutamine 17 of the BPV1 E5 protein is required for interaction with the 16K protein and that the E5 protein, the PDGF β-receptor, and the 16K protein exist in a ternary complex with the 16K protein bridging the viral protein and the receptor, more recent results do not confirm these findings (Adduci and Schlegel, 1999; Conrad et al., 1993; Goldstein et al., 1992a, b). The BPV1 and the HPV16 E5 proteins appear to associate with the highly conserved fourth membrane-spanning segment of the 16K protein, and glutamic acid 143 in this segment is important for the interaction with the BPV1 E5 protein (Andresson et al., 1995). However, many HPV16 E5 deletion mutants as well as substitution mutants containing random amino acid sequences bind the

16K protein, and different groups have mapped the 16K binding site to different segments of the HPV16 E5 protein (Adam et al., 2000; Gieswein et al., 2003; Rodriguez et al., 2000). These results raise the possibility that this interaction may be due to relatively nonspecific hydrophobic interactions.

The V-ATPase is a complex, multi-subunit enzyme that pumps hydrogen ions into organelles such as the Golgi apparatus and endosomes, thereby lowering luminal pH in these compartments (Stevens and Forgac, 1997). By using pH-sensitive probes and other methods, it was shown that acidification of the lumen of endosomes and the Golgi apparatus is impaired in cells expressing the HPV16 and BPV1 E5 proteins, respectively, presumably as a consequence of E5-mediated inhibition of V-ATPase activity (Disbrow et al., 2005; Schapiro et al., 2000; Straight et al., 1995). Indeed, physiological experiments conducted in NIH3T3 fibroblasts expressing the BPV1 E5 protein indicate that it inhibits V-ATPase activity without affecting the levels or localization of the enzyme (Schapiro et al., 2000). Furthermore, expression of the HPV16 E5 protein in yeast decreases the assembly and stability of the V-ATPase and results in growth and metabolic defects consistent, at least in part, with inhibition of V-ATPase activity (Briggs et al., 2001). Inhibition of endosome acidification by the HPV16 E5 protein may account for the reduction in EGF receptor degradation, because degradation normally occurs in acidic endosomes and lysosomes. The inhibition of EGF receptor degradation by chloroquine and balfinomycin A1, agents that increase endosomal pH, is consistent with this notion (Straight et al., 1995). However, these compounds do not stimulate EGF receptor recycling, suggesting that increased recycling in response to E5 expression is not simply due to alterations of endosomal pH. In addition, some HPV16 E5 mutants bind the 16K protein without affecting EGF receptor signaling (Adam et al., 2000; Briggs, 2001; Rodriguez et al., 2000).

Despite the results summarized above, there is no clear agreement on the effect of E5 proteins on V-ATPase activity. For example, HPV16 and BPV4 E5 mutants have been reported that bind the 16K protein without affecting V-ATPase activity (Adam et al., 2000; Ashrafi et al., 2000; Briggs et al., 2001), and there is a report that the apparent effects of the HPV16 E5 protein on the pH of late endosomes are not due to decreased acidification, but rather to inhibition of endocytic trafficking (Thomsen et al., 2000). Finally, a second study in yeast that directly measured V-ATPase activity in isolated vacuoles failed to demonstrate inhibition by BPV1, BPV4, or HPV16 E5 proteins (Ashby et al., 2001). Based on this latter finding, it was proposed that the E5 protein/16K protein interaction may primarily serve to deliver the E5 protein to the Golgi apparatus. A recent report demonstrated that the HPV16 E5 protein altered the pH of early endosomes even though the protein is localized largely to the endoplasmic reticulum (Disbrow et al., 2005), suggesting that the E5 protein may affect organellar pH by interfering with assembly of active V-ATPase (which occurs in the endoplasmic reticulum (Stevens and Forgac, 1997)) or through some other indirect mechanism.

It was also proposed that the E5 protein/16K protein interaction plays a role in BPV1 E5-mediated cell transformation (Goldstein et al., 1991). This proposal was suggested by the finding that a 16K mutant induces anchorage independence in NIH3T3 cells, a finding which indicates that altered V-ATPase function can elicit transformation (Andresson et al., 1995). However, although the ability of a small panel of BPV1 E5 mutants to impair Golgi acidification correlates with their ability to transform cells (Schapiro et al., 2000), other studies found no apparent correlation between 16K binding and transformation by BPV1, BPV4, and HPV11 E5 mutants (Adduci and Schlegel, 1999; Ashrafi et al., 2000; Chen et al., 1996b; O'Brien et al., 1999).

The E5 proteins of BPV4 and HPV16 inhibit gap junctional communication in keratinocytes, an effect that correlates with reduced phosphorylation or expression of connexin 43, a major component of gap junctions (Faccini et al., 1996; Oelze et al., 1995; Tomakidi et al., 2000). Gap junctions are sites of cell-to-cell movement of low molecular weight molecules, and gap junctional communication is frequently reduced in transformed cells. Interestingly, the 16K V-ATPase subunit is also a component of gap junctions, and 16K mutants can inhibit gap junctional communication (Finbow et al., 1994; Saito et al., 1998). These results led to the suggestion that interaction with the 16K protein is responsible for reduced gap junction communication in cells expressing the E5 protein, but mutational analysis of BPV4 E5 dissociated inhibition of gap junction activity from 16K protein binding and cell transformation (Ashrafi et al., 2000). Growth factor receptor signaling can also impair gap junctional communication e.g. (Hossain et al., 1998), providing another potential mechanism for the effect of the E5 protein.

In summary, the interaction between various E5 proteins and the V-ATPase appears well conserved and has been implicated in regulation of V-ATPase activity and acidification of organelles, cell transformation, EGF receptor metabolism, gap junction communication, and immune evasion (see below) (Fig. 9.3). However, a clear consensus regarding the relevance of the E5/V-ATPase interaction to these diverse activities has yet to emerge.

9.6.   E5 Proteins and Major Histocompatibility Antigen Expression

E5 proteins can impair the expression and activity of major histocompatibility (MHC) antigens. Because these antigens play central roles in antigen presentation and immune recognition, inhibition of MHC expression and function can impair host immune defenses against virally infected cells. Cell surface expression of class I MHC antigens is reduced in cultured cells expressing the E5 proteins of several papillomavirus types (Ashrafi et al., 2005; Ashrafi et al., 2002; Cartin and Alonso, 2003; Marchetti et al., 2002). The HPV16 E5 protein down-regulates cell surface expression of HLA-A and HLA-B (Ashrafi et al., 2005), which present viral peptides to cytotoxic T lymphocytes (CTLs). Thus, the E5 protein may allow HPV-infected cells to avoid CTL-mediated killing. In contrast, non-classical HLA-C and HLA-E were not down-regulated by the E5 protein, which may allow the cells to escape natural killer (NK) cells as well. Reduced cell surface expression of MHC antigens in E5-expressing cells appears to be due in large part to their retention within the Golgi apparatus, although antigen degradation and inhibition of MHC gene transcription can also occur (Marchetti et al., 2002). Reduced cell surface MHC class I expression also occurs in naturally occurring papillomas in animals and in cervical dysplasia (Araibi et al., 2004). The HPV16 E5 protein also interferes with cell surface expression of mature, peptide-loaded MHC class II antigens, which normally present viral peptides during the generation of a CD4 helper T cell response (Zhang et al., 2003).

Although MHC antigens are transmembrane proteins, the effects of the E5 proteins on MHC function are not necessarily due to direct physical inter-action with MHC proteins. Rather, it has been speculated that the interaction between the E5 protein and V-ATPase and the resulting impaired Golgi acidi-fication is responsible for altered class I trafficking, because similar effects are elicited by treatment of cells with monensin, a V-ATPase inhibitor (Marchetti et al., 2002). Similarly, maturation of class II antigens is pH-dependent, suggesting that impaired acidification in response to E5 expression and V-ATPase inhibition may also contribute to decreased assembly of functional class II dimers. Indeed, inhibition of V-ATPase activity with concanamycin B prevents maturation of MHC class II molecules (Zhang et al., 2003). However, there are recent reports that E5 proteins physically interact with MHC class I heavy chains (Marchetti et al., 2005; Ashrafi et al., 2006).

9.7.   Effects of E5 Proteins on Apoptosis

E5 proteins also affect apoptosis. The HPV16 E5 protein protects primary human keratinocytes from apoptosis induced by ultraviolet B irradiation (Zhang et al., 2002). This protective effect appears mediated by phosphoinositol 3′ kinase and MAP kinase survival pathways. Similarly, the HPV16 E5 protein inhibits apoptosis induced by Fas ligand and tumor necrosis factor-related apoptosis inducing ligand (TRAIL) in monolayer and organotypic cultures of keratinocyte cell lines (Kabsch and Alonso, 2002b; Kabsch et al., 2004). On the other hand, the HPV16 E5 protein sensitizes HaCaT human keratinocytes to apoptosis induced by hyperosmotic stress (Kabsch and Alonso, 2002a). The basis for these diverse effects on apoptosis are not clear, but they may be related to modulation of growth factor receptor-mediated survival signals or the colocalization of the HPV16 E5 protein with the anti-apoptotic protein, Bcl-2 (Auvinen et al., 2004).

BPV1 E5-transformed primary human fibroblasts undergo massive caspase 3-independent, Bcl-2-resistant apoptosis after they have been maintained at confluence without medium change for 2 weeks, whereas non-transformed cells survive much longer under these conditions (Zhang et al., 2002). Apoptosis in this setting requires PDGF receptor-mediated signaling, because it is blocked by specific inhibitors of the receptor and it is induced under the same conditions by v-sis, a homologue of PDGF that specifically activates the PDGF receptor.

In this system, apoptosis appears to be induced by an as-yet-uncharacterized secreted small peptide factor.

9.8.   Role of E5 Proteins in the Virus Life Cycle and Carcinogenesis

It is thought that the fibropapillomavirus E5 proteins are responsible for the ability of these viruses to induce fibroblastic tumors in natural and experimental animal hosts, but experiments to test this possibility have not been performed. Similarly, the role of the E5 proteins during the virus life cycle is unclear. The BPV1 E5 protein is present in basal epithelial cells and the differentiated outer layer of warts, and the HPV31b E5 protein localizes to epithelial cells in the basal and granular layer in organotypic keratinocyte raft cultures that promote cell differentiation and stratification (Bohl et al., 2001; Burnett et al., 1992; Mayer and Meyers, 1998). Abundant mRNA capable of encoding E5 is expressed in differentiated keratinocytes.

To explore the role of the E5 protein in the HPV life cycle, null mutations were constructed in E5 genes in the context of the full-length HPV genome. An HPV16 E5 mutant displays a modest reduction in the number of suprabasal host cells undergoing DNA synthesis in organotypic keratinocyte cultures (Genther et al., 2003). In related experiments, an HPV31b E5 mutant is impaired in its ability to undergo viral genome amplification and differentiation-dependent late gene expression in differentiating keratinocytes (Fehrmann et al., 2003). Taken together, these results suggest that the high-risk HPV E5 protein may play a role in generating the optimal permissive host cell for virus replication. Although the underlying biochemical basis for these effects is not known, the ability of the E5 protein to stimulate growth factor receptor activity may contribute to the ability of suprabasal keratinocytes to support viral DNA replication. In addition, induction of cellular transcription factors by the HPV16 E5 protein suggests that E5 activity may also stimulate expression of viral or cellular genes required for papillomavirus replication. For example, the HPV16 E5 protein induces expression of c-jun and c-fos, which may activate the AP1-responsive viral long control region (Bouvard et al., 1994). However, the existence of papillomavirus types that lack an E5 gene indicates that E5 function is not absolutely required for papillomavirus replication, or that essential E5 activities are carried out by other viral proteins.

Unlike the HPV E6 and E7 oncogenes, the E5 gene is frequently lost during integration of the viral genome into cellular DNA during carcinogenic progression, although the E5 gene is expressed in a substantial fraction of cervical carcinomas containing extrachromosomal HPV16 DNA (Borzacchiello et al., 2003; Chang et al., 2000). Thus, the E5 protein does not appear to be required for the later stages of carcinogenesis or maintenance of the carcinogenic phenotype, but it may be involved in the initial stages. Indeed, sequence comparisons of E5 genes from different HPV types and from cervical carcinomas suggest that the

E5 protein may play a role in carcinogenesis (Bible et al., 2000; Bravo et al., 2005; Schiffman et al., 2005). The nature of this role is not known, but it may reflect the ability of the HPV E5 protein to inhibit immune function, apoptosis, or gap-junctional communication. It is possible that vaccination could elicit E5-specific cytolytic T-lymphocyte-mediated killing of precancerous cells or even carcinoma cells that express the E5 protein (Chen et al., 2004, Liu et al., 2000).

9.9. Conclusions

Because of their small size and unusual sequence, study of the E5 proteins may reveal new insights into membrane proteins and the cellular processes they control. It will be of interest to determine whether the diverse activities of the E5 proteins reflect a multitude of cellular targets, or whether there are functional interactions with only one or a few targets, such as the PDGF β receptor and the vacuolar ATPase. Although substantial progress has been made in elucidating the mechanism of action of the BPV1 E5 protein, important and interesting questions remain. Biophysical and structural experiments must be conducted to determine the molecular structure of the BPV1 E5 protein/PDGF β-receptor complex, and it should be resolved whether the BPV1 E5 protein can transform cells in a PDGF β-receptor-independent fashion. Using the BPV1 E5 protein as a model to construct small transmembrane proteins with novel properties may be a new approach to manipulate cell physiology. Despite considerable effort devoted to studies of the HPV E5 proteins, the biological activities and biochemical properties of these proteins remain important areas of investigation. The role of the V-ATPase in the activities of the BPV and HPV E5 proteins needs to be clarified. Finally, the role of the E5 proteins in the papillomavirus life cycle, carcinogenesis, and immune evasion is largely unexplored and demands further analysis.

References

Adam, J.L., Briggs, M.W., and McCance, D.J. (2000). A mutagenic analysis of the E5 protein of human papillomavirus type 16 reveals that E5 binding to the vacuolar H+-ATPase is not sufficient for biological activity, using mammalian and yeast expression systems. *Virology* 272:315–325.

Adduci, A.J., andSchlegel, R. (1999). The transmembrane domain of the E5 oncoprotein contains functionally discrete helical faces. *J. Biol. Chem.* 274:10249–10258.

Andresson, T., Sparkowski, J., Goldstein, D.J., and Schlegel, R. (1995). Vacuolar H+-ATPase mutants transform cells and define a binding site for the papillomavirus E5 oncoprotein. *J. Biol. Chem.* 270:6830–6837.

Araibi, E.H., Marchetti, B., Ashrafi, G.H., and Campo, M.S. (2004). Downregulation of major histocompatibility complex class I in bovine papillomas. *J. Gen. Virol.* 85:2809–2814.

Ashby, A.D.M., Meagher, L., Campo, M.S., and Finbow, M.E. (2001). E5 transforming proteins of papillomaviruses do not disturb the activity of the vacuolar H+-ATPase. *J. Gen. Virol.* 82:2353–2362.

Ashrafi, G.H., Haghshenas, M., Marchetti, B., and Campo, M.S. (2006). E5 protein of human papillomavirus 16 downregulates HLA class I and interacts with the heavy chain *via* its first hydrophobic domain. *Int. J. Cancer* 119:2105–2112.

Ashrafi, G.H., Haghshenas, M.R, Marchetti, B., O'Brien, P.M., and Campo, M.S. (2005). E5 protein of human papillomavirus type 16 selectively downregulates surface HLA class *I Int. J. Cancer* 113:276–283.

Ashrafi, G.H., Pitts, J.D., Faccini, A., McLean, P., O'Brien, P.M., Finbow, M.E., and Campo, M.S. (2000). Binding of bovine papillomavirus type 4 E8 to ductin (16K proteolipid), down-regulation of gap junction intercellular communication and full cell transformation are independent events. *J. Gen. Virol.* 81:689–694.

Ashrafi, G.H., Tsirimonaki, E., Marchetti, B., O'Brien, P.M., Sibbet, G.J., Andrew, L., and Campo, M.S. (2002). Down-regulation of MHC class I by bovine papillomavirus E5 oncoproteins. *Oncogene* 21:248–259.

Auvinen, E., Alonso, A., and Auvinen, P. (2004). Human papillomavirus type 16 E5 protein colocalizes with the antiapoptotic Bcl-2 protein. *Arch. Virol.* 149:1745–1759.

Bergman, P., Ustav, M., Sedman, J., Moreno-Lopez, J., Vennstrom, B., and Pettersson, U. (1988). The E5 gene of bovine papillomavirus type 1 is sufficient for complete oncogenic transformation of mouse fibroblasts. *Oncogene* 2:453–459.

Bible, J.M., Mant, C., Best, J.M., Kell, B., Starkey, W.G., Raju, K.S., Seed, P., Biswas, C., Muir, P., Banatvala, J.E., and Cason, J. (2000). Cervical lesions are associated with human papillomavirus type 16 intratypic variants that have high transcriptional activity and increased usage of common mammalian codons. *J. Gen. Virol.* 8:1517–1527.

Bohl, J., Hull, B., and Vande Pol, S.B. (2001). Cooperative transformation and coexpression of bovine papillomavirus type 1 E5 and E7 proteins. *J. Virol.* 75:513–521.

Borzacchiello, G., Iovane, G., Marcante, M.L., Poggiali, F., Roperto, F., Roperto, S., and Venuti, A. (2003). Presence of bovine papillomavirus type 2 DNA and expression of the viral oncoprotein E5 in naturally occurring urinary bladder tumours in cows. *J. Gen. Virol.* 84:2921–2926.

Borzacchiello, G., Russo, V., Gentile, F., Roperto, F., Venuti, A., Nitsch, L., Campo, M.S., and Roperto, S. (2006). Bovine papillomavirus E5 oncoprotein binds to the activated form of the platelet-derived growth factor beta receptor in naturally occurring bovine urinary blader tumours. *Oncogene* 25:1251–1260.

Bouvard, V., Matlashewski, G., Gu, Z.-M., Storey, A., and Banks, L. (1994). The human papillomavirus type 16 E5 gene cooperates with the E7 gene to stimulate proliferation of primary cells and increases viral gene expression. *Virology* 203:73–80.

Bravo, I.G., and Alonso, A. (2004). Mucosal human papillomaviruses encode four different E5 proteins whose chemistry and phylogeny correlate with malignant or benign growth. *J. Virol.* 78:13613–13626.

Bravo, I.G., Crusius, K., and Alonso, A. (2005). The E5 protein of the human papillomavirus type 16 modulates composition and dynamics of membrane lipids in keratinocytes. *Arch. Virol.* 150:231–246.

Briggs, M.W., Adam, J.L., and McCance, D.J. (2001). The human papillomavirus type 16 E5 protein alters vacuolar H^+-ATPase function and stability in *Saccharomyces cerevisiae*. *Virology* 280:169–175.

Burkhardt, A., DiMaio, D., and Schlegel, R. (1987). Genetic and biochemical definition of the bovine papillomavirus E5 transforming protein. *EMBO J.* 6(8):2381–2385.

Burkhardt, A., Willingham, M., Gay, C., Jeang, K.-T., and Schlegel, R. (1989). The E5 oncoprotein of bovine papillomavirus is oriented asymmetrically in Golgi and plasma membranes. *Virology* 170:334–339.

Burnett, S., Jareborg, N., and DiMaio, D. (1992). Localization of bovine papillomavirus type 1 E5 protein to transformed basal keratinocytes and permissive differentiated cells in fibropapilloma tissue. *Proc. Natl. Acad. Sci. U S A* 89(12):5665–5669.

Cartin, W., and Alonso, A. (2003). The human papillomavirus HPV2a E5 protein localizes to the Golgi apparatus and modulates signal transduction. *Virology* 314:572–579.

Chang, J.-L., Tsao, Y.-P., Liu, D.-W., Huang, S.-J., Lee, W.-H., and Chen, S.-L. (2000). The expression of HPV-16 E5 protein in squamous neoplastic changes in the uterine cervix. *J. Biomed. Sci.* 8:206–213.

Chen, S.L., Huang, C.H., Tsai, T.C., Lu, K.Y., and Tsao, Y.P. (1996a). The regulation mechanism of c-jun and junB by human papillomavirus type 16 E5 oncoprotein. *Arch. Virol.* 141:791–800.

Chen, S.L., and Mounts, P. (1990). Transforming activity of E5a protein of human papillomavirus type 6 in NIH 3T3 and C127 cells. *J. Virol.* 64:3226–3233.

Chen, S.-L., Tsai, T.-C., Han, C.-P., and Tsao, Y.-P. (1996b). Mutational analysis of human papillomavirus type 11 E5a oncoprotein. *J. Virol.* 70:3502–3508.

Chen, Y.F., Lin, C.W., Tsao, Y.P., and Chen, S.L. (2004). Cytotoxic-T-lymphocyte human papillomavirus type 16 E5 peptide with CpG-oligodeoxynucleotide can eliminate tumor growth in C57BL/6 mice. *J. Virol.* 78:1333–1343.

Cohen, B.D., Goldstein, D.J., Rutledge, L., Vass, W.C., Lowy, D.R., Schlegel, R., and Schiller, J.T. (1993). Transformation-specific interaction of the bovine papillomavirus E5 oncoprotein with the platelet-derived growth factor receptor transmembrane domain and the epidermal growth factor receptor cytoplasmic domain. *J. Virol.* 67:5303–5311.

Conrad, M., Bubb, V.J., and Schlegel, R. (1993). The human papillomavirus type 6 and 16 E5 proteins are membrane-associated proteins which associate with the 16-kilodalton pore-forming protein. *J. Virol.* 67:6170–6178.

Conrad, M., Goldstein, D., Andresson, T., and Schlegel, R. (1994). The E5 protein of HPV-6, but not HPV-16, associates efficiently with cellular growth factor receptors. *Virology* 200:796–800.

Conrad-Stoppler, M., Straight, S.W., Tsao, G., Schlegel, R., and McCance, D.J. (1996). The E5 gene of HPV-16 enhances keratinocyte immortalization by full-length DNA. *Virology* 223:251–254.

Constantinescu, S.N., Liu, X., Beyer, W., Fallon, A., Shekar, S., Henis, Y.I., Smith, S.O., and Lodish, H.F. (1999). Activation of the erythropoietin receptor by the gp^{55-P} viral envelope protein is determined by a single amino acid in its transmembrane domain. *EMBO J.* 18:3334–3347.

Crusius, K., Auvinen, E., and Alonso, A. (1997). Enhancement of EGF- and PMA-mediated MAP kinase activation in cells expressing the human papillomavirus type 16 E5 protein. *Oncogene* 15:1437–1444.

Crusius, K., Auvinen, E., Steuer, B., Gaissert, H., and Alonso, A. (1998). The human papillomavirus type 16 E5-protein modulates ligand-dependent activation of the EGF receptor family in the human epithelial cell line HaCaT. *Exp. Cell Res.* 241: 76–83.

Crusius, K., Kaszkin, M., Kinzel, V., and Alonso, A. (1999). The human papillomavirus type 16 E5 protein modulates phospholipase C-γ-1 activity and phosphatidyl inositol turnover in mouse fibroblasts. *Oncogene* 18:6714–6718.

Crusius, K., Rodriguez, I., and Alonso, A. (2000). The human papillomavirus type 16 E5 protein modulates ERK1/2 and p38 MAP kinase activation by an EGFR-independent process in stressed human keratinocytes. *Virus Genes* 20:65–69.

DiMaio, D., Guralski, D., and Schiller, J.T. (1986). Translation of open reading frame E5 of bovine papillomavirus is required for its transforming activity. *Proc. Natl. Acad. Sci. U S A* 83(6):1797–801.

Disbrow, G.L., Hanover, J.A., and Schlegel, R. (2005). Endoplasmic reticulum-localized human papillomavirus type 16 E5 protein alters endosomal pH but not trans-Golgi pH. *J. Virol.* 79:5839–5846.

Drummond-Barbosa, D.A., Vaillancourt, R.R., Kazlauskas, A., and DiMaio, D. (1995). Ligand-independent activation of the platelet-derived growth factor beta receptor: requirements for bovine papillomavirus E5-induced mitogenic signaling. *Mol. Cell. Biol.* 15(5):2570–2581.

Faccini, A.M., Cairney, M., Ashrafi, G.H., Finbow, M.E., Campo, M.S., and Pitts, J.D. (1996). The bovine papillomavirus type 4 E8 protein binds to ductin and causes loss of gap junctional intercellular communication in primary fibroblasts. *J. Virol.* 70:9041–9045.

Fehrmann, F., Klumpp, D.J., and Laimins, L.A. (2003). Human papillomavirus type 31 E5 protein supports cell cycle progression and activates late viral functions upon epithelial differentiation. *J. Virol.* 77:2819–2831.

Finbow, M.E., Goodwin, S.F., Meagher, L., Lane, N.J., Keen, J., Findlay, J.B., and Kaiser, K. (1994). Evidence that the 16 kDa proteolipid (subunit c) of the vacuolar H(+)-ATPase and ductin from gap junctions are the same polypeptide in Drosophila and Manduca: molecular cloning of the *Vha16k* gene from Drosophila. *J. Cell Sci.* 107:1817–1824.

Freeman-Cook, L., Dixon, A.M., Frank, J.B., Xia, Y., Ely, L., Gerstein, M., Engelman, D.M., and DiMaio, D. (2004). Selection and characterization of small random transmembrane proteins that bind and activate the platelet-derived growth factor β receptor. *J. Mol. Biol.* 338:907–920.

Freeman-Cook, L.L., and DiMaio, D. (2005). Modulation of cell function by small transmembrane proteins modeled on the bovine papillomavirus E5 protein. Oncogene 24:7756–7762.

Freeman-Cook, L.L., Edwards, A.P.B., Dixon, A.M., Yates, K.E., Ely, L., Engelman, D.M., and DiMaio, D. (2005). Specific locations of hydrophilic amino acids in constructed transmembrane ligands of the platelet-derived growth factor β receptor. *J. Mol. Biol.* 345:907–921.

Genther, S.M., Sterling, S., Duensing, S., Munger, K., Sattler, C., and Lambert, P.F. (2003). Quantitative role of the human papillomavirus type 16 E5 gene during the productive stage of the viral life cycle. *J. Virol.* 77:2832–2842.

Genther-Williams, S.M., Disbrow, G.L., Schlegel, R., Lee, D., Threadgill, D.W., and Lambert, P.F. (2005). Requirement of epidermal growth factor receptor for hyperplasia induced by E5, a high-risk human papillomavirus oncogene. *Cancer Res.* 65:6534–6542.

Gieswein, C.E., Sharom, F.J., and Wildeman, A.G. (2003). Oligomerization of the E5 protein of human papillomavirus type 16 occurs through multiple hydrophobic regions. *Virology* 313:415–426.

Goldstein, D.J., Andresson, T., Sparkowski, J.J., and Schlegel, R. (1992a). The BPV-1 E5 protein, the 16 kDa membrane pore-forming protein and the PDGF receptor exist in a complex that is dependent on hydrophobic transmembrane interactions. *EMBO J.* 11:4851–4859.

Goldstein, D.J., Finbow, M.E., Andresson, T., McLean, P., Smith, K., Bubb, V., and Schlegel, R. (1991). Bovine papillomavirus E5 oncoprotein binds to the 16K component of vacuolar H(+)-ATPases. Nature 352:347–349.

Goldstein, D.J., Kulke, R., DiMaio, D., and Schlegel, R. (1992b). A glutamine residue in the membrane-associating domain of the bovine papillomavirus type 1 E5 oncoprotein mediates its binding to a transmembrane component of the vacuolar H(+)-ATPase. *J. Virol.* 66(1):405–413.

Goldstein, D.J., Li, W., Wang, L.-M., Heidaran, M.A., Aaronson, S.A., Shinn, R., Schlegel, R., and Pierce, J.H. (1994). The bovine papillomavirus type 1 E5 transforming protein specifically binds and activates the beta-type receptor for platelet-derived growth factor but not other tyrosine kinase-containing receptors to induce cellular transformation. *J. Virol.* 68:4432–4441.

Grindlay, G.J., Campo, M.S., and O'Brien, V. (2005). Transactivation of the cyclin A promoter by bovine papillomavirus type 4 E5 protein. *Virus Res.* 108:29–38.

Gu, Z.-M., and Matlashewski, G. (1995). Effect of human papillomavirus type 16 oncogenes on MAP kinase activity. *J. Virol.* 69:8051–8056.

Horwitz, B.H., Burkhardt, A.L., Schlegel, R., and DiMaio, D. (1988). 44-amino-acid E5 transforming protein of bovine papillomavirus requires a hydrophobic core and specific carboxyl-terminal amino acids. *Mol. Cell. Biol.* 8(10):4071–4078.

Horwitz, B.H., Weinstat, D.L., and DiMaio, D. (1989). Transforming activity of a 16-amino-acid segment of the bovine papillomavirus E5 protein linked to random sequences of hydrophobic amino acids. *J. Virol.* 63(11):4515–4519.

Hossain, M.Z., Ao, P., and Boynton, A.L. (1998). Rapid disruption of Gap junctional communication and phosphorylation of connexin43 by platelet-derived growth factor in T51B rat liver epithelial cells expressing platelet-derived growth factor receptor. *J. Cell. Physiol.* 174:66–77.

Hwang, E.S., Nottoli, T., and DiMaio, D. (1995). The HPV16 E5 protein: expression, detection, and stable complex formation with transmembrane proteins in COS cells. *Virology* 211(1):227–233.

Kabsch, K., and Alonso, A. (2002a). The human papillomavirus type 16 (HPV-16) E5 protein sensitizes human keratinocytes to apoptosis induced by osmotic stress. *Oncogene* 21:947–953.

Kabsch, K., and Alonso, A. (2002b). The human papillomavirus type 16 E5 protein impairs TRAIL- and FasL-mediated apoptosis in HaCaT cells by different mechanisms. *J. Virol.* 76:12162–12172.

Kabsch, K., Mossadegh, N., Kohl, A., Komposch, G., Schenkel, J., Alonso, A., and Tomakidi, P. (2004). The HPV-16 E5 protein inhibits TRAIL- and FasL-mediated apoptosis in human keratinocyte raft cultures. *Intervirology* 47:48–56.

Kim, S.H., Juhnn, Y.S., Kang, S., Park, S.W., Sung, M.W., Bang, Y.J., and Song, Y.S. (2006). Human papillomavirus 16 E5 up-regulates the expression of vascular endothelial growth factor through the activation of epidermal growth factor receptor, MEK/ ERK1,2 and PI3K/Akt. *Cell. Mol. Life Sci.* 63:930–938.

Klein, O., Kegler-Ebo, D., Su, J., Smith, S., and DiMaio, D. (1999). The bovine papillomavirus E5 protein requires a juxtamembrane negative charge for activation of the platelet-derived growth factor beta receptor and transformation of C127 cells. *J. Virol.* 73(4):3264–3272.

Klein, O., Polack, G. W., Surti, T., Kegler-Ebo, D., Smith, S. O., and DiMaio, D. (1998). Role of glutamine 17 of the bovine papillomavirus E5 protein in platelet-derived growth factor beta receptor activation and cell transformation. *J. Virol.* 72(11): 8921–8932.

Lai, C.-C., Edwards, A.P.B., and DiMaio, D. (2005). Productive interaction between transmembrane mutants of the bovine papillomavirus E5 protein and the platelet-derived growth factor β receptor. *J. Virol.* 79:1924–1929.

Lai, C.C., Henningson, C., and DiMaio, D. (1998). Bovine papillomavirus E5 protein induces oligomerization and trans-phosphorylation of the platelet-derived growth factor beta receptor. *Proc. Natl. Acad. Sci. U S A* 95(26):15241–15246.

Lai, C.C., Henningson, C., and DiMaio, D. (2000). Bovine papillomavirus E5 protein induces the formation of signal transduction complexes containing dimeric activated platelet-derived growth factor β receptor and associated signaling proteins. *J. Biol. Chem.* 275:9832–9840.

Leechanachai, P., Banks, L., Moreau, F., and Matlashewski, G. (1992). The E5 gene from human papillomavirus type 16 is an oncogene which enhances growth factor-mediated signal transduction to the nucleus. *Oncogene* 7:19–25.

Leptak, C., Ramon y Cajal, S., Kulke, R., Horwitz, B.H., Riese, D.J., II, Dotto, G.P., and DiMaio, D. (1991). Tumorigenic transformation of murine keratinocytes by the E5 genes of bovine papillomavirus type 1 and human papillomavirus type 16. *J Virol* 65(12):7078–7083.

Liu, D.W., Tsao, Y.P., Hsieh, C.H., Hsieh, J.T., Kung, J.T., Chiang, C.L., Huang, S.J., and Chen, S.L. (2000). Induction of CD8 T cells by vaccination with recombinant adenovirus expressing human papillomavirus type 16 E5 gene reduces tumor growth. *J. Virol.* 74:9083–9089.

Marchetti, B., Ashrafi, G.H., Dornan, E.S., Araibi, E.H., Ellis, S.A., and Campo, M.S. (2006). The E5 protein of BPV-4 interacts with the heavy chain of MHC class I and irreversibly retains the MHC complex in the Golgi apparatus. *Oncogene* 25:2254–2263.

Marchetti, B., Ashrafi, G.H., Tsirimonaki, E., O'Brien, P.M., and Campo, M.S. (2002). The bovine papillomavirus oncoprotein E5 retains MHC class I molecules in the Golgi apparatus and prevents their transport to the cell surface. *Oncogene* 21: 7808–7816.

Martin, P., Vass, W.C., Schiller, J.T., Lowy, D.R., and Velu, T.J. (1989). The bovine papillomavirus E5 transforming protein can stimulate the transforming activity of EGF and CSF-1 receptors. *Cell* 59:21–32.

Mattoon, D., Gupta, K., Doyon, J., Loll, P.J., and DiMaio, D. (2001). Identification of the transmembrane dimer interface of the bovine papillomavirus E5 protein. *Oncogene* 20:3824–3834.

Mayer, T.J., Frauenhoffer, E.E., and Meyers, A.C. (2000). Expression of epidermal growth factor and platelet-derived growth factor receptors during cervical carcinogenesis. *In Vitro Cell Dev. Biol. Anim.* 36:667–676.

Mayer, T.J., and Meyers, C. (1998). Temporal and spatial expression of the E5a protein during the differentiation-dependent life cycle of human papillomavirus type 31b. *Virology* 248:208–217.

Meyer, A.N., Xu, Y.-F., Webster, M.K., Smith, A.S., and Donoghue, D.J. (1994). Cellular transformation by a transmembrane peptide: structural requirements for the bovine papillomavirus E5 oncoprotein. *Proc. Natl. Acad. Sci. USA* 91:4634–4638.

Nappi, V.M., and Petti, L.M. (2002). Multiple transmembrane amino acid requirements suggest a highly specific interaction between the bovine papillomavirus E5 oncoprotein and the platelet-derived growth factor beta receptor. *J. Virol.* 76:7976–7986.

Nappi, V.M., Schaefer, J.A., and Petti, L.M. (2002). Molecular examination of the transmembrane requirements of the platelet-derived growth factor beta receptor for a productive interaction with the bovine papillomavirus E5 oncoprotein. *J. Biol. Chem.* 277:47149–47159.

Nilson, L.A., and DiMaio, D. (1993). Platelet-derived growth factor receptor can mediate tumorigenic transformation by the bovine papillomavirus E5 protein. *Mol. Cell. Biol.* 13(7):4137–4145.

Nilson, L.A., Gottlieb, R.L., Polack, G.W., and DiMaio, D. (1995). Mutational analysis of the interaction between the bovine papillomavirus E5 transforming protein and the endogenous beta receptor for platelet-derived growth factor in mouse C127 cells. *J. Virol.* 69(9):5869–5874.

O'Brien, P.M., Ashrafi, G.H., Grindlay, G.J., Anderson, R., and Campo, M.S. (1999). A mutational analysis of the transforming functions of the E8 protein of bovine papillomavirus type 4. *Virology* 255:385–394.

O'Brien, P.M., Grindlay, G.J., and Campo, M.S. (2001). Cell transformation by the E5/E8 protein of bovine papillomavirus type 4. *J. Biol. Chem.* 276:33861–33868.

O'Brien, V., and Campo, M.S. (1998). BPV-4 E8 transforms NIH 3T3 cells, up-regulates cyclin A and cyclin A-associated kinase activity and de-regulates expression of the cdk inhibitor p27^{KIP1}. *Oncogene* 17:293–301.

Oelze, I., Kartenbeck, J., Crusius, K., and Alonso, A. (1995). Human papillomavirus type 16 E5 protein affects cell–cell communication in an epithelial cell line. *J. Virol.* 69:4489–4494.

Petti, L., and DiMaio, D. (1992). Stable association between the bovine papillomavirus E5 transforming protein and activated platelet-derived growth factor receptor in transformed mouse cells. *Proc. Natl. Acad. Sci U S A* 89(15):6736–6740.

Petti, L., and DiMaio, D. (1994). Specific interaction between the bovine papillomavirus E5 transforming protein and the beta receptor for platelet-derived growth factor in stably transformed and acutely transfected cells. *J. Virol.* 68(6):3582–3592.

Petti, L., Nilson, L.A., and DiMaio, D. (1991). Activation of the platelet-derived growth factor receptor by the bovine papillomavirus E5 transforming protein. *EMBO J.* 10(4):845–855.

Petti, L.M., and Ray, F.A. (2000). Transformation of mortal human fibroblasts and activation of a growth inhibitory pathway by the bovine papillomavirus E5 oncoprotein. *Cell Growth Differ.* 11:395–408.

Petti, L.M., Reddy, V., Smith, S.O., and DiMaio, D. (1997). Identification of amino acids in the transmembrane and juxtamembrane domains of the platelet-derived growth factor receptor required for productive interaction with the bovine papillomavirus E5 protein. *J. Virol.* 71(10):7318–7327.

Pim, D., Collins, M., and Banks, L. (1992). Human papillomavirus type 16 E5 gene stimulates the transforming activity of the epidermal growth factor receptor. *Oncogene* 7:27–32.

Riese, D.J., II, and DiMaio, D. (1995). An intact PDGF signaling pathway is required for efficient growth transformation of mouse C127 cells by the bovine papillomavirus E5 protein. *Oncogene* 10(7):1431–1439.

Rodriguez, M.I., Finbow, M.E., and Alonso, A. (2000). Binding of human papillomavirus 16 E5 to the 16 kDa subunit c (proteolipid) of the vacuolar H$^+$-ATPase can be dissociated from the E5-mediated epidermal growth fator receptor overactivation. *Oncogene* 19:3727–3732.

Saito, T., Schlegel, R., Andresson, T., Yuge, L., Yamamoto, M., and Yamasaki, H. (1998). Induction of cell transformation by mutated 16K vacuolar H+-ATPase (ductin) is accompanied by down-regulation of gap junctional intercellular communication and translocation of connexin 43 in NIH3T3 cells. *Oncogene* 17:1673–1680.

Schapiro, F., Sparkowski, J., Adduci, A., Suprynowicz, F.A., Schlegel, R., and Grinstein, S. (2000). Golgi alkalinization by the papillomavirus E5 oncoprotein. *J. Cell. Biol.* 148:305–315.

Schiffman, M., Herrero, R., Desalle, R., Hildesheim, A., Wacholder, S., Rodriguez, A.C., Bratti, M.C., Sherman, M.E., Morales, J., Guillen, D., Alfaro, M., Hutchinson, M., Wright, T. C., Solomon, D., Chen, Z., Schussler, J., Castle, P.E., and Burk, R.D. (2005). The carcinogenicity of human papillomavirus types reflects viral evolution. *Virology* 337:76–84.

Schiller, J.T., Vass, W.C., Vousden, K.H., and Lowy, D.R. (1986). E5 open reading frame of bovine papillomavirus type 1 encodes a transforming gene. *J. Virol.* 57:1–6.

Schlegel, R., Wade-Glass, M., Rabson, M.S., and Yang, Y.-C. (1986). The E5 transforming gene of bovine papillomavirus encodes a small hydrophobic protein. *Science* 233:464–467.

Settleman, J., Fazeli, A., Malicki, J., Horwitz, B.H., and DiMaio, D. (1989). Genetic evidence that acute morphologic transformation, induction of cellular DNA synthesis, and focus formation are mediated by a single activity of the bovine papillomavirus E5 protein. *Mol Cell Biol* 9(12):5563–5572.

Sparkowski, J., Anders, J., and Schlegel, R. (1995). E5 oncoprotein retained in the endoplasmic reticulum/*cis* Golgi still induces PDGF receptor autophosphorylation but does not transform cells. *EMBO J.* 14:3055–3063.

Sparkowski, J., Mense, M., Anders, M., and Schlegel, R. (1996). E5 oncoprotein transmembrane mutants dissociate fibroblast transforming activity from 16-kilodalton protein binding and platelet-derived growth factor receptor binding and phosphorylation. *J. Virol.* 70:2420–2430.

Staebler, A., Pierce, J.H., Brazinski, S., Heidaran, M.A., Li, W., Schlegel, R., and Goldstein, D. J. (1995). Mutational analysis of the beta-type platelet-derived growth factor receptor defines the site of interaction with the bovine papillomavirus type 1 E5 transforming protein. *J. Virol.* 69:6507–6517.

Stevens, T.H., and Forgac, M. (1997). Structure, function and regulation of the vacuolar (H$^+$)-ATPase. *Ann. Rev. Cell Dev. Biol.* 13:779–808.

Straight, S.W., Herman, B., and McCance, D.J. (1995). The E5 oncoprotein of human papillomavirus type 16 inhibits the acidification of endosomes in human keratinocytes. *J. Virol.* 69:3185–3192.

Straight, S.W., Hinkle, P.M., Jewers, R.J., and McCance, D.J. (1993). The E5 oncoprotein of human papillomavirus type 16 transforms fibroblasts and effects the downregulation of the epidermal growth factor receptor in keratinocytes. *J. Virol.* 67:4521–4532.

Suprynowicz, F.A., Baege, A., Sunitha, I., and Schlegel, R. (2002). c-Src activation by the E5 oncoprotein enables transformation independently of PDGF receptor activation. *Oncogene* 21:1695–1706.

Suprynowicz, F.A., Disbrow, G.L., Simic, V., and Schlegel, R. (2005). Are transforming properties of the bovine papillomavirus E5 protein shared by E5 from high-risk human papillomavirus type 16? *Virology* 332:102–113.

Suprynowicz, F.A., Sparkowski, J., Baege, A., and Schlegel, R. (2000). E5 oncoprotein mutants activate phosphoinositide 3'-kinase independently of platelet-derived growth factor receptor activation. *J. Biol. Chem.* 275:5111–5119.

Surti, T., Klein, O., Aschheim, K., DiMaio, D., and Smith, S.O. (1998). Structural models of the bovine papillomavirus E5 protein. *Proteins* 33(4):601–612.

Thomsen, P., van Deurs, B., Norrild, B., and Kayser, L. (2000). The HPV16 E5 oncogene inhibits endocytic trafficking. *Oncogene* 19:6023–6032.

Tomakidi, P., Cheng, H., Kohl, A., Komposch, G., and Alonso, A. (2000). Connexin 43 expression is downregulated in raft cultures of human keratinocytes expressing the human papillomavirus type 16 E5 protein. *Cell Tissue Res.* 301:323–327.

Tsao, Y.-P., Li, L.-Y., Tsai, T.-C., and Chen, S.-L. (1996). Human papillomavirus type 11 and 16 E5 represses p21$^{WafI/SdiI/CipI}$ gene expression in fibroblasts and keratinocytes. *J. Virol.* 70:7535–7539.

Vallee, G.F., and Banks, L. (1995). The human papillomavirus (HPV)-6 and HPV-16 proteins co-operate with HPV-16 E7 in the transformation of primary rodent cells. *J. Gen. Virol.* 76:1239–1245.

Venuti, A., Salani, D., Poggiali, F., Manni, V., and Bagnato, A. (1998). The E5 oncoprotein of human papillomavirus type 16 enhances endothelin-1-induced keratinocyte growth. *Virology* 248:1–5.

Zago, M., Campo, M. S., and O'Brien, V. (2004). Cyclin A expression and growth in suspension can be uncoupled from p27 de-regulation and ERK activity in cells transformed by BPV-4 E5. *J. Gen. Virol.* 85:3585–3595.

Zhang, B., Li, P., Wang, E., Brahmi, Z., Dunn, K.W., Blum, J.S., and Roman, A. (2003). The E5 protein of human papillomavirus type 16 perturbs MHC class II antigen maturation in human foreskin keratinocytes treated with interferon-γ. *Virology* 310:100–108.

Zhang, B., Spandau, D.F., and Roman, A. (2002). E5 protein of human papillomavirus type 16 protects human foreskin keratinocytes from UV B-irradiation-induced apoptosis. *J. Virol.* 76:220–231.

Zhang, B., Srirangam, A., Potter, D.A., and Roman, A. (2005). HPV16 E5 protein disrupts the c-Cbl-EGFR interaction and EGFR ubiquitination in human foreskin keratinocytes. *Oncogene.* 24:2585–2588.

Zhang, Y., Lehman, J.M., and Petti, L.M. (2002). Apoptosis of mortal human fibroblasts transformed by the bovine papillomavirus E5 oncoprotein. *Mol. Cancer Res.* 1:122–136.

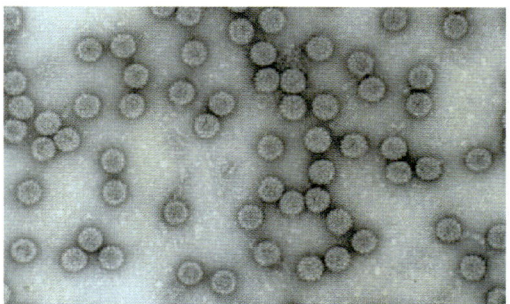

PLATE 1. FIGURE 5.2. Electron micrograph of bovine papillomavirus.

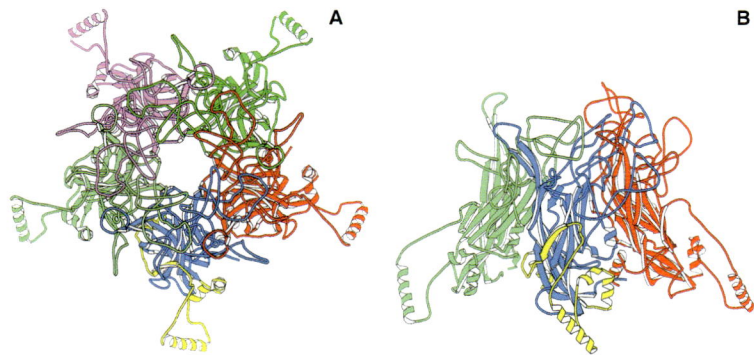

PLATE 2. FIGURE 5.3. An HPV L1 pentamer viewing from the top (**A**) and from the side (**B**). The side view in panel B only show the three monomers in the front, with the two monomers in the back taken away to give a clearer view (Chen et al., 2000).

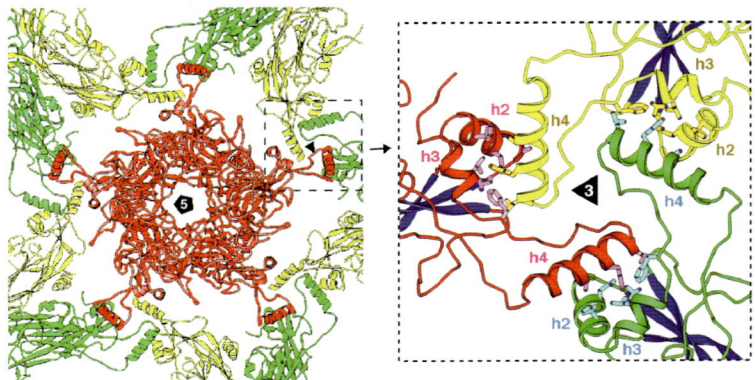

PLATE 3. FIGURE 5.4. The threefold interactions in a T1 particle assembled with HPV16 L1 pentamers (Modis et al., 2002). In a T = 1 particle all pentamers of L1 have five neighboring pentamers, so that the inter-pentamer bonds are all equivalent. These bonds may be distinct from those seen in the complete T = 7 virion.

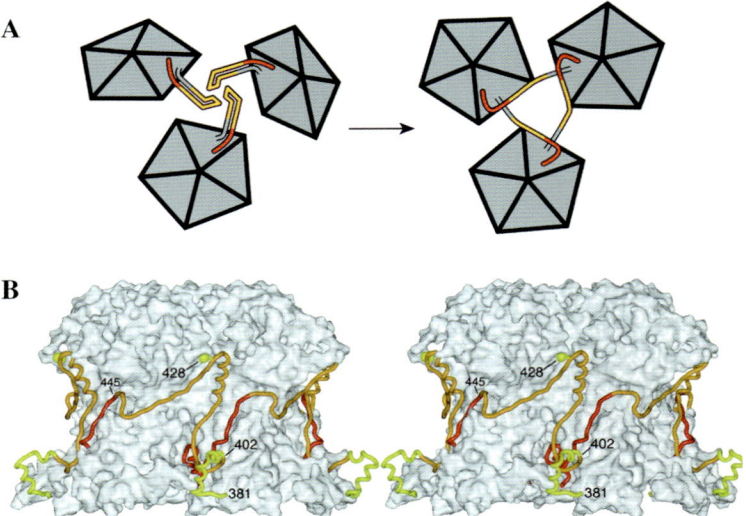

PLATE 4. FIGURE 5.5. Schematic showing the bonding interactions modeled by Modis et al. (2002) in the complete T = 7 papilloma virion. A: The threefold pentamer–pentamer interactions observed in the T1 structure on the left and the possible C-terminal arm conformation at the same threefold location in a T7 particle. B: The invading C-terminal arms of adjacent pentamers wrapping around their neighboring pentamers allowing important juxtaposition of cysteine residues (e.g., cys428 HPV16L1) for disulfide bond formation.

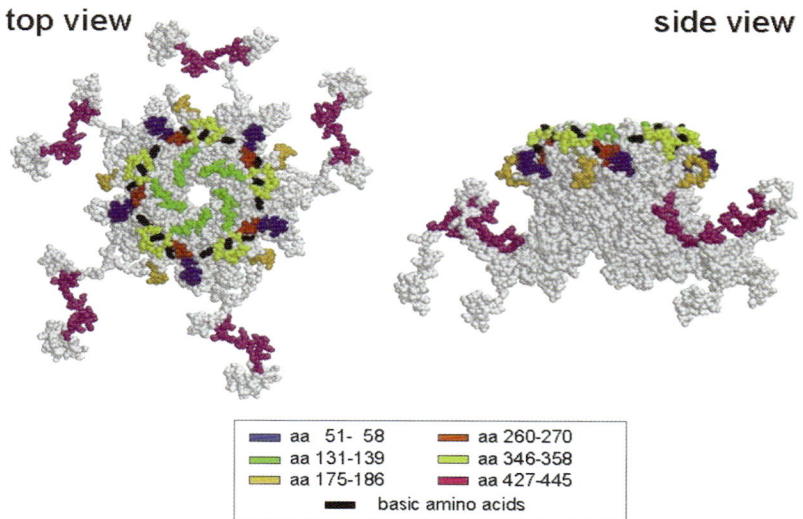

top view   side view

aa 51- 58	aa 260-270
aa 131-139	aa 346-358
aa 175-186	aa 427-445
basic amino acids	

PLATE 5. FIGURE 6.2. Location of neutralization epitopes and surface-exposed basic amino acids. The structure of the HPV16 capsomere is taken from Chen et al. (2000). References for the neutralization epitopes are Roden et al. (1997), McClements et al. (2001), and Carter et al. (2003).

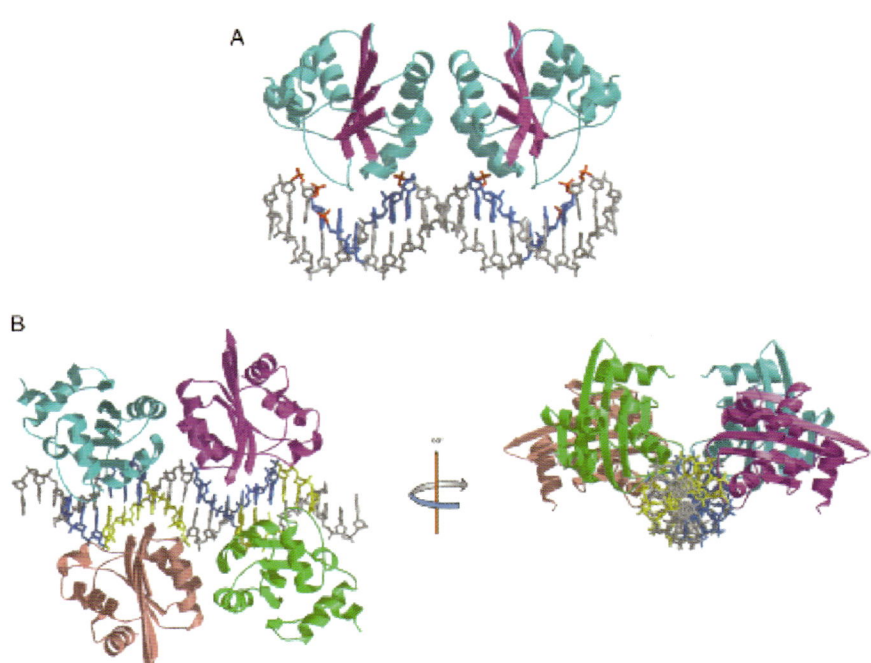

PLATE 6. FIGURE 8.3. **A:** Structure of the E1 DBD dimer bound to E1 BS 2 and 4. The binding site nucleotides identified by mutagenesis are colored in blue and phosphate contacts identified by ethylation interference are colored red. **B:** Structure of the E1 DBD tetramer bound to E1 binding sites 1–4 viewed perpendicular and parallel to the DNA helical axis. The four monomers are individually colored: one dimer is colored cyan and purple; the other, dimer pink and green; and the two sets of binding sites are colored blue (E1 BS 2 and 4) and yellow (E1 BS 1 and 3). Adapted from Enemark et al. (2002). Crystal structures of two intermediates in the assembly of the papillomavirus replication initiation complex. *Embo. J.* 21:1487–1496.

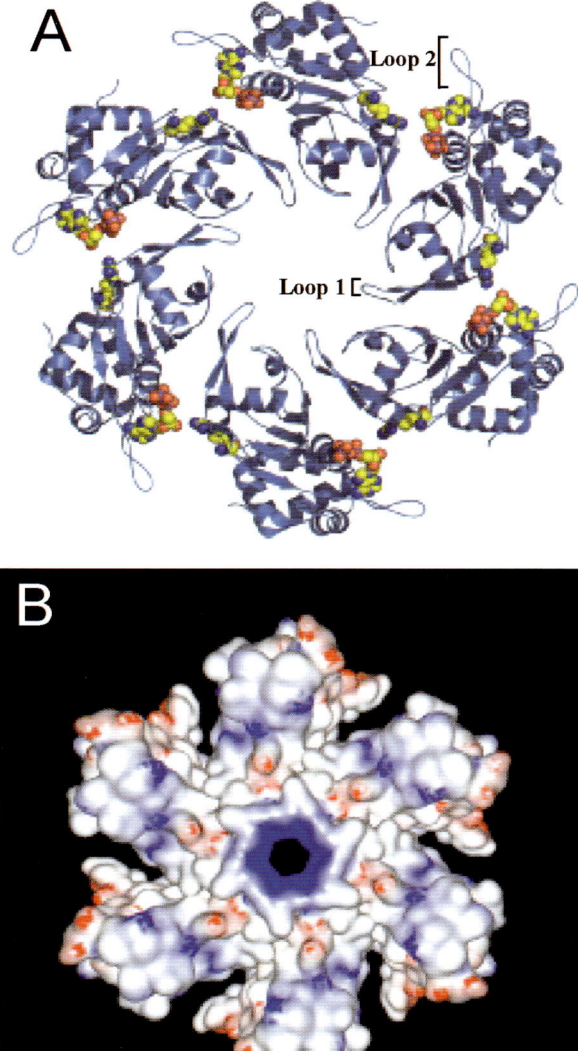

PLATE 7. FIGURE 8.4. Structure of the helicase domain from HPV-18 E1. **A**: The X-ray crystal structure of a fragment (aa 428–631) corresponding to the helicase domain of HPV-18 E1 was solved, and the resulting structure was modeled on the hexameric structure of the SV40 T-ag helicase domain. **B**: Electrostatic potential surface representation (GRASP) of the E1 hexamer. Areas of positive potential are blue and negative potential red. Adapted from Abbate et al. (2004). The X-ray structure of the papillomavirus helicase in complex with its molecular matchmaker E2. *Genes Dev.* 18:1981–1996.

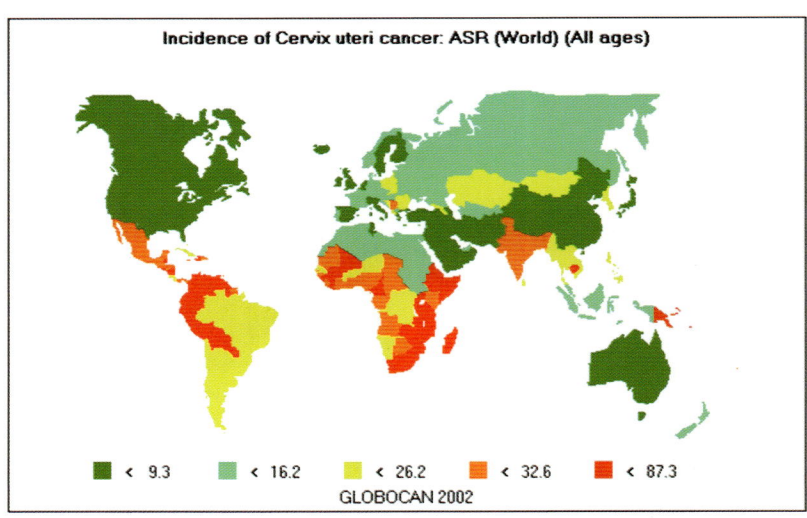

PLATE 8. FIGURE 16.2. Estimated age-standardized incidence rates of cervical cancer by country as color-coded from highest (red) to lowest (dark green) per 100,000 women. Available at www.iarc.fr.

10
Human Papillomavirus E6 and E7 Oncogenes

Karl Münger[1], Peter Howley[2], and Daniel DiMaio[3]

[1] *Brigham & Women's Hospital, Channing Labs, Boston, MA 02115-5701*
[2] *Department of Pathology, Harvard Medical School Boston, MA 02115*
[3] *Departments of Genetics and Therapeutic Radiology, Yale University School of Medicine, New Haven, CT 06520-8005*

10.1.   Introduction

Like most other DNA tumor viruses, papillomaviruses encode multiple oncogenes. In epithelial cells and transgenic mice, the E6 and E7 genes of the high-risk HPV display a number of growth promoting activities, including the ability to cooperate to immortalize primary cultured human keratinocytes. The E6 and E7 proteins affect cell growth by binding to and modulating the activity of cellular proteins involved in growth control, including p53, retinoblastoma family members, and PDZ domain-containing proteins. These same cellular proteins are also modulated by DNA tumor viruses of entirely different virus families, suggesting that there are relatively few intracellular growth regulatory pathways that viruses target to facilitate virus replication. In general the E6 and E7 proteins from low-risk HPV display reduced ability to perturb these cellular targets. Thus, these *in vitro* measures of activity reflect the underlying biology of cervical carcinogenesis. Strikingly, repression of HPV E6 and E7 expression in cervical cancer cells inhibits the growth of these cells, validating the viral oncogenes as potential therapeutic targets in HPV-associated precancerous and cancerous lesions.

10.2.   Papillomavirus E6 Proteins

The papillomavirus E6 gene is encoded at the 5′ end of the viral early region and is well conserved among all of the papillomaviruses. The E6 protein has multiple functions important for interactions of the virus with the host cell. These functions have been best studied for the cervical cancer–associated high-risk HPV types and for BPV1. The role of E6 in cellular transformation is better understood at this time than its role in the papillomavirus life cycle, where it appears to have a role in viral DNA replication and countering cellular antiviral defenses.

The E6 protein is required for the productive life cycle of the papillomaviruses and has been implicated in supporting the stable maintenance of viral DNA in cultured keratinocytes (Park and Androphy, 2002; Thomas et al., 1999; Wu et al., 1994). In transgenic mice, high-risk HPV E6 genes can induce skin hyperplasia and carcinoma, primarily by affecting the progression stage, which involves the malignant conversion of benign tumors (Song et al., 1999, 2000). In contrast, the E6 gene on its own does not induce cervical cancer in transgenic mice, but it is able to cooperate with the E7 gene to induce large, invasive cervical cancers (Riley et al., 2003).

The E6 proteins also have transforming activities in cultured cells, which were first described for the bovine papillomavirus type 1. In mouse fibroblasts the full transformed phenotype requires the expression of the E6 and E7 genes as well as E5 (Neary and DiMaio, 1989; Sarver et al., 1984). In addition, BPV1 E6 expressed from a strong heterologous promoter is sufficient for transformation of mouse C127 cells (Schiller et al., 1984). In contrast, the E6 proteins encoded by the high-risk HPV types are weakly oncogenic by themselves in rodent cell transformation assays. The major transforming activity for the high-risk HPV E6 protein in cultured cells is its ability to enhance immortalization of human keratinocytes in collaboration with the E7 protein (Hawley-Nelson et al., 1989; Hudson et al., 1990; Munger et al., 1989a). In addition, high-risk E6 can efficiently immortalize certain populations of human mammary epithelial cells (Band et al., 1991). The ability of E6 to immortalize cells is mediated through a number of its activities, including its ability to block p53 activity and to activate cellular telomerase. These functions as well as other cellular targets and activities of the papillomavirus E6 proteins are described in detail below. It should be noted that no intrinsic enzymatic activities have been identified for the E6 proteins, and they are believed to function exclusively through protein:protein interactions with cellular targets.

10.2.1.   Biochemical Characterization of Papillomavirus E6 Proteins

The E6 and E7 proteins encoded by the papillomaviruses are structurally related and display some level of conservation among all of the papillomaviruses. These proteins have conserved structural motifs. Both E6 and E7 contain domains of almost identically spaced CYS-X-X-CYS motifs (four in E6 and two in the carboxy-terminal half of E7). This structural similarity suggests that the E6 and E7 genes may have arisen from duplication events involving a core sequence containing one of these motifs (Cole and Danos, 1987). The CYS-X-X-CYS motifs found in a number of nucleic acid binding proteins are characteristic of zinc binding proteins, and indeed, the E6 and E7 proteins bind zinc through these cysteine residues (Barbosa et al., 1989; Grossman and Laimins, 1989).

The E6 proteins have been difficult to study biochemically because they are poorly folded and largely insoluble when expressed at high levels. Recently, an NMR analysis of the C-terminal domain of HPV16 E6 has been published

(Nomine et al., 2005). Despite their small size (generally less than 170 amino acids), there have been no published structural studies of a full-length E6 protein.

Although most functional studies have been performed on the full-length E6 proteins or genes, shorter spliced variants of E6 and fusion proteins generated by fusion of E6 with other viral ORFs may also exist. These putative alternate E6 proteins are predicted by viral RNA transcript analyses, but it should be noted that no E6 proteins other than those encoded by the full-length ORFs have been detected. Several shorter forms of the HPV E6 proteins (referred to as the E6* proteins) are predicted by cDNA analyses. These spliced RNAs might serve as transcripts for the E7 gene rather than as transcripts for these shortened E6* protein variants. In addition to mRNAs that encode the full-length E6 protein, BPV1-transformed cells express mRNAs with the potential to encode fusion proteins consisting of the amino terminal half of E6 fused to portions of downstream proteins (Yang et al., 1985). No studies have been performed to analyze these potential BPV1 E6 fusion proteins.

10.2.2. Human Papillomavirus E6 Proteins

10.2.2.1. HPV E6 and p53

Most studies on the human papillomavirus E6 proteins have focused on the high-risk HPV types associated with human cervical cancer. The first activity identified for the high-risk HPV E6 proteins was the ability to form a complex with p53 (Werness et al., 1990), a property not shared with the low-risk HPV E6 proteins. Through this interaction, high-risk E6 can block the ability of p53 to transcriptionally activate p53-responsive promoters (Mietz et al., 1992). The steady state levels of p53 are generally quite low in HPV-positive carcinoma cell lines and in cells immortalized by the HPV oncoproteins (Scheffner et al., 1991). The observation that E6 proteins of high-risk HPV types 16 and 18 promote the ubiquitin-dependent degradation of the p53 protein in vitro (Scheffner et al., 1990) led to the hypothesis that E6-mediated ubiquitylation of p53 and its subsequent degradation accounted for the low steady-state levels of p53 in cells. Indeed, the half-life of p53 is dramatically decreased in E6-expressing cells, and E6 can prevent the increase in p53 levels when cells are challenged with genotoxic agents (Hubbert et al., 1992; Kessis et al., 1993; Scheffner et al., 1991). Expression of the high-risk E7 proteins and their engagement of pRB can result in an increase in the levels of p53 within cells, which in turn transcriptionally activates the expression of proapoptotic genes or genes such as p21 that can arrest the cell cycle (Fig. 10.1) (Jones et al., 1997b). Thus a primary function of E6 may be to counteract the E7-induced increase in p53. By targeting p53, the high-risk HPV E6 proteins inhibit DNA damage and oncogene-mediated cell death and growth-inhibitory signals in cultured cells and animals (Eichten et al., 2004; Jones and Munger, 1997; Kessis et al., 1993; Song et al., 1998); induction of genomic instability, as evidenced by the long-term development of chromosome translocations and aneuploidy (Reznikoff et al., 1994; White et al., 1994); and immortalization of human mammary epithelial cells (Band et al., 1991; Shamanin and Androphy, 2004). However, some experiments indicate that

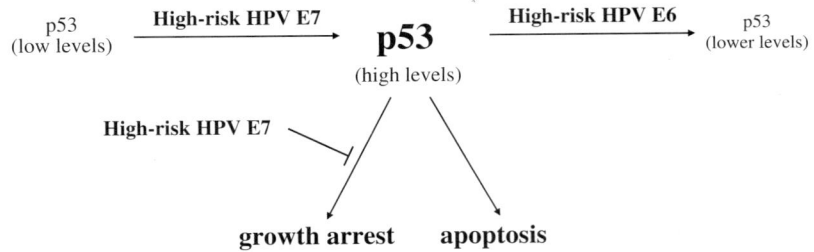

FIGURE 10.1. HPV E6 and E7 regulation of p53 levels. The level of p53 in keratinocytes is quite low. DNA-damaging agents, viral infection, and expression of the high-risk HPV E7 proteins increase the level of p53 through a combination of mechanisms, including increased protein stability and translation. Elevated levels of p53 in turn result in the transcriptional activation of genes that can lead to apoptosis or a cell cycle checkpoint arrest in the G1 phase of the cell cycle, respectively. The high-risk HPV E6 proteins deal with this negative growth regulatory function of p53 by targeting it for ubiquitylation and accelerated proteolysis.

high-risk E6 mitigates, but does not abolish, p53-dependent cell cycle effects (Butz et al., 1995).

If the ability of high-risk E6 to induce p53 degradation is important for carcinogenesis, then it is possible that p53 variants with altered binding to E6 may affect cancer risk. Indeed, Storey et al. (1998) reported that p53 variants containing arginine at position 72 (in comparison to the more common version containing proline) are more susceptible to E6-mediated degradation in vitro and in cultured cells. Furthermore, case control studies suggested that individuals homozygous for the arginine variant displayed a severalfold increased risk of developing cervical and skin carcinoma. However, this finding has been confirmed by some, but not all, subsequent studies (e.g. Cenci et al., 2003; Madeleine et al., 2000; Ojeda et al., 2003), which may reflect differences in sample populations or analysis (Makni et al., 2000). Clearly, the association between p53 status and the risk of HPV-associated cancer deserves further study.

10.2.2.2.   The E6-Associated Protein (E6AP)

E6 does not bind directly to p53. Rather, the binding of E6 to p53 is mediated by a 100-kDa cellular protein, E6AP (E6 associated protein) (Huibregtse et al., 1995). E6AP plays a direct role in p53 ubiquitylation by acting as an E3 ubiquitin protein ligase (Huibregtse et al., 1993a,b; Scheffner et al., 1993). Ubiquitin-dependent proteolysis is a multistep process that results in degradation of a targeted protein. The ubiquitylation of protein substrates is carried out by a series of cellular enzymes known as E1, E2, and E3. Initially, the 76-amino-acid ubiquitin moiety is activated by the E1 ubiquitin-activating enzyme (UBA) in an ATP-dependent manner. The activation of ubiquitin involves the formation of a thiolester bond between the carboxy-terminal glycine of ubiquitin and the active site cysteine residue of E1. This activated ubiquitin moiety is then transferred to one of a family of E2 proteins (ubiquitin-conjugating enzymes, or UBCs) characterized by a highly conserved catalytic site. E2 enzymes catalyze the formation of an isopeptide bond between the C-terminal glycine of ubiquitin and the ε-amino group on a lysine residue of the protein substrate, either directly or in conjunction with an E3 (ubiquitin protein ligase). The 26S proteasome complex recognizes the polyubiquitylated protein, which is then deubiquitylated and degraded.

Structure-function studies of E6AP revealed that it contains an 18-amino-acid region (amino acids 391–408) that is sufficient for binding E6. The E6-dependent binding of p53 involves amino acids 280–781 of E6AP, a domain which encompasses the E6 binding region. In addition to the sequences necessary for p53 binding, an intact C-terminus of E6AP is necessary for E6-mediated p53 ubiquitylation (Huibregtse et al., 1995b). The C-terminal 350 amino acids comprise the HECT (homologous to the E6AP C-terminus) domain, a region of homology shared by several proteins structurally and functionally related to E6AP (Huibregtse et al., 1995). These proteins form the HECT domain family of E3 ubiquitin protein ligases, which exist in yeast, *Drosophila*, and *C. elegans* as well as in vertebrates. HECT E3 ligases participate directly in the transfer of ubiquitin from E2 ubiquitin-conjugating enzymes to their substrates. The biochemical pathway by which E6 mediates E6AP-dependent ubiquitylation of p53 is depicted in Figure 10.2. The X-ray structure of the E6AP HECT domain complexed to UbcH7 has been determined (Huang et al., 1999). In addition, the NMR structure of a peptide containing the E6 binding region of E6AP has been determined to be α-helical and contains leucines conserved in other E6 binding proteins described below (Be et al., 2001). The structure of the ternary complex of E6:E6AP:p53 has not been determined.

It should be noted that E6AP does not regulate p53 ubiquitylation in the absence of E6. By binding E6AP, E6 directs E6AP to p53, allowing it to form a ternary complex active in p53 ubiquitylation. Studies using antisense oligonucleotides or using a catalytically inactive, dominant-negative mutant form of E6AP established that E6AP is not involved in targeting the ubiquitylation of p53 in cells that do not contain E6 (Beer-Romano et al., 1997; Talis et al., 1998). Instead, mdm2 is the major E3 ubiquitin ligase responsible for p53 degradation in the absence of E6 (Haupt et al., 1997; Kubbutat et al., 1997). Thus the high-risk E6 protein has adopted the unusual strategy of retargeting a component of the cellular protein degradation machinery to a novel substrate, p53, which regulates many important cellular processes.

E6AP has been implicated in the human genetic disorder Angelman syndrome (AS) (Kishino et al., 1997; Matsuura et al., 1997), a rare, maternally imprinted, inherited disorder characterized by severe mental retardation, seizures, frequent out-of-context laughter, and abnormal gait. This association strongly suggests a role for E6AP and the ubiquitin system in human brain development. It is postulated that AS may result from the absence of E6AP from neurons of the hippocampus and the resulting failure of normal E6AP substrates to be ubiquitylated and degraded. To date, a few normal, E6-independent substrates of E6AP have been identified, including the human homologue of the yeast Rad23,

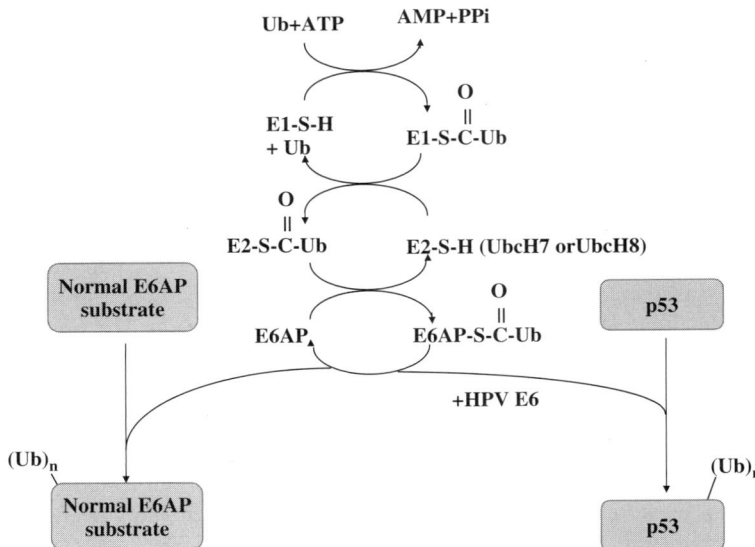

FIGURE 10.2. E6/E6AP/p53 ubiquitylation pathway: HPV16 E6–dependent ubiquitylation of p53. E6 binds the cellular protein E6AP and directs it to function as an E3 ubiquitin protein ligase in the ubiquitylation of p53 (Scheffner et al., 1993). The ubiquitylation of a protein involves three cellular activities: E1 (ubiquitin-activating enzyme), E2 (ubiquitin-conjugating enzyme), and E3 (ubiquitin protein ligase). Ubiquitin (Ub) is activated in an ATP-dependent manner and is bound to E1 through a high-energy thiolester bond. It is then transferred to the E2 ubiquitin-conjugating enzyme, again through a thiolester linkage. Ubiquitin can then be transferred to a cysteine within the HECT domain of E6AP, through a similar thiolester linkage (Scheffner et al., 1995), through the direct binding of E6AP with UbcH7 or UbcH8 (Kumar et al., 1997). In conjunction with HPV-16 E6, E6AP then recognizes p53 and catalyses the formation of an isopeptide bond between the carboxy-terminal glycine of ubiquitin and a lysine side chain of p53.

HHR23A (Kumar et al., 1999), the Src-family kinase Blk (Oda et al., 1999), and the MCM7 subunit of replication licensing factor (Kuhne and Banks, 1998). However, the relevance of these proteins to HPV transformation or to AS, if any, has not been determined. It is possible that HPV E6, through promoting the self-ubiquitylation of E6AP (Kao et al., 2000) and redirecting E6AP to p53, could stabilize the normal cellular substrates of E6AP. Alternatively, E6 binding to E6AP might activate its overall E3 ligase activity and enhance the ubiquitin-mediated proteolysis of the normal substrates of E6AP.

10.2.2.3.   PDZ Domain-Containing Proteins

In addition to targeting p53 for degradation, E6 can mediate the E6AP-dependent degradation of a number of cellular PDZ domain proteins. PDZ domains are approximately 90-amino acid sequence homology regions first noted within a number of signaling molecules. The name PDZ is derived from the first three proteins in which these domains were identified: PSD-95 (a 95-kDa protein involved in signaling in the postsynaptic density), Dlg (the *Drosophila* disc large protein), and ZO1 (the zonula occludens 1 protein with functions in epithelial cell polarity). PDZ domain proteins are scaffolds and cellular organizing centers that can affect a number of cellular processes, including signal transduction, transcriptional regulation, receptor assembly, and cell polarity. Specific PDZ domain proteins have been shown to be important in keratinocytes.

A number of different PDZ domain proteins have been shown to be able to associate with E6, which serves as a molecular bridge between these PDZ domain proteins and E6AP, facilitating their ubiquitylation and proteolysis. The binding of E6 to PDZ domains is mediated by the C-terminal five amino acids of the high-risk HPV E6 proteins. This sequence, (S/T)-X-V-I-L, is present at the C-terminus of all high-risk E6 proteins but not at the C-terminus of the low-risk E6 proteins. Among the PDZ domain proteins implicated as E6 targets are hDlg, the human homologue of the *Drosophila* melanogaster discs large tumor suppressor, and hScrib, the human homologue of the *Drosophila* Scribble tumor suppressor (Gardiol et al., 1999; Nakagawa and Huibregtse, 2000). Additional PDZ domain proteins capable of binding to E6 are MAGI-1, MAGI-2, MAGI-3, MUPP1, and TIP-2/GIPC (Favre-Bonvin et al., 2005; Glaunsinger et al., 2000; Lee et al., 2000; Thomas et al., 2002).

Several groups have analyzed the consequences of mutating the PDZ binding domain of the high-risk E6 proteins. E6 mutants deleted of the C-terminal five amino acids are unable to transform rodent cells in vitro (Kiyono et al., 1997). In organotypic raft cultures of keratinocytes this domain of HPV31 E6 is required for normal levels of viral DNA replication and normal HPV-mediated effects on tissue morphology (Lee and Laimins, 2004). Furthermore, the ability of the HPV16 E6 protein to induce epithelial hyperplasia in transgenic mice is abrogated by mutations that prevent PDZ protein binding, even though these mutations do not appear to affect p53 inactivation (Nguyen et al., 2003). These results imply that interactions of PDZ proteins with the E6 protein are important for several E6 activities, but the relative importance of specific members of the large PDZ domain protein family to the biological activities of the E6 proteins remains to be defined.

10.2.2.4. HPV E6 Activation of Cellular Telomerase

E6 displays a number of p53-independent functions that are relevant to cellular transformation and immortalization. Indeed, there are HPV16 E6 mutations that separate p53 degradation from cellular immortalization (Kiyono et al., 1998; Liu et al., 1999), although another study indicated that p53 degradation was required for immortalization (McMurray and McCance, 2004). One such function in keratinocytes is the ability of E6 to activate cellular telomerase through the transcriptional upregulation of the rate-limiting catalytic subunit of human telomerase (hTERT) (Kiyono et al., 1998; Klingelhutz et al., 1996; Oh et al.,

2001; Veldman et al., 2001). Maintenance of telomere length is an important step in cellular immortalization and transformation and occurs either through transcriptional activation of hTERT expression or through the activation of the ALT recombination pathway. Activation of hTERT is observed in most human cancers, including HPV-positive cervical cancers. In addition, the expression of hTERT is sufficient for the immortalization of many different cell types (Hahn et al., 1999). The mechanism by which E6 activates the hTERT promoter has not been yet fully elucidated, but two E6-related mechanisms of hTERT promoter activation have been proposed. Interestingly, both involve the interaction of E6AP. One involves the binding of E6 to Myc, with the activation of the hTERT promoter being dependent on E6AP activity and the Myc binding sites within the promoter (Liu et al., 2005; McMurray and McCance, 2003). The second involves the E6:E6AP-dependent degradation of a putative transcriptional repressor of the hTERT promoter, NFX1-91 (Gewin and Galloway, 2001; Gewin et al., 2004). Although it is somewhat difficult to reconcile the data of these two groups, the proposed mechanisms involving E6AP-mediated activation are not necessarily mutually exclusive.

10.2.2.5.   Other Cellular Targets of High-Risk HPV E6 Proteins

Because the high-risk HPV E6 proteins have cellular transformation and immortalization activities that are p53-independent, a variety of approaches have been used to identify additional E6 interacting proteins. Figure 10.3 shows a number of the proteins that have been identified as E6 interaction partners. One important partner is Interferon Regulatory Factor-3 (IRF-3), which was identified as an HPV16 E6 binding protein in a yeast two-hybrid screen (Ronco et al., 1998). IRF-3 plays a key role in innate immunity and in mediating apoptosis following viral infection. IRF-3 is activated by viral infection to form a stable complex with other transcriptional regulators that bind to the regulatory element of the interferon-ß promoter (Wathelet et al., 1998), and it is a potent transcriptional activator. HPV16 E6 inhibits the transactivation function of IRF-3, but it should be noted that the binding of HPV16 E6 does not result in IRF-3 ubiquitylation or degradation (Ronco et al., 1998). The E6 binding region of IRF3 is quite similar to the E6 binding region of E6AP, so it is possible that the binding of IRF3 and E6AP are mutually exclusive. High-risk E6 proteins also affect other aspects of interferon signaling. HPV16 E6 can inhibit the ability of IL-18 to induce interferon-γ signaling in human immune cells, and HPV18 E6 associates with Tyk2 and impairs Jak-STAT activation during interferon α-signaling (Lee et al., 2001; Li et al., 1999). Finally, HPV16 E6-mediated downregulation of E-cadherin depletes Langerhans antigen-presenting cells in infected skin (Matthews et al., 2003). Thus through multiple mechanisms, expression of high-risk E6 in primary keratinocytes may play a critical role in allowing the virus to evade the host innate immune and apoptotic responses during the normal virus life cycle.

The high-risk E6 proteins can also induce cellular mutagenesis (Havre et al., 1995). In part, this may be due to interference with p53 function and abrogation

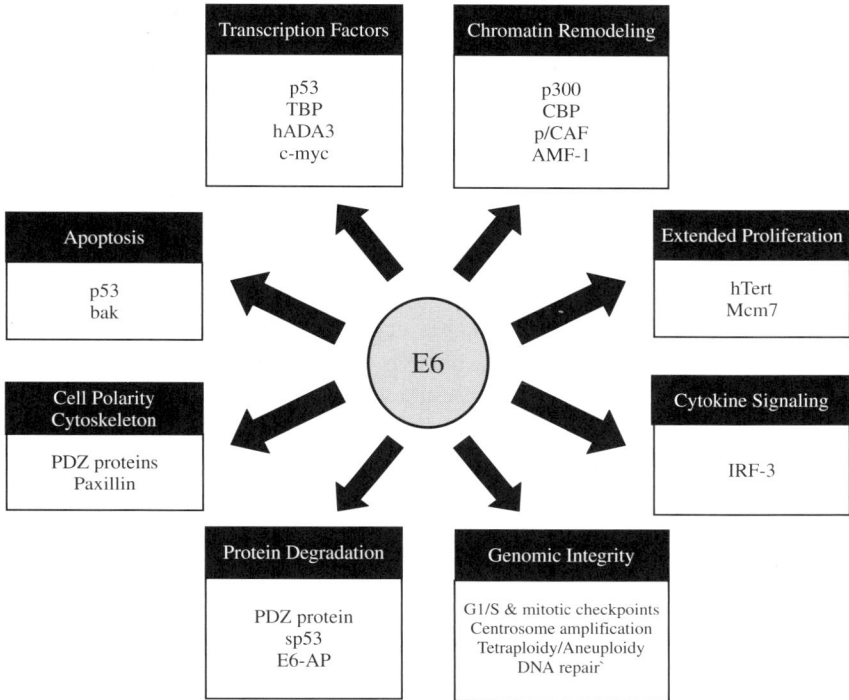

FIGURE 10.3. Papillomavirus E6 cellular targets and activities. Schematic of the various cellular processes and specific cellular proteins that high-risk HPV and BPV E6 proteins have been reported to target. It should be noted that some of the E6 cellular targets (e.g., some of the PDZ domain proteins) may be targeted for ubiquitylation through the additional interaction of E6AP and E6. See text for details and references.

of DNA damage (Kessis et al., 1993) and mitotic (Thomas and Banks, 1998; Thompson and Belinsky, 1997) checkpoint control. However, E6 has also been shown to interact with cellular proteins directly involved in DNA repair. E6 binds and destabilizes O(6)-methylguanine DNA methyltransferase, which plays a role in base excision repair, and binds and inhibits the activity of the XRCC1 protein, a scaffolding protein involved in single-strand break repair (Iftner et al., 2002; Srivenugopal and Ali-Osman, 2002). Mutagenesis induced by the E6 protein presumably plays an important role in the accumulation of mutations required for carcinogenic progression.

Other cellular proteins that form a complex with the high-risk E6 proteins include E6TP1, which exhibits homology to GTPase activating proteins (GAPs) (Gao et al., 1999, 2002; Singh et al., 2003), the proapoptotic protein Bak (Thomas et al., 1999), and the transcriptional coactivators p300 and ADA3 (Kumar et al., 2002; Patel et al., 1999; Zimmermann et al., 1999). In several cases these interactions result in the inactivation or degradation of the cellular protein. Finally, the cellular protein, Gps2, binds high- and low-risk HPV E6 proteins (Degenhardt and Silverstein, 2001a). The relevance of many of these E6 binding proteins to the oncogenic potential of HPV and to virus life cycle remain to be determined, but mutational analysis suggests that at least some of these partners are required for E6-mediated oncogenesis in transgenic mice (Nguyen et al., 2002).

10.2.2.6. Low-Risk HPV E6 Proteins

Several of the cellular proteins found to bind the high-risk HPV E6 proteins also bind the low-risk HPV E6 proteins. These include Bak, Gps2, and E6AP. It should be noted, however, that although the low-risk HPV E6 proteins bind E6AP, they fail to engage p53 as part of a ternary complex and do not cause the ubiquitylation or degradation of p53. In addition, zylin, a focal adhesion protein, has been identified in a yeast two-hybrid screen as an HPV6 E6 binding protein (Degenhardt and Silverstein, 2001b).

10.2.3. Bovine Papillomavirus E6 Protein

Although the E5 gene is the major transforming gene of BPV1, E6 has independent transforming activity (Schiller et al., 1984; Yang et al., 1985). As is the case for the high-risk HPV E6 proteins, no intrinsic enzymatic activities have been identified for BPV1 E6, and it too is believed to function through binding cellular targets. Several different cellular proteins have been found to interact with BPV1 E6: E6AP, E6BP (also known as ERC55), the γ subunit of the AP1 clatherin adaptor complex, and the focal adhesion protein paxillin (Chen et al., 1995; Tong et al., 1998, 1997; Vande Pol et al., 1998). Unlike the high-risk HPV E6 proteins, the BPV1 E6 interaction with E6AP does not result in the binding and degradation of p53. Indeed, the physiologic consequences resulting from the interaction of BPV1 E6 with E6AP have not been determined. It is possible that BPV1 E6 may direct E6AP to other targets or affect the ubiquitylation of normal E6AP substrates. Similarly, no in vivo consequences of the interaction of BPV1 E6 with either ERC55 or the γ subunit of the AP1 have been determined. Analysis of a series of BPV1 E6 mutants showed a good correlation between the binding to paxillin and fibroblast transformation (Das et al., 2000; Tong and Howley, 1997; Vande Pol et al., 1998). E6/paxillin binding correlates with the disruption of the cellular actin cytoskeleton, a characteristic of transformed cells (Tong and Howley, 1997). E6 binds to α-helical charged leucine motifs in paxillin known as LD motifs and in doing so competes with the ability of paxillin to bind to vinculin and the focal adhesion kinase (Tong et al., 1997; Vande Pol et al., 1998). The LD motifs have sequence similarity to the E6 binding domains in E6AP, IRF-3, and E6PB (Bohl et al., 2000). At this point, however, a detailed understanding of the mechanisms by which BPV1 E6 transforms cells remains to be achieved.

10.3.   Papillomavirus E7 Proteins

10.3.1.   Biochemical Characterization of HPV E7 Proteins

The E7 proteins are encoded by an open reading frame immediately downstream of the E6 gene and are small, rather acidic polypeptides of approximately 100 amino acid residues (Figure 10.4). The amino terminal third of the HPV E7 protein sequence is homologous to a portion of conserved region (CR) 1 and the entire CR2 of adenovirus (Ad) E1A proteins and related sequences in the large tumor antigens (T Ag) of polyomaviruses, including SV40 (Figge et al., 1988; Phelps et al., 1998; Vousden et al., 1988). The CR1 homology domain of low-risk and high-risk HPV E7 proteins has been identified as a binding site for p600 (Huh et al., 2005), a 5,183-amino-acid protein that has been implicated as a regulator of apoptosis (Nakatani et al., 2005) and a ubiquitin ligase of N-end rule pathway (Tasaki et al., 2005). The BPV1 E7 protein also associates with p600 through its CR1 homology domain (DeMasi et al., 2005). The CR2 homology domain includes a Leu-X-Cys-X-Glu (LXCXE) motif, which constitutes the core binding site for the pRB, the retinoblastoma tumor suppressor (Munger et al., 1989b) and the related "pocket proteins" p107 and p130 (Dyson et al., 1992). Immediately adjacent to the LXCXE motif is a consensus phosphorylation site for casein kinase II (CK II) (Barbosa et al., 1989; Firzlaff et al., 1991). The E7 carboxyl terminus contains a zinc-binding domain consisting of two Cys-X-X-Cys sequence motifs (Barbosa et al., 1989; McIntyre et al., 1993). This domain is often referred to as a "zinc finger," but the spacing between the Cys-X-X-Cys motifs is larger (29 amino acid residues) than in standard zinc finger proteins and forms a novel zinc-binding structure (Liu et al., 2006). The E7 carboxyl terminus is vital for the structural integrity of the protein and can function as a dimerization/multimerization domain (Clemens et al., 1995; Clements et al., 2000; Liu et al., 2006; McIntyre et al., 1993). Biophysical investigations have detected multimeric E7 complexes (Alonso et al., 2004). In

FIGURE 10.4. Schematic of the HPV-16 E7 oncoprotein. The amino terminal portion of E7 contains homology to a portion of conserved region 1 (CR1 (white)) and CR2 of the adenovirus E1A oncoprotein. The HPV-16 E7 CR2 homology sequence (gray) encompasses the core binding site (LXCXE) for the retinoblastoma tumor suppressor pRB and the related proteins p107 and p130 as well as a casein kinase II consensus phosphorylation site (CK II). The carboxyl terminal E7 sequence (black) contains a cysteine-rich zinc-binding motif. The HPV E6 proteins contain two copies of a related zinc-binding module.

addition, the HPV16 E7 carboxyl terminus contains a site that is phosphorylated by an as yet unidentified protein kinase (Massimi and Banks, 2000; Clements et al., 2000).

HPV16 E7 migrates aberrantly on SDS PAGE with an apparent molecular size of 18–20 kDa rather than the predicted molecular size of ∼11kDa (Smotkin and Wettstein, 1986). This abnormal electrophoretic mobility is mainly mediated by the amino terminal CR1 homology domain (Heck et al., 1992; Munger et al., 1991) and may be due to its high content of acidic residues (Armstrong and Roman, 1993). Even though HPV16 E7 lacks a recognizable nuclear targeting sequence, it is actively imported into the nucleus. Nuclear transport of E7 is Ran-dependent but does not involve the classical cytosolic Kap ß import receptors (Angeline et al., 2003). In vivo, there is evidence for nuclear as well as cytoplasmic pools of E7 (Greenfield et al., 1991; Sato et al., 1989; Smith-McCune et al., 1999; Smotkin and Wettstein, 1987), and nucleolar E7 localization has also been reported (Zatsepina et al., 1997). Like the E6 protein, E7 lacks intrinsic enzymatic activities and does not appear to directly interact with specific DNA sequences. It is now well accepted that the biological activities of E7 are due to its ability to associate with and subvert the normal activities of cellular regulatory complexes (Figure 10.5).

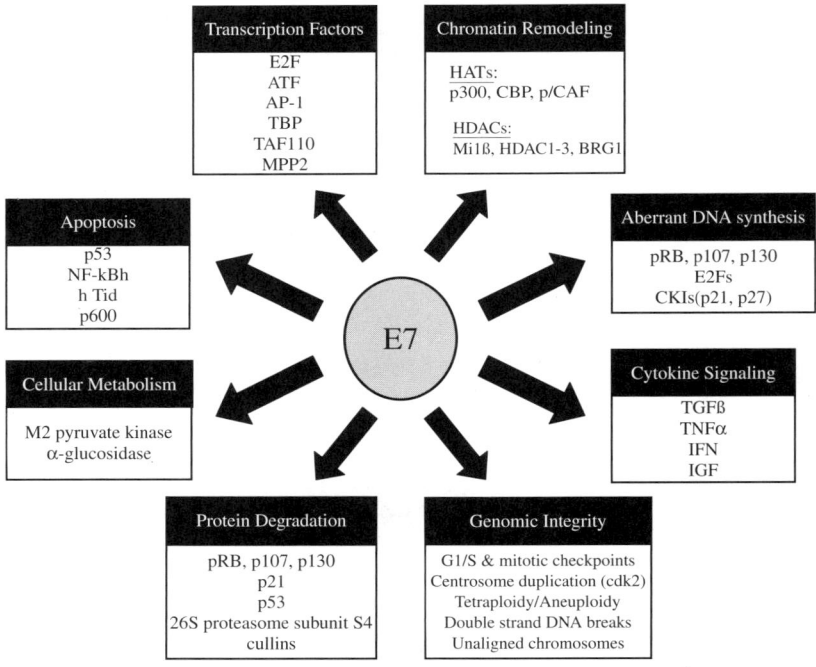

FIGURE 10.5. Papillomavirus E7 cellular targerts and activities. Schematic of the various cellular processes and specific cellular proteins that high-risk HPV E7 proteins have been reported to target. See text for details and references.

10.3.2. Biological Activities of the E7 Protein

The HPV16 E7 protein has been studied most extensively as it was the first high-risk HPV oncogene product to be discovered (Kanda et al., 1988; Phelps et al., 1988; Vousden et al., 1988; Yutsudo et al., 1988). HPV16 E7 is sufficient for morphological transformation of rodent fibroblast cell lines, such as mouse NIH3T3 or rat 3Y1 cells (Bedell et al., 1989; Phelps et al., 1988; Vousden et al., 1988; Yasumoto et al., 1986), and can cooperate with the *ras* oncogene to transform primary baby rat kidney cells (Matlashewski et al., 1987; Phelps et al., 1988). HPV16 E7 expression interferes with senescence signaling (Bischof et al., 2005). It also extends the life span of primary human genital epithelial cells (Foster and Galloway, 1996; Halbert et al., 1991) and facilitates their immortalization in combination with E6 (Hawley-Nelson et al., 1989; Hudson et al., 1990; Munger et al., 1989a). In some populations of human mammary epithelial cells, E7 expression is sufficient to induce immortalization (Wazer et al., 1995). The low-risk HPV E7 proteins have much lower transforming and immortalizing activities than the high-risk HPV E7 proteins (Halbert et al., 1992).

Even though HPV E7 proteins do not directly bind to specific DNA sequences, they can modulate gene transcription through cellular E2F and ATF binding sites (Phelps et al., 1988, 1991). HPV E7 proteins also associate with other transcription factors, including the forkhead family member MPP2 (Luscher-Firzlaff et al., 1999) and AP1 (Antinore et al., 1996) as well as components of the basic transcription factor machinery, including TBP (Massimi et al., 1996; Phillips and Vousden, 1997) and the 110-kDa TBP-associated factor (TAF110) (Mazzarelli et al., 1995).

Expression of high-risk HPV E7 proteins causes genomic instability in normal human cells (White et al., 1994). HPV16 E7 induces G1/S and mitotic cell-cycle checkpoint defects (Martin et al., 1998; Peacock et al., 1995; Thomas and Banks, 1998) and uncouples synthesis of centrosomes from the cell division cycle. This latter activity can cause chromosome missegregation and aneuploidy (Duensing et al., 2000) and is described in detail in a later section. Malignant progression of high-risk HPV-associated lesions is frequently associated with integration of the viral genome into a host cellular chromosome and termination of the viral life cycle (Schneider-Maunoury et al., 1987). The frequent loss of the viral transcriptional repressor E2 on integration (Schneider-Gädicke and Schwarz, 1986) results in increased expression of E6/E7 mRNAs and the generation of E6/E7 mRNAs fused to genomic sequences, events which stabilize the viral mRNAs (Jeon et al., 1995). This contributes to increased E6/E7 protein levels and enhanced cell growth after HPV genome integration.

The oncogenic activities of high-risk HPV E7 proteins are manifestations of their functions during the viral life cycle. Expression of E7 is necessary for the virus life cycle and the stable maintenance of HPV episomes in epithelial cells (Flores et al., 2000; Thomas et al., 1999), and E7 sequences that contribute to cellular transformation are important during the life cycle of high-risk and low-risk HPVs (Longworth and Laimins, 2004a,b;

Oh et al., 2004b; Thomas et al., 1999). The replicative life cycle of HPVs is tightly coupled to the differentiation state of the host keratinocyte, with high-level viral genome replication occurring in differentiated epithelial cells. HPV genome synthesis requires the host cellular DNA replication machinery, but these differentiated cells do not intrinsically support DNA replication (reviewed by Stubenrauch and Laimins (1999)). HPV E7 retains differentiating keratinocytes in a DNA replication competent state (Cheng et al., 1995; Woodworth et al., 1992). Therefore the ability of E7 proteins to induce unscheduled DNA replication through the mechanisms described in later sections is not only relevant for cellular transformation, but it is a vital component of the HPV replication strategy.

In transgenic mice the E7 protein can stimulate apoptosis (Pan and Griep, 1994). E7 can also induce the formation of epidermal hyperplasia and locally invasive skin carcinomas, activities that require conserved regions 1 and 2 (Gulliver et al., 1997; Herber et al., 1996). E7 acts primarily at the promotion stage in the two-step model for skin carcinogenesis (Song et al., 2000). In trans-genic mice chronically treated with estrogen the HPV16 E7 gene induced the formation of microinvasive cervical tumors and cooperated with the E6 gene to induce the formation of large, invasive cervical carcinomas (Riley et al., 2003).

The transforming activities of E7 proteins are not limited to the human papillomaviruses. The bovine papillomavirus type 1 (BPV1) causes benign fibropapillomas in cattle, and BPV-1 E7 is weakly transforming and enhances anchorage-independent growth of murine C127 cells induced by BPV-1 E5 or E6 (Bohl et al., 2001; Neary and DiMaio, 1989). The transforming activities of BPV1 E7 in this assay have been linked to p600 binding (DeMasi et al., 2005). In addition, the BPV4 E7 protein can cooperate with the *ras* oncogene in transformation of primary bovine palate fibroblasts (Pennie et al., 1993), and the E7 protein encoded by cottontail rabbit papillomavirus (CRPV) can transform murine NIH3T3 and cottontail rabbit skin epithelial cells (Meyers et al., 1992).

10.3.3. Association of E7 with the Retinoblastoma Tumor Suppressor pRB and the Related p107 and p130 Pocket Proteins

The HPV E7 oncoprotein associates with pRB, p107, and p130 through the LXCXE motif that is shared with Ad E1A and SV40 T Ag (Figge et al., 1988; Phelps et al., 1988; Vousden and Jat, 1989). Similar to SV40 T Ag, HPV-16 E7 preferentially associates with the growth suppressive, hypophosphorylated form of pRB (Dyson et al., 1992). pRB is an important signal integrator that regulates cell-cycle progression, differentiation, senescence, and cell death (Thomas et al., 2003). The p107 and p130 proteins have overlapping activities, particularly in cell-cycle regulation, and studies with cells derived from knockout mice have documented functional overlap and compensation in response to loss of specific pocket protein family members (Mulligan et al., 1998). However, the ability of pocket proteins to regulate differentiation is specific to distinct family members. The best-studied pRB/p107/p130 function is its ability to regulate cell-cycle progression by modulating the activities of E2F transcription factors (reviewed by Frolov and Dyson (2004)).

Members of the E2F transcription factor family bind directly to sites in cellular promoters and control expression of several rate-limiting factors for cellular DNA synthesis, including DNA polymerase α; enzymes involved in nucleotide biosynthesis; and cyclin E, an activator of cyclin-dependent kinase (CDK) 2 that controls S-phase progression (reviewed by Dyson (1998)). E2F transcription factors also modulate expression of genes that control mitosis, DNA damage responses, and many other key cellular processes (Hernando et al., 2004; Ren et al., 2002; Weinmann et al., 2002). Eight distinct E2F family members have been identified thus far. With the exception of E2F-7 they form heterodimeric complexes with one of two "dimerization partner" (DP) subunits (Bandara et al., 1993; Helin et al., 1993). The discovery of E2F-1 as a pRB-associated protein provided the mechanistic link between the ability of viral oncoproteins to associate with pRB family members and induction of aberrant S-phase entry (Helin et al., 1992; Kaelin et al., 1992). E2Fs 1–5 associate with and are regulated by pRB family members, whereas E2Fs 6–8 lack conserved pRB interaction domains. E2Fs 1–5 are often grouped into "activator" and "repressor" E2Fs based on their differential ability to activate transcription of reporters and/or to induce quiescent cells to enter the cell-cycle (reviewed by Trimarchi and Lees (2002)). Activator E2Fs include E2F-1, E2F-2, and E2F-3 and preferentially associate with pRB, whereas the repressor E2F-4 and E2F-5 proteins primarily interact with p107 and p130.

Association with pRB family members regulates the transcriptional activities of E2F family members through multiple mechanisms, but typically, this association inhibits E2F-dependent transcription. The pRB binding site is within the transcriptional activation domain of E2F, and pRB binding can block E2F transactivation (Flemington et al., 1993). E2F-bound pRB also interferes with assembly of transcription preinitiation complexes (Ross et al., 1999). The pRB family members associate with histone-modifying enzymes (see below) through a binding site that is distinct from the E2F binding site, and hence the pocket proteins may function as adapters for recruitment of histone-modifying enzymes to E2F-regulated promoters (reviewed by Frolov and Dyson (2004)). Cell-cycle-dependent phosphorylation of pRB in normal cells displaces E2F. Similarly, interaction of E7 with pRB can dissociate pRB/E2F complexes. E2F and E7 associate with distinct domains of pRB (Dick and Dyson, 2002; Lee et al., 1998, 2002a; Wu et al., 1993; Xiao et al., 2003), and disruption of E2F/pRB complexes requires carboxyl terminal E7 sequences in addition to the LXCXE motif (Helt and Galloway, 2001; Huang et al., 1993; Wu et al., 1993), which may be an independent, low-affinity pRB binding site (Patrick et al., 1994).

It was initially believed that high-risk HPV E7 expression caused cellular transformation through stoichiometric interaction with pRB/p107/p130, thereby displacing and aberrantly "activating" E2F. In support of this model, pRB binding-deficient HPV16 E7 mutants are transformation defective (reviewed by Phelps et al. (1992)). In addition, low-risk HPV E7 proteins bind to pRB less efficiently than high-risk HPV E7 proteins (Gage et al., 1990; Munger et al., 1989b; Schmitt et al., 1994), and the transforming activities of low-risk E7 proteins are enhanced by mutations that cause increased pRB binding (Heck et al., 1992; Sang and Barbosa, 1992). Several lines of evidence suggest, however, that this model is overly simplistic. The low-risk HPV6 E7 protein, which binds pRB poorly, can activate E2F-dependent promoters as efficiently as the high-risk HPV16 E7 protein in some cell types (Armstrong and Roman, 1997; Munger et al., 1991). Similarly, studies with E7 mutants where the carboxyl terminus was replaced by structurally related HPV E6-derived sequences demonstrated that the ability of E7 to disrupt E2F/pRB complexes is not required for transformation (Braspenning et al., 1998; Mavromatis et al., 1997). Moreover, the ability of E7 to efficiently associate with pRB and activate E2F-responsive promoters is not correlated to cellular transformation. The cutaneous low-risk HPV1 E7 protein is transformation negative, even though it efficiently binds pRB and potently activates E2F-responsive promoters (Ciccolini et al., 1994; Schmitt et al., 1994). In addition, several nontransforming HPV16 E7 mutants efficiently interact with pRB and activate transcription of E2F-responsive promoters (Banks et al., 1990; Brokaw et al., 1994; Edmonds and Vousden, 1989; Jewers et al., 1992; Phelps et al., 1992). Mutations within the amino terminal CR1 homology domain or the CKII site of E7 can interfere with cellular transformation but do not affect pRB binding (Barbosa et al., 1990; Firzlaff et al., 1991). Integrity of the CKII site is critical for the ability of HPV-18 E7 to induce cell-cycle entry in keratinocytes (Chien et al., 2000) and may modulate association of E7 to the TATA box binding protein TBP (Massimi et al., 1996; Phillips and Vousden, 1997) and F-actin (Rey et al., 2000). Carboxyl terminal E7 mutants also exhibit defects in cellular immortalization and transformation that are independent of pRB binding (Helt and Galloway, 2001; Jewers et al., 1992).

10.3.4. Association of E7 with Histone-Modifying Enzymes

The exposed amino terminal tails of histones can undergo a number of reversible posttranslational modifications, including phosphorylation, methylation, acetylation, and ubiquitination. The nature and extent of these modifications determines nucleosome packaging and thus accessibility of chromosomal loci. The most thoroughly studied mechanism is histone acetylation. Histone hyperacetylation occurs within the regulatory regions of genes that are actively transcribed, whereas histones at promoter regions of nontranscribed genes are generally hypoacetylated. The level of histone acetylation, and hence the transcriptional status of a given gene, is determined by the dynamic interplay of two opposing families of enzymes: histone acetyl transferases (HATs) and histone deacetylases (HDACs) (reviewed by Jenuwein and Allis (2001)). E7 proteins associate directly and indirectly with both classes of these enzymes. The transcriptional repression activity of pRB and the pocket proteins, p107 and p130, is related to their ability to associate with HDACs (Brehm et al., 1998; De Luca et al., 1998;

Ferreira et al., 1998; Luo et al., 1998; Magnaghi-Jaulin et al., 1998) through sequences other than the E7 binding domain (Dick et al., 2000). It is not clear whether E7 association displaces pRB-bound HDACs or if E7/pRB/E2F/HDAC complexes are formed. In either event, E7 binding may interfere with the biological activities of pRB/E2F/HDAC repressor complexes. In addition, the carboxyl terminal sequences of E7 can associate with HDACs (Brehm et al., 1999), and an HPV16 E7 mutant where leucine 67 is changed to arginine cannot (Longworth and Laimins, 2004b). The ability of E7 to associate with HDACs appears to be important for its biological activity because the L67R E7 mutant is defective for activating an E2F responsive reporter, exhibits a defect in trans-formation (Phelps et al., 1992), and does not extend the life span of primary human foreskin keratinocytes nor support the maintenance HPV episomes in keratinocytes (Longworth and Laimins, 2004b).

HPV16 E7 also associates with the HATs, p300, CBP, and the p300/CBP-associated factor (p/CAF) (Avvakumov et al., 2003; Bernat et al., 2003; Huang and McCance, 2002), which contribute to E2F-mediated transcriptional activation (reviewed by Frolov and Dyson (2004)), and acetylated histone H3 levels were increased at E2F-responsive promoters in HPV16 E7 expressing human foreskin keratinocytes (Zhang et al., 2004). HPV16 E7 can associate with E2F-1 in a pRB-independent manner through carboxyl terminal sequences (Hwang et al., 2002). E7 expression increased transcriptional activity of a pRB binding-defective E2F-1 mutant, suggesting that association with E7 may alter the transcriptional activity of E2F through pRB/p107/p130-independent mechanisms (Hwang et al., 2002).

10.3.5. Destabilization of pRB, p107, and p130

Binding of the high-risk HPV E7 proteins results in destabilization of pRB, p107, and p130 (Berezutskaya et al., 1997; Boyer et al., 1996; Giarre et al., 2001; Gonzalez et al., 2001; Helt and Galloway, 2001; Jones et al., 1997b; Smith-McCune et al., 1999) via a proteasome-dependent pathway (Boyer et al., 1996; Gonzalez et al., 2001). Interestingly, the pp71 protein of the human cytomegalovirus (CMV) and the Epstein-Barr virus nuclear antigen 3c (EBV EBNA 3C) also induce degradation of pRB family members (Kalejta and Shenk, 2003; Knight et al., 2005). Association of E7 with pRB is necessary but not suffi-cient for pRB degradation since transformation-defective amino terminal CR1 homology domain E7 mutants efficiently associate with pRB but are defective for pRB degradation (Edmonds and Vousden, 1989; Gonzalez et al., 2001; Helt and Galloway, 2001; Jones et al., 1997b; Phelps et al., 1992) and maintenance of HPV episomes in keratinocytes (Thomas et al., 1999). Similarly, the low-risk HPV1 E7 protein efficiently binds pRB but is unable to destabilize pRB and is transformation-deficient (Giarre et al., 2001; Gonzalez et al., 2001). The ability to destabilize p130 is shared by mucosal high-risk and low-risk HPV E7 proteins, suggesting that this activity contributes to the role of E7 in the viral life cycle (Zhang et al., 2006).

HPV-16 E7 has a short half-life of less than 2 hours (Smotkin and Wettstein, 1987) and is itself degraded through a proteasome-mediated mechanism (Reinstein et al., 2000), potentially involving a cullin 1-containing ubiquitin ligase complex that associates with E7 (Oh et al., 2004a). There is no evidence, however, that E7-induced pRB degradation is linked to ubiquitin-mediated degradation of E7 as a degradation-resistant version of E7 retains the capacity to target pRB for degradation (Gonzalez et al., 2001).

The detailed mechanism that E7 employs to stimulate pRB degradation remains elusive. HPV16 E7 binds the S4 subunit of the 26S proteasome (Berezutskaya and Bagchi, 1998), but S4 binding does not appear necessary for E7-mediated pRB degradation (Gonzalez et al., 2001). Some results suggest that E7 may accelerate pRB degradation through association with a cellular factor. Indeed, ectopic expression of a pRB-binding deficient E7 mutant interferes with pRB destabilization mediated by wild-type E7 (Gonzalez et al., 2001).

The ability of high-risk HPV E7 to destabilize pRB is not shared by the Ad E1A and SV40 T Ag oncoproteins, illustrating that these viruses have developed divergent strategies to inactivate this vital cellular regulatory protein. In the case of SV40 T-antigen an amino terminal J-domain (Kelley and Georgopoulos, 1997) is necessary to inactivate pRB (Sheng et al., 1997; Zalvide et al., 1998) and to alter steady state levels of phosphorylated p107 and p130 (Stubdal et al., 1996, 1997), even though this domain does not directly contribute to pocket protein binding. J-domains are signature sequences of the DnaJ/hsp40 family of proteins, specificity subunits of DnaJ-DnaK chaperone complexes (reviewed by Silver and Way (1993)). HPV16 E7-mediated activation of E2F-responsive promoters may also involve a J-domain protein (Sheng et al., 1997). Although HPV16 E7 does not contain a J-domain, it binds to the cellular DnaJ protein, hTid1 (Schilling et al., 1998).

The ability of HPV E7 to destabilize pocket proteins provides an effective strategy for viral oncoproteins that are expressed at low levels to inactivate cellular target proteins that are expressed at much higher levels. Through reduction of pRB steady state levels these viral oncoproteins can impair the entire spectrum of biological activities of the pocket proteins (Sellers et al., 1998). It is conceivable that the carcinogenic potential of high-risk HPVs may be related, at least in part, to the ability of high-risk HPV E7 proteins to subvert biological activities of pRB other than cell-cycle regulation, such as the ability of the pRB tumor suppressor to regulate differentiation. This activity is unique to pRB and is not shared by p107 and p130, which are not recognized as bona fide human tumor suppressors (Nguyen and McCance, 2005; Thomas et al., 2003). Low-risk HPV E7 proteins do not destabilize pRB and may not be able to modify these putative tumor suppressive activities.

10.3.6.   HPV E7 and the p53 Tumor Suppressor

Primary human cells that express high-risk HPV E7 are predisposed to apoptosis, particularly under conditions of growth factor deprivation. This intrinsic tumor suppression mechanism, the "trophic sentinel response" to E7 expression, is p53-dependent (Eichten et al., 2002). Levels of p53 are increased in E7-expressing cells (Demers et al., 1994b), suggesting that E7 perturbs p53 degradation (Jones et al., 1997b). The mechanism of E7-mediated p53 stabilization is not known, but it is independent of p14^{ARF}, an E2F-regulated inhibitor of mdm2-mediated p53 degradation (Bates et al., 1998; Seavey et al., 1999; Stott et al., 1998; Zhang et al., 1998). Nevertheless, normal mdm2-mediated p53 turnover does not occur in cervical carcinoma cell lines (Hengstermann et al., 2001), and mdm2 may be less tightly associated with p53 in E7-expressing cells than in normal cells (Seavey et al., 1999). This is interesting since mdm2 has also been implicated in pRB degradation (Uchida et al., 2005), and there is a tight correlation between E7-mediated pRB degradation and p53 stabilization (Jones et al., 1997b). The increased levels of p53 in E7-expressing cells do not appear to be transcriptionally active (Eichten et al., 2004), and E7 interferes with the activation of p53-responsive reporter constructs in transient transfection assays, possibly through association with TBP (Massimi and Banks, 1997). Steady state levels of p21^{CIP1} are also augmented in E7-expressing cells (Thomas and Banks, 1998) as a result of protein stabilization (Jones et al., 1999; Noya et al., 2001).

HPV E7 can also interfere with p53-mediated G1 growth arrest signaling (Demers et al., 1994a; Hickman et al., 1994; Slebos et al., 1994; Vousden et al., 1993). This may occur through multiple mechanisms, including aberrant expression of E2F transcriptional targets such as cyclin E, cyclin A (Hickman et al., 1994), and cdc25A (Katich et al., 2001); inactivation of p21^{CIP1} (Funk et al., 1997; Jones et al., 1997a); and destabilization of pRB (Funk et al., 1997; Jones and Munger, 1997).

10.3.7. Interactions of HPV E7 with Components of the Cell-Cycle Machinery

HPV genome replication critically depends on the host cell DNA synthesis machinery and occurs in terminally differentiated epithelial strata that normally do not support DNA synthesis. The E7 protein plays a critical role in inducing and/or sustaining a replication-competent cellular milieu in infected, differentiated host keratinocytes, and it has been shown to induce DNA synthesis in differentiated keratinocytes (Blanton et al., 1992; Cheng et al., 1995).

Cdk2 plays a central role in initiation, progression, and completion of S-phase. Cdk2 activity is stimulated by association with cyclin E and cyclin A. Regulation is also provided by inhibitory as well as activating phosphorylations of cdk2 and through association of cdk2 complexes with the cdk inhibitors (CKIs) p21^{CIP1}, p27^{KIP1}, and p57^{KIP2} (reviewed by Massague (2004)). Through inactivation of the pocket proteins, HPV E7 expression causes increased E2F-mediated transcription of the cdk2 activators cyclin E and cyclin A (Martin et al., 1998; Zerfass et al., 1995) as well as of the Cdc25A phosphatase, which activates cdk2 through dephosphorylation of an inhibitory tyrosine residue (Katich et al., 2001).

The HPV16 E7 oncoprotein is associated with cdk activity and interacts with cdk2/cyclin A and cdk2/cyclin E complexes (Davies et al., 1993; McIntyre et al., 1996; Tommasino et al., 1993). These interactions may be indirect and mediated by p107 and p130, each of which have been reported to associate with cdk2 complexes (Hannon et al., 1993; Lees et al., 1992). HPV16 E7 can also associate with and abrogate inhibition of cdk2 activity by $p27^{KIP1}$ and $p21^{CIP1}$ (Funk et al., 1997; Jian et al., 1998; Jones et al., 1997b; Morozov et al., 1997; Ruesch and Laimins, 1997; Zerfass-Thome et al., 1996). These inhibitors have been implicated in coupling cdk2 inhibition and cell-cycle withdrawal to differentiation in keratinocytes (Alani et al., 1998; DiCunto et al., 1998; Missero et al., 1996; Ruesch and Laimins, 1998). Therefore the ability of E7 to interact with $p21^{CIP1}$ and/or $p27^{KIP1}$ may be related to its ability to maintain a replication-competent cellular milieu in HPV-infected cells (Blanton et al., 1992; Cheng et al., 1995). Interestingly, $p21^{CIP1}$ levels are increased in E7 expressing cells, at least in part through protein stabilization (Jian et al., 1998; Jones et al., 1999).

The HPV E1 replication protein is also a phosphorylation target of cdk2/cyclin E complexes (Lin et al., 2000), and nuclear localization of the HPV E1 protein is dependent on phosphorylation by cdk2/cyclin E (Deng et al., 2004; Lin et al., 2000). Aberrant cyclin E and cdc25A expression also causes chromosome instability (Mailand et al., 2000; Molinari et al., 2000; Spruck et al., 1999), and the ability of HPV-16 E7 to induce aberrant centrosome duplication (see below) depends on cdk2 activity (Duensing and Munger, 2004). Hence the functional and biochemical interactions of high-risk HPV E7 with components of the cell-cycle machinery may also contribute directly to activation of the viral replication machinery and to induction of numerical and structural chromosome aberrations (Duensing and Munger, 2002).

HPV E7-expressing cells also express increased levels of $p16^{INK4A}$, an inhibitor of cdk4/cyclin D and cdk6/cyclin D (Khleif et al., 1996). Since the major cdk4/cdk6 targets, pRB, p107, and p130 (Bruce et al., 2000), are rendered dysfunctional through HPV E7-mediated destabilization, HPV E7-expressing cells are resistant to $p16^{INK4A}$-mediated G1 growth arrest, and high-level $p16^{INK4A}$ expression serves as a specific diagnostic biomarker for cells infected with high-risk HPV (Klaes et al., 2001).

10.3.8. Modulation of Cytostatic Cytokine Signaling by HPV E7

Transforming growth factor ß (TGF-ß) is a potent inhibitor of epithelial cell growth, but most squamous cell carcinoma cell lines, including cervical cancer cell lines, are resistant to TGF-ß growth suppression (reviewed by Polyak (1996)). High-risk HPV E7 proteins have the intrinsic capacity to interfere with TGF-ß growth inhibition (Pietenpol et al., 1990). Both $p21^{CIP1}$ and $p27^{KIP1}$ (Datto et al., 1995; Elbendary et al., 1994; Polyak et al., 1994) have been implicated in TGF-ß mediated growth inhibition, and the ability of HPV E7 to inactivate these CKIs may contribute to its ability to override growth inhibition by TGF-ß. TGF-ß also induces expression of p15^{ink4B}, a cdk4/cdk6-specific CKI (Hannon and Beach, 1994). Similar to p16^{ink4A}, growth suppression induced by p15^{ink4B} may require pRB, p107, and p130 (Bruce et al., 2000), which are functionally impaired in E7-expressing cells. The acquisition of TGF-ß resistance in HPV expressing cervical cancer lines, however, is complex. TGF-ß treatment leads to decreased expression of the HPV E6 and E7 oncoproteins through a mechanism that involves the *ski* oncogene and a nuclear factor I (NFI) site in the viral regulatory region (Baldwin et al., 2004). As a consequence, freshly derived HPV-containing epithelial cell lines are sensitive to TGF-ß (Creek et al., 1995). Acquisition of TGF-ß resistance coincides with a loss of transcriptional sensitivity of HPV oncogene expression to TGF-ß and occurs in parallel with resistance to differentiation cues, such as retinoic acid treatment (Borger et al., 2000) and decreased TGF-ß receptor 1 expression (Mi et al., 2000).

High-risk HPV-positive cervical cancer cell lines also acquire resistance to tumor necrosis factor α (TNF) (Villa et al., 1992), an important immune response mediator that is produced by cytotoxic T cells in response to viral infection. Primary human keratinocytes undergo G1 growth arrest and differentiation in response to TNF (Basile et al., 2001) through a pathway that involves NF-κB-mediated induction of p21^{CIP1} (Basile et al., 2003). Expression of high-risk HPV E7 proteins renders keratinocytes resistant to TNF in a process that correlates with the ability of E7 to inactivate pocket protein functions (Basile et al., 2001; Boccardo et al., 2004). TNF treatment triggers apoptosis on coadministration with the protein synthesis inhibitor cycloheximide. HPV E7-expressing keratinocytes show increased apoptosis in response to TNF/cycloheximide (Basile et al., 2001; Stoppler et al., 1998), whereas caspase 8 activation and apoptosis is inhibited in HPV E7-expressing normal human fibroblasts (Thompson et al., 2001).

HPV E7 has also been reported to compromise interferon (IFN) signaling by associating with and inhibiting IFN-α-induced nuclear translocation of p48, the DNA binding component of ISGF-3 (Barnard and McMillan, 1999; Barnard et al., 2000). In addition, E7 binds and inactivates the IFN-γ-induced transcription factor IRF-1, which regulates IFN-ß expression (Park et al., 2000; Perea et al., 2000). However, there is also evidence that IFN-γ can inhibit HPV16 E7 expression, and the IFN-γ-induced suppressor of cytokine signaling-1 (SOCS-1)/JAB can associate with E7 and induce its ubiquitin-mediated degradation (Kamio et al., 2004).

Many cytokine signaling pathways involve NF-κB activation, and HPV16 gene expression is negatively regulated by NF-κB (Fontaine et al., 2000). HPV16 E7 interferes with NF-κB activation by interaction with the IκB kinase complex (Spitkovsky et al., 2002), resulting in increased cytoplasmic levels of the NF-κB precursor proteins p100 and p105 in keratinocytes expressing E7 (Havard et al., 2005).

HPV E7 may also affect cellular survival signaling by insulin-like growth factors (IGFs) and associates with insulin-like growth factor binding protein-3

(IGFBP-3) (Mannhardt et al., 2000). IGFBP-3 is a transcriptional target of p53 (Bertherat, 1996) and a marker of cellular senescence that increases IGF-induced mitogenic signaling. Somewhat surprisingly, however, IGFBP-3 expression is increased in cervical cancer cell lines and in keratinocytes expressing high-risk HPV (Baege et al., 2004; Berger et al., 2002).

10.3.9.   Effects of HPV E7 on Cellular Metabolism

Cancer cells efficiently derive energy through glycolytic processes rather than mitochondrial respiration (reviewed by Aisenberg (1961) and Warburg (1936)). Cells expressing high-risk HPV E7 undergo similar metabolic alterations. Pyruvate kinase (PK) exists in a highly active tetrameric form and a less active dimeric form. Fructose 1,6 bisphosphate levels regulate the transition between these two forms, which determines glycolytic flux. Like many cancer cell lines, HPV E7-transformed cells contain mostly dimeric PK, despite high levels of fructose 1,6 bisphosphate, which would normally favor the more active form. HPV16 E7 associates with PK, and it has been hypothesized that this interaction may interfere with physiological regulation of PK (Mazurek et al., 2001; Zwerschke et al., 1999). HPV-16 E7 also allosterically activates α-glucosidase (Zwerschke et al., 2000), which may cause depletion of glycogen stores, which has been observed in cervical and other human cancers (Bannasch et al., 1997; Pedersen, 1975).

Transformed cells generally have a high intracellular pH. HPV E7-expressing NIH 3T3 cells showed increased activity of the Na^+/H^+ exchanger protein, resulting in cytoplasmic alkalinization. Inhibition of E7-induced alkalinization caused alterations in glycolytic metabolism and reduced several parameters of cellular transformation, including proliferation, growth in low serum, or anchorage-independent growth (Reshkin et al., 2000). These physiological abnormalities in E7-expressing cells may directly contribute to induction of cellular transformation, or they may reflect necessary metabolic adaptations to the altered energy requirements of rapidly proliferating transformed cells.

10.3.10.   HPV E7 and Chromosomal Instability

Loss of genomic integrity is a defining hallmark of human tumors, and cancer has been described as a disease of genomic instability (Klausner, 2002). Most human solid tumors, including cervical cancers, contain substantial numerical chromosome abnormalities and are aneuploid. This is in part a consequence of the loss of cell-cycle checkpoint functions, which allows cells with genomic alterations to remain in the replicative pool. DNA repair pathways are also compromised in human tumors (reviewed by Khanna and Jackson (2001)). Increased genomic plasticity of tumor cells may be necessary for survival and clonal expansion of an emerging neoplasm (reviewed by Cahill et al. (1999)). Genomic instability has been noted even in early premalignant high-risk HPV-associated lesions (Heselmeyer et al., 1996), and high-risk, but not low-risk,

HPV E6 and E7 oncoproteins can each independently render normal human cells genomically unstable (White et al., 1994).

Multipolar mitoses are a distinguishing feature of high-risk HPV-positive cancers (Winkler et al., 1984). Such abnormal mitotic processes can lead to asymmetrical chromosome segregation, thereby contributing to the induction of aneuploidy. The characteristic multipolar mitoses in cervical lesions are caused by supernumerary centrosomes, and the high-risk HPV E6 and E7 proteins cooperate to generate these defects (Duensing et al., 2000). In contrast, cells expressing low-risk HPV E6/E7 proteins do not exhibit centrosome abnormalities. Centrosome abnormalities also arise in cervical cancers (Balsitis et al., 2003; Reznikoff et al., 1994) and skin lesions that develop in HPV-16 E6- and/or E7-expressing transgenic mice (Schaeffer et al., 2004).

Centrosome synthesis is tightly linked to the cell division cycle. After cell division, each daughter cell contains a single centrosome, which consists of two centrioles surrounded by pericentriolar material. On S-phase entry the centrosome is licensed for duplication, which involves separation of the two centrioles. Each of the separated centrioles then serves as a template for synthesis of a daughter centriole. Centriole synthesis is completed at the end of S-phase, and cells contain two centrosomes that form the poles of the bipolar mitotic spindle, ensuring equal and symmetric chromosome segregation during cell division (reviewed by Stearns (2001) and Storchova and Pellman (2004)). Expression HPV E7 induces centrosome abnormalities in normal diploid cells by causing aberrant centriole synthesis, presumably by uncoupling the process of centrosome duplication from the cell division cycle (Duensing et al., 2001; Guarguaglini and Duncan, 2005) through a pathway that is at least in part independent of its ability to target pRB family members (Duensing and Munger, 2003). Inhibition of cdk2 specifically abrogates E7-induced centrosome abnormalities and aneuploidy without affecting normal centrosome duplication (Duensing and Munger, 2004). Hence HPV16 E7 may act as a mitotic mutator that increases the probability of mitotic errors during each round of cell division. This may generate the necessary genomic plasticity for acquisition of additional cellular mutations that contribute to malignant progression (reviewed by Duensing and Munger (2004)).

HPV16 E7 expression also causes centrosome-independent manifestations of chromosomal instability, including lagging chromosomal material and anaphase bridges that may represent chromosomal fusions resulting from double-strand DNA breaks (Duensing and Munger, 2002). The presence of DNA repair foci in HPV16 E7-expressing cells suggests that E7 may induce double-strand DNA breaks or interfere with break repair, thereby facilitating viral genome integration. Indeed, HPV16-expressing cells have a higher propensity for integration of exogenous plasmid DNA (Kessis et al., 1996). Consequently, E7 may facilitate integration of high-risk HPV genomes into host cellular chromosomes, an event that frequently accompanies malignant progression of high-risk HPV-associated lesions.

Expression of HPV-18 E7 enhances the degree of tetraploidy in suprabasal epithelial cells (Southern et al., 2001) through a pathway that is independent of pRB inactivation (Southern et al., 2004). Tetraploidy is often regarded as a precursor to aneuploidy (reviewed by Storchova and Pellman (2004)), and high-risk HPV E7 proteins can subvert mitotic checkpoints and/or the tetraploidy checkpoint that normally blocks tetraploid cells from reentering the cell division cycle (Thomas and Banks, 1998). Genomic analyses provided additional evidence for dysregulation of mitotic pathways in cervical cancer and high-risk HPV-expressing cell lines (Thierry et al., 2004; Patel et al., 2004).

10.4.   Role of the HPV E6 and E7 Genes in Survival and Proliferation of Cancer Cells

Human epithelial cells immortalized by the high-risk HPV E6 and E7 proteins are not tumorigenic in experimental animals, but tumorigenic variants arise with the continued passage of these cells. As discussed in previous sections, cells expressing the high-risk HPV E6 and E7 proteins are genetically unstable and accumulate mutations. These mutations are thought to underlie the acquisition of tumorigenicity and, in patients, carcinogenic progression. However, the continued expression of the E6 and E7 proteins in cervical cancer cells for many passages in culture suggests that the viral proteins also confer an ongoing growth advantage to the cells. A number of experimental approaches have been used to show that the activity of the E6 and E7 proteins is continuously required for the survival and proliferation of cervical cancer cell lines.

10.4.1.   Antisense and Related Technologies That Reduce HPV RNA in Cervical Cancer Cells

The role of HPV proteins in the proliferation of cervical cancer cells was first tested by the expression of antisense RNA to the E6 and E7 genes in the HPV18-positive C4-1 cervical carcinoma cell line (von Knebel Doeberitz et al., 1988). Expression of antisense RNA reduced colony formation by about one-half, suggesting that the viral gene products are required for optimal growth in culture. Numerous subsequent experiments that employed a variety of different plasmid or viral antisense expression constructs, small interfering RNAs, antisense oligodeoxyribonucleotides, or HPV E6/E7-specific ribozymes demonstrated similar growth inhibition or induction of apoptosis in HPV16 or HPV18-positive cervical and oral carcinoma cell lines (Alvarez-Salas et al., 1998, 1999; Butz et al., 2003; Cho et al., 2002; Choo et al., 2000; Hall and Alexander, 2003; Hamada et al., 1996b; Hu et al., 1995; Jiang and Milner, 2002; Steele et al., 1992, 1993; Tan and Ting, 1995; Venturini et al., 1999, Watanabe et al., 1993). In addition, several studies showed that HPV antisense treatment inhibited the tumorigenicity of cervical carcinoma cell lines in nude mice (Alvarez-Salas et al., 1998, 1999; Choo et al., 2000; Hamada et al., 1996a,b; Hu et al., 1995; Tan and Ting, 1995; von Knebel Doeberitz et al., 1992).

Although HPV antisense RNA can exert nonspecific effects (Storey et al., 1991), the activity displayed by a wide variety of reagents that target HPV mRNA provides confidence that these phenotypic responses were due to HPV repression. Furthermore, in several cases, antisense treatment was shown to downregulate E6/E7 expression and increase p53 levels, and growth inhibition was impaired by constitutive expression of the E6/E7 region from an ectopic gene or by expression of a dominant-negative p53 mutant.

Cellular genes required for HPV gene expression are also potential antisense targets. The transcription factor Brn3a is required for HPV16 E6/E7 expression in SiHa cervical cancer cells, and antisense Brn3a RNA repressed E6 expression and inhibited cell growth in vitro and tumorigenicity in nude mice (Ndisang et al., 1999, 2001). Similarly, nucleolin is required for HPV18 E6/E7 expression, and antisense inhibition of nucleolin expression blocked HPV18 E6/E7 expression in HeLa and SW756 cells and inhibited their growth (Grinstein et al., 2002). The expression of the HPV control region is also repressed by other cellular proteins, including TGF-β, γ-interferon, leukoregulin, TNF-α, serpinB2, the retinoblastoma protein, and interleukin-1 (Darnell et al., 2005; Kyo et al., 1994; Woodworth et al., 1990, 1992). However, modulating cellular genes to influence viral transcription seems likely to be a less specific approach than strategies that directly target viral RNA because the cellular proteins presumably control expression of various cellular genes as well as the integrated HPV genes.

10.4.2. Small Molecules and Peptides That Inhibit HPV Expression or Activity in Cervical Carcinoma Cells

Small molecules can also repress expression of the HPV oncogenes. Treatment with 5-azacytidine, a DNA demethylating agent, repressed HPV18 E6/E7 expression and reduced colony formation in nontumorigenic hybrids between HeLa cells and normal cells (Rosl et al., 1988). The steroid dexamethasone blocked transcription of the endogenous HPV18 E6/E7 genes in SW 756 cells, resulting in p53 induction and growth inhibition, unless exogenous copies of HPV16 E6/E7 genes were constitutively expressed (von Knebel Doeberitz et al., 1991,1994). Growth inhibition did not occur in cervical carcinoma cell lines in which dexamethasone did not repress HPV transcription. An inhibitor of the glycolytic pathway and protein glycosylation, 2-deoxyglucose, repressed expression of the HPV18 E6 and E7 genes, likely through inactivation of Sp1, an activator of the HPV promoter (Kang et al., 2003; Maehama et al., 1998). This resulted in activation of p53 and apoptosis in HeLa cells. Similarly, all-trans and 9-cis retinoic acids, which inhibit HPV16 expression in HPV-immortalized human keratinocytes (Creek et al., 1994), also greatly reduced levels of HPV16 E6/E7 RNA, increased p53 levels, and induced cell-cycle arrest in CaSki cells (Narayanan et al., 1998), and the antiviral drug Cidofovir repressed HPV E6/E7 expression in cervical and head and neck cancer cell lines, resulting in accumulation of p53 and Rb, impaired cell proliferation, and reduced tumorigenicity in nude mice (Abdulkarim et al., 2002). Finally, combined treatment of several cervical carcinoma cell lines with leptomycin B, a nuclear export inhibitor, and actinomycin D, which inhibits transcription, resulted in elimination of E6/E7 RNA, induction of p53, and p53-dependent apoptosis (Hietanen et al., 2000). However, it is unclear if the primary target of this treatment is E6/E7 transcription.

Chemicals and small peptides have also been used to inhibit the action of the E6 or E7 protein. The chemical 4, 4′-dithiodimorpholine, which causes the ejection of zinc from the zinc fingers of the E6 protein, induced p53 expression and apoptosis in SiHa and CaSki cells (Beerheide et al., 1999). In a novel approach, small peptide aptamers that bind the HPV16 E6 protein were isolated by genetic selection in yeast (Butz et al., 2000). Treatment of HPV16-positive SiHa and CaSki cervical carcinoma cells with some, but not all, of these aptamers resulted in p53 induction and apoptosis, whereas cells containing heterologous HPV18 DNA did not die. Similarly, peptide aptamers that bind the HPV16 E7 protein induced apoptosis of CaSki cells (Nauenburg et al., 2001). More recently, phage display libraries have been used to select a single-chain antibody that binds HPV16 E6 with high affinity (Griffin et al., 2006). Expression of this antibody in cervical cancer cells that contain HPV16 DNA caused p53 induction and apoptosis. Finally, peptides that contain an α-helical E6 binding motif have also been described that bind to the E6 protein and inhibit E6/E6AP-mediated in vitro degradation of various proteins including p53 (Sterlinko et al., 2004). This α-helical E6 protein binding motif has also been grafted onto protein scaffolds and shown to inhibit E6-E6AP complex formation, but the effect of these chimeric proteins on cells has not been reported (Liu et al., 2004).

10.4.3.   Papillomavirus E2 Protein as a Repressor of HPV Expression in Cervical Carcinoma Cells

In 1987, Thierry (1987) reported that transient transfection of the bovine papillomavirus type 1 (BPV1) E2 gene repressed transcription of reporter genes linked to the E6/E7 promoter from high-risk HPV genomes. Repression was exerted by full-length E2 proteins derived from BPV1, HPV11, and HPV18 and was due to the binding of the E2 protein to specific binding sites in the HPV long control region immediately upstream of the E6 and E7 genes (Bernard et al., 1989; Chin et al., 1996; Dostatni et al., 1991; Romanczuk et al., 1990; Thierry and Howley, 1991; Thierry and Yaniv, 1987). These findings led to the suggestion that loss of E2 expression on integration of HPV DNA into the cellular genome de-represses expression of the E6 and E7 genes, resulting in an enhanced proliferative stimulus and further progression toward cancer. In the initial report that the BPV1 E2 protein represses the HPV E6/E7 promoter, the authors noted that they were not able to establish HeLa cells expressing the BPV1 E2 protein (Thierry and Yaniv, 1987).

To determine the effect of the E2 protein on the transcription of the endogenous HPV genomes in cervical cancer cells, a replication-defective recombinant SV40-based viral vector was used to deliver the BPV1 E2 protein into cervical cancer cell lines (Hwang et al., 1993). Expression of the E2 protein in HeLa cells and other cervical carcinoma cell lines caused dramatic repression of the resident HPV E6/E7 oncogenes and rapid inhibition of cell proliferation (Hwang et al., 1993). In these experiments, E2 caused growth inhibition of HT-3 cervical carcinoma cells, which were initially thought to be HPV-negative but which were later shown to contain HPV30 DNA (Naeger et al., 1999). Subsequent experiments showed that expression of the HPV18, HPV16, and BPV1 E2 proteins by transfection or viral infection, or by shifting a temperature-sensitive E2 mutant to the permissive temperature, inhibited E6/E7 expression in cervical carcinoma cell lines and caused a dramatic inhibition of colony formation (Desaintes et al., 1997; Dowhanick et al., 1995; Wells et al., 2000). Growth inhibition occurred in cervical carcinoma cell lines expressing different high-risk HPV types as well as in HPV16-immortalized nontumorigenic cervical keratinocytes (Dowhanick et al., 1995; Hwang et al., 1993; Lee et al., 2002b; Moon et al., 2001). Thus susceptibility to growth inhibition does not require long-term passage of the cell lines or additional genetic events that occurred during carcinogenic progression. In contrast to these results, there is a report that the HPV16 E2 protein *increased* the expression of the endogenous HPV16 E6/E7 genes in SiHa cells (Sanchez-Perez et al., 1997), consistent with the ability of this E2 protein to activate the HPV16 early promoter in some transient transfection experiments (Bouvard et al., 1994; Kovelman et al., 1996).

In many experiments the E2 protein did not exert growth-inhibitory effects in cells lacking HPV DNA, and growth suppression was inhibited by mutations in the E2 protein that impaired DNA binding or E6/E7 repression (Desaintes et al., 1999; Dowhanick et al., 1995; Francis et al., 2000; Goodwin et al., 1998; Hwang et al., 1993; Nishimura et al., 2000). Most importantly, these growth-inhibitory effects (and the biochemical changes described below) were greatly attenuated when the cells were engineered to express copies of the E6 and E7 genes that were not repressed by the E2 protein (DeFilippis et al., 2003; Francis et al., 2000; Kang et al., 2004; Wells et al., 2000). Thus repression of HPV gene expression is required for growth inhibition. However, these experiments did not rule out the possibility that the E2 protein must repress HPV expression *and* affect an unidentified cellular target to induce growth arrest.

The BPV1 E2 protein and E2-mediated E6/E7 repression also caused dramatic biochemical changes in cells. The levels of p53 and the hypophosphorylated form of pRB increased posttranscriptionally, and the downstream p53 and pRB growth-inhibitory pathways were mobilized (Dowhanick et al., 1995; Goodwin and DiMaio, 2000; Hwang et al., 1993, 1996; Lee et al., 2002c; Moon et al., 2001; Naeger et al., 1999; Wu et al., 2000). Microarray analysis of cervical carcinoma cells after HPV E6/E7 repression revealed that a substantial fraction of the genome remains under immediate control of the viral oncogenes, even in advanced cancer (Thierry et al., 2004; Wells et al., 2003). The transcriptional response was dominated by induction of p53-responsive genes and repression of E2F-responsive genes, including those involved in mitosis and other aspects of cell-cycle progression. Additional experiments demonstrated that DNA binding and activation of tumor suppressor pathways were required for efficient growth inhibition by the E2 protein (Goodwin et al., 1998; Horner et al., 2004; Psyrri et al., 2004; Thierry et al., 2004; Wells et al., 2003) and that p53 levels remained under the control of the E6 protein and pRB levels remained under the control of the E7 protein in HeLa cells (DeFilippis et al., 2003). Taken together, these results indicate that proliferation of cervical cancer cells requires continuous expression of the HPV oncogenes to maintain tumor suppressor pathways in their inactive state. As noted in the final section, this finding validates the E6 and E7 proteins as rational targets for therapy. It also implies that the E6/E7 proteins may be good targets for therapeutic vaccination strategies against HPV-associated cancers because the cancer cells cannot escape immune killing simply by downregulating expression of these essential viral proteins.

Cells forced to undergo growth arrest by HPV repression display the same phenotype as primary cells that have undergone replicative senescence (Goodwin and DiMaio, 2000; Wells et al., 2000). Senescence, a permanent growth-arrested state attained by normal somatic cells after extended continuous passage in culture, is regarded as an important tumor suppressor mechanism and model of cellular aging (Campisi, 2005). Senescence induced by E2 expression is rapid, synchronous, and uniform (Goodwin and DiMaio, 2000); does not require short telomeres or low telomerase activity (Goodwin and DiMaio, 2001); and is impaired by antisense p21 oligonucleotides (Wells et al., 2000). Although induced senescence can be reversed after a few days by extinction of E2 expression or by reexpression of E7, after this time, it becomes irreversible (Kang et al., 2004; Wells et al., 2003). Senescence was also induced in HeLa cells by small interfering RNAs that target the HPV18 E6/E7 genes, demonstrating that the E2 protein per se is not required for this phenotype (Hall and Alexander, 2003). Analysis of HeLa cells engineered to separately repress E6 or E7 demonstrated that E7 repression activated pRB-dependent signaling that resulted in senescence, whereas E6 repression activated p53-dependent signaling that caused senescence or delayed apoptosis (DeFilippis et al., 2003; Horner et al., 2004; Psyrri et al., 2004). E7 repression inhibits the expression of the DEK protooncogene, an effect which appears to contribute to the senescent phenotype (Wise-Draper et al., 2005).

10.4.4.   HPV-Independent Effects of the E2 Protein

In addition to the effects described above that require HPV repression, E2 proteins can also exert HPV-independent effects. E2 expression can induce apoptosis or perturbation of the cell cycle in HPV-negative cells, and E2 mutants defective for DNA binding can elicit HPV-independent effects in HPV-positive cells. For example, delivery of the HPV31 E2 gene on an adenovirus vector caused normal foreskin keratinocytes to arrest in S-phase with ongoing cellular DNA replication, resulting in the generation of cells with greater than 4N DNA content and, eventually, the induction of apoptosis (Frattini et al., 1997). The HPV18 and BPV1 E2 proteins expressed from adenovirus vectors induced rapid p53-independent apoptosis in the absence of E2 DNA binding in HPV-positive or HPV-negative cells, effects which appeared to require high-level E2 expression (Desaintes et al., 1997, 1999). Expression of an ectopic HPV16 E2 gene by induction from a metallothionein promoter by treatment with heavy metals or by transfection induced p53-dependent apoptosis in the absence of serum growth factors, even in cells lacking endogenous HPV genomes (Sanchez-Perez et al., 1997; Webster et al., 2000). Finally, fusion of the HPV16 E2 protein to the herpes simplex virus VP22 delivery domain generated a protein that caused apoptosis in a variety of p53-positive cells, including those that lacked HPV DNA (Roeder et al., 2004).

The generation of HPV-independent responses in these experiments may be due to technical factors, such as E2 overexpression or the coexpression of adenovirus proteins from expression vectors. A more interesting possibility is that E2 proteins encoded by different papillomaviruses have different biological activities. In general, the response to the BPV1 E2 protein appears to reflect primarily its repressor activity at the HPV promoter, whereas HPV-independent effects are more commonly reported for HPV E2 proteins. In fact, high-multiplicity infection with an adenovirus vector expressing the transactivation domain of the HPV18 E2 protein fused to the green fluorescent protein induced caspase-8-dependent, p53-independent apoptosis (Demeret et al., 2003). This response may involve a caspase-8 cleavage site that is present in the high-risk HPV E2 proteins but lacking from the BPV1 E2 protein. A number of other biological activities have been attributed to high-risk but not low-risk E2 proteins (Blachon et al., 2005; Bellanger et al., 2005). Because E2 proteins can bind to a number of cellular proteins and genes, including some involved in transcriptional regulation, it is not surprising that in some settings they exert effects in addition to HPV repression (e.g. Grm et al., 2005; Li et al., 1991; Lee et al., 2002b; Oldak et al., 2004; Rehtanz et al., 2004; Steger and Corbach, 1997; You et al., 2004).

10.5. HPV E6 and E7 Proteins as Potential Therapeutic Targets

Each of the approaches used to modulate the expression or activity of the E6 and E7 proteins in cervical cancer cells has the potential to exert nonspecific or unanticipated effects. Nevertheless, the ability of numerous unrelated approaches to repress HPV function and inhibit the growth or survival of cervical carcinoma cells provides convincing evidence that HPV gene expression is continuously required in cervical carcinoma cell lines. If similar effects occur in primary cancer cells, treatments that inhibit the expression or action of the E6 and E7 proteins may be therapeutically useful by inducing senescence or apoptosis, even in advanced cancer. The binding of E6 to E6AP and E6-dependent ubiquitylation of p53 as well as the binding of E7 to pocket proteins and other targets are therefore validated targets for the development of screens to identify small molecule inhibitors that might serve as drug development candidates for HPV-associated cancers. Because HPV genes and gene products are missing from uninfected cells, it may be possible to develop agents that are highly specific for the viral targets with minimal effects on normal tissue.

Many of the approaches described above cause coordinate repression of both the E6 and the E7 proteins, which are expressed from a single primary transcript. In targeting the ubiquitylation and degradation of p53 through E6AP the oncogenic HPV E6 proteins protect the cells from the p53-dependent proapoptotic signals created by the engagement of high-risk HPV E7 proteins with the pRB family of pocket proteins. Thus treatments that interfere with E6 activity and its ability to mediate the proteolysis of p53, while leaving E7 activity intact, are more likely to induce apoptosis (Butz et al., 2000, 2003; DeFilippis et al., 2003; Griffin et al., 2006), which may be a more desirable therapeutic outcome than growth inhibition or senescence. Small molecules that target the E6 protein directly may be more specific in this regard than are agents that target expression of the viral genes, which tend to affect E6 and E7 expression coordinately.

Studies analyzing the effects of the HPV E6 and E7 proteins on normal cells, combined with the results of removing these proteins from cancer cells, have revealed the role these proteins play in inducing and maintaining the cancer state. These studies have also validated these viral proteins as promising targets in treating cervical neoplastic disease. The era of effective antiviral treatment against human cancer is in sight.

References

Abdulkarim, B., Sabri, S., Deutsch, E., Chagraoui, H., Maggiorella, L., Thierry, J., Eschwege, F., Vainchenker, W., Chouaib, S., and Bourhis, J. (2002). Antiviral agent Cidofovir restores p53 function and enhances the radiosensitivity in HPV-associated cancers. *Oncogene* 21:2334–2346.

Aisenberg, A.C. (1961). *The Glycolysis and Respiration of Tumors*. New York: Academic.

Alani, R.M., Hasskarl, J., and Munger, K. (1998). Alterations in cyclin-dependent kinase 2 function during differentiation of primary human keratinocytes. *Mol. Carcinog.* 23:226–233.

Alonso, L.G., Garcia-Alai, M.M., Smal, C., Centeno, J.M., Iacono, R., Castano, E., Gualfetti, P., and de Prat-Gay, G. (2004). The HPV16 E7 viral oncoprotein self-assembles into defined spherical oligomers. *Biochemistry* 43:3310–3317.

Alvarez-Salas, L.M., Arpawong, T.E., and DiPaolo, J.A. (1999). Growth inhibition of cervical tumor cells by antisense oligodeoxynucleotides directed to the human papillomavirus type 16 E6 gene. *Antisense Nucleic Acid Drug Dev.* 9:441–450.

Alvarez-Salas, L.M., Cullinan, A.E., Siwkowski, A., Hampel, A., and DiPaolo, J.A. (1998). Inhibition of *HPV-16 E6/E7* immortalization of normal keratinocytes by hairpin ribozymes. *Proc. Natl. Acad. Sci. U. S. A.* 95:1189–1194.

Angeline, M., Merle, E., and Moroianu, J. (2003). The E7 oncoprotein of high-risk human papillomavirus type 16 enters the nucleus via a nonclassical Ran-dependent pathway. *Virology* 317:13–23.

Antinore, M.J., Birrer, M.J., Patel, D., Nader, L., and McCance, D.J. (1996). The human papillomavirus type 16 E7 gene product interacts with and trans-activates the AP1 family of transcription factors. *EMBO J.* 15:1950–1960.

Armstrong, D.J., and Roman, A. (1993). The anomalous electrophoretic behavior of the human papillomavirus type-16 E7-protein is due to the high content of acidic amino acid residues. *Biochem. Biophys. Res. Commun.* 192:1380–1387.

Armstrong, D.J., and Roman, A. (1997). The relative ability of human papillomavirus type 6 and human papillomavirus type 16 E7 proteins to transactivate E2F-responsive elements is promoter- and cell-dependent. *Virology* 239:238–246.

Avvakumov, N., Torchia, J., and Mymryk, J.S. (2003). Interaction of the HPV E7 proteins with the pCAF acetyltransferase. *Oncogene* 22:3833–3841.

Baege, A.C., Disbrow, G.L., and Schlegel, R. (2004). IGFBP-3, a marker of cellular senescence, is overexpressed in human papillomavirus-immortalized cervical cells and enhances IGF-1-induced mitogenesis. *J. Virol.* 78:5720–5727.

Baldwin, A., Pirisi, L., and Creek, K.E. (2004). NFI-Ski interactions mediate transforming growth factor beta modulation of human papillomavirus type 16 early gene expression. *J. Virol.* 78:3953–3964.

Balsitis, S.J., Sage, J., Duensing, S., Munger, K., Jacks, T., and Lambert, P.F. (2003). Recapitulation of the effects of the human papillomavirus type 16 E7 oncogene on mouse epithelium by somatic Rb deletion and detection of pRb-independent effects of E7 *in vivo*. *Mol. Cell. Biol.* 23:9094–9103.

Band, V., DeCaprio, J.A., Delmolino, L., Kulesa, V., and Sager, R. (1991). Loss of p53 protein in human papillomavirus type 16 E6-immortalized human mammary epithelial cells. *J. Virol.* 65:6671–6676.

Bandara, L.R., Buck, V.M., Zamanian, M., Johnston, L.H., and La Thangue, N.B. (1993). Functional synergy between DP-1 and E2F-1 in the cell cycle-regulating transcription factor DRTF1/E2F. *EMBO J.* 12:4317–4324.

Banks, L., Edmonds, C., and Vousden, K.H. (1990). Ability of the HPV16 E7 protein to bind RB and induce DNA synthesis is not sufficient for efficient transforming activity in NIH3T3 cells. *Oncogene* 5:1383–1389.

Bannasch, P., Klimek, F., and Mayer, D. (1997). Early bioenergetic changes in hepatocarcinogenesis: preneoplastic phenotypes mimic responses to insulin and thyroid hormone. *J. Bioenerg. Biomembr.* 29:303–313.

Barbosa, M.S., Edmonds, C., Fisher, C., Schiller, J.T., Lowy, D.R., and Vousden, K.H. (1990). The region of the HPV E7 oncoprotein homologous to adenovirus E1a and SV40 large T antigen contains separate domains for Rb binding and casein kinase II. *EMBO J.* 9:153–160.

Barbosa, M.S., Lowy, D.R., and Schiller, J.T. (1989). Papillomavirus polypeptides E6 and E7 are zinc binding proteins. *J. Virol.* 63:1404–1407.

Barnard, P., and McMillan, N.A. (1999). The human papillomavirus E7 oncoprotein abrogates signaling mediated by interferon-alpha. *Virology* 259:305–313.

Barnard, P., Payne, E., and McMillan, N.A. (2000). The human papillomavirus E7 protein is able to inhibit the antiviral and anti-growth functions of interferon-alpha. *Virology* 277:411–419.

Basile, J.R., Eichten, A., Zacny, V., and Munger, K. (2003). NK-kB-mediated induction of p21[cip1/Waf1] by tumor necrosis factor a induces growth arrest and cytoprotection in normal human keratinocytes. *Mol. Cancer Res.* 1:262–270.

Basile, J.R., Zacny, V., and Munger, K. (2001). The cytokines tumor necrosis factor-alpha (TNF-alpha) and TNF-related apoptosis-inducing ligand differentially modulate proliferation and apoptotic pathways in human keratinocytes expressing the human papillomavirus-16 E7 oncoprotein. *J. Biol. Chem.* 276:22522–22528.

Bates, S., Phillips, A.C., Clark, P.A., Stott, F., Peters, G., Ludwig, R.L., and Vousden, K.H. (1998). p14ARF links the tumour suppressors RB and p53. *Nature* 395:124–125.

Be, X., Hong, Y., Wei, J., Androphy, E.J., Chen, J.J., and Baleja, J.D. (2001). Solution structure determination and mutational analysis of the papillomavirus E6 interacting peptide of E6AP. *Biochemistry* 40:1293–1299.

Bedell, M.A., Jones, K.H., Grossman, S.R., and Laimins, L.A. (1989). Identification of human papillomavirus type 18 transforming genes in immortalized and primary cells. *J. Virol.* 63:1247–1255.

Beerheide, W., Bernard, H.U., Tan, Y.J., Ganesan, A., Rice, W.G., and Ting, A.E. (1999). Potential drugs against cervical cancer: zinc-ejecting inhibitors of the human papillomavirus type 16 E6 oncoprotein. *J. Natl. Cancer Inst.* 91:1211–1220.

Beer-Romano, P., Glass, S., and Rolfe, M. (1997). Antisense targeting of E6AP elevates p53 in HPV-infected cells but not in normal cells. *Oncogene* 14:595–602.

Bellanger, S., Blachon, S., Mechali, F., Bonne-Andrea, C., and Thierry, F. (2005). High-risk but not low-risk HPV E2 proteins bind to the APC activators Cdh1 and Cdc20 and cause genomic instability. *Cell Cycle* 4:1608–1615.

Berezutskaya, E., and Bagchi, S. (1998). The human papillomavirus E7 oncoprotein functionally interacts with the S4 subunit of the 26 S proteasome. *J. Biol. Chem.* 272:30135–30140.

Berezutskaya, E., Yu, B., Morozov, A., Raychaudhuri, P., and Bagchi, S. (1997). Differential regulation of the pocket domains of the retinoblastoma family proteins by the HPV16 E7 oncoprotein. *Cell Growth Differ.* 8:1277–1286.

Berger, A.J., Baege, A., Guillemette, T., Deeds, J., Meyer, R., Disbrow, G., and Schlegel, R. (2002). Insulin-like growth factor-binding protein 3 expression increases during immortalization of cervical keratinocytes by human papillomavirus type 16 E6 and E7 proteins. *Am. J. Pathol.* 161:603–610.

Bernard, B.A., Bailly, C., Lenoir, M.-C., Darmon, M., Thierry, F., and Yaniv, M. (1989). The human papillomavirus type 18 (HPV18) E2 gene product is a repressor of the HPV18 regulatory region in human keratinocytes. *J. Virol.* 63:4317–4324.

Bernat, A., Avvakumov, N., Mymryk, J.S., and Banks, L. (2003). Interaction between the HPV E7 oncoprotein and the transcriptional coactivator p300. *Oncogene* 22:7871–7881.

Bertherat, J. (1996). Insulin-like growth factor binding protein 3 (IGFBP-3): a novel target of the tumor suppressor p53 inhibiting cell growth. *Eur. J. Endocrinol.* 134:426–427.

Bischof, O., Nacerddine, K., and Dejean, A. (2005). Human papillomavirus oncoprotein E7 targets the promyelocytic leukemia protein and circumvents cellular senescence via the Rb and p53 tumor suppressor pathways. *Mol. Cell. Biol.* 25:1013–1024.

Blachon, S., Bellanger, S., Demeret, C., and Thierry, F. (2005). Nucleo-cytoplasmic shuttling of high-risk human papillomavirus E2 proteins induces apoptosis. *J. Biol. Chem.* 280:36088–36098.

Blanton, R.A., Coltrera, M.D., Gown, A.M., Halbert, C.L., and McDougall, J.K. (1992). Expression of the HPV16 E7 gene generates proliferation in stratified squamous cell cultures which is independent of endogenous p53 levels. *Cell Growth Differ.* 3:791–802.

Boccardo, E., Noya, F., Broker, T.R., Chow, L.T., and Villa, L.L. (2004). HPV-18 confers resistance to TNF-alpha in organotypic cultures of human keratinocytes. *Virology* 328:233–243.

Bohl, J., Das, K., Dasgupta, B., and Vande Pol, S.B. (2000). Competitive binding to a charged leucine motif represses transformation by a papillomavirus E6 oncoprotein. *Virology* 271:163–170.

Bohl, J., Hull, B., and Vande Pol, S.B. (2001). Cooperative transformation and coexpression of bovine papillomavirus type 1 E5 and E7 proteins. *J. Virol.* 75:513–521.

Borger, D.R., Mi, Y., Geslani, G., Zyzak, L.L., Batova, A., Engin, T.S., Pirisi, L., and Creek, K.E. (2000). Retinoic acid resistance at late stages of human papillomavirus type 16- mediated transformation of human keratinocytes arises despite intact retinoid signaling and is due to a loss of sensitivity to transforming growth factor-beta. *Virology* 270:397–407.

Bouvard, V., Storey, A., Pim, D., and Banks, L. (1994). Characterization of the human papillomavirus E2 protein: evidence of trans-activation and trans-repression in cervical keratinocytes. *EMBO J.* 13:5451–5459.

Boyer, S.N., Wazer, D.E., and Band, V. (1996). E7 protein of human papilloma virus-16 induces degradation of retinoblastoma protein through the ubiquitin-proteosome pathway. *Cancer Res.* 56:4620–4624.

Braspenning, J., Marchini, A., Albarani, V., Levy, L., Ciccolini, F., Cremonesi, C., Ralston, R., Gissmann, L., and Tommasino, M. (1998). The CXXC Zn binding motifs of the human papillomavirus type 16 E7 oncoprotein are not required for its in vitro transforming activity in rodent cells. *Oncogene* 16:1085–1089.

Brehm, A., Miska, E.A., McCance, D.J., Reid, J.L., Bannister, A.J., and Kouzarides, T. (1998). Retinoblastoma protein recruits histone deacetylase to repress transcription. *Nature* 391:597–601.

Brehm, A., Nielsen, S.J., Miska, E.A., McCance, D.J., Reid, J.L., Bannister, A.J., and Kouzarides, T. (1999). The E7 oncoprotein associates with Mi2 and histone deacetylase activity to promote cell growth. *EMBO J.* 18:2449–2458.

Brokaw, J.L., Yee, C.L., and Munger, K. (1994). A mutational analysis of the amino terminal domain of the human papillomavirus type 16 E7 oncoprotein. *Virology* 205:603–607.

Bruce, J.L., Hurford, R.K., Classon, M., Koh, H., and Dyson, N.J. (2000). Requirements for cell cycle arrest by p16^{ink4a}. *Mol. Cell* 6:737–742.

Butz, K., Denk, C., Ullmann, A., Scheffner, M., and Hoppe-Seyler, F. (2000). Induction of apoptosis in human papillomavirus-positive cancer cells by peptide aptamers targeting the viral E6 oncoprotein. *Proc. Natl. Acad. Sci. U. S. A.* 97:6693–6697.

Butz, K., Ristriani, T., Hengstermann, A., Denk, C., Scheffner, M., and Hoppe-Seyler, F. (2003). siRNA targeting of the viral E6 oncogene efficiently kills human papillomavirus-positive cancer cells. *Oncogene* 22:5938–5945.

Butz, K., Shahabeddin, L., Geisen, C., Spitkovsky, D., Ullmann, A., and Hoppe-Seyler, F. (1995). Functional p53 protein in human papillomavirus-positive cancer cells. *Oncogene* 10:927–936.

Cahill, D.P., Kinzler, K.W., Vogelstein, B., and Lengauer, C. (1999). Genetic instability and darwinian selection in tumours. *Trends Cell Biol.* 9:M57–60.

Campisi, J. (2005). Senescent cells, tumor suppression, and organismal aging: good citizens, bad neighbors. *Cell* 120:513–522.

Cenci, M., French, D., Pisani, T., Alderisio, M., Lombardi, A.M., Marchese, R., Colelli, F., and Vecchione, A. (2003). p53 polymorphism at codon 72 is not a risk factor for cervical carcinogenesis in central Italy. *Anticancer Res.* 23:1385–1387.

Chen, J.J., Reid, C.E., Band, V., and Androphy, E.J. (1995). Interaction of papillomavirus E6 oncoproteins with a putative calcium-binding protein. *Science* 269:529–531.

Cheng, S., Schmidt-Grimminger, D.C., Murant, T., Broker, T.R., and Chow, L.T. (1995). Differentiation-dependent up-regulation of the human papillomavirus E7 gene reactivates cellular DNA replication in suprabasal differentiated keratinocytes. *Genes Dev.* 9:2335–2349.

Chien, W.M., Parker, J.N., Schmidt-Grimminger, D.C., Broker, T.R., and Chow, L.T. (2000). Casein kinase II phosphorylation of the human papillomavirus-18 E7 protein is critical for promoting S-phase entry. *Cell Growth Differ.* 11:425–435.

Chin, Y.E., Kitagawa, M., Su, W.C., You, Z.H., Iwamoto, Y., and Fu, X.Y. (1996). Cell growth arrest and induction of cyclin-dependent kinase inhibitor p21 WAF1/CIP1 mediated by STAT1. *Science* 272:719–722.

Cho, C.W., Poo, H., Cho, Y.S., Cho, M.C., Lee, K.A., Lee, S.J., Park, S.N., Kim, I.K., Jung, Y.K., Choe, Y.K., Yeom, Y.I., Choe, I.S., and Yoon, D.Y. (2002). HPV E6 antisense induces apoptosis in CaSki cells via suppression of E6 splicing. *Exp. Mol. Med.* 34:159–166.

Choo, C.K., Ling, M.T., Suen, C.K., Chan, K.W., and Kwong, Y.L. (2000). Retrovirus-mediated delivery of HPV16 E7 antisense RNA inhibited tumorigenicity of CaSki cells. *Gynecol. Oncol.* 78:293–301.

Ciccolini, F., Dipasquale, G., Carlotti, F., Crawford, L., and Tommasino, M. (1994). Functional studies of E7 proteins from different HPV types. *Oncogene* 9:2633–2638.

Clemens, K.E., Brent, R., Gyuris, J., and Munger, K. (1995). Dimerization of the human papillomavirus E7 oncoprotein *in vivo*. *Virology* 214:289–293.

Clements, A., Johnston, K., Mazzarelli, J.M., Ricciardi, R.P., and Marmorstein, R. (2000). Oligomerization properties of the viral oncoproteins adenovirus E1A and human papillomavirus E7 and their complexes with the retinoblastoma protein. *Biochemistry* 39:16033–16045.

Cole, S.T., and Danos, O. (1987). Nucleotide sequence and comparative analysis of the human papillomavirus type 18 genome: phylogeny of papillomaviruses and repeated structure of the E6 and E7 gene products. *J. Mol. Biol.* 193:599–608.

Creek, K.E., Geslani, G., Batova, A., and Pirisi, L. (1995). Progressive loss of sensitivity to growth control by retinoic acid and transforming growth factor-beta at late stages of human papillomavirus type 16-initiated transformation of human keratinocytes. *Adv. Exp. Med. Biol.* 375:117–135.

Creek, K.E., Jenkins, G.R., Khan, M.A., Batova, A., Hodam, J.R., Tolleson, W.H., and Pirisi, L. (1994). Retinoic acid suppresses human papillomavirus type 16 (HPV16)-mediated transformation of human keratinocytes and inhibits the expression of the HPV16 oncogenes. *Adv. Exp. Med. Biol.* 354:19–35.

Darnell, G.A., Antalis, T.M., Rose, B.R., and Suhrbier, A. (2005). Silencing of integrated human papillomavirus type 18 oncogene transcription in cells expressing SerpinB2. *J. Virol.* 79:4246–4256.

Das, K., Bohl, J., and Vande Pol, S.B. (2000). Identification of a second transforming function in bovine papillomavirus type 1 E6 and the role of E6 interactions with paxillin, E6BP, and E6AP. *J. Virol.* 74:812–816.

Datto, M.B., Li, Y., Panus, J.F., Howe, D.J., Xiong, Y., and Wang, X.F. (1995). Transforming growth factor beta induces the cyclin- dependent kinase inhibitor p21 through a p53-independent mechanism. *Proc. Natl. Acad. Sci. U. S. A.* 92:5545–5549.

Davies, R., Hicks, R., Crook, T., Morris, J., and Vousden, K. (1993). Human papillomavirus type 16 E7 associates with a histone H1 kinase and with p107 through sequences necessary for transformation. *J. Virol.* 67:2521–2528.

DeFilippis, R.A., Goodwin, E.C., Wu, L., and DiMaio, D. (2003). Endogenous human papillomavirus E6 and E7 proteins differentially regulate proliferation, senescence, and apoptosis in HeLa cervical carcinoma cells. *J. Virol.* 77:1551–1563.

Degenhardt, Y.Y., and Silverstein, S. (2001a). Interaction of zyxin, a focal adhesion protein, with the E6 protein from human papillomavirus type 6 results in its nuclear translocation. *J. Virol.* 75:11791–11802.

Degenhardt, Y.Y., and Silverstein, S.J. (2001b). Gps2, a protein partner for human papillomavirus E6 proteins. *J. Virol.* 75:151–160.

De Luca, P., Majello, B., and Lania, L. (1998). Retinoblastoma protein tethered to promoter DNA represses TBP-mediated transcription. *J. Cell. Biochem.* 70:281–287.

DeMasi, J., Huh, K.W., Nakatani, Y., Munger, K., and Howley, P.M. (2005). Bovine papillomavirus E7 transformation function correlates with cellular p600 protein binding. *Proc. Natl. Acad. Sci. U. S. A.* 102:11486–11491.

Demeret, C., Garcia-Carranca, A., and Thierry, F. (2003). Transcription-independent triggering of the extrinsic pathway of apoptosis by human papillomavirus 18 E2 protein. *Oncogene* 22:168–175.

Demers, G.W., Foster, S.A., Halbert, C.L., and Galloway, D.A. (1994a). Growth arrest by induction of p53 in DNA damaged keratinocytes is bypasses by human papillomavirus 16 E7. *Proc. Natl. Acad. Sci. U. S. A.* 91:4382–4386.

Demers, G.W., Halbert, C.L., and Galloway, D.A. (1994b). Elevated wild-type p53 protein levels in human epithelial cell lines immortalized by the human papillomavirus type 16 E7 gene. *Virology* 198:169–174.

Deng, W., Lin, B.Y., Jin, G., Wheeler, C.G., Ma, T., Harper, J.W., Broker, T.R., and Chow, L.T. (2004). Cyclin/CDK regulates the nucleocytoplasmic localization of the human papillomavirus E1 DNA helicase. *J. Virol.* 78:13954–13965.

Desaintes, C., Demeret, C., Goyat, S., Yaniv, M., and Thierry, F. (1997). Expression of the papillomavirus E2 protein in HeLa cells leads to apoptosis. *EMBO J.* 16:504–514.

Desaintes, C., Goyat, S., Garbay, S., Yaniv, M., and Thierry, F. (1999). Papillomavirus E2 induces p53-independent apoptosis in HeLa cells. *Oncogene* 18:4538–4545.

Dick, F.A., and Dyson, N.J. (2002). Three regions of the pRB pocket domain affect its inactivation by human papillomavirus E7 proteins. *J. Virol.* 76:6224–6234.

Dick, F.A., Sailhamer, E., and Dyson, N.J. (2000). Mutagenesis of the pRB pocket reveals that cell cycle arrest functions are separable from binding to viral oncoproteins. *Mol. Cell. Biol.* 20:3715–3727.

DiCunto, F., Topley, G., Calautti, E., Hsiao, J., Ong, L., Seth, P.K., and Dotto, G.P. (1998). Inhibitory function of p21^{Cip1}/WAF1 in differentiation of primary mouse keratinocytes independent of cell cycle control. *Science* 280:1069–1072.

Dostatni, N., Lambert, P.F., Sousa, R., Ham, J., Howley, P.M., and Yaniv, M. (1991). The functional BPV-1 E2 trans-activating protein can act as a repressor by preventing formation of the initiation complex. *Genes Dev.* 5:1657–1671.

Dowhanick, J.J., McBride, A.A., and Howley, P.M. (1995). Suppression of cellular proliferation by the papillomavirus E2 protein. *J. Virol.* 69:7791–7799.

Duensing, S., Duensing, A., Crum, C.P., and Munger, K. (2001). Human papillomavirus type 16 E7 oncoprotein-induced abnormal centrosome synthesis is an early event in the evolving malignant phenotype. *Cancer Res.* 61:2356–2360.

Duensing, S., Lee, L.Y., Duensing, A., Basile, J., Piboonniyom, S.O., Gonzalez, S.L., Crum, C.P., and Munger, K. (2000). The human papillomavirus type 16 E6 and E7 oncoproteins cooperate to induce mitotic defects and genomic instability by uncoupling centrosome duplication from the cell division cycle. *Proc. Natl. Acad. Sci. U. S. A.* 97:10002–10007.

Duensing, S., and Munger, K. (2002). The human papillomavirus type 16 E6 and E7 oncoproteins independently induce numerical and structural chromosome instability. *Cancer Res.* 62:7075–7082.

Duensing, S., and Munger, K. (2003). Human papillomavirus type 16 E7 oncoprotein can induce abnormal centrosome duplication through a mechanism independent of retinoblastoma protein family members. *J. Virol.* 77:12331–12335.

Duensing, S., and Munger, K. (2004). Mechanisms of genomic instability in human cancer: insights from studies with human papillomavirus oncoproteins. *Int. J. Cancer* 109:157–162.

Dyson, N. (1998). The regulation of E2F by pRB-family proteins. *Genes Dev.* 12:2245–2262.

Dyson, N., Guida, P., Munger, K., and Harlow, E. (1992). Homologous sequences in adenovirus E1A and human papillomavirus E7 proteins mediate interaction with the same set of cellular proteins. *J. Virol.* 66:6893–6902.

Edmonds, C., and Vousden, K.H. (1989). A point mutational analysis of human papillomavirus type 16 E7 protein. *J. Virol.* 63:2650–2656.

Eichten, A., Rud, D.S., Grace, M., Piboonniyom, S.O., Zacny, V., and Munger, K. (2004). Molecular pathways executing the "trophic sentinel" response in HPV-16 E7-expressing normal human diploid fibroblasts upon growth factor deprivation. *Virology* 319:81–93.

Eichten, A., Westfall, M., Pietenpol, J.A., and Munger, K. (2002). Stabilization and functional impairment of the tumor suppressor p53 by the human papillomavirus type 16 E7 oncoprotein. *Virology* 295:74–95.

Elbendary, A., Berchuck, A., Davis, P., Havrilesky, L., Bast, R.C., Jr., Iglehart, J.D., and Marks, J.R. (1994). Transforming growth factor beta 1 can induce CIP1/WAF1 expression independent of the p53 pathway in ovarian cancer cells. *Cell Growth Differ.* 5:1301–1307.

Favre-Bonvin, A., Reynaud, C., Kretz-Remy, C., and Jalinot, P. (2005). Human papillomavirus type 18 E6 protein binds the cellular PDZ protein TIP-2/GIPC, which is involved in transforming growth factor beta signaling and triggers its degradation by the proteasome. *J. Virol.* 79:4229–4237.

Ferreira, R., Magnaghi-Jaulin, L., Robin, P., Harel-Bellan, A., and Trouche, D. (1998). The three members of the pocket proteins family share the ability to repress E2F activity through recruitment of a histone deacetylase. *Proc. Natl. Acad. Sci. U. S. A.* 95:10493–10498.

Figge, J., Webster, T., Smith, T.F., and Paucha, E. (1988). Prediction of similar transforming regions in Simian virus 40 large T, adenovirus E1A and myc oncoproteins. *J. Virol.* 62:1814–1818.

Firzlaff, J.M., Luscher, B., and Eisenman, R.N. (1991). Negative charge at the casein kinase II phosphorylation site is important for transformation but not for Rb protein binding by the E7 protein of human papillomavirus type 16. *Proc. Natl. Acad. Sci. U. S. A.* 88:5187–5191.

Flemington, E.K., Speck, S.H., and Kaelin, W.G., Jr. (1993). E2F-1-mediated transactivation is inhibited by complex formation with the retinoblastoma susceptibility gene product. *Proc. Natl. Acad. Sci. U. S. A.* 90:6914–6918.

Flores, E.R., Allen-Hoffmann, B.L., Lee, D., and Lambert, P.F. (2000). The human papillomavirus type 16 E7 oncogene is required for the productive stage of the viral life cycle. *J. Virol.* 74:6622–6631.

Fontaine, V., van der Meijden, E., de Graaf, J., ter Schegget, J., and Struyk, L. (2000). A functional NF-kappaB binding site in the human papillomavirus type 16 long control region. *Virology* 272:40–49.

Foster, S.A., and Galloway, D.A. (1996). Human papillomavirus type 16 E7 alleviates a proliferation block in early passage human mammary epithelial cells. *Oncogene* 12:1773–1779.

Francis, D.A., Schmid, S.I., and Howley, P.M. (2000). Repression of the integrated papillomavirus E6/E7 promoter is required for growth suppression of cervical cancer cells. *J. Virol.* 74:2679–2686.

Frattini, M.G., Hurst, S.D., Lim, H.B., Swaminathan, S., and Laimins, L.A. (1997). Abrogation of a mitotic checkpoint by E2 proteins from oncogenic human papillomaviruses correlates with increased turnover of the p53 tumor suppressor protein. *EMBO J.* 16:318–331.

Frolov, M.V., and Dyson, N.J. (2004). Molecular mechanisms of E2F-dependent activation and pRB-mediated repression. *J. Cell Sci.* 117:2173–2181.

Funk, J.O., Waga, S., Harry, J.B., Espling, E., Stillman, B., and Galloway, D.A. (1997). Inhibition of CDK activity and PCNA-dependent DNA replication by p21 is blocked by interaction with the HPV-16 E7 oncoprotein. *Genes Dev.* 11:2090–2100.

Gage, J.R., Meyers, C., and Wettstein, F.O. (1990). The E7 proteins of the nononcogenic human papillomavirus type 6b (HPV-6b) and of the oncogenic HPV-16 differ in retinoblastoma protein binding and other properties. *J. Virol.* 64:723–730.

Gao, Q., Kumar, A., Singh, L., Huibregtse, J.M., Beaudenon, S., Srinivasan, S., Wazer, D.E., Band, H., and Band, V. (2002). Human papillomavirus E6-induced degradation of E6TP1 is mediated by E6AP ubiquitin ligase. *Cancer Res.* 62:3315–3321.

Gao, Q., Srinivasan, S., Boyer, S.N., Wazer, D.E., and Band, V. (1999). The E6 oncoproteins of high-risk papillomaviruses bind to a novel putative GAP protein, E6TP1, and target it for degradation. *Mol. Cell. Biol.* 19:733–744.

Gardiol, D., Kuhne, C., Glaunsinger, B., Lee, S.S., Javier, R., and Banks, L. (1999). Oncogenic human papillomavirus E6 proteins target the discs large tumor suppressor for proteasome-mediated degradation. *Oncogene* 18:5487–5496.

Gewin, L., and Galloway, D.A. (2001). E box-dependent activation of telomerase by human papillomavirus type 16 E6 does not require induction of c-myc. *J. Virol.* 75:7198–7201.

Gewin, L., Myers, H., Kiyono, T., and Galloway, D.A. (2004). Identification of a novel telomerase repressor that interacts with the human papillomavirus type-16 E6/E6-AP complex. *Genes Dev.* 18:2269–2282.

Giarre, M., Caldeira, S., Malanchi, I., Ciccolini, F., Leao, M.J., and Tommasino, M. (2001). Induction of pRb degradation by the human papillomavirus type 16 E7 protein is essential to efficiently overcome p16^{INK4a}-imposed G_1 cell cycle arrest. *J. Virol.* 75:4705–4712.

Glaunsinger, B.A., Lee, S.S., Thomas, M., Banks, L., and Javier, R. (2000). Interactions of the PDZ-protein MAGI-1 with adenovirus E4-ORF1 and high-risk papillomavirus E6 oncoproteins. *Oncogene* 19:5270–5280.

Gonzalez, S.L., Stremlau, M., He, X., Basile, J.R., and Munger, K. (2001). Degradation of the retinoblastoma tumor suppressor by the human papillomavirus type 16 E7 oncoprotein is important for functional inactivation and is separable from proteasomal degradation of E7. *J. Virol.* 75:7583–7591.

Goodwin, E.C., and DiMaio, D. (2000). Repression of human papillomavirus oncogenes in HeLa cervical carcinoma cells causes the orderly reactivation of dormant tumor suppressor pathways. *Proc. Natl. Acad. Sci. U. S. A.* 97:12513–12518.

Goodwin, E.C., and DiMaio, D. (2001). Induced senescence in HeLa cervical carcinoma cells containing elevated telomerase activity and extended telomeres. *Cell Growth Differ.* 12:525–534.

Goodwin, E.C., Naeger, L.K., Breiding, D.E., Androphy, E.J., and DiMaio, D. (1998). Transactivation-competent bovine papillomavirus E2 protein is specifically required for efficient repression of human papillomavirus oncogene expression and for acute growth inhibition of cervical carcinoma cell lines. *J. Virol.* 72:3925–3934.

Goodwin, E.C., Yang, E., Lee, C.-J., Lee, H.-W., DiMaio, D., and Hwang, E.-S. (2000). Rapid induction of senescence in human cervical carcinoma cells. *Proc. Natl. Acad. Sci. U. S. A.* 97:10978–10983.

Greenfield, I., Nickerson, J., Penman, S., and Stanley, M. (1991). Human papillomavirus 16 E7 protein is associated with the nuclear matrix. *Proc. Natl. Acad. Sci. U. S. A.* 88:11217–11221.

Griffin, H., Elston, R., Jackson, D., Ansell, K., Coleman, M., Winter, G., and Doorbar, J. (2006). Inhibition of papillomavirus protein function in cervical cancer cells by intrabody targeting. *J. Mol. Biol.* 355:360–378.

Grinstein, E., Wernet, P., Snijders, P.J., Rosl, F., Weinert, I., Jian, W., Kraft, R., Schewe, C., Schwabe, M., Hauptmann, S., Dietel, M., Meijer, C.J., and Royer, H.D. (2002). Nucleolin as activator of human papillomavirus type 18 oncogene transcription in cervical cancer. *J. Exp. Med.* 196:1067–1078.

Grm, H.S., Massimi, P., Gammoh, N., and Banks, L. (2005). Crosstalk between the human papillomavirus E2 transcriptional activator and the E6 oncoprotein. *Oncogene* 24:5149–5164.

Grossman, S.R., and Laimins, L.A. (1989). E6 protein of human papillomavirus type 18 binds zinc. *Oncogene* 4:1089–1093.

Guarguaglini, G., and Duncan, P.I. (2005). The forkhead-associated domain protein Cep170 interacts with Polo-like kinase 1 and serves as a marker for mature centrioles. *Mol. Biol. Cell* 16:1095–1107.

Gulliver, G.A., Herber, R.L., Liem, A., and Lambert, P.F. (1997). Both conserved region 1 (CR1) and CR2 of the human papillomavirus type 16 E7 oncogene are required for induction of epidermal hyperplasia and tumor formation in transgenic mice. *J. Virol.* 71:5905–5914.

Hahn, W.C., Stewart, S.A., Brooks, M.W., York, S.G., Eaton, E., Kurachi, A., Beijersbergen, R.L., Knoll, J.H.M., Meyerson, M., and Weinberg, R.A. (1999). Inhibition of telomerase limits the growth of human cancer cells. *Nature Med.* 5:1164–1170.

Halbert, C.L., Demers, G.W., and Galloway, D.A. (1991). The E7 gene of human papillomavirus type 16 is sufficient for immortalization of human epithelial cells. *J. Virol.* 65:473–478.

Halbert, C.L., Demers, G.W., and Galloway, D.A. (1992). The E6 and E7 genes of human papillomavirus type 6 have weak immortalizing activity in human epithelial cells. *J. Virol.* 66:2125–2134.

Hall, A.H.S., and Alexander, K.A. (2003). RNA interference of human papillomavirus type 18 E6 and E7 induces senescence in HeLa cells. *J. Virol.* 77:6066–6069.

Hamada, K., Alemany, R., Zhang, W.W., Hittelman, W.N., Lotan, R., Roth, J.A., and Mitchell, M.F. (1996a). Adenovirus-mediated transfer of a wild-type p53 gene and induction of apoptosis in cervical cancer. *Cancer Res.* 56:3047–3054.

Hamada, K., Sakaue, M., Alemany, R., Zhang, W.W., Horio, Y., Roth, J.A., and Mitchelle, M.F. (1996b). Adenovirus-mediated transfer of HPV 16 E6/E7 antisense RNA to human cervical cancer cells. *Gynecol. Oncol.* 63:219–227.

Hannon, G.J., and Beach, D. (1994). p15(INK4B) is a potential effector of TGF-beta-induced cell cycle arrest. *Nature* 371:257–261.

Hannon, G.J., Demetrick, D., and Beach, D. (1993). Isolation of the RB-related p130 through its interaction with CDK2 and cyclins. *Genes Dev.* 7:2378–2391.

Haupt, Y., Maya, R., Kazaz, A., and Oren, M. (1997). Mdm2 promotes the rapid degradation of p53. *Nature* 387:296–299.

Havard, L., Rahmouni, S., Boniver, J., and Delvenne, P. (2005). High levels of p105 (NFKB1) and p100 (NFKB2) proteins in HPV16-transformed keratinocytes: role of E6 and E7 oncoproteins. *Virology* 331:357–366.

Havre, P.A., Yuan, J., Hedrick, L., Cho, K.R., and Glazer, P.M. (1995). p53 inactivation by HPV16 E6 results in increased mutagenesis in human cells. *Cancer Res.* 55:4420–4424.

Hawley-Nelson, P., Vousden, K.H., Hubbert, N.L., Lowy, D.R., and Schiller, J.T. (1989). HPV16 E6 and E7 proteins cooperate to immortalize human foreskin keratinocytes. *EMBO J.* 8:3905–3910.

Heck, D.V., Yee, C.L., Howley, P.M., and Munger, K. (1992). Efficiency of binding the retinoblastoma protein correlates with the transforming capacity of the E7 oncoproteins of the human papillomaviruses. *Proc. Natl. Acad. Sci. U. S. A.* 89:4442–4446.

Helin, K., Lees, J.A., Vidal, M., Dyson, N., Harlow, E., and Fattaey, A. (1992). A cDNA encoding a pRB-binding protein with properties of the transcription factor E2F. *Cell* 70:337–350.

Helin, K., Wu, C.-L., Fattaey, A., Lees, J.A., Dynlacht, B., Ngwu, C., and Harlow, E. (1993). Heterodimerization of the transcription factors E2F-1 and DP-1 leads to cooperative trans-activation. *Genes Dev.* 7:1850–1861.

Helt, A.M., and Galloway, D.A. (2001). Destabilization of the retinoblastoma tumor suppressor by human papillomavirus type 16 E7 is not sufficient to overcome cell cycle arrest in human keratinocytes. *J. Virol.* 75:6737–6747.

Hengstermann, A., Linares, L.K., Ciechanover, A., Whitaker, N.J., and Scheffner, M. (2001). Complete switch from Mdm2 to human papillomavirus E6-mediated degradation of p53 in cervical cancer cells. *Proc. Natl. Acad. Sci. U. S. A.* 98:1218–1223.

Herber, R., Liem, A., Pitot, H.C., and Lambert, P.F. (1996). Squamous epithelial hyperplasia and carcinoma in mice transgenic for the human papillomavirus type 16 E7 oncogene. *J. Virol.* 70:1873–1881.

Hernando, E., Nahle, Z., Juan, G., Diaz-Rodriguez, E., Alaminos, M., Hermann, M., Michel, L., Mittal, V., Gerald, W., Benezra, R., Lowe, S.W., and Cordon-Cardo, C. (2004). Rb inactivation promotes genomic instability by uncoupling cell cycle progression from mitotic control. *Nature* 430:797–802.

Heselmeyer, K., Schrock, E., du Manoir, S., Blegen, H., Shah, K., Steinbeck, R., Auer, G., and Ried, T. (1996). Gain of chromosome 3q defines the transition from severe dysplasia to invasive carcinoma of the uterine cervix. *Proc. Natl. Acad. Sci. U. S. A.* 93:479–484.

Hickman, E.S., Picksley, S.M., and Vousden, K.H. (1994). Cells expressing HPV16 E7 continue cell cycle progression following DNA damage induced p53 activation. *Oncogene* 9:2177–2181.

Hietanen, S., Lain, S., Krausz, E., Blattner, C., and Lane, D.P. (2000). Activation of p53 in cervical carcinoma cells by small molecules. *Proc. Natl. Acad. Sci. U. S. A.* 97:8501–8506.

Horner, S.M., DeFilippis, R.A., Manuelidis, L., and DiMaio, D. (2004). Repression of the human papillomavirus E6 gene initiates p53-dependent, telomerase-independent senescence and apoptosis in HeLa cervical carcinoma cells. *J. Virol.* 78:4063–4073.

Hu, G., Liu, W., Hanania, E.G., Fu, S., Wang, T., and Deisseroth, A.B. (1995). Suppression of tumorigenesis by transcription units expressing the antisense E6 and E7 messenger RNA (mRNA) for the transforming proteins of the human papilloma virus and the sense mRNA for the retinoblastoma gene in cervical carcinoma cells. *Cancer Gene Ther.* 2:19–32.

Huang, L., Kinnucan, E., Wang, G., Beaudenon, S., Howley, P.M., Huibregtse, J.M., and Pavletich, N.P. (1999). Structure of an E6AP-UbcH7 complex: insights into ubiquitination by the E2-E3 enzyme cascade. *Science* 286:1321–1326.

Huang, P.S., Patrick, D.R., Edwards, G., Goodhart, P.J., Huber, H.E., Miles, L., Garsky, V.M., Oliff, A., and Heimbrook, D.C. (1993). Protein domains governing interactions between E2F, the retinoblastoma gene product, and human papillomavirus type 16 E7 protein. *Mol. Cell. Biol.* 13:953–960.

Huang, S.M., and McCance, D.J. (2002). Down regulation of the interleukin-8 promoter by human papillomavirus type 16 E6 and E7 through effects on CREB binding protein/p300 and P/CAF. *J. Virol.* 76:8710–8721.

Hubbert, N.L., Sedman, S.A., and Schiller, J.T. (1992). Human papillomavirus type 16 E6 increases the degradation rate of p53 in human keratinocytes. *J. Virol.* 66:6237–6241.

Hudson, J.B., Bedell, M.A., McCance, D.J., and Laimins, L.A. (1990). Immortalization and altered differentiation of human keratinocytes *in vitro* by the E6 and E7 open reading frames of human papillomavirus type 18. *J. Virol.* 64:519–526.

Huh, K.W., DeMasi, J., Ogawa, H., Nakatani, Y., Howley, P.M., and Munger, K. (2005). Association of the human papillomavirus type 16 E7 oncoprotein with the 600-kDa retinoblastoma protein-associated factor, p600. *Proc. Natl. Acad. Sci. U. S. A.* 102:11492–11497.

Huibregtse, J.M., Scheffner, M., Beaudenon, S., and Howley, P.M. (1995). A family of proteins structurally and functionally related to the E6-AP ubiquitin-protein ligase. *Proc. Natl. Acad. Sci. U. S. A.* 92:2563–2567.

Huibregtse, J.M., Scheffner, M., and Howley, P.M. (1991). A cellular protein mediates association of p53 with the E6 oncoprotein of human papillomavirus types 16 or 18. *EMBO J.* 10:4129–4135.

Huibregtse, J.M., Scheffner, M., and Howley, P.M. (1993a). Cloning and expression of the cDNA for E6-AP: a protein that mediates the interaction of the human papillomavirus E6 oncoprotein with p53. *Mol. Cell. Biol.* 13:775–784.

Huibregtse, J.M., Scheffner, M., and Howley, P.M. (1993b). Localization of the E6-AP regions that direct HPV E6 binding, association with p53, and ubiquination of associated proteins. *Mol. Cell. Biol.* 13:4918–4927.

Hwang, E.-S., Naeger, L.K., and DiMaio, D. (1996). Activation of the endogenous p53 growth inhibitory pathway in HeLa cervical carcinoma cells by expression of the bovine papillomavirus E2 gene. *Oncogene* 12:795–803.

Hwang, E.-S., Riese, D.J., II, Settleman, J., Nilson, L.A., Honig, J., Flynn, S., and DiMaio, D. (1993). Inhibition of cervical carcinoma cell line proliferation by introduction of a bovine papillomavirus regulatory gene. *J. Virol.* 67:3720–3729.

Hwang, S.G., Lee, D., Kim, J., Seo, T., and Choe, J. (2002). Human papillomavirus type 16 E7 binds to E2F1 and activates E2F1-driven transcription in a retinoblastoma protein-independent manner. *J. Biol. Chem.* 277:2923–2930.

Iftner, T., Elbel, M., Schopp, B., Hiller, T., Loizou, J.I., Caldecott, K.W., and Stubenrauch, F. (2002). Interference of papillomavirus E6 protein with single-strand break repair by interaction with XRCC1. *EMBO J.* 21:4741–4748.

Jenuwein, T., and Allis, C.D. (2001). Translating the histone code. *Science* 293:1074–1080.

Jeon, S., Allen, H.B., and Lambert, P.F. (1995). Integration of human papillomavirus type 16 into the human genome correlates with a selective growth advantage of cells. *J. Virol.* 69:2989–2997.

Jewers, R.J., Hildebrandt, P., Ludlow, J.W., Kell, B., and McCance, D.J. (1992). Regions of human papillomavirus type 16 E7 oncoprotein required for immortalization of human keratinocytes. *J. Virol.* 66:1329–1335.

Jian, Y., Schmidt-Grimminger, D.-C., Chien, W.-M., Wu, X., Broker, T.R., and Chow, L.T. (1998). Post-transcriptional induction of p21cip1 protein by human papillomavirus E7 inhibits unscheduled DNA synthesis reactivated in differentiated keratinocytes. *Oncogene* 17:2027–2038.

Jiang, M., and Milner, J. (2002). Selective silencing of viral gene expression in HPV-positive human cervical carcinoma cells treated with siRNA, a primer of RNA interference. *Oncogene* 21:6041–6048.

Jones, D.L., Alani, R.M., and Munger, K. (1997a). The human papillomavirus E7 oncoprotein can uncouple cellular differentiation and proliferation in human keratinocytes by abrogating p21^{Cip1}-mediated inhibition of cdk2. *Genes Dev.* 11:2101–2111.

Jones, D.L., and Munger, K. (1997). Analysis of the p53-mediated G_1 growth arrest pathway in cells expressing the human papillomavirus type 16 E7 oncoprotein. *J. Virol.* 71:2905–2912.

Jones, D.L., Thompson, D.A., and Munger, K. (1997b). Destabilization of the RB tumor suppressor protein and stabilization of p53 contribute to HPV type 16 E7-induced apoptosis. *Virology* 239:97–107.

Jones, D.L., Thompson, D.A., Suh-Burgmann, E., Grace, M., and Munger, K. (1999). Expression of the HPV E7 oncoprotein mimics but does not evoke a p53-dependent cellular DNA damage response pathway. *Virology* 258:406–414.

Kaelin, W.G., Jr., Krek, W., Sellers, W.R., DeCaprio, J.A., Ajchenbaum, F., Fuchs, C.S., Chittenden, T., Li, Y., Farnham, P.J., Blanar, M.A., Livingston, D.M., and Flemington, E.K. (1992). Expression cloning of a cDNA encoding a retinoblastoma-binding protein with E2F-like properties. *Cell* 70:351–364.

Kalejta, R.F., and Shenk, T. (2003). Proteasome-dependent, ubiquitin-independent degradation of the Rb family of tumor suppressors by the human cytomegalovirus pp71 protein. *Proc. Natl. Acad. Sci. U. S. A.* 100:3263–3268.

Kamio, M., Yoshida, T., Ogata, H., Douchi, T., Nagata, Y., Inoue, M., Hasegawa, M., Yonemitsu, Y., and Yoshimura, A. (2004). SOC1 inhibits HPV-E7-mediated transformation by inducing degradation of E7 protein. *Oncogene* 23:3107–3115.

Kanda, T., Furuno, A., and Yoshiike, K. (1988). Human papillomavirus type 16 open reading frame E7 encodes a transforming gene for rat 3Y1 cells. *J. Virol.* 62:610–613.

Kang, H.T., Ju, J.W., Cho, J.W., and Hwang, E.S. (2003). Down-regulation of Sp1 activity through modulation of *O*-glycosylation by treatment with a low glucose mimetic, 2-deoxyglucose. *J. Biol. Chem.* 278:51223–51231.

Kang, H.T., Lee, C.J., Seo, E.J., Bahn, Y.J., Kim, H.J., and Hwang, E.S. (2004). Transition to an irreversible state of senescence in HeLa cells arrested by repression of HPV E6 and E7 genes. *Mech. Ageing Dev.* 125:31–40.

Kao, W.H., Beaudenon, S.L., Talis, A.L., Huibregtse, J.M., and Howley, P.M. (2000). Human papillomavirus type 16 E6 induces self-ubiquitination of the E6AP ubiquitin-protein ligase. *J. Virol.* 74:6408–6417.

Katich, S.C., Zerfass-Thome, K., and Hoffmann, I. (2001). Regulation of the Cdc25A gene by the human papillomavirus type 16 E7 oncogene. *Oncogene* 20:543–550.

Kelley, W.L., and Georgopoulos, C. (1997). The T/t common exon of simian virus 40, JC, and BK polyomavirus T antigens can functionally replace the J-domain of the Escherichia coli DnaJ molecular chaperone. *Proc. Natl. Acad. Sci. U. S. A.* 94:3679–3684.

Kessis, T.D., Connolly, D.C., Hedrick, L., and Cho, K.R. (1996). Expression of HPV16 E6 or E7 increases integration of foreign DNA. *Oncogene* 13:427–431.

Kessis, T.D., Slebos, R.J.C., Nelson, W.G., Kastan, M.B., Plunkett, B.S., Han, S.M., Lorincz, A.T., Hedrick, L., and Cho, K.R. (1993). Human papillomavirus 16 E6 expression disrupts the p53-mediated cellular response to DNA damage. *Proc. Natl. Acad. Sci. U. S. A.* 90:3988–3992.

Khanna, K.K., and Jackson, S.P. (2001). DNA double-strand breaks: signaling, repair and the cancer connection. *Nature Genet.* 27:247–254.

Khleif, S.N., DeGregori, J., Yee, C.L., Otterson, G.A., Kaye, F.J., Nevins, J.R., and Howley, P.M. (1996). Inhibition of cyclin D-CDK4/CDK6 activity is associated with an E2F-mediated induction of cyclin kinase inhibitor activity. *Proc. Natl. Acad. Sci. U. S. A.* 93:4350–4354.

Kishino, T., Lalande, M., and Wagstaff, J. (1997). UBE3A/E6-AP mutations cause Andelman syndrome. *Nature Genet.* 15:70–73.

Kiyono, T., Foster, S.A., Koop, J.I., McDougall, J.K., Galloway, D.A., and Klein-gelhutz, A.J. (1998). Both Rb/p16^{INK4a} inactivation and telomerase activity are required to immortalize human epithelial cells. *Nature* 396:84–88.

Kiyono, T., Hiraiwa, A., Fujita, M., Hayashi, Y., Akiyama, T., and Ishibashi, M. (1997). Binding of high-risk human papillomavirus E6 oncoproteins to the human homologue of the Drosophila discs large tumor suppressor protein. *Proc. Natl. Acad. Sci. U. S. A.* 94:11612–11616.

Klaes, R., Friedrich, T., Spitkovsky, D., Ridder, R., Rudy, W., Petry, U., Dallenbach-Hellweg, G., Schmidt, D., and von Knebel Doeberitz, M. (2001). Overexpression of p16^{INK4A} as a specific marker for dysplastic and neoplastic epithelial cells of the cervix uteri. *Int. J. Cancer* 92:276–284.

Klausner, R.D. (2002). The fabric of cancer cell biology—weaving together the strands. *Cancer Cell* 1:3–10.

Klingelhutz, A.J., Foster, S.A., and McDougall, J.K. (1996). Telomerase activation by the E6 gene product of human papillomavirus type 16. *Nature* 380:79–82.

Knight, J.S., Sharma, N., and Robertson, E.S. (2005). Epstein-Barr virus latent antigen 3C can mediate the degradation of the retinoblastoma protein through an SCF cellular ubiquitin ligase. *Proc. Natl. Acad. Sci. U. S. A.* 102:18562–18566.

Kovelman, R., Bilter, G.K., Glezer, E., Tsou, A.Y., and Barbosa, M.S. (1996). Enhanced transcriptional activation by E2 proteins from the oncogenic human papillomaviruses. *J. Virol.* 70:7549–7560.

Kubbutat, M.H.G., Jones, S.N., and Vousden, K.H. (1997). Regulation of p53 stability by Mdm2. *Nature* 387:299–303.

Kühne, C., and Banks, L. (1998). E3-Ubiquitin Ligase/E6-AP links multicopy maintenance protein 7 to the ubiquitination pathway by a novel motif, the L2G Box. *J.Biol. Chem.* 273:34302–34309.

Kumar, A., Zhao, Y., Meng, G., Zeng, M., Srinivasan, S., Delmolino, L.M., Gao, Q., Dimri, G.P., Weber, G.F., Wazer, D.E., Band, H., and Band, V. (2002). Human papillomavirus oncoprotein E6 inactivates the transcriptional coactivator human ADA3. *Mol. Cell. Biol.* 22:5801–5812.

Kumar, S., Kao, W.H., and Howley, P.M. (1997). Physical interaction between specific E2 and HECT E3 enzymes determines functional cooperativity. *J. Biol. Chem.* 272:13548–13554.

Kumar, S., Talis, A.L., and Howley, P.M. (1999). Identification of HHR23A as a substrate for E6-associated protein-mediated ubiquitination. *J. Biol. Chem.* 274:18785–18792.

Kyo, S., Inoue, M., Hayasaka, N., Inoue, T., Yutsudo, M., Tanizawa, O., and Hakura, A. (1994). Regulation of early gene expression of human papillomavirus type 16 by inflammatory cytokines. *Virology* 200:130–139.

Lee, C., Chang, J.H., Lee, H.S., and Cho, Y.S. (2002a). Structural basis for the recognition of the E2F transactivation domain by the retinoblastoma tumor suppressor. *Genes Dev.* 16:3199–3212.

Lee, C., and Laimins, L.A. (2004). Role of the PDZ domain-binding motif of the oncoprotein E6 in the pathogenesis of human papillomavirus type 31. *J. Virol.* 78:12366–12377.

Lee, C.J., Suh, E.J., Kang, H.T., Im, J.S., Um, S.J., Park, J.S., and Hwang, E.S. (2002b). Induction of senescence-like state and suppression of telomerase activity through inhibition of HPV E6/E7 gene expression in cells immortalized by HPV16 DNA. *Exp. Cell Res.* 277:173–182.

Lee, D., Kim, H.-Z., Jeong, K.W., Shim, Y.S., Horikawa, I., Barrett, J.C., and Choe, J. (2002c). Human papillomavirus E2 down-regulates the human telomerase reverse transcriptase promoter. *J. Biol. Chem.* 277:27748–27756.

Lee, J.O., Russo, A.A., and Pavletich, N.P. (1998). Structure of the retinoblastoma tumour-suppressor pocket domain bound to a peptide from HPV E7. *Nature* 391:859–865.

Lee, S.J., Cho, Y.S., Cho, M.C., Shim, J.H., Lee, K.A., Ko, K.K., Choe, Y.K., Park, S.N., Hoshino, T., Kim, S., Dinarello, C.A., and Yoon, D.Y. (2001). Both E6 and E7 oncoproteins of human papillomavirus 16 inhibit IL-18-induced IFN-gamma production in human peripheral blood mononuclear and NK cells. *J. Immunol.* 167:497–504.

Lee, S.S., Glaunsinger, B., Mantovani, F., Banks, L., and Javier, R.T. (2000). Multi-PDZ domain protein MUPP1 is a cellular target for both adenovirus E4-ORF1 and high-risk papillomavirus type 18 E6 oncoproteins. *J. Virol.* 74:9680–9693.

Lees, E., Faha, B., Dulic, V., Reed, S.L., and Harlow, E. (1992). Cyclin E/cdk2 and cyclin A/cdk2 kinases associate with p107 and E2F in a temporally distinct manner. *Genes Dev.* 6:1874–1885.

Li, R., Knight, J.D., Jackson, S.P., Tjian, R., and Botchan, M.R. (1991). Direct interaction between Sp1 and the BPV enhancer E2 protein mediates synergistic activation of transcription. *Cell* 65:493–505.

Li, S., Labrecque, S., Gauzzi, M.C., Cuddihy, A.R., Wong, A.H., Pellegrini, S., Matlashewski, G., and Koromilas, A.E. (1999). The human papilloma virus (HPV)-18 E6 oncoprotein physically associates with Tyk2 and impairs Jak-STAT activation by interferon-alpha. *Oncogene* 18:5727–5737.

Lin, B.Y., Ma, T., Liu, J.S., Kuo, S.R., Jin, G., Broker, T.R., Harper, J.W., and Chow, L.T. (2000). HeLa cells are phenotypically limiting in cyclin E/CDK2 for efficient human papillomavirus DNA replication. *J. Biol. Chem.* 275:6167–6174.

Liu, X., Clements, A., Zhao, K., and Marmorstein, R. (2006). Structure of the human papillomavirus E7 oncoprotein and its mechanism for inactivation of the retinoblastoma tumor suppressor. *J. Biol. Chem.* 281:578–586.

Liu, X., Yuan, H., Fu, B., Disbrow, G.L., Apolinario, T., Tomaic, V., Kelley, M.L., Baker, C.C., Huibregtse, J., and Schlegel, R. (2005). The E6AP ubiquitin ligase is required for transactivation of the hTERT promoter by the human papillomavirus E6 oncoprotein. *J. Biol. Chem.* 280:10807–10816.

Liu, Y., Chen, J.J., Gao, Q., Dalal, S., Hong, Y., Mansur, C.P., Band, V., and Androphy, E.J. (1999). Multiple functions of human papillomavirus type 16 E6 contribute to the immortalization of mammary epithelial cells. *J. Virol.* 73:7297–7307.

Liu, Y., Liu, Z., Androphy, E.J., Chen, J., and Baleja, J.D. (2004). Design and characterization of helical peptides that inhibit the E6 protein of papillomavirus. *Biochemistry* 43:7421–7431.

Longworth, M.S., and Laimins, L.A. (2004a). The binding of histone deacetylases and the integrity of zinc finger-like motifs of the E7 protein are essential for the life cycle of human papillomavirus type 31. *J. Virol.* 78:3533–3541.

Longworth, M.S., and Laimins, L.A. (2004b). Pathogenesis of human papillomaviruses in differentiating epithelia. *Microbiol. Mol. Biol. Rev.* 68:362–372.

Luo, R.X., Postigo, A.A., and Dean, D.C. (1998). Rb interacts with histone deacetylase to repress transcription. *Cell* 92:463–473.

Luscher-Firzlaff, J.M., Westendorf, J.M., Zwicker, J., Burkhardt, H., Henriksson, M., Muller, R., Pirollet, F., and Luscher, B. (1999). Interaction of the fork head domain transcription factor MPP2 with the human papilloma virus 16 E7 protein: enhancement of transformation and transactivation. *Oncogene* 18:5620–5630.

Madeleine, M.M., Shera, K., Schwartz, S.M., Daling, J.R., Galloway, D.A., Wipf, G.C., Carter, J.J., McKnight, B., and McDougall, J.K. (2000). The p53 Arg72Pro polymorphism, human papillomavirus, and invasive squamous cell cervical cancer. *Cancer Epidemiol. Biomarkers Prevent.* 9:225–227.

Maehama, T., Patzelt, A., Lengert, M., Hutter, K.J., Kanazawa, K., Hausen, H., and Rosl, F. (1998). Selective down-regulation of human papillomavirus transcription by 2-deoxyglucose. *Int. J. Cancer* 76:639–646.

Magnaghi-Jaulin, L., Groisman, R., Naguibneva, I., Robin, P., Lorain, S., Le Villain, J.P., Troalen, F., Trouche, D., and Harel-Bellan, A. (1998). Retinoblastoma protein represses transcription by recruiting a histone deacetylase. *Nature* 391:601–605.

Mailand, N., Falck, J., Lukas, C., Syljuasen, R.G., Welcker, M., Bartek, J., and Lukas, J. (2000). Rapid destruction of human Cdc25A in response to DNA damage. *Science* 288:1425–1429.

Makni, H., Franco, E.L., Kaiano, J., Villa, L.L., Labrecque, S., Dudley, R., Storey, A., and Matlashewski, G. (2000). p53 polymorphism in codon 72 and risk of human papillomavirus induced cervical cancer: effect of inter-laboratory variation. *Int. J. Cancer* 87:528–533.

Mannhardt, B., Weinzimer, S.A., Wagner, M., Fiedler, M., Cohen, P., Jansen-Durr, P., and Zwerschke, W. (2000). Human papillomavirus type 16 E7 oncoprotein binds and inactivates growth-inhibitory insulin-like growth factor binding protein 3. *Mol. Cell. Biol.* 20:6483–6495.

Martin, L.G., Demers, G.W., and Galloway, D.A. (1998). Disruption of the G_1/S transition in human papillomavirus type 16 E7-expressing human cells is associated with altered regulation of cyclin E. *J. Virol.* 72:975–985.

Massague, J. (2004). G1 cell-cycle control and cancer. *Nature* 432:298–306.

Massimi, P., and Banks, L. (1997). Repression of p53 transcriptional activity by the HPV E7 proteins. *Virology* 227:255–259.

Massimi, P., and Banks, L. (2000). Differential phosphorylation of the HPV-16 E7 oncoprotein during the cell cycle. *Virology* 276:388–394.

Massimi, P., Pim, D., Storey, A., and Banks, L. (1996). HPV-16 E7 and adenovirus E1a complex formation with TATA box binding protein is enhanced by casein kinase II phosphorylation. *Oncogene* 12:2325–2330.

Matlashewski, G., Schneider, J., Banks, L., Jones, N., Murray, A., and Crawford, L. (1987). Human papillomavirus type 16 cooperates with activated *ras* in transforming primary cells. *EMBO J.* 6:1741–1746.

Matsuura, T., Sutcliffe, J.S., Fang, P., Galjaard, R.J., Jiang, Y.H., Benton, C.S., Rommens, J.M., and Beaudet, A.L. (1997). *De novo* truncating mutations in E6AP ubiquitin-protein ligase gene (UBE3A) in Angelman syndrome. *Nature Genet.* 15:74–77.

Matthews, K., Leong, C.M., Baxter, L., Inglis, E., Yun, K., Backstrom, B.T., Doorbar, J., and Hibma, M. (2003). Depletion of Langerhans cells in human papillomavirus type 16-infected skin is associated with E6-mediated down regulation of E-cadherin. *J. Virol.* 77:8378–8385.

Mavromatis, K.O., Jones, D.L., Mukherjee, R., Yee, C., Grace, M., and Münger, K. (1997). The carboxyl-terminal zinc-binding domain of the human papillomavirus E7 protein can be functionally replaced by the homologous sequences of the E6 protein. *Virus Res.* 52:109–118.

Mazurek, S., Zwerschke, W., Jansen-Durr, P., and Eigenbrodt, E. (2001). Effects of the human papilloma virus HPV-16 E7 oncoprotein on glycolysis and glutaminolysis: role of pyruvate kinase type M2 and the glycolytic-enzyme complex. *Biochem. J.* 356:247–256.

Mazzarelli, J.M., Atkins, G.B., Geisberg, J.V., and Ricciardi, R.P. (1995). The viral oncoproteins Ad5 E1A, HPV16 E7 and SV40 TAg bind a common region of the TBP-associated factor-110. *Oncogene* 11:1859–1864.

McIntyre, M.C., Frattini, M.G., Grossman, S.R., and Laimins, L.A. (1993). Human papillomavirus type 18 E7 protein requires intact Cys-X-X-Cys motifs for zinc binding, dimerization, and transformation but not for Rb binding. *J. Virol.* 67:3142–3150.

McIntyre, M.C., Ruesch, M.N., and Laimins, L.A. (1996). Human papillomavirus E7 oncoproteins bind a single form of cyclin E in a complex with cdk2 and p107. *Virology* 215:73–82.

McMurray, H.R., and McCance, D.J. (2003). Human papillomavirus type 16 E6 activates TERT gene transcription through induction of c-Myc and release of USF-mediated repression. *J. Virol.* 77:9852–9861.

McMurray, H.R., and McCance, D.J. (2004). Degradation of p53, not telomerase activation, by E6 is required for bypass of crisis and immortalization by human papillomavirus type 16 E6/E7. *J. Virol.* 78:5698–5706.

Meyers, C., Frattini, M.G., Hudson, J.B., and Laimins, L.A. (1992). Biosynthesis of human papillomavirus from a continuous cell line upon epithelial differentiation. *Science* 257:971–973.

Mi, Y., Borger, D.R., Fernandes, P.R., Pirisi, L., and Creek, K.E. (2000). Loss of transforming growth factor-beta (TGF-beta) receptor type I mediates TGF-beta resistance in human papillomavirus type 16- transformed human keratinocytes at late stages of in vitro progression. *Virology* 270:408–416.

Mietz, J.A., Unger, T., Huibregtse, J.M., and Howley, P.M. (1992). The transcriptional transactivation function of wild-type p53 is inhibited by SV40 large T-antigen and by HPV-16 oncoprotein. *EMBO J.* 11:5013–5020.

Missero, C., DiCunto, F., Kiyokawa, H., Koff, A., and Dotto, G.P. (1996). The absence of p21[Cip1]/WAF1 alters keratinocytes growth and differentiation and promotes ras-tumor progression. *Genes Dev.* 10:3065–3075.

Molinari, M., Mercurio, C., Dominiguez, J., Goubin, F., and Draetta, G.F. (2000). Human Cdc25 A inactivation in response to S phase inhibition and its role in preventing premature mitosis. *EMBO Rep.* 1:71–79.

Moon, M.S., Lee, C.J., Um, S.J., Park, J.S., Yang, J.M., and Hwang, E.S. (2001). Effect of BPV1 E2-mediated inhibition of E6/E7 expression in HPV16-positive cervical carcinoma cells. *Gynecol. Oncol.* 80:168–175.

Morozov, A., Shiyanov, P., Barr, E., Leiden, J.M., and Raychaudhuri, P. (1997). Accumulation of human papillomavirus type 16 E7 protein bypasses G_1 arrest induced by serum deprivation and by the cell cycle inhibitor p21. *J. Virol.* 71:3451–3457.

Mulligan, G.J., Wong, J., and Jacks, T. (1998). p130 is dispensable in peripheral T lymphocytes: evidence for functional compensation by p107 and pRB. *Mol. Cell. Biol.* 18:206–220.

Munger, K., Phelps, W.C., Bubb, V., Howley, P.M., and Schlegel, R. (1989a). The E6 and E7 genes of the human papillomavirus type 16 together are necessary and sufficient for transformation of primary human keratinocytes. *J. Virol.* 63:4417–4421.

Munger, K., Werness, B.A., Dyson, N., Phelps, W.C., Harlow, E., and Howley, P.M. (1989b). Complex formation of human papillomavirus E7 proteins with the retinoblastoma tumor suppressor gene product. *EMBO J.* 8:4099–4105.

Munger, K., Yee, C.L., Phelps, W.C., Pietenpol, J.A., Moses, H.L., and Howley, P.M. (1991). Biochemical and biological differences between E7 oncoproteins of the high- and low-risk human papillomavirus types are determined by amino-terminal sequences. *J. Virol.* 65:3943–3948.

Naeger, L.K., Goodwin, E.C., Hwang, E.-S., DeFilippis, R.A., Zhang, H., and DiMaio, D. (1999). Bovine papillomavirus E2 protein activates a complex growth-inhibitory program in p53-negative HT-3 cervical carcinoma cells that includes repression of cyclin A and cdc25A phosphatase genes and accumulation of hypophosphorylated retinoblastoma protein. *Cell Growth Differ.* 10:413–422.

Nakagawa, S., and Huibregtse, J.M. (2000). Human scribble (vartul) is targeted for ubiquitin-mediated degradation by the high-risk papillomavirus E6 proteins and the E6AP ubiquitin-protein ligase. *Mol. Cell. Biol.* 20:8244–8253.

Nakatani, Y., Konishi, H., Vassilev, A., Kurooka, H., Ishiguro, K., Sawada, J., Ikura, T., Korsmeyer, S.J., Qin, J., and Herlitz, A.M. (2005). p600, a unique protein required for membrane morphogenesis and cell survival. *Proc. Natl. Acad. Sci. U. S. A.* 102:15093–15098.

Narayanan, B.A., Holladay, E.B., Nixon, D.W., and Mauro, C.T. (1998). The effect of all-trans and 9-cis retinoic acid on the steady state level of HPV16 E6/E7 mRNA and cell cycle in cervical carcinoma cells. *Life Sci.* 63:565–573.

Nauenburg, S., Zwerschke, W., and Jansen-Durr, P. (2001). Induction of apoptosis in cervical carcinoma cells by peptide aptamers that bind to the HPV-16 E7 oncoprotein. *FASEB J.* 15:592–594.

Ndisang, D., Budhram-Mahadeo, V., and Latchman, D.S. (1999). The Brn-3a transcription factor plays a critical role in regulating human papilloma virus gene expression and determining the growth characteristics of cervical cancer cells. *J. Biol. Chem.* 274:28521–28527.

Ndisang, D., Budhram-Mahadeo, V., Pedley, B., and Latchman, D.S. (2001). The Brn-3a transcriptin factor plays a key role in regulating the growth of cervical cancer cells *in vivo. Oncogene* 20:4899–4903.

Neary, K., and DiMaio, D. (1989). Open reading frames E6 and E7 of bovine papillomavirus type 1 are both required for full transformation of mouse C127 cells. *J. Virol.* 63:259–266.

Nguyen, D.X., and McCance, D.J. (2005). Role of the retinoblastoma tumor suppressor protein in cellular differentiation. *J. Cell. Biochem.* 94:870–879.

Nguyen, M.L., Nguyen, M.M., Lee, D., Griep, A.E., and Lambert, P.F. (2003). The PDZ ligand domain of the human papillomavirus type 16 E6 protein is required for E6's induction of epithelial hyperplasia *in vivo. J. Virol.*77:6957–6964.

Nguyen, M., Song, S., Liem, A., Androphy, E., Liu, Y., and Lambert, P.F. (2002). A mutant of human papillomavirus type 16 E6 deficient in binding alpha-helix partners displays reduced oncogenic potential *in vivo. J. Virol.* 76:13039–13048.

Nishimura, A., Ono, T., Ishimoto, A., Dowhanick, J.J., Frizzell, M.A., Howley, P.M., and Sakai, H. (2000). Mechanisms of human papillomavirus E2-mediated repression of viral oncogene expression and cervical cancer cell growth inhibition. *J. Virol.* 74:3752–3760.

Nomine, Y., Charbonnier, S., Miguet, L., Potier, N., Van Dorsselaer, A., Atkinson, R.A., Trave, G., and Kieffer, B. (2005). 1H and 15N resonance assignment, secondary structure and dynamic behaviour of the C-terminal domain of human papillomavirus oncoprotein E6. *J. Biomol. NMR* 31:129–141.

Noya, F., Chien, W.-M., Broker, T.R., and Chow, L.T. (2001). p21cip1 degradation in differentiated keratinocytes is abrogated by co-stabilization with cyclin E induced by HPV E7. *J. Virol.* 75:6121–6134.

Oda, H., Kumar, S., and Howley, P.M. (1999). Regulation of the Src family tyrosine kinase Blk through E6AP-mediated ubiquitination. *Proc. Natl. Acad. Sci. U. S. A.* 96:9557–9562.

Oh, K.J., Kalinina, A., Wang, J., Nakayama, K., Nakayama, K.I., and Bagchi, S. (2004a). The papillomavirus E7 oncoprotein is ubiquitinated by UbcH7 and Cullin 1- and Skp2-containing E3 ligase. *J. Virol.* 78:5338–5346.

Oh, S.T., Kyo, S., and Laimins, L.A. (2001). Telomerase activation by human papillomavirus type 16 E6 protein: induction of human telomerase reverse transcriptase expression through Myc and GC-rich Sp1 binding sites. *J. Virol.* 75:5559–5566.

Oh, S.T., Longworth, M.S., and Laimins, L.A. (2004b). Roles of the E6 and E7 proteins in the life cycle of low-risk human papillomavirus type 11. *J. Virol.* 78:2620–2626.

Ojeda, J.M., Ampuero, S., Rojas, P., Prado, R., Allende, J.E., Barton, S.A., Chakraborty, R., and Rothhammer, F. (2003). p53 codon 72 polymorphism and risk of cervical cancer. *Biol. Res.* 36:279–283.

Oldak, M., Smola, H., Aumailley, M., Rivero, F., Pfister, H., and Smola-Hess, S. (2004). The human papillomavirus type 8 E2 protein suppresses b4-integrin expression in primary human keratinocytes. *J. Virol.* 78:10738–10746.

Pan, H., and Griep, A.E. (1994). Altered cell cycle regulation in the lens of HPV-16 E6 or E7 transgenic mice: implications for tumor suppressor gene function in development. *Genes Dev.* 8:1285–1299.

Park, J.S., Kim, E.J., Kwon, H.J., Hwang, E.S., Namkoong, S.E., and Um, S.J. (2000). Inactivation of interferon regulatory factor-1 tumor suppressor protein by HPV E7 oncoprotein. Implication for the E7-mediated immune evasion mechanism in cervical carcinogenesis. *J. Biol. Chem.* 275:6764–6769.

Park, R.B., and Androphy, E.J. (2002). Genetic analysis of high-risk E6 in episomal maintenance of human papillomavirus genomes in primary human keratinocytes. *J. Virol.* 76:11359–11364.

Patel, D., Huang, S.M., Baglia, L.A., and McCance, D.J. (1999). The E6 protein of human papillomavirus type 16 binds to and inhibits co-activation by CBP and p300. *EMBO J.* 18:5061–5072.

Patel, D. Incassati, A., Wang, N., and McCance, D.J. (2004). Human papillomavirus type 16 E6 and E7 cause polyploidy in human keratinocytes and up-regulation of G2-M-phase proteins. *Cancer Res.* 64:1299–1306.

Patrick, D.R., Oliff, A., and Heimbrook, D.C. (1994). Identification of a novel retinoblastoma gene product binding site on human papillomavirus type 16 E7 protein. *J. Biol. Chem.* 269:6842–6850.

Peacock, J.W., Chung, S., Bristow, R.G., Hill, R.P., and Benchimol, S. (1995). The p53-mediated G(1) checkpoint is retained in tumorigenic rat embryo fibroblast clones transformed by the human papillomavirus type 16 E7 gene and EJ-ras. *Mol. Cell. Biol.* 15:1446–1454.

Pedersen, S.N. (1975). Enzymatic studies of glycogen metabolism in nonmalignant and malignant biopsies from the human uterine cervix. *Acta Obstet. Gynecol. Scand.* 54:443–448.

Pennie, W.D., Grindlay, G.J., Cairney, M., and Campo, M.S. (1993). Analysis of the transforming functions of bovine papillomavirus type 4. *Virology* 193:614–620.

Perea, S.E., Massimi, P., and Banks, L. (2000). Human papillomavirus type 16 E7 impairs the activation of the interferon regulatory factor-1. *Int. J. Mol. Med.* 5:661–666.

Phelps, W.C., Bagchi, S., Barnes, J.A., Raychaudhuri, P., Kraus, V.B., Munger, K., Howley, P.M., and Nevins, J.R. (1991). Analysis of trans activation by human papillomavirus type 16 E7 and adenovirus 12S E1A suggests a common mechanism. *J. Virol.* 65:6922–6930.

Phelps, W.C., Munger, K., Yee, C.L., Barnes, J.A., and Howley, P.M. (1992). Structure-function analysis of the human papillomavirus type 16 E7 oncoprotein. *J. Virol.* 66:2418–2427.

Phelps, W.C., Yee, C.L., Munger, K., and Howley, P.M. (1988). The human papillomavirus type 16 E7 gene encodes transactivation and transformation functions similar to adenovirus E1a. *Cell* 53:539–547.

Phillips, A.C., and Vousden, K.H. (1997). Analysis of the interaction between human papillomavirus type 16 E7 and the TATA-binding protein, TBP. *J. Gen. Virol.* 78:905–909.

Pietenpol, J.A., Stein, R.W., Moran, E., Yaciuk, P., Schlegel, R., Lyons, R.M., Pittelkow, M.R., Münger, K., Howley, P.M., and Moses, H.L. (1990). TGFβ1 inhibition of c-myc transcription and growth in keratinocytes is abrogated by viral transforming proteins with pRB binding domains. *Cell* 61:777–785.

Polyak, K. (1996). Negative regulation of cell growth by TGF beta. *Biochim. Biophys. Acta Rev. Cancer* 1242:185–199.

Polyak, K., Kato, J.Y., Solomon, M.J., Sherr, C.J., Massague, J., Roberts, J.M., and Koff, A. (1994). p27(kip1), a cyclin-cdk inhibitor, links transforming growth factor-beta and contact inhibition to cell cycle arrest. *Genes Dev.* 8:9–22.

Psyrri, A., DeFilippis, R.A., Edwards, A.P.B., Yates, K.E., Manuelidis, L., and DiMaio, D. (2004). Role of the retinoblastoma pathway in senescence triggered by repression of the human papillomavirus E7 protein in cervical carcinoma cells. *Cancer Res.* 64:3079–3086.

Rehtanz, M., Schmidt, H.M., Warthorst, U., and Steger, G. (2004). Direct interaction between nucleosome assembly protein 1 and the papillomavirus E2 proteins involved in activation of transcription. *Mol. Cell. Biol.* 24:2153–2168.

Reinstein, E., Scheffner, M., Oren, M., Ciechanover, A., and Schwartz, A. (2000). Degradation of the E7 human papillomavirus oncoprotein by the ubiquitin- proteasome system: targeting via ubiquitination of the N-terminal residue. *Oncogene* 19:5944–5950.

Ren, B., Cam, H., Takahashi, Y., Volkert, T., Terragni, J., Young, R.A., and Dynlacht, B.D. (2002). E2F integrates cell cycle progression with DNA repair, replication, and G(2)/M checkpoints. *Genes Dev.* 16:245–256.

Reshkin, S.J., Bellizzi, A., Caldeira, S., Albarani, V., Malanchi, I., Poignee, M., Alunni-Fabbroni, M., Casavola, V., and Tommasino, M. (2000). Na + /H+ exchanger-dependent intracellular alkalinization is an early event in malignant transformation and plays an essential role in the development of subsequent transformation-associated phenotypes. *FASEB J.* 14:2185–2197.

Rey, O., Lee, S., Baluda, M.A., Swee, J., Ackerson, B., Chiu, R., and Park, N.H. (2000). The E7 oncoprotein of human papillomavirus type 16 interacts with F-actin *in vitro* and *in vivo*. *Virology* 268:372–381.

Reznikoff, C.A., Belair, C., Savelieva, E., Zhai, Y., Pfeifer, K., Yeager, T., Thompson, K.J., DeVries, S., Bindley, C., and Newton, M.A. (1994). Long-term genome stability and minimal genotypic and phenotypic alterations in HPV-16 E7-, but not E6-immortalized human uroepithelial cells. *Genes Dev.* 8:2227–2240.

Riley, R.R., Duensing, S., Brake, T., Munger, K., Lambert, P.F., and Arbeit, J.M. (2003). Dissection of human papillomavirus E6 and E7 function in transgenic mouse models of cervical carcinogenesis. *Cancer Res.* 63:4862–4871.

Roeder, G.E., Parish, J.L., Stern, P.L., and Gaston, K. (2004). Herpes simplex virus VP22-human papillomavirus E2 fusion proteins produced in mammalian or bacterial cells enter mammalian cells and induce apoptotic cell death. *Biotechnol. Appl. Biochem.* 40:157–165.

Romanczuk, H., Thierry, F., and Howley, P.M. (1990). Mutational analysis of *cis* elements involved in E2 modulation of human papillomavirus type 16 P_{97} and type 18 P_{105} promoters. *J. Virol.* 64:2849–2859.

Ronco, L.V., Karpova, A.Y., Vidal, M., and Howley, P.M. (1998). Human papillomavirus 16 E6 oncoprotein binds to interferon regulatory factor-3 and inhibits its transcriptional activity. *Genes Dev.* 12:2061–2072.

Rosl, F., Durst, M., and zur Hausen, H. (1988). Selective suppresion of human papillomavirus transcription in non tumorigenic cells by S-azacytidine. *EMBO J.* 7:1321–1328.

Ross, J.F., Liu, X., and Dynlacht, B.D. (1999). Mechanism of transcriptional repression of E2F by the retinoblastoma tumor suppressor protein. *Mol. Cell.* 3:195–205.

Ruesch, M.N., and Laimins, L.A. (1997). Initiation of DNA synthesis by human papillomavirus E7 oncoproteins is resistant to p21-mediated inhibition of cyclin E-cdk2 activity. *J. Virol.* 71:5570–5578.

Ruesch, M.N., and Laimins, L.A. (1998). Human papillomavirus oncoproteins alter differentiation-dependent cell cycle exit on suspension in semisolid medium. *Virology* 250:19–29.

Sanchez-Perez, A.-M., Soriano, S., Clarke, A.R., and Gaston, K. (1997). Disruption of the human papillomavirus type 16 E2 gene protects cervical carcinoma cells from E2F-induced apoptosis. *J. Gen. Virol.* 78:3009–3018.

Sang, B.-C., and Barbosa, M.S. (1992). Single amino acid substitutions in "low-risk" human papillomavirus (HPV) type 6 E7 protein enhance features characteristic of the "high-risk" HPV E7 oncoprotein. *Proc. Natl. Acad. Sci. U. S. A.* 89:8063–8067.

Sarver, N., Rabson, M.S., Yang, Y.C., Byrne, J.C., and Howley, P.M. (1984). Localization and analysis of bovine papillomavirus type 1 transforming functions. *J. Virol.* 52:377–388.

Sato, H., Watanabe, S., Furuno, A., and Yoshiike, K. (1989). Human papillomavirus type 16 E7 protein expressed in Eschericia coli and monkey COS-1 cells: immunofluorescence detection of the nuclear E7 protein. *Virology* 170:311–315.

Schaeffer, A.J., Nguyen, M., Liem, A., Lee, D., Montagna, C., Lambert, P.F., Ried, T., and Difilippantonio, M.J. (2004). E6 and E7 oncoproteins induce distinct patterns of chromosomal aneuploidy in skin tumors from transgenic mice. *Cancer Res.* 64:538–546.

Scheffner, M., Huibregtse, J.M., Vierstra, R.D., and Howley, P.M. (1993). The HPV-16 E6 and E6-AP complex functions as a ubiquitin-protein ligase in the ubiquitination of p53. *Cell* 75:495–505.

Scheffner, M., Munger, K., Byrne, J.C., and Howley, P.M. (1991). The state of the p53 and retinoblastoma genes in human cervical carcinoma cell lines. *Proc. Natl. Acad. Sci. U. S. A.* 88:5523–5527.

Scheffner, M., Nuber, U., and Huibregtse, J. (1995). Protein ubiquitination involving an E1-E2-E3 enzyme ubiquitin thioester cascade. *Nature* 373:81–83.

Scheffner, M., Werness, B.A., Huibregtse, J.M., Levine, A.J., and Howley, P.M. (1990). The E6 oncoprotein encoded by human papillomavirus type 16 and 18 promotes the degradation of p53. *Cell* 63:1129–1136.

Schiller, J.T., Vass, W.C., and Lowy, D.R. (1984). Identification of a second transforming region in bovine papillomavirus DNA. *Proc. Natl. Acad. Sci. U. S. A.* 81:7880–7884.

Schilling, B., De-Medina, T., Syken, J., Vidal, M., and Munger, K. (1998). A novel human DnaJ protein, hTid-1, a homolog of the Drosophila tumor suppressor protein Tid56, can interact with the human papillomavirus type 16 E7 oncoprotein. *Virology* 247:74–85.

Schmitt, A., Harry, J.B., Rapp, B., Wettstein, F.O., and Iftner, T. (1994). Comparison of the properties of the E6 and E7 genes of low- and high-risk cutaneous papillomaviruses reveals strongly transforming and high RB-binding activity for the E7 protein of the low-risk human papillomavirus type 1. *J. Virol.* 68:7051–7059.

Schneider-Gädicke, A., and Schwarz, E. (1986). Different human cervical carcinoma cell lines show similar transcription patterns of human papillomavirus type 18 early genes. *EMBO J.* 5:2285–2292.

Schneider-Maunoury, S., Croissant, O., and Orth, G. (1987). Integration of human papillomavirus type 16 DNA sequences: a possible early event in the progression of genital tumors. *J. Virol.* 61:3295–3298.

Seavey, S.E., Holubar, M., Saucedo, L.J., and Perry, M.E. (1999). The E7 oncoprotein of human papillomavirus type 16 stabilizes p53 through a mechanism independent of p19(ARF). *J. Virol.* 73:7590–7598.

Sellers, W.R., Novitch, B.G., Miyake, S., Heith, A., Otterson, G.A., Kaye, F.J., Lassar, A.B., and Kaelin, W.G., Jr. (1998). Stable binding to E2F is not required for the retinoblastoma protein to activate transcription, promote differentiation, and suppress tumor cell growth. *Genes Dev.* 12:95–106.

Shamanin, V.A., and Androphy, E.J. (2004). Immortalization of human mammary epithelial cells is associated with inactivation of the p14ARF-p53 pathway. *Mol. Cell. Biol.* 24:2144–2152.

Sheng, Q., Denis, D., Ratnofsky, M., Roberts, T.M., DeCaprio, J.A., and Schaffhausen, B. (1997). The DnaJ domain of polyomavirus large T antigen is required to regulate Rb family tumor suppressor function. *J. Virol.* 71:9410–9416.

Silver, P.A., and Way, J.C. (1993). Eukaryotic dnaJ homologs and the specificity of hsp70 activity. *Cell* 74:5–6.

Singh, L., Gao, Q., Kumar, A., Gotoh, T., Wazer, D.E., Band, H., Feig, L.A., and Band, V. (2003). The high-risk human papillomavirus type 16 E6 counters the GAP function of E6TP1 toward small Rap G proteins. *J. Virol.* 77:1614–1620.

Slebos, R.J.C., Lee, M.H., Plunkett, B.S., Kessis, T.D., Williams, B.O., Jacks, T., Hedrick, L., Kastan, M.B., and Cho, K.R. (1994). p53-dependent G(1) arrest involves pRB-related proteins and is disrupted by the human papillomavirus 16 E7 oncoprotein. *Proc. Natl. Acad. Sci. U. S. A.* 91:5320–5324.

Smith-McCune, K., Kalman, D., Robbins, C., Shivakumar, S., Yuschenkoff, L., and Bishop, J.M. (1999). Intranuclear localization of human papillomavirus 16 E7 during transformation and preferential binding of E7 to the Rb family member p130. *Proc. Natl. Acad. Sci. U. S. A.* 96:6999–7004.

Smotkin, D., and Wettstein, F.O. (1986). Transcription of human papillomavirus type 16 early genes in cervical cancer and a cerivcal cancer derived cell line and identification of the E7 protein. *Proc. Natl. Acad. Sci. U. S. A.* 83:4680–4684.

Smotkin, D., and Wettstein, F.O. (1987). The major human papillomavirus protein in cervical cancers is a cytoplasmic phosphoprotein. *J. Virol.* 61:1686–1689.

Song, S., Gulliver, G.A., and Lambert, P.F. (1998). Human papillomavirus type 16 E6 and E7 oncogenes abrogate radiation-induced DNA damage responses *in vivo* through p53-dependent and p53-independent pathways. *Proc. Natl. Acad. Sci. U. S. A.* 95:2290–2295.

Song, S., Liem, A., Miller, J.A., and Lambert, P.F. (2000). Human papillomavirus types 16 E6 and E7 contribute differently to carcinogenesis. *Virology* 267:141–150.

Song, S., Pitot, H.C., and Lambert, P.F. (1999). The human papillomavirus type 16 E6 alone is sufficient to induce carcinomas in transgenic animals. *J. Virol.* 73:5887–5893.

Southern, S.A., Lewis, M.H., and Herrington, C.S. (2004). Induction of tetrasomy by human papillomavirus type 16 E7 protein is independent of pRb binding and disruption of differentiation. *Br. J. Cancer* 90:1949–1954.

Southern, S.A., Noya, F., Meyers, C., Broker, T.R., Chow, L.T., and Herrington, C.S. (2001). Tetrasomy is induced by human papillomavirus type 18 E7 gene expression in keratinocyte raft cultures. *Cancer Res.* 61:4858–4863.

Spitkovsky, D., Hehner, S.P., Hofmann, T.G., Moller, A., and Schmitz, M.L. (2002). The human papillomavirus oncoprotein E7 attenuates NF-kappa B activation by targeting the Ikappa B kinase complex. *J. Biol. Chem.* 277:25576–25582.

Spruck, C.H., Won, K.A., and Reed, S.I. (1999). Deregulated cyclin E induces chromosome instability. *Nature* 401:297–300.

Srivenugopal, K.S., and Ali-Osman, F. (2002). The DNA repair protein, O(6)-methylguanine-DNA methyltransferase is a proteolytic target for the E6 human papillomavirus oncoprotein. *Oncogene* 21:5940–5945.

Stearns, T. (2001). Centrosome duplication: a centriolar pas de deux. *Cell* 105:417–420.

Steele, C., Cowsert, L.M., and Shillitoe, E.J. (1993). Effects of human papillomavirus type 18-specific antisense oligonucleotides on the transformed phenotype of human carcinoma cell lines. *Cancer Res.* 53:2330–2337.

Steele, C., Sacks, P.G., Adler-Storthz, K., and Shillitoe, E.J. (1992). Effect on cancer cells of plasmids that express antisense RNA of human papillomavirus type 18. *Cancer Res.* 52:4706–4711.

Steger, G., and Corbach, S. (1997). Dose-dependent regulation of the early promoter of human papillomavirus type 18 by the viral E2 protein. *J. Virol.* 71:50–58.

Sterlinko, H., Weber, M., Elston, R., McIntosh, P., Griffin, H., Banks, L., and Doorbar, J. (2004). Inhibition of E6-induced degradation of its cellular substrates by novel blocking peptides. *J. Mol. Biol.* 335:971–985.

Stoppler, H., Conrad Stoppler, M., Johnson, E., Simbulan-Rosenthal, C.M., Smulson, M.E., Iyer, S., Rosenthal, D.S., and Schlegel, R. (1998). The E7 protein of human papillomavirus type 16 sensitizes primary human keratinocytes to apoptosis. *Oncogene* 17:1207–1214.

Storchova, Z., and Pellman, D. (2004). From polyploidy to aneuploidy, genome instability and cancer. *Nature Rev. Mol. Cell. Biol.* 5:45–54.

Storey, A., Oates, D., Banks, L., Crawford, L., and Crook, T. (1991). Anti-sense phosphorothioate oligonucleotides have both specific and non-specific effects on cells containing human papillomavirus type 16. *Nucleic Acids Res.* 19:4109–4114.

Storey, A., Thomas, M., Kalita, A., Harwood, C., Gardiol, D., Mantovani, F., Breuer, J., Leigh, I.M., Matlashewski, G., and Banks, L. (1998). Role of a p53 polymorphism in the development of human papillomavirus-assocated cancer. *Nature* 393:229–234.

Stott, F.J., Bates, S., James, M.C., McConnell, B.B., Starborg, M., Brookes, S., Palmero, I., Ryan, K., Hara, E., Vousden, K.H., and Peters, G. (1998). The alternative product from the human CDKN2A locus, p14(ARF), participates in a regulatory feedback loop with p53 and MDM2. *EMBO J.* 17:5001–5014.

Stubdal, H., Zalvide, J., Campbell, K.S., Schweitzer, C., Roberts, T.M., and DeCaprio, J.A. (1997). Inactivation of pRB-related proteins p130 and p107 mediated by the J domain of simian virus 40 large T antigen. *Mol. Cell. Biol.* 17:4979–4990.

Stubdal, H., Zalvide, J., and Decaprio, J.A. (1996). Simian virus 40 large T antigen alters the phosphorylation state of the RB-related proteins p130 and p107. *J. Virol.* 70:2781–2788.

Stubenrauch, F., and Laimins, L.A. (1999). Human papillomavirus life cycle: active and latent phases. *Sem. Cancer Biol.* 9:379–386.

Talis, A.L., Huibregtse, J.M., and Howley, P.M. (1998). The role of E6AP in the regulation of p53 protein levels in human papillomavirus (HPV) positive and HPV negative cells. *J. Biol. Chem.* 273:6439–6445.

Tan, T.M., and Ting, R.C. (1995). In vitro and in vivo inhibition of human papillomavirus type 16 E6 and E7 genes. *Cancer Res.* 55:4599–4605.

Tasaki, T., Mulder, L.C., Iwamatsu, A., Lee, M.J., Davydov, I.V., Varshavsky, A., Muesing, M., and Kwon, Y.T. (2005). A family of mammalian E3 ubiquitin ligases that contain the UBR box motif and recognize N-degrons. *Mol. Cell. Biol.* 25:7120–7136.

Thierry, F., Benotmane, M.A., Demeret, C., Mori, M., Teissier, S., and Desaintes, C. (2004). A genomic approach reveals a novel mitotic pathway in papillomavirus carcinogenesis. *Cancer Res.* 64:895–903.

Thierry, F., and Howley, P.M. (1991). Functional analysis of E2-mediated repression of the HPV18 P105 promoter. *New Biol.* 3:90–100.

Thierry, F., and Yaniv, M. (1987). The BPV1-E2 trans-acting protein can be either an activator or a repressor of the HPV18 regulatory region. *EMBO J.* 6:3391–3397.

Thomas, D.M., Yang, H.S., Alexander, K.A., and Hinds, P.W. (2003). Role of the retinoblastoma protein in differentiation and senescence. *Cancer Biol. Ther.* 2:124–130.

Thomas, J.T., Hubert, W.G., Ruesch, M.N., and Laimins, L.A. (1999). Human papillomavirus type 31 oncoproteins E6 and E7 are required for the maintenance of episomes during the viral life cycle in normal human keratinocytes. *Proc. Natl. Acad. Sci. U. S. A.* 96:8449–8454.

Thomas, J.T., and Laimins, L.A. (1998). Human papillomavirus oncoproteins E6 and E7 independently abrogate the mitotic spindle checkpoint. *J. Virol.* 72:1131–1137.

Thomas, M., and Banks, L. (1998). Inhibition of Bak-induced apoptosis by HPV-18 E6. *Oncogene* 17:2943–2954.

Thomas, M., and Banks, L. (1999). Human papillomavirus (HPV) E6 interactions with Bak are conserved amongst E6 proteins from high and low risk HPV types. *J. Gen. Virol.* 80:1513–1517.

Thomas, M., Laura, R., Hepner, K., Guccione, E., Sawyers, C., Lasky, L., and Banks, L. (2002). Oncogenic human papillomavirus E6 proteins target the MAGI-2 and MAGI-3 proteins for degradation. *Oncogene* 21:5088–5096.

Thompson, D.A., and Belinsky, G. (1997). The human papillomavirus-16 E6 oncoprotein decreases the vigilance of mitotic checkpoints. *Oncogene* 15:3025–3036.

Thompson, D.A., Zacny, V., Belinsky, G.S., Clason, M., Jones, D.L., Schlegel, R., and Münger, K. (2001). The HPV E7 oncoprotein E7 inhibits tumor necrosis factor a-mediated apoptosis in normal human fibroblasts. *Oncogene* 20:3629–3640.

Tommasino, M., Adamczewski, J.P., Carlotti, F., Barth, C.F., Manetti, R., Contorni, M., Cavalieri, F., Hunt, T., and Crawford, L. (1993). HPV16 E7 protein associates with the protein kinase p33CDK2 and cyclin A. *Oncogene* 8:195–202.

Tong, X., Boll, W., Kirschhausen, T., and Howley, P.M. (1998). Interaction of the bovine papillomavirus E6 protein with the clatherin adaptor complex AP-1. *J. Virol.* 72:476–482.

Tong, X., and Howley, P.M. (1997). The bovine papillomavirus E6 oncoprotein interacts with paxillin and disrupts the actin cytoskeleton. *Proc. Natl. Acad. Sci. U. S. A.* 94:4412–4417.

Tong, X., Salgia, R., Li, J.-L., Griffin, J.D., and Howley, P.M. (1997). The bovine papillomavirus E6 protein binds to the LD motif repeats of paxillin and blocks its interaction with vinculin and the focal adhesion kinase. *J. Biol. Chem.* 272:33373–33376.

Trimarchi, J.M., and Lees, J.A. (2002). Sibling rivalry in the E2F family. *Nature Rev. Mol. Cell. Biol.* 3:11–20.

Uchida, C., Miwa, S., Kitagawa, K., Hattori, T., Isobe, T., Otani, S., Oda, T., Sugimura, H., Kamijo, T., Ookawa, K., Yasuda, H., and Kitagawa, M. (2005). Enhanced Mdm2 activity inhibits pRB function via ubiquitin-dependent degradation. *EMBO J.* 24:160–169.

Vande Pol, S.B., Brown, M.C., and Turner, C.E. (1998). Association of bovine papillomavirus type 1 E6 oncoprotein with the focal adhesion protein paxillin through a conserved protein interaction motif. *Oncogene* 16:43–52.

Veldman, T., Horikawa, I., Barrett, J.C., and Schlegel, R. (2001). Transcriptional activation of the telomerase hTERT gene by human papillomavirus type 16 E6 oncoprotein. *J. Virol.* 75:4467–4472.

Venturini, F., Braspenning, J., Homann, M., Gissmann, L., and Sczakiel, G. (1999). Kinetic selection of HPV 16 E6/E7-directed antisense nucleic acids: anti-proliferative effects on HPV 16-transformed cells. *Nucleic Acids Res.* 27:1585–1592.

Villa, L.L., Vieira, K.-B.-L., Pei, X.F., and Schlegel, R. (1992). Differential effect of tumor necrosis factor on proliferation of primary human keratinocytes and cell lines containing human papillomavirus types 16 and 18. *Mol. Carcinog.* 6:5–9.

von Knebel Doeberitz, M., Bauknecht, T., Bartsch, D., and zur Hausen, H. (1991). Influence of chromosomal integration on glucocorticoid-regulated transcription of growth-stimulating papillomavirus genes E6 and E7 in cervical carcinoma cells. *Proc. Natl. Acad. Sci. U. S. A.* 88:1411–1415.

von Knebel Doeberitz, M., Oltersdorf, T., Schwarz, E., and Gissmann, L. (1998). Correlation of modified human papilloma virus early gene expression with altered growth properties in C4-1 cervical carcinoma cells. *Cancer Res.* 48:3780–3786.

von Knebel Doeberitz, M., Rittmuller, C., Aengeneyndt, F., Jansen-Durr, P., and Spitkovsky, D. (1994). Reversible repression of papillomavirus oncogene expression in cervical carcinoma cells: consequences for the phenotype and E6-p53 and E7-pRB interactions. *J. Virol.* 68:2811–2821.

von Knebel Doeberitz, M., Rittmuller, C., zur Hausen, H., and Durst, M. (1992). Inhibition of tumorigenicity of cervical cancer cells in nude mice by HPV E6-E7 anti-sense RNA. *Int. J. Cancer* 51:831–834.

Vousden, K.H., Doninger, J., DiPaolo, J.A., and Lowy, D.R. (1988). The E7 open reading frame of human papillomavirus type 16 encodes a transforming gene. *Oncogene Res.* 3:167–175.

Vousden, K.H., and Jat, P.S. (1989). Functional similarity between HPV16 E7, SV40 large T and adenovirus E1a proteins. *Oncogene* 4:153–158.

Vousden, K.H., Vojtesek, B., Fisher, C., and Lane, D. (1993). HPV-16 E7 or adenovirus E1A can overcome the growth arrest of cells immortalized with a temperature-sensitive p53. *Oncogene* 8:1697–1702.

Warburg, O. (1936). *Ueber den Stoffwechsel der Tumoren.* Berlin: Springer.

Watanabe, S., Kanda, T., and Yoshiike, K. (1993). Growth dependence of human papillomavirus 16 DNA-positive cervical cancer cell lines and human papillomavirus 16-transformed human and rat cells on the viral oncoproteins. *Jpn. J. Cancer Res.* 84:1043–1049.

Wathelet, M., Lin, C.H., Parekh, B., Ronco, L.V., Howley, P.M., and Maniatis, T. (1998). Virus infection induces the assembly of coordinately activated transcription factors on the IFN-b enhancer *in vivo. Mol. Cell* 1:507–518.

Wazer, D.E., Liu, X.L., Chu, Q., Gao, Q., and Band, V. (1995). Immortalization of distinct human mammary epithelial cell types by human papilloma virus 16 E6 or E7. *Proc. Natl. Acad. Sci. U. S. A.* 92:3687–3691.

Webster, K., Parish, J., Pandya, M., Stern, P.L., Clarke, A.R., and Gaston, K. (2000). The human papillomavirus (HPV) 16 E2 protein induces apoptosis in the absence of other HPV proteins and via a p53-dependent pathway. *J. Biol. Chem.* 275:87–94.

Weinmann, A.S., Yan, P.S., Oberley, M.J., Huang, T.H., and Farnham, P.J. (2002). Isolating human transcription factor targets by coupling chromatin immunoprecipitation and CpG island microarray analysis. *Genes Dev.* 16:235–244.

Wells, S.I., Aronow, B.J., Wise, T.M., Williams, S.S., Couget, J.A., and Howley, P.M. (2003). Transcriptome signature of irreversible senescence in human papillomavirus-positive cervical cancer cells. *Proc. Natl. Acad. Sci. U. S. A.* 100:7093–7098.

Wells, S.I., Francis, D.A., Karpova, A.Y., Dowhanick, J.J., Benson, J.D., and Howley, P.M. (2000). Papillomavirus E2 induces senescence in HPV-positive cells via pRB- and p21^{CIP}-dependent pathways. *EMBO J.* 19:5762–5771.

Werness, B.A., Levine, A.J., and Howley, P.M. (1990). Association of human papillomavirus types 16 and 18 E6 proteins with p53. *Science* 248:76–79.

White, A.E., Livanos, E.M., and Tlsty, T.D. (1994). Differential disruption of genomic integrity and cell cycle regulation in normal human fibroblasts by the HPV oncoproteins. *Genes Dev.* 8:666–677.

Winkler, B., Crum, C.P., Fujii, T., Ferenczy, A., Boon, M., Braun, L., Lancaster, W.D., and Richart, R.M. (1984). Koilocytotic lesions of the cervix: the relationship of mitotic abnormalities to the presence of papillomavirus antigens and nuclear DNA content. *Cancer* 53:1081–1087.

Wise-Draper, T.M., Allen, H.V., Thobe, M.N., Jones, E.E., Habash, K.B., Munger, K., and Wells, S.I. (2005). The human DEK proto-oncogene is a senescence inhibitor and an upregulated target of high-risk human papillomavirus E7. *J. Virol.* 79:14309–14317.

Woodworth, C.D., Lichti, U., Simpson, S., Evans, C.H., and DiPaolo, J.A. (1992). Leukoregulin and gamma-interferon inhibit human papillomavirus type 16 gene transcription in human papillomavirus-immortalized human cervical cells. *Cancer Res.* 52:456–463.

Woodworth, C.D., Notario, V., and DiPaolo, J.A. (1990). Transforming growth factors beta 1 and 2 transcriptionally regulate human papillomavirus (HPV) type 16 early gene expression in HPV-immortalized human genital epithelial cells. *J. Virol.* 64:4767–4775.

Wu, E.W., Clemens, K.E., Heck, D.V., and Munger, K. (1993). The human papillomavirus E7 oncoprotein and the cellular transcription factor E2F bind to separate sites on the retinoblastoma tumor suppressor protein. *J. Virol.* 67:2402–2407.

Wu, L., Goodwin, E.C., Naeger, L.K., Vigo, E., Galaktionov, K., Helin, K., and DiMaio, D. (2000). E2F-Rb complexes assemble and inhibit cdc25A transcription in cervical carcinoma cells following repression of human papillomavirus oncogene expression. *Mol. Cell. Biol.* 20:7059–7067.

Wu, X., Xiao, W., and Brandsma, J.L. (1994). Papilloma formation by cottontail rabbit papillomavirus requires E1 and E2 regulatory genes in addition to E6 and E7 transforming genes. *J. Virol.* 68:6097–6102.

Xiao, B., Spencer, J., Clements, A., Ali-Khan, N., Mittnacht, S., Bronceno, C., Burghammer, M., Perrakis, A., Marmorstein, R., and Gamblin, S.J. (2003). Crystal structure of the retinoblastoma tumor suppressor protein bound to E2F and the molecular basis of its regulation. *Proc. Natl. Acad. Sci. U. S. A.* 100:2363–2368.

Yang, Y.C., Okayama, H., and Howley, P.M. (1985). Bovine papillomavirus contains multiple transforming genes. *Proc. Natl. Acad. Sci. U. S. A.* 82:1030–1034.

Yasumoto, S., Burkhardt, A.L., Doninger, J., and DiPaolo, J. (1986). Human papillomavirus type 16 DNA-induced malignant transformation of NIH 3T3 cells. *J. Virol.* 57:572–577.

You, J., Croyle, J.L., Nishimura, A., Ozato, K., and Howley, P.M. (2004). Interaction of the bovine papillomavirus E2 protein with Brd4 tethers the viral DNA to host mitotic chromosomes. *Cell* 117:349–360.

Yutsudo, M., Okamoto, Y., and Hakura, A. (1988). Functional dissociation of transforming genes of human papillomavirus type 16. *Virology* 166:594–597.

Zalvide, J., Stubdal, H., and DeCaprio, J.A. (1998). The J domain of simian virus 40 large T antigen is required to functionally inactivate RB family proteins. *Mol. Cell. Biol.* 18:1408–1415.

Zatsepina, O., Braspenning, J., Robberson, D., Hajibagheri, M.A., Blight, K.J., Ely, S., Hibma, M., Spitkovsky, D., Trendelenburg, M., Crawford, L., and Tommasino, M. (1997). The human papillomavirus type 16 E7 protein is associated with the nucleolus in mammalian and yeast cells. *Oncogene* 14:1137–1145.

Zerfass, K., Schulze, A., Spitkovsky, D., Friedman, V., Henglein, B., and Jansen-Durr, P. (1995). Sequential activation of cyclin E and cyclin A gene expression by human papillomavirus type 16 E7 through sequences necessary for transformation. *J. Gen. Virol.* 69:6389–6399.

Zerfass-Thome, K., Zwerschke, W., Mannhardt, B., Tindle, R., Botz, J.W., and Jansen-Durr, P. (1996). Inactivation of the cdk inhibitor p27KIP1 by the human papillomavirus type 16 E7 oncoprotein. *Oncogene* 13:2323–2330.

Zhang, B., Chen, W., and Roman, A. (2006). The E7 proteins of low- and high-risk human papillomaviruses share the ability to target the pRB family member p130 for degradation. *Proc. Natl. Acad. Sci. U. S. A.* 103:437–442.

Zhang, B., Laribee, R.N., Klemsz, M.J., and Roman, A. (2004). Human papillomavirus type 16 E7 protein increases acetylation of histone H3 in human foreskin keratinocytes. *Virology* 329:189–198.

Zhang, Y., Xiong, Y., and Yarbrough, W.G. (1998). ARF promotes MDM2 degradation and stabilizes p53: ARF-INK4a locus deletion impairs both the Rb and p53 tumor suppression pathways. *Cell* 92:725–734.

Zimmermann, H., Degenkolbe, R., Bernard, H.U., and O'Connor, M.J. (1999). The human papillomavirus type 16 E6 oncoprotein can down-regulate p53 activity by targeting the transcriptional coactivator CBP/p300. *J. Virol.* 73:6209–6219.

Zwerschke, W., Mannhardt, B., Massimi, P., Nauenburg, S., Pim, D., Nickel, W., Banks, L., Reuser, A.J., and Jansen-Durr, P. (2000). Allosteric activation of acid alpha-glucosidase by the human papillomavirus E7 protein. *J. Biol. Chem.* 275:9534–9541.

Zwerschke, W., Mazurek, S., Massimi, P., Banks, L., Eigenbrodt, E., and Jansen-Durr, P. (1999). Modulation of type M2 pyruvate kinase activity by the human papillomavirus type 16 E7 oncoprotein. *Proc. Natl. Acad. Sci. U. S. A.* 96:1291–1296.

11
In Vivo Models for the Study of Animal and Human Papillomaviruses

Paul F. Lambert[1] and Anne E. Griep[2]

[1] McArdle Laboratory for Cancer Research, Department of Oncology, and
[2] Department of Anatomy, University of Wisconsin, Madison, WI 53706

11.1. Introduction

Much of our understanding of papillomaviruses, their life cycle, and their oncogenic potential arose from studies of animal papillomaviruses beginning in the early 1900s with the identification of a transmissible agent responsible for tumors in wild rabbits. These studies led to the identification of the cottontail rabbit papillomavirus (CRPV) or Shope papillomavirus, followed by the identification of bovine papillomaviruses, most notably BPV1, and its study in tissue culture in the 1970s and 1980s. In these studies, the transforming potential of papillomaviruses in tissue culture was first elucidated. These viruses and the canine oral papillomavirus became important animal virus models for the development of prophylactic vaccines for human papillomaviruses. With the advent of transgenic mouse technologies in the 1980s and the recognition that HPVs are associated with human cancer, studies were undertaken to establish in vivo animal models of the role of HPVs in cancer. In this chapter, we review the literature pertinent to these areas of study, with focus given to studies in the natural hosts in the case of animal papillomaviruses and transgenic mice in the context of human papillomaviruses.

11.2. Animal Papillomaviruses

This section provides a brief overview of animal papillomaviruses, the diseases they cause in their natural hosts, and their investigation in the laboratory. Many prior reviews have been written on these viruses and many different aspects of these viruses are covered in other chapters of this book. The description given here focuses on in vivo studies and is meant to provide a basis for comparison to studies described later in this chapter regarding transgenic mouse models for human papillomaviruses.

11.2.1.   Rabbit Papillomaviruses

In 1935, Shope discovered a transmissible agent, subsequently identified as the cottontail rabbit papillomavirus (CRPV), which was responsible for warts arising in cottontail rabbits and domestic rabbits (Shope, 1935). Rous and Beard discovered that these warts can progress to squamous carcinomas, providing the first demonstration that DNA viruses can induce cancer (Rous and Beard, 1934). CRPV infects the epidermis where it induces benign papillomas in the vast majority of infected cottontail rabbits. In a quarter of these animals, progression to squamous cell carcinoma is observed by 18 months. In contrast, 80 percent or greater of domestic rabbits develop frank cancers within 12 months, suggestive of a host genetic contribution to cancer progression (Jeckel et al., 2002). Furthermore, little to no infectious virus is produced in lesions arising in domestic rabbits, in contrast to abundant virus production in lesions in the cottontail rabbit natural hosts (Nasseri and Wettstein, 1984; Zeltner et al., 1994). The use of in vivo DNA transfection methodologies using DNA gun technologies allowed the dissection of the individual roles of viral genes in the induction of disease in the context of the domestic rabbit laboratory host. Based on these studies, it was found that the E6, E7, E4, E2, and L1 ORFs were required for the induction of warts, whereas, the E5 and L2 ORFs were dispensible (Brandsma et al., 1992; Nasseri et al., 1989; Peh et al., 2004; Meyers and Wettstein, 1991; Wu et al., 1994). In addition to the genetic analysis of gene function in vivo, investigators made use of CRPV for preclinical trials of prophylactic vaccines (see Section 1.4).

11.2.2.   Bovine Papillomaviruses

There are six BPV genotypes that can be subgrouped based upon their ability to induce fibropapillomas (BPV1 and -2) versus papillomas that are entirely epithelial in origin (BPV3, -4, and -6). BPV5 is unique in that it can induce both fibropapillomas and papillomas. The genomic organization of BPV1, -2, and -5 are similar to most other papillomaviruses, whereas BPV3, -4, and -6 have the E5 gene (originally called the E8 gene) located where normally E6 is found. Another difference is the location of twelve E2 binding sites within the LCR in the former group versus four E2 binding sites in the latter.

   BPV1 induces fibropapillomas of the penis of bulls and the teats and udders of cows (Campo et al., 1981). In addition, meningiomas arose in a large majority of calves injected in the brain with the virus (Gordon and Olson, 1968). In hamsters, BPV1 causes fibrosarcomas and fibromas of the skin, chondromas of the ear and meningiomas (Olson et al., 1969). Much of our basic understanding regarding the biological properties of papillomaviral genes came from the study of BPV1 in tissue culture. In the late 1970s it was discovered that infection with BPV1 virus or transfection with recombinant BPV1 genomes led to the morphologic and tumorigenic transformation of mouse C127 cells, an immortalized mouse fibroblast line (Dvoretzky et al., 1980). In these cells, the viral genome was maintained as a nuclear plasmid much like that observed in natural infections, and a subset of viral genes were expressed that are also expressed in the poorly differentiated basal epithelial cells within a wart (Law et al., 1981). Thus, the BPV1 transformed mouse C127 cells appeared to mimic the nonproductive infective state seen in the lower compartment of epithelia within a wart. The 1980s saw the comprehensive genetic analysis of the BPV1 *trans* and *cis* elements involved in viral transcription, DNA replication and cellular transformation (reviewed in Lambert et al., 1988). Some of the highlights include identification of the E2 family of transcriptional regulators, viral promoters and transcript maps, identification of the E1 DNA helicase and its cognate E1-dependent replication origin, and determination of the mechanism of action of the major BPV1 oncogene, E5. Other chapters in this book cover in greater detail the many facets of papillomavirus biology elucidated through the study of BPV1 transformed mouse C127 cells. An important point considering the focus of this chapter is that BPV1 was the first papillomavirus to be studied in the context of transgenic mice, where it was found to induce fibropapillomas (see BPV Transgenic Mice).

BPV2 is etiologically associated with squamous cell carcinoma of the gastrointestinal tract and bladder tumors in cattle. A cofactor in these tumors is bracken fern, which both induces immunosuppression and contains mutagens necessary for malignant progression. BPV2 efficiently induced fibromas and polyps of the skin, vagina, and bladder in the majority of infected animals (Olson et al., 1959). BPV2 also induced fibropapillomas in experimentally infected hamsters (Moar et al., 1981). BPV3 was isolated from a bovine skin papilloma, but no experimental studies have been carried out with this genotype (Pfister et al., 1979). BPV4 has been ascribed a role in gastrointestinal papillomas and carcinomas in cattle, and it synergizes with bracken fern to induce bladder cancer, but the viral genome cannot be found in the cancers (Campo, 1987).

11.2.3. Other Animal Papillomavirus

Another animal papillomavirus of importance to experimental research is the canine oral papillomavirus (COPV). COPV induces oral papillomas in dogs, particularly those kept in kennels. COPV-induced lesions normally regress spontaneously with only rare progression to carcinomas in immune-competent animals (Bregman et al., 1987). COPV became an important animal papillomavirus for preclinical trials on prophylactic vaccines (see Use of Animal Papillomaviruses in the Analysis of Prophylactic Vaccines). In addition to CRPV, BPVs, and COPV, many other animal papillomaviruses have been described (Table 11.1).

11.2.4. Use of Animal Papillomaviruses in the Analysis of Prophylactic Vaccines

Vaccination studies on BPVs, CRPV, and COPV were highly instrumental in defining the requirements for an effective prophylactic vaccine against HPVs. Summarized below are studies based upon papillomavirus type. Studies of HPV vaccines are reviewed in Chapter 14.

TABLE 11.1. Animal Papillomaviruses

Natural Host	Virus Name	Site of Infection	Type of Disease
Cottontail rabbits	Cottontail rabbit PV (CRPV)	Skin, ears	Papilloma, carcinoma
Domestic rabbits	Rabbit oral PV (ROPV)	Oral cavity	Papilloma
Cattle	Bovine PV (BPV)-1	Skin, penis, teats	Fibropapilloma, penile carcinoma
Cattle	BPV-2	Skin and oral cavity	Fibropapilloma, bladder cancer
Cattle	BPV-3	Skin	Papilloma
Cattle	BPV-4	Gastrointestinal tract	Papilloma, carcinomas
Cattle	BPV-5	Teats, udder, face	Papilloma and fibropapilloma
Cattle	BPV-6	Teats and udder	Papilloma
Canine	Canine oral PV	Mouth	Papilloma, carcinomas in immunosuppressed animals
Feline	*Felis domesticus* PV (FdPV)	Skin	Papilloma
Elk	(European) Elk PV EEPV or EPV	Skin	Fibropapilloma
Reindeer	Reindeer PV (RPV)	Skin	Fibropapilloma
Red deer	Red deer PV (RDPV)	Skin	Fibropapilloma
Mule and white-tailed deer	Deer PV (DPV)	Skin	Fibropapilloma
Sheep	Ovine PV (OPV)	Skin	Papilloma, carcinoma
European harvest mice	*Micromys minutus* PV (MmPV)	Skin	Papilloma, carcinoma
African rodent	*Mastomys natalensis* PV (MnPV)	Skin, stomach	Papilloma, carcinoma
Chaffinches (avian)	*Fringilla* PV (FPV)	Legs	Papilloma
African Grey Parrot	*Psittacus erithacus* PV (PePV)	Skin	Papilloma
Colubus monkey	*Colubus guereza* PV (CgPV-1)	Penis	Papilloma
Colubus monkey	CgPV-2	Skin	Papilloma
Pigmy chimpanzees	*Pan paniscus* PV (PCPV)	Skin	Papilloma
Rhesus monkey	Rhesus PV (RhPV)	Penis, cervix	Papilloma

11.2.4.1.   BPVs

Genotype specific prophylactic vaccination was achieved with intramuscular injection of BPV1 virus (Campo et al., 1997). BPV2 L1 but not L2 was found to elicit protection against infection, though the L2 vaccine induced more rapid regression of preformed lesions (Jarrett et al., 1991). In contrast, BPV4 L2 did afford protection against challenge with the cognate virus (Campo et al., 1993) and a single epitope on L2 was defined to be sufficient for this protective effect (Campo et al., 1997). BPV4 L1 or L1/L2 virus-like particles (VLPs) were very potent prophylactic vaccines (Kirnbauer et al., 1996).

11.2.4.2.   CRPV

Studies performed in the 1930s first demonstrated that sera from rabbits with papillomas contain antibodies that inhibit infection with CRPV (Shope, 1937). Subsequently, neutralizing antibodies were identified for both CRPV L1 and L2 (Christensen et al., 1991; Lin et al., 1992, 1993). Protection by L1 required full-length intact L1 or intact L1 VLPs (Christensen et al., 1991; Lin et al., 1993). Long-term protection was demonstrated with CRPV VLPs (Christensen et al., 1996). Incorporation of L2 into VLPs increased the effectiveness of VLPs as immunogens (Breitburd et al., 1995).

11.2.4.3.   COPV

Canine oral papillomas induced by COPV are highly predictable with respect to the time required for lesions to form (4–8 weeks) and then to regress (an additional 4–8 weeks). Dogs previously exposed to COPV, in which papillomas arose and regressed, failed to develop papillomas upon re-exposure to COPV, indicating long-term immunity to the virus (Chambers and Evans, 1959). Immunization of dogs with an COPV L1 VLP led to complete prevention of infection by COPV (Suzich et al., 1995). Bacterially synthesized GST-L1 fusions of COPV L1, which cannot assemble into VLPs but can form pentamers, also provided protection (Yuan et al., 2001). This finding is of potential importance to the development of inexpensive HPV vaccines that might be accessible to developing countries.

11.3.   Transgenic Mouse Models for the Study of Human Papillomaviruses

In the mid-1980s, Dr. Harold zur Hausen and colleagues (reviewed in Gissmann et al., 1984) made the seminal discovery that certain anogenital HPV genotypes, termed the high-risk HPVs, were present in cervical cancers and their derived cell lines. This finding was followed soon thereafter with the recognition that in these cancers, the HPV genomes were commonly found integrated with a subset of viral genes, most notably E6 and E7, selectively expressed

(Schwarz et al., 1985; Yee et al., 1985). This second observation indicated the likely importance of E6 and E7 in cervical cancer. Integration was found to lead not only to the selective expression but also to the increased expression of these two viral genes (Jeon et al., 1995). Many investigators carried out studies that established that E6 and E7 genes of high-risk HPVs possess oncogenic properties in tissue culture (reviewed in Munger, 2002). Furthermore, E6 and E7 proteins were found to bind and inactivate the cellular tumor suppressors p53 and pRb, respectively (Dyson et al., 1989; Werness et al., 1990). These cellular genes are often mutated or their pathways disrupted in a variety of human cancers, but not in cervical cancers (Crook et al., 1991; Scheffner et al., 1991), implying a role of HPV E6 and E7 in this cancer type. Later studies demonstrated that the continued expression of E6 and E7 was critical for the growth of cervical cancer-derived cell lines (Goodwin and DiMaio, 2000; Goodwin et al., 2000; Hwang et al., 1993; Wells et al., 2000). In sum, these studies pointed to an etiological role of HPVs in cervical cancer. High-risk HPVs are also implicated in other anogenital cancers, as well as approximately 20 percent of human head-and-neck cancers, most notably those of the oropharynx including the tonsils.

11.3.1. BPV Transgenic Mice

Given the recognition that HPVs contribute to human cancer, the species-specific tropism of papillomaviruses and the need for an in vivo model to study their contribution to cancer, multiple investigators initiated studies using transgenic mouse technologies to investigate the role of HPVs in cancer. The potential value of transgenic mice in studying the oncogenic potential of papillomaviruses was first demonstrated by studies on BPV1. Hanahan and colleagues generated mice carrying 1.6 tandem copies of the BPV1 genome integrated into the mouse genome. These mice developed fibrosarcomas (Lacey et al., 1986). In these fibrosarcomas the integrated viral genome was found to have undergone recombination and to exist as a nuclear plasmid. The E5 gene, which was demonstrated in earlier tissue culture studies to be the major transforming protein of BPV1, was expressed in these fibrosarcomas (Sippola et al., 1989). The expression of the E5 protein along with cytogenetic changes in the mouse genome were determined to contribute to this tumorigenic phenotype (Lindgren et al., 1989). While a similar approach to the study of human papillomaviuses proved less fruitful, the recognition that genetically engineered mice might unveil the roles of HPV oncogenes in human cancer was quickly recognized and led to a number of studies in which the entire HPV genome or segments thereof were introduced into mice under the control of heterologous transcriptional promoters.

11.3.2. HPV Transgenic Mice

A number of groups generated HPV transgenic mice with which to study the contributions of high-risk HPVs in cancer. The first HPV transgenic mice described in the literature expressed both the E6 and E7 oncogenes of HPV16 from the mouse mammary tumor virus (MMTV) enhancer/promoter region (Kondoh et al., 1991). Male transgenic mice spontaneously developed testicular tumors, which were subsequently characterized to be Leydig cell in origin (Kondoh et al., 1994) and to contain activated c-kit receptor (Kondoh et al., 1995). Female mice from this line developed salivary gland carcinomas, lymphomas, and skin histiocytomas (Sasagawa et al., 1994). In addition, some of the female mice developed dysplastic and/or hyperplastic changes in the cervix and vagina (Sasagawa et al., 1994). Expression of HPV16 E6 and E7 from the human beta-actin promoter led to neuroepithelial carcinomas (Arbeit et al., 1993). Expression of HPV16 E6 and E7 from the αA crystallin promoter, which normally directs expression of genes to the lens, was found to cause not only epithelial tumors in the lens (Griep et al., 1993), but also, due to ectopic transgene expression in one line, squamous cell carcinomas in the skin (Lambert et al., 1993) and retinoblastomas (Griep et al., 1998). This mouse model, in which the individual or combinatorial effect of HPV16 E6 and E7 were characterized in the lens, led to important new insights regarding the biological activities of E6 and E7. E7 was discovered to induce apoptosis (Pan and Griep, 1994b), in part through its inactivation of pRb and the latter's immediate downstream target E2Fs (McCaffrey et al., 1999). E6 was found to inhibit this process, through both p53-dependent and p53-independent processes (Pan and Griep, 1995b). E6 also was found to inhibit the normal differentiation process in the lens and this correlated with the inability of cells to undergo intracellular organelle loss (Pan and Griep, 1994b), a process that involves the activation of caspase 6 or caspase 6 like activities (Foley et al., 2004).

HPVs normally infect stratified epithelia lining the skin, anogenital tract, and oral cavity. By the mid 1990s, transcriptional promoters, such as ones derived from keratin genes, were used to direct expression of the HPV oncogenes to stratified squamous epithelia. These mouse models provided insight into the effects of high-risk HPV E6 and E7 genes in more relevant cell types, in particular the ability of these genes to induce the proliferation of cells within stratified epithelia, a property that likely contributes to the oncogenic potential of the HPV in humans. Transgenic expression of the HPV16 E6 and E7 genes in the suprabasal compartment of the epidermis, achieved using the bovine keratin 10 promoter, led to hyperplasia, but not tumorigenesis (Auewarakul et al., 1994). In contrast, expression of the high-risk HPV18 E6 and E7 genes in the same compartment, by using the human keratin 1 (K1) promoter, led to wart-like lesions in the skin of adult mice, and these papillomas carried activating mutations in H-ras (Greenhalgh et al., 1994). Expression of the entire early region of HPV16 genome from the human keratin 14 promoter, which directs expression to the basal compartment of stratified squamous epithelia, led to epithelial hyperplasia (Arbeit et al., 1994), and squamous cell carcinomas were found to arise once this transgene was crossed onto the FVB/n genetic background, again arguing a role of host genes in determining risk for HPV associated cancer (Coussens et al., 1996). In another study, transgenic mice expressing the HPV16 E6 and E7 genes from the keratin 14 promoter on the C57Bl/6 genetic background failed to develop skin tumors (Melero et al., 1997), but did so when crossed to FVB/n mice carrying an activated ras oncogene (Schreiber et al., 2004). When HPV16 E6 and E7 were expressed in the basal compartment from the mouse keratin 5 promoter, the resulting mice developed lung carcinomas and thymic hyperplasia (Carraresi et al., 2001). Expression of HPV16 E6 and E7 genes from the tyrosinase promoter induced epidermal hyperplasia and susceptibility to chemically induced skin tumors (Kang et al., 2000), whereas their expression from the bovine keratin 6 promoter altered the growth phase of hair follicles (Escalante-Alcalde et al., 2000).

11.4. Mechanistic Studies of E6 in Transgenic Mouse Models

The biological properties of HPV16 E6 in stratified squamous epithelia such as the epidermis and the cervical epithelium were elucidated from the study of K14E6 transgenic mice, in which the HPV16 E6 gene was under the transcriptional control of the keratin 14 (K14) promoter, targeting expression to the basal compartment of stratified squamous epithelium (Song et al., 1999). The study of K14E6 mice provided several important insights into the biological properties of E6. First it was observed that E6 could induce epithelial hyperplasia in the mouse epidermis and the lens epithelium (Song et al., 1999; Nguyen et al., 2002b). This hyperplasia was associated with a delay in normal differentiation as evidenced by a delay in keratin 10 expression in the suprabasal compartment of the epidermis, and a delay in expression of markers of fiber cell differentiation in the lens. Hyperplasia was also associated with the presence of suprabasal epidermal cells undergoing DNA synthesis. Normally only cells within the basal compartment, where cell proliferation is normally restricted, undergo new rounds of DNA synthesis. Because p53-null mice do not display the induction of epithelial hyperplasia in the epidermis or lens, this phenotype in the K14E6 mice is not a consequence of the inactivation of p53 by E6 (Nguyen et al., 2002b; Pan and Griep, 1995a; Song et al., 1999). Furthermore, K14E6[I128T] mice, expressing a mutant of HPV16 E6 that fails to inactivate p53, retains an ability to induce epithelial hyperplasia in the epidermis and the lens; and this phenotype is not further accentuated when the transgene is placed on a p53-deficient background (Nguyen et al., 2002a,b). The ability of the E6 protein to induce epithelial hyperplasia, instead, correlates with its binding to a class of cellular proteins called PDZ domain proteins (Nguyen et al., 2003a,b). This conclusion was based upon the characterization of K14E6[Δ146–151] mice which express a mutant version of HPV16 E6 missing amino acids at the C terminus that mediate wild-type E6's interaction with the PDZ domain proteins. PDZ domain partners of E6 include the mammalian homologs of two *Drosophila* proteins, *Dlg* and *Scribble*, considered to be tumor suppressors because their disruption in *Drosophila* leads to hyperplastic and disorganized epithelial sheets such as the imaginal discs during development.

Another important insight gained from the study of the K14E6 mice came from an analysis of DNA damage responses. It had earlier been demonstrated that HPV16 E6 can inhibit DNA damage responses in tissue culture (Foster et al., 1994; Kessis et al., 1993). As seen in tissue culture, expression of E6 in vivo led to a complete abrogation of DNA damage responses in the epidermis of K14E6 transgenic mice following ionizing radiation (Song et al., 1998). The normal response to ionizing radiation in the epidermis is a transient inhibition of DNA synthesis, which is mediated through the cdk inhibitor p21 (Song and Lambert, 1999). Surprisingly, this response was inhibited more completely in E6 transgenic mice than in p53-null mice, arguing that E6 must abrogate DNA damage responses through p53-independent as well as p53-dependent pathways (Song et al., 1998). This p53-independent pathway has not been identified. Whether it relates to the p53-independent pathway by which E6 inhibits apoptosis in the lens fiber cells is also unknown.

Perhaps the most significant insight gained from the analysis of the K14E6 mice is that E6 can contribute to multiple stages of carcinogenesis in the skin. K14E6 mice spontaneously develop squamous cell carcinomas (Song et al., 1999). Through the analysis of these mice in studies involving their treatment with chemical carcinogens (Song et al., 2000), it was learned that E6 contributes to the "promotion" stage of tumorigenesis that is involved in the formation of benign tumors or papillomas in the skin, and also to the progression stage, which defines the process by which a benign tumor progresses to a malignant carcinoma. The mechanisms by which HPV16 E6 contributes to these stages of carcinogenesis are distinct. This conclusion was drawn from the observation that, in K14E6$^{\Delta146-151}$ mice, E6 is defective for contributing to the promotion stage of carcinogenesis but retains its ability to contribute to progression (Simonson and Lambert, manuscript submitted). These same mice are defective for E6-induced epidermal hyperplasia, indicating a correlation between an ability of E6 to contribute to promotion and induce hyperplasia. Furthermore, this observation indicates that the interaction of E6 with PDZ domain partners likely contributes to promotion. How E6 contributes to tumor progression has yet to be defined unambiguously. However, K14E6^{I128T} mice, which encode a form of HPV16 E6 protein that is unable to inactivate p53 in vivo, display reduced contribution to progression compared to mice encoding wild-type E6 protein, a result which supports the hypothesis that the inactivation of p53 by E6 contributes to its role in tumor progression (Nguyen et al., 2002a,b).

11.5.   Mechanistic Studies of E7 in Transgenic Mouse Models

The activities of E7 in stratified squamous epithelia, likewise, were elucidated from the study of transgenic mice in which E7 was placed under the control of the keratin 14 promoter. One of the most striking consequences of E7 expression is induction of DNA synthesis at a higher rate than in control mice in differentiating, post-mitotic cells. Targeting E7 to the basal layer of stratified squamous epithelia using the K14 promoter resulted in hyperplasia of the ear and trunk skin, mouth palate, esophagus, forestomach, and exocervix (Herber et al., 1996). Assays for the presence of the DNA synthesis marker PCNA or incorporation of the nucleoside analog BrdU showed that hyperplasia coincided with an increased rate of DNA synthesis in the epithelium. Furthermore, DNA synthesis occurred not only in the normally proliferative basal epithelial layer, but also in the normally quiescent suprabasal epithelial layers of ear and trunk skin from E7 transgenic mice (Gulliver et al., 1997; Herber et al., 1996).

In many epithelial tissues, the process of differentiation is linked to cell-cycle withdrawal. Thus, it is perhaps not surprising that E7-induced cell cycle progression is accompanied by alteration of the normal differentiation program of the affected tissue. This effect is evidenced not only by the induction of DNA synthesis in normally postmitotic cells, but also by alterations in the morphology of differentiating cells in transgenic lens and skin (Gulliver et al., 1997; Herber et al., 1996; Pan and Griep, 1994a) and altered expression patterns of differentiation-specific keratins (Balsitis et al., 2003; Griep et al., 1993; Gulliver et al., 1997; Nguyen et al., 2002b).

The study of the K14E7 mice also revealed that HPV16 E7 abrogated DNA damage-induced responses. When nontransgenic mice are exposed to 5 Gy ionizing radiation, the cell cycle pauses to avoid replication of damaged DNA and allow time for DNA repair to occur. Consequently, little-to-no DNA synthesis is detectable in the epidermis 24 h after irradiation (Song et al., 1998). By contrast, K14E7 mice show no decrease in DNA synthesis in the epidermis after exposure to ionizing radiation. This effect may accelerate the accumulation of additional mutations necessary for a high-risk HPV-infected cell to become malignant.

A second activity of E7 has also been proposed to accelerate the accumulation of mutations in E7-expressing cells. In cultured cells in vitro, E7 is capable of inducing supernumerary centrosomes (more than 2 centrosomes per cell), which can result in multipolar mitoses (Duensing et al., 2001a; Duensing and Munger, 2003). These centrosome abnormalities occur prior to the development of aneuploidy, and it has been hypothesized that they may contribute to tumor development by promoting chromosome mis-segregation or breakage during multipolar cell divisions (Duensing et al., 2001a). Examination of K14E7 transgenic mice demonstrated that E7 induces centrosome abnormalities in vivo (Balsitis et al., 2003). However, analysis of multiple tumor cell lines derived from K14E7 mice showed no correlation between frequency of centrosome abnormalities and severity of karyotypic instability (Schaeffer et al., 2004). Thus, while E7 clearly induced supernumerary centrosomes both in vitro and in vivo, the significance of this effect in tumor development requires further investigation.

The role of E7 in skin tumorigenesis was elucidated by treating mice with the tumor-initiating agent DMBA and tumor-promoting agent (TPA). In these studies, E7 was found to increase the formation of benign tumors in DMBA/TPA-treated mice or mice treated with DMBA alone. E7, however, was not able to initiate tumors in mice treated with TPA alone, or able to induce progression to malignancy at a rate higher than seen in control mice. These observations show that E7 acts primarily at the tumor promotion stage in mouse skin, rather than at the tumor initiation or progression stages (Song et al., 2000). A subsequent study found that expression of dominant-negative c-*jun* in the skin of K14E7 mice inhibits E7's promotion activity in DMBA/TPA carcinogensis studies, but not its ability to induce DNA synthesis or hyperplasia or disrupt DNA damage responses. These observations indicate that c-*jun* does not play a role in the acute effects of E7 expression, but c-*jun* activation may be an important component of E7's tumor promotion activity (Young et al., 2002).

Germline inactivation of pRb recapitulates the apoptotic effect of E7 in the mouse lens, and somatic pRb inactivation in the epidermis reproduces E7's effects on DNA synthesis, differentiation, centrosome abnormalities, and DNA damage responses (Balsitis et al., 2003; Morgenbesser et al., 1994; Pan and Griep, 1995a), suggesting that Rb plays a role in mediating E7's effects. Consistent with this interpretation, the biological activities of E7 were greatly attenuated when αAcryE7 transgenic mice were placed on a genetic background nulligenic for E2F1, a transcription factor normally regulated by pRb and deregulated in the presence of E7 (McCaffrey et al., 1999). While these correlations support a major role for pRb inactivation in the phenotypes of E7, E7 was also shown to produce additional phenotypes when expressed in pRb-deficient skin (Balsitis et al., 2003). Combined, these observations suggest that pRb inactivation is very important in the functioning of E7 in these tissues, but other molecular targets of E7 are likely to play additional roles.

11.6.   Transgenic Mouse Studies on the Role of HPV in Cervical Cancer in Transgenic Mice

Models for specifically studying cervical cancer improved with the generation of the K14 HPV16 transgenic mouse (Arbeit et al., 1994). In this model, the human K14 promoter targets HPV gene expression to the basal layer of the squamous epithelium including the uterine cervix. These mice express the entire early region of the HPV genome minus the LCR region. Virgin female mice were treated with 17-β-estradiol chronically in the form of a 60 day continual release pellet starting at 4 weeks of age for a period of up to 6 months (Arbeit et al., 1996). Chronic estrogen treatment places the mouse in constant estrus and prevents it from estrogen cycling. Estrogen treatment of nontransgenic mice increased the proliferative capacity in the epithelium of the reproductive tract, resulting in a thickened or hyperplastic epithelia that retained the ability to differentiate normally. Cervical lesions in estrogen-treated K14HPV16 transgenic mice progressed from low-grade dysplasia to high-grade dysplasia, or carcinoma in situ (CIS), by 6 months of treatment with a majority of cancers occurring in the transformation zone, the region in which cervical cancer commonly arises in women (Elson et al., 2000). Untreated transgenic mice never developed spontaneous cervical cancer when monitored up to 1.5 years of age. The levels of transgene expression were monitored throughout the course of estrogen treatment and neoplastic progression. E6 and E7 mRNA levels remained constant in both cases and did not increase in frank cancer.

The individual roles of HPV16 E6 and E7 in cervical carcinogenesis were elucidated by the further study of K14E6 (Song et al., 1999) and K14E7 (Herber et al., 1996) mice. These transgenic mice were treated with estrogen for six months in order to identify the individual roles of E6 and E7 (Riley et al., 2003). A similar percentage of K14E7 and K14HPV16 mice developed cervical tumors, with all of the K14E7 mice developing high-grade dysplasias or microinvasive cancers (MIC). K14E6 mice, however, did not develop any cancer and only displayed CIN I (cervical intraepithelial neoplasia) to CIN II lesions. E6 function was confirmed by scoring for centrosome abnormalities and p53 protein expression. Tumors were larger in estrogen treated K14E6/K14E7 doubly transgenic mice compared to K14E7 mice, leading to the conclusion that E6 contributes late in cervical carcinogenesis. The results from the cervical carcinogenesis studies with K14E6 and K14E7 mice differ from the findings made in the skin of these same mice. In the cervix, E7 clearly is the more potent oncogene based upon the above described findings. In contrast, E6 is the more potent oncogene in the skin, as K14E6 mice spontaneously develop primarily skin carcinomas, whereas K14E7 mice develop primarily benign skin papillomas (Song et al., 2000).

In a follow-up study, the importance of E7's inactivation of pRb in mediating E7's oncogenic activities in the cervix was investigated. Mice deficient for Rb in the cervical epithelium failed to develop cervical cancers, when treated with estrogen for six months (Balsitis et al., 2006). Furthermore, K14E7 mice on a genetic background encoding a mutant form of pRB unable to be bound by E7 developed cervical cancers when treated with estrogen for six months (Balsitis et al., 2006). These studies demonstrate that activities other than E7's inactivation of pRb contribute to its oncogenic activities in this relevant tissue.

The effect of prolonged treatment with estrogen was also investigated (Brake and Lambert, 2005). Here, K14E7 and K14E6/K14E7 mice were treated with the same dose of estrogen as used in the prior study, but for 9 months instead of 6. Results from this study reconfirmed that the presence of E6 greatly contributed to increased tumor size and severity. Interestingly, when these mice treated for 6 months with exogenous estrogen were aged for an additional 3 months in the absence of any exogenous estrogen, the percentage of mice developing cervical cancer was reduced from 100 percent in both E7 and E6E7 mice with 9 months of estrogen to 38 percent and 82 percent, respectively. In addition, the area of tumor invasion markedly decreased, by approximately 85 percent in the highest case. These results argue for the necessity of estrogen in the induction, maintenance and progression of cervical carcinogenesis in combination with HPV oncogenes.

Studies using K14E6 and K14E7 mice led to the discovery of a new biomarker for murine and human cervical cancer. Current biomarkers for cervical cancer include Ki67 and cyclin E (Keating et al., 2001). Immunohistological analysis of the estrogen treated HPV transgenic mice with antibodies to these known biomarkers for cervical cancer showed similar staining patterns as in human cervical cancer, thus further confirming the validity of this mouse model for human cervical cancer. Life cycle studies with HPV16 E7 revealed that MCM7, an E2F-responsive gene, is up-regulated with the inactivation of pRb. Staining of cervical lesions in mice showed that MCM7 staining correlated with neoplastic progression, as was also true for human cervical cancer. Thus, these mouse models allowed the investigation of a new biomarker for human cervical cancer (Brake et al., 2003).

11.7. Use of HPV Transgenic Mice as Preclinical Models for Testing New Modalities for Preventing or Treating Cervical Cancer

One goal in developing animal models for cervical cancer is to identify new drugs for either chemoprevention or treatment of cervical cancer. One potential therapeutic stemming from HPV transgenic mouse studies is the anti-estrogen phytochemical, indole-3-carbinol (I3C), found in cruciferous vegetables or in the form of supplements. Tumor incidence was drastically decreased from 76 percent to 8 percent when transgenic mice were fed a diet supplemented with I3C, and the reproductive epithelium of estrogen-treated mice fed I3C displayed less PCNA staining, a marker indicative of proliferation. (Jin et al., 1999). Thus, I3C was effective in opposing the effects of estrogen. Similarly, although estrogen did not have an effect on ear epithelium, transgenic mice on the I3C diet showed a reduction in the severity of skin dysplasias and cancer.

    I3C is metabolized into diindolylmethane (DIM) by acid-catalyzed condensation in the stomach (Grose and Bjeldanes, 1992) after a meal of cruciferous vegetables. DIM was also tested for its effectiveness in reducing the severity of lesions. DIM was more potent and effective at inducing apoptosis in cervical cancer cell lines than I3C (Chen et al., 2001). These chemicals induced apoptosis via DNA strand breaks specifically in transformed cells in the cervical epithelium. This specificity, however, did not necessitate the presence of HPV, as both chemicals were effective in inducing apoptosis in HPV negative cervical cancer cell lines and MCF breast carcinoma cell lines. The mechanism of I3C/DIM action is still somewhat unclear, but several possibilities exist. I3C has been shown to disrupt transcription (Cram et al., 2001) by interfering with cdk6 expression (Cover et al., 1998), resulting in cell cycle inhibition. Furthermore, the condensation products of I3C, including DIM, are ligands for the aryl hydrocarbon receptor (Chen et al., 1996) and have been shown to alter the metabolism of estrogen in both breast and cervical cancer cell lines. In cervical cancer cell lines, I3C influences estrogen metabolism to produce an anti-estrogen derivative, 2-hydroxyestrone (Yuan et al., 1999) and thus could compete with estrogen for receptor binding and reduce the production of 16α-hydroxyestrone, a metabolite that has been linked to carcinogenic activity

(Auborn et al., 1991; Yuan et al., 1999). Based on these in vivo studies, I3C could potentially serve as a therapeutic for cervical cancer.

Because HPV transgenic mouse models express viral antigens in the skin, grafting studies can be used to investigate immune recognition of these antigens, when expressed in clinically relevant amounts in keratinocytes, in naïve immunologically competent animals. Surprisingly, E7 was found not to function as a classical minor transplantation antigen in the context of its expression in mouse epidermis. Skin from K14E7 mice, when grafted onto nontransgenic mice, was not rejected even when these mice were immunized against E7 and developed E7-specific CTL responses (Dunn et al., 1997; Frazer et al., 2001). Similarly, grafts in which the E6 protein is expressed from the K14 promoter were also accepted (Matsumoto et al., 2004). However, recipient animals could be induced to reject the skin grafts in an E7-specific manner by stimulating systemic proinflammatory responses (Frazer et al., 2001), or through passive transfer of E7-specific CTLs in combination with E7-specific immunization (Matsumoto et al, 2004). Why it is so difficult for the immune system to recognize E7 antigen in the context of the skin remains unclear. Perhaps E7 can inhibit host immune responses. One transgenic mouse study demonstrated that E7 can suppress innate immune responses, including induction of IFN-responsive genes, through its interaction with IRF1 (Um et al., 2000). Another transgenic mouse study, however, noted no effect of E7 on the induction of MHC class I protein levels by interferon (Leggatt et al., 2002). Alternatively, it is possible that E7, a nonsecreted antigen, is poorly cross-presented by professional antigen presenting cells and/or that the local immune environment of the skin is not conducive to the recognition of keratinocyte-expressed antigens by the host immune system. Regardless of the mechanism, the insights gained from these grafting studies have clear implications regarding the design of effective therapeutic vaccines for treating patients infected with HPVs.

11.8. Additional Mouse Models for HPV-Associated Neoplasia

High risk mucosal HPVs, particularly HPV16, are etiologically associated with a subset of human head and neck squamous cell carcinoma (HNSCC). A new model for assessing HPVs role in HNSCC has been developed using K14E6/K14E7 double transgenic mice. When these mice expressing both the HPV16 E6 and E7 oncogenes were treated with a synthetic carcinogen 4-nitroquinoline N oxide (4NQO); they developed high grade cancers at a much higher frequency than like-treated nontransgenic mice (Strati et al., 2006). The HNSCC arising in the 4NQO-treated HPV transgenic mice, as compared to those arising in the 4NQO-treated nontransgenic mice, were more basaloid in their differentiative state and over-expressed p16, as seen in HPV-positive HNSCC in humans. Using this mouse model MCM7 was identified as a potentially useful biomarker for HPV-associated HNSCC (Strati et al., 2006), as had been demonstrated to be the case in the mouse model for HPV-associated cervical cancers, and in human cervical cancers (Brake et al., 2003).

Cutaneous HPVs are associated with a subset of skin cancers. Two laboratories recently developed transgenic mouse models for investigating the role of cutaneous HPVs that cause epidermodysplasia verruciformis (EV). Patients who develop EV are genetically predisposed to the disease and are at an increased risk of developing squamous cell carcinomas. Both labs found that mice transgenic for either of two EV-associated HPVs are more prone to skin cancers, supporting the role of these cutaneous papillomaviruses in cancer. The Pfister lab generated transgenic mice expressing the early genes of HPV8 behind the keratin 14 promoter. These mice developed a progressive skin disease characterized by acanthosis, hyperkeratosis, dysplasia, frequent induction of papillomas, and in some animals squamous cell carcinoma (Schaper et al., 2005). They utilized their HPV8 transgenic mice to evaluate expression of metalloproteinases altered in their expression by the viral early genes (Akgul et al., 2006). The Tommasino lab generated transgenic mice in which the E6 and E7 genes of HPV38 were directed in their expression from the bovine equivalent of the human keratin 10 transcriptional promoter. These mice displayed localized areas of epithelial hyperplasia and an increased sensitivity to chemical carcinogen-induced skin tumors (Dong et al., 2005). Utilizing their HPV38 transgenic mice, the Tommasino lab discovered a novel means by which cutaneous HPV regulate p53 function by inducing expression of a dominant negative isoform of p73, which suppresses the transcriptional transactivation activity of p53 (Accardi et al., 2006).

11.9. Summary

This chapter provides an overview of the knowledge gained from the study of animal papillomviruses in their natural hosts and the study of human papillomaviruses in transgenic mice. HPV transgenic mice provide tractable model systems with which to define the mechanisms of action of HPV oncogenes in carcinogenesis. These mice also provide preclinical models in which to test new modalities for treating or preventing HPV-associated cancers. While the development of effective prophylactic HPV vaccines has great potential for reducing the worldwide incidence of HPV-associated cancers, there remains a critical need to find better means for treating those who do acquire such cancers or who are at increased risk due to persistent high-risk HPV infections. The World Health Organization predicts that these prophylactic vaccines will not cause a significant reduction in HPV-associated cancers before 2040. In the interim, it is predicted that over 15 million women will contract cervical cancer. Furthermore, these vaccines will only be able to protect against HPV16 and HPV18, which at most account for three quarters of the HPV-associated cervical cancers. Thus, even if the vaccine is administered to all women and is 100 percent effective, over 100,000 new cases of cervical cancer will still arise annually due to other high-risk HPVs. Therefore, the existence of preclinical models for testing new approaches for preventing or treating cervical cancer is critical.

References

Accardi R, Dong W, Smet A, Cui R, Hautefeuille A, Gabet AS, Sylla BS, Gissmann L, Hainaut P, Tommasino M. (2006). Skin human papillomavirus type 38 alters p53 functions by accumulation of deltaNp73. EMBO Rep. 7:334–340.

Akgul B, Pfefferle R, Marcuzzi GP, Zigrino P, Krieg T, Pfister H, Mauch C. (2006). Expression of matrix metalloproteinase (MMP)-2, MMP-9, MMP-13, and MT1-MMP in skin tumors of human papillomavirus type 8 transgenic mice. *Exp. Dermatol.* 15:35–42.

Arbeit, J., Howley, P.M., and Hanahan, D. (1996). Chronic estrogen-induced cervical and vaginal squamous carcinogenesis in human papillomavirus type 16 transgenic mice. *Proc. Natl. Acad. Sci. USA* 93:2930–2935.

Arbeit, J., Munger, K., Howley, P.M., and Hanahan, D. (1993). Neurorpithelial carcinomas in mice transgenic with human papillomavirus type 16 E6/E7 ORFs. *Am. J. Pathol.* 142: 1187–1197.

Arbeit, J., Munger, K., Howley, P.M., and Hanahan, D. (1994). Progressive squamous epithelial neoplasia in K14-Human papillomavirus type 16 transgenic mice. *J. Virol.* 68:4358–4368.

Auborn, K.J., Woodworth, C., DiPaolo, J.A., and Bradlow, H.L. (1991). The interaction between HPV infection and estrogen metabolism in cervical carcinogenesis. *International Journal of Cancer* 49:867–869.

Auewarakul, P., Gissmann, L., and Cid, A.A. (1994). Targeted expression of the E6 and E7 oncogenes of human papillomavirus type 16 in the epidermis of transgenic mice elicits generalized epidermal hyperplasia involving autocrine factors. *Mol. Cell. Biol.* 14:8250–8258.

Balsitis, S., Dick, F., Dyson, N., and Lambert, P.f.(2006). Critical roles for non-pRb targets of human papillomavirus type 16 E7 in cervical carcinogenesis. *Cancer res.,* (66):9393–9400.

Balsitis, S.J., Sage, J., Duensing, S., Munger, K., Jacks, T., and Lambert, P.F. (2003). Recapitulation of the effects of the HPV-16 E7 oncogene on mouse epithelium by somatic Rb deletion and detection of pRb-independent effects of E7 *in vivo. Mol. Cell. Biol.* 23:9094–9103.

Brake, T., Connor, J.P., Petereit, D.G., and Lambert, P.F. (2003). Comparative analysis of cervical cancer in women and in a human papillomavirus-transgenic mouse model: Identification of minichromosome maintenance protein 7 as an informative biomarker for human cervical cancer. *Cancer Res* 63:8173–8180.

Brake, T., and Lambert, P.F. (2005). Estrogen contributes to the onset, persistence, and malignant progression of cervical cancer in human papillomavirus-transgenic mouse model. *Proc. Natl. Acad. Sci. U.S.A.* 102 (7):2490–2495.

Brandsma, J.L., Yang, Z.H., DiMaio, D., Barthold, S.W., Johnson, E., and Xiao, W. (1992). The putative E5 open reading frame of cottontail rabbit papillomavirus is dispensable for papilloma formation in domestic rabbits. *J. Virol.* 66:6204–6207.

Bregman, C.L., Hirth, R.S., Sundberg, J.P., and Christensen, E.F. (1987). Cutaneous neoplasms in dogs associated with canine oral papillomavirus vaccine. *Veterinary Pathology* 24:477–487.

Breitburd, F., Kirnbauer, R., Hubbert, N.L., Nonnenmacher, B., Trin-Dinh-Desmarquet,C., Orth, G., Schiller, J.T., and Lowy, D.R. (1995). Immunization with viruslike particles from cottontail rabbit papillomavirus (CRPV) can protect against experimental CRPV infection. *J. Virol.* 69:3959–3963.

Campo, M.S. (1987). Papillomas and cancer in cattle. *Cancer Surveys* 6:39–54.

Campo, M.S. (1997). Vaccination against papillomavirus in cattle. *Clin. Dermatol.* 15:275–283.

Campo, M.S., Grindlay, G.J., O'Neil, B.W., Chandrachud, L.M., McGarvie, G.M., and Jarrett, W.F. (1993). Prophylactic and therapeutic vaccination against a mucosal papillomavirus. *J. Gen. Virol.* 74:945–953.

Campo, M.S., Moar, M.H., Laird, H.M., and Jarrett, W.F. (1981). Molecular heterogeneity and lesion site specificity of cutaneous bovine papillomaviruses. *Virology* 113:323–335.

Campo, M.S., O'Neil, B.W., Grindlay, G.J., Curtis, F., Knowles, G., and Chandrachud, L. (1997). A peptide encoding a B-cell epitope from the N-terminus of the capsid protein L2 of bovine papillomavirus-4 prevents disease. *Virology* 234:261–266.

Carraresi, L., Tripodi, S.A., Mulder, L.C., Bertini, S., Nuti, S., Schuerfeld, K., Cintorino, M., Bensi, G., Rossini, M., and Mora, M. (2001). Thymic hyperplasia and lung carcinomas in a line of mice transgenic for keratin 5-driven HPV16 E6/E7 oncogenes. *Oncogene* 20:8148–8153.

Chambers, V.C., and Evans, C.A. (1959). Canine oral papillomavirus. 1. Virus assay and observations on the various stages of the experimental infection. *Cancer Res.* 19:1188–1195.

Chen, D.Z., Qi, M., Auborn, K.J., and Carter, T.H. (2001). Indole-3-carbinol and diindolylmethane induce apoptosis of human cervical cancer cells and in murine HPV16-transgenic preneoplastic cervical epithelium. *J. Nutr.* 131:3294–3302.

Chen, I., Safe, S., and Bjeldanes, L. (1996). Indole-3-carbinol and diindolylmethane as aryl hydrocarbon (Ah) receptor agonists and antagonists in T47D human breast cancer cells. *Biochem. Pharmacol.* 51:1069–1076.

Christensen, N.D., Kreider, J.W., Kan, N.C., and DiAngelo, S.L. (1991). The open reading frame L2 of cottontail rabbit papillomavirus contains antibody-inducing neutralizing epitopes. *Virology* 181:572–579.

Christensen, N.D., Reed, C.A., Cladel, N.M., Han, R., and Kreider, J.W. (1996). Immunization with viruslike particles induces long-term protection of rabbits against challenge with cottontail rabbit papillomavirus. *J. Virol.* 70:960–965.

Coussens, L., Hanahan, D., and Arbeit, J. (1996). Genetic predisposition and parameters of malignant progression in K14-HPV16 transgenic mice. *Am. J. Pathol.* 149:1899–1917.

Cover, C.M., Hsieh, S.J., Tran, S.H., Hallden, G., Kim, G.S., Bjeldanes, L.F., and Firestone, G.L. (1998). Indole-3-carbinol inhibits the expression of cyclin-dependent kinase-6 and induces a G1 cell cycle arrest of human breast cancer cells independent of estrogen receptor signaling. *J. Biol. Chem.* 273:3838–3847.

Cram, E.J., Liu, B.D., Bjeldanes, L.F., and Firestone, G.L. (2001). Indole-3-carbinol inhibits CDK6 expression in human MCF-7 breast cancer cells by disrupting Sp1 transcription factor interactions with a composite element in the CDK6 gene promoter. *J. Biol. Chem.* 276:22332–22340.

Crook, T., Wrede, D., and Vousden, K.H. (1991). p53 point mutations in HPV negative human cervical carcinoma cell lines. *Oncogene* 6:873–875.

Dong W, Kloz U, Accardi R, Caldeira S, Tong WM, Wang ZQ, Jansen L, Durst M, Sylla BS, Gissmann L, Tommasino M. (2005). Skin hyperproliferation and susceptibility to chemical carcinogenesis in transgenic mice expressing E6 and E7 of human papillomavirus type 38. *J. Virol.* 79:14899–14908.

Duensing, S., Duensing, A., Crum, C.P., and Munger, K. (2001a). Human papillomavirus type 16 E7 oncoprotein-induced abnormal centrosome synthesis is an early event in the evolving malignant phenotype. *Cancer Res.* 61:2356–2360.

Duensing, S., and Munger, K. (2003). Human Papillomavirus Type 16 E7 oncoprotein can induce abnormal centrosome duplication through a mechanism independent of inactivation of retinoblastoma protein family members. *J. Virol.* 77:12331–12335.

Dunn, L.A., Evander, M., Tindle, R.W., Bulloch, A.L., de Kluyver, R.L., Fernando, G.J., Lambert, P.F., and Frazer, I.H. (1997). Presentation of the HPV16E7 protein by skin grafts is insufficient to allow graft rejection in an E7-primed animal. *Virology* 235:94–103.

Dvoretzky, I., Shober, R., Chattopadhyay, S.K., and Lowy, D.R. (1980). A quantitative *in vitro* focus assay for bovine papilloma virus. *Virology* 103:369–375.

Dyson, N., Howley, P.M., Munger, K., and Harlow, E. (1989). The human papilloma virus-16 E7 oncoprotein is able to bind to the retinoblastoma gene product. *Science* 243:934–937.

Elson, D., Riley, R., Lacey, A., Thordarson, G., Talamantes, F., and Arbeit, J. (2000). Sensitivity of the cervical transformation zone to estrogen-induced squamous carcinogenesis. *Cancer Res.* 60:1267–1275.

Escalante-Alcalde, D., Recillas-Targa, F., Valencia, C., Santa-Olalla, J., Chavez, P., Marroquin, A., Gutierrez, X., Gariglio, P., and Covarrubias, L. (2000). Expression of E6 and E7 papillomavirus oncogenes in the outer root sheath of hair follicles extends the growth phase and bypasses resting at telogen. *Cell Growth Differ.* 11:527–539.

Foley, J.D., Rosenbaum, H., and Griep, A.E. (2004). Temporal regulation of VEID-7-amino-4-trifluoromethylcoumarin cleavage activity and caspase-6 correlates with organelle loss during lens development. *J. Biol. Chem.* 279:32142–32150.

Foster, S.A., Demers, G.W., Etscheid, B.G., and Galloway, D.A. (1994). The ability of human papillomavirus E6 proteins to target p53 for degradation *in vivo* correlates with their ability to abrogate actinomycin D-induced growth arrest. *J. Virol.* 68:5698–5705.

Frazer, I.H., Kluyver, R.D., Leggatt, G.R., Yang Guo, H., Dunn, L., White, O., Harris, C., Liem, A., and Lambert, P. (2001). Tolerance or immunity to a tumor antigen expressed in somatic cells can be determined by systemic proinflammatory signals at the time of first antigen exposure. *J. Immunol.* 167:6180–6187.

Gissmann, L., Boshart, M., Durst, M., Ikenberg, H., Wagner, D., and zur Hausen, H. (1984). Presence of human papillomavirus in genital tumors. *Journal of Investigative Dermatology* 83:26s–28s.

Goodwin, E.C., and DiMaio, D. (2000). Repression of human papillomavirus oncogenes in HeLa cervical carcinoma cells causes the orderly reactivation of dormant tumor suppressor pathways. *Proc. Natl. Acad. Sci. USA* 97:12513–12518.

Goodwin, E.C., Yang, E., Lee, C.J., Lee, H.W., DiMaio, D., and Hwang, E.S. (2000). Rapid induction of senescence in human cervical carcinoma cells. *Proc. Natl. Acad. Sci. USA* 97:10978–10983.

Gordon, D.E., and Olson, C. (1968). Meningiomas and fibroblastic neoplasia in calves induced with the bovine papilloma virus. *Cancer Research* 28:2423–2431.

Greenhalgh, D.A., Wang, X.J., Rothnagel, J.A., Eckhardt, J.N., Quintanilla, M.I., Barber, J.L., Bundman, D.S., Longley, M.A., Schlegel, R., and Roop, D.R. (1994). Transgenic mice expressing targeted HPV-18 E6 and E7 oncogenes in the epidermis develop verrucous lesions and spontaneous, rasHa-activated papillomas. *Cell Growth & Differentiation* 5:667–675.

Griep, A., Herber, R., Jeon, S., Lohse, J., Dubielzig, R., and Lambert, P.F. (1993). Tumorigenicity by human papillomavirus type 16 E6 and E7 in transgenic mice correlates with alterations in epithelial cell growth and differentiation. *J. Virol.* 67:1373–1384.

Griep, A., Krawcek, J., Lee, D., Liem, A., Albert, D., Carabeo, R., Drinkwater, N., McCall, M., Sattler, C., Lasudry, J., and Lambert, P.F. (1998). Multiple genetic loci modify risk for retinoblastoma in transgenic mice. *Invest. Ophth. Vis. Sci.* 39:2723–2732.

Grose, K.R., and Bjeldanes, L.F. (1992). Oligomerization of indole-3-carbinol in aqueous acid. *Chem. Res. Toxicol.* 5:188–193.

Gulliver, G.A., Herber, R.L., Liem, A., and Lambert, P.F. (1997). Both conserved region 1 (CR1) and CR2 of the human papillomavirus type 16 E7 oncogene are required for induction of epidermal hyperplasia and tumor formation in transgenic mice. *J. Virol.* 71:5905–5914.

Herber, R., Liem, A., Pitot, H., and Lambert, P.F. (1996). Squamous epithelial hyperplasia and carcinoma in mice transgenic for the human papillomavirus type 16 E7 oncogene. *J. Virol.* 70:1873–1881.

Hwang, E.S., Riese, D.D., Settleman, J., Nilson, L.A., Honig, J., Flynn, S., and DiMaio, D. (1993). Inhibition of cervical carcinoma cell line proliferation by the introduction of a bovine papillomavirus regulatory gene. *J. Virol.* 67:3720–3729.

Jarrett, W.F., Smith, K.T., O'Neil, B.W., Gaukroger, J.M., Chandrachud, L.M., Grindlay, G.J., McGarvie, G.M., and Campo, M.S. (1991). Studies on vaccination against papillomaviruses: Prophylactic and therapeutic vaccination with recombinant structural proteins. *Virology* 184:33–42.

Jeckel, S., Huber, E., Stubenrauch, F., and Iftner, T. (2002). A transactivator function of cottontail rabbit papillomavirus E2 is essential for tumor induction in rabbits. *J. Virol.* 76:11209–11215.

Jeon, S., Allen, H.B., and Lambert, P.F. (1995). Integration of human papillomavirus type 16 into the human genome correlates with a selective growth advantage of cells. *J. Virol.* 69:2989–2997.

Jin, L., Qi, M., Chen, D.Z., Anderson, A., Yang, G.Y., Arbeit, J.M., and Auborn, K.J. (1999). Indole-3-carbinol prevents cervical cancer in human papilloma virus type 16 (HPV16) transgenic mice. *Cancer. Res.* 59:3991–3997.

Kang, J.K., Kim, J.H., Lee, S.H., Kim, D.H., Kim, H.S., Lee, J.E., and Seo, J.S. (2000). Development of spontaneous hyperplastic skin lesions and chemically induced skin papillomas in transgenic mice expressing human papillomavirus type 16 E6/E7 genes. *Cancer Lett.* 160:177–183.

Keating, J.T., Ince, T., and Crum, C.P. (2001). Surrogate biomarkers of HPV infection in cervical neoplasia screening and diagnosis. *Adv. Anat. Pathol.* 8:83–92.

Kessis, T.D., Slebos, R.J., Nelson, W.G., Kastan, M.B., Plunkett, B.S., Han, S.M., Lorincz, A.T., Hedrick, L., and Cho, K.R. (1993). Human papillomavirus 16 E6 expression disrupts the p53-mediated cellular response to DNA damage. *Proceedings of the National Academy of Sciences of the United States of America* 90:3988–3992.

Kirnbauer, R., Chandrachud, L.M., O'Neil, B.W., Wagner, E.R., Grindlay, G.J., Armstrong, A., McGarvie, G.M., Schiller, J.T., Lowy, D.R., and Campo, M.S. (1996). Virus-like particles of bovine papillomavirus type 4 in prophylactic and therapeutic immunization. *Virology* 219:37–44.

Kondoh, G., Hayasaka, N., Li, Q., Nishimune, Y., and Hakura, A. (1995). An *in vivo* model for receptor tyrosine kinase autocrine/paracrine activation: Auto-stimulated KIT receptor acts as a tumor promoting factor in papillomavirus-induced tumorigenesis. *Oncogene* 10:341–347.

Kondoh, G., Murata, Y., Aozasa, K., Yutsudo, M., and Hakura, A. (1991). Very high incidence of germ cell tumorigenesis (seminomagenesis) in human papillomavirus type 16 transgenic mice. *Journal of Virology* 65:3335–3339.

Kondoh, G., Nishimune, Y., Nishizawa, Y., Hayasaka, N., Matsumoto, K., and Hakura, A. (1994). Establishment and further characterization of a line of transgenic mice showing testicular tumorigenesis at 100% incidence. *Journal of Urology.* 152:2151–2154

Lacey, M., Alpert, S., and Hanahan, D. (1986). Bovine papillomavirus genome elicits skin tumours in transgenic mice. *Nature* 322:609–612.

Lambert, P.F., Baker, C.C., and Howley, P.M. (1988). The genetics of bovine papillomavirus type 1. *Annual Review of Genetics* 22:235–258.

Lambert, P.F., Pan, H., Pitot, H., Liem, A., Jackson, M., and Griep, A. (1993). Epidermal cancer associated with expression of human papillomavirus type 16 E6 and E7 oncogenes in the skin of transgenic mice. *Proc. Natl. Acad. Sci. USA* 90:5583–5587.

Law, M.F., Lowy, D.R., Dvoretzky, I., and Howley, P.M. (1981). Mouse cells transformed by bovine papillomavirus contain only extrachromosomal viral DNA sequences. *Proceedings of the National Academy of Sciences of the United States of America* 78:2727–2731.

Leggatt, G.R., Dunn, L.A., De Kluyver, R.L., Stewart, T., and Frazer, I.H. (2002). Interferon-gamma enhances cytotoxic T lymphocyte recognition of endogenous peptide in keratinocytes without lowering the requirement for surface peptide. *Immunol. Cell. Biol.* 80:415–424.

Lin, Y.L., Borenstein, L.A., Ahmed, R., and Wettstein, F.O. (1993). Cottontail rabbit papillomavirus L1 protein-based vaccines: Protection is achieved only with a full-length, nondenatured product. *J. Virol.* 67:4154–4162.

Lin, Y.L., Borenstein, L.A., Selvakumar, R., Ahmed, R., and Wettstein, F.O. (1992). Effective vaccination against papilloma development by immunization with L1 or L2 structural protein of cottontail rabbit papillomavirus. *Virology* 187:612–619.

Lindgren, V., Sippola, T.M., Skowronski, J., Wetzel, E., Howley, P.M., and Hanahan, D. (1989). Specific chromosomal abnormalities characterize fibrosarcomas of bovine papillomavirus type 1 transgenic mice. *Proceedings of the National Academy of Sciences of the United States of America* 86:5025–5029.

Matsumoto, K., Leggatt, G.R., Zhong, J., Liu, X., de Kluyver, R.L., Peters, T., Fernando, G.J., Liem, A., Lambert, P.F., and Frazer, I.H. (2004). Impaired antigen presentation and effectiveness of combined active/passive immunotherapy for epithelial tumors. *J. Natl. Cancer Inst.* 96:1611–1619.

McCaffrey, J., Yamasaki, L., Dyson, N.J., Harlow, E., and Griep, A.E. (1999). Disruption of retinoblastoma protein family function by human papillomavirus type 16 E7 oncoprotein inhibits lens development in part through E2F-1. *Mol. Cell. Biol.* 19:6458–6468.

Melero, I., Singhal, M.C., McGowan, P., Haugen, H.S., Blake, J., Hellstrom, K.E., Yang, G., Clegg, C.H., and Chen, L. (1997). Immunological ignorance of an E7-encoded cytolytic T-lymphocyte epitope in transgenic mice expressing the E7 and E6 oncogenes of human papillomavirus type 16. *J. Virol.* 71:3998–4004.

Meyers, C., and Wettstein, F.O. (1991). The late region differentially regulates the in vitro transformation by cottontail rabbit papillomavirus DNA in different cell types. *Virology* 181:637–646.

Moar, M.H., Campo, M.S., Laird, H.M., and Jarrett, W.F. (1981). Unintegrated viral DNA sequences in a hamster tumor induced by bovine papilloma virus. *J. Virol.* 39: 945–949.

Morgenbesser, S.D., Williams, B.O., Jacks, T., and DePinho, R.A. (1994). p53-dependent apoptosis produced by Rb-deficiency in the developing mouse lens. *Nature* 371:72–74.

Munger, K. (2002) The role of human papillomaviruses in human cancers. *Front. Biosci.* 7:d641–649.

Nasseri, M., Meyers, C., and Wettstein, F.O. (1989). Genetic analysis of CRPV pathogenesis: The L1 open reading frame is dispensable for cellular transformation but is required for papilloma formation. *Virology* 170:321–325.

Nasseri, M., and Wettstein, F.O. (1984). Differences exist between viral transcripts in cottontail rabbit papillomavirus-induced benign and malignant tumors as well as non-virus-producing and virus-producing tumors. *J. Virol.* 51:706–712.

Nguyen, M., Song, S., Liem, A., Androphy, E., Liu, Y., and Lambert, P.F. (2002a). A mutant of human papillomavirus type 16 E6 deficient in binding alpha-helix partners displays reduced oncogenic potential *in vivo*. *J. Virol.* 76:13039–13048.

Nguyen, M.L., Nguyen, M.M., Lee, D., Griep, A.E., and Lambert, P.F. (2003a). The PDZ ligand domain of the human papillomavirus type 16 E6 protein is required for E6's induction of epithelial hyperplasia *in vivo*. *J. Virol.* 77:6957–6964.

Nguyen, M.M., Nguyen, M.L., Caruana, G., Bernstein, A., Lambert, P.F., and Griep, A.E. (2003b). Requirement of PDZ-containing proteins for cell cycle regulation and differentiation in the mouse lens epithelium. *Mol. Cell. Biol.* 23:8970–8981.

Nguyen, M.M., Potter, S.J., and Griep, A.E. (2002b). Deregulated cell cycle control in lens epithelial cells by expression of inhibitors of tumor suppressor function. *Mech. Dev.* 112:101–113.

Olson, C., Gordon, D.E., Robl, M.G., and Lee, K.P. (1969). Oncogenicity of bovine papilloma virus. *Archives of Environmental Health* 19:827–837.

Olson, C., Pamukcu, A.M., Brobst, D.F., Kowolczk, T., Satter, E.J., and Proce, J.M. (1959). A urinary bladder tumor induced by a bovine cutaneous papilloma agent. *Cancer Res.* 19:779–783

Pan, H., and Griep, A. (1994a). Altered cell cycle regultion in the lens of HPV-16 E6 or E7 transgenic mice: Implications for tumor suppressor gene function in development. *Genes Dev.* 8:1285–1299.

Pan, H., and Griep, A. (1995a). Temporally distinct patterns of p53-dependent and p53-independent apoptosis during mouse lens development. *Genes Dev.* 9:2157–2169.

Pan, H., and Griep, A.E. (1994b). Altered cell cycle regulation in the lens of HPV-16 E6 or E7 transgenic mice: Implications for tumor suppressor gene function in development. *Genes Dev.* 8:1285–1299.

Pan, H., and Griep, A.E. (1995b). Temporally distinct patterns of p53-dependent and p53-independent apoptosis during mouse lens development. *Genes Dev.* 9:2157–2169.

Peh, W.L., Brandsma, J.L., Christensen, N.D., Cladel, N.M., Wu, X., and Doorbar, J. (2004). The viral E4 protein is required for the completion of the cottontail rabbit papillomavirus productive cycle *in vivo*. *J. Virol.* 78:2142–2151.

Pfister, H., Gross, G., and Hagedorn, M. (1979). Characterization of human papillomavirus 3 in warts of a renal allograft patient. *Journal of Investigative Dermatology* 73:349–353.

Riley, R., Duensing, S., Brake, T., Munger, K., Lambert, P.F., and Arbeit, J. (2003). Dissection of human papillomavirus E6 and E7 function in transgenic mouse models of cervical carcinogenesis. *Cancer Res.* 63:4862–4871.

Rous, P., and Beard, J.W. (1934). A virus induced mammalian growth with the characteristics of a tumor (the Shope rabbit papillomavirus). 1. The growth on implantation within a favorable host. *J. Exp. Med.* 60:701–722.

Sasagawa, T., Kondoh, G., Inoue, M., Yutsudo, M., and Hakura, A. (1994). Cervical/vaginal dysplasias of transgenic mice harbouring human papillomavirus type 16 E6-E7 genes. *Journal of General Virology.*

Schaeffer, A., Nguyen, M., Liem, A., Lee, D., Montagna, C., Lambert, P.F., Ried, T., and Difilippantonio, M. (2004). E7 and E7 oncoproteins induce distinct patterns of chromosomal aneuploidy in skin tumors from transgenic mice. *Cancer Res.* 64:538–546.

Schaper ID, Marcuzzi GP, Weissenborn SJ, Kasper HU, Dries V, Smyth N, Fuchs P, Pfister H. (2005). Development of skin tumors in mice transgenic for early genes of human papillomavirus type 8. *Cancer Res.* 65:1394–400.

Scheffner, M., Munger, K., Byrne, J.C., and Howley, P.M. (1991). The state of the p53 and retinoblastoma genes in human cervical carcinoma cell lines. *Proceedings of the National Academy of Sciences of the United States of America* 88:5523–5527.

Schreiber, K., Cannon, R.E., Karrison, T., Beck-Engeser, G., Huo, D., Tennant, R.W., Jensen, H., Kast, W.M., Krausz, T., Meredith, S.C., et al. (2004). Strong synergy between mutant ras and HPV16 E6/E7 in the development of primary tumors. *Oncogene* 23:3972–3979.

Schwarz, E., Freese, U.K., Gissmann, L., Mayer, W., Roggenbuck, B., Stremlau, A., and zur Hausen. H. (1985). Structure and transcription of human papillomavirus sequences in cervical carcinoma cells. *Nature* 314:111–114.

Shope, R.E. (1935). Serial transmission of the virus of infectious papillomatosis in domestic rabbits. *Proc. Soc. Exp. Biol. Med.* 32:830–832.

Shope, R.E. (1937). Immunization of rabbits to infectious papillomatosis. *J. Exp. Med.* 65:607–624.

Sippola, T.M., Hanahan, D., and Howley, P.M. (1989). Cell-heritable stages of tumor progression in transgenic mice harboring the bovine papillomavirus type 1 genome. *Molecular & Cellular Biology* 9:925–934.

Song, S., Gulliver, G.A., and Lambert, P.F. (1998). Human papillomavirus type 16 E6 and E7 oncogenes abrogate radiation-induced DNA damage responses *in vivo* through p53-dependent and p53-independent pathways. *Proc. Natl. Acad. Sci. USA* 95:2290–2295.

Song, S., and Lambert, P.F. (1999). Different responses of epidermal and hair follicular cells to radiation correlate with distinct patterns of p53 and p21 induction. *Am. J. Pathol.* 155:1121–1127.

Song, S., Liem, A., Miller, J.A., and Lambert, P.F. (2000). Human papillomavirus types 16 E6 and E7 contribute differently to carcinogenesis. *Virology* 267:141–150.

Song, S., Pitot, H.C., and Lambert, P.F. (1999). The human papillomavirus type 16 E6 gene alone is sufficient to induce carcinomas in transgenic animals. *J. Virol.* 73:5887–5893.

Strati, K., Pitot, H.C. and Lambert, P.F. Identification of Biomarkers that Distinguish HPV-positive versus HPV-negative Head and Neck Cancers in a Mouse Model. (2006). *Proc. Natl. Acad. Sci., U.S.A.,*103:14152–14157.

Suzich, J.A., Ghim, S.J., Palmer, H.F., White, W.I., Tamura, J.K., Bell, J.A., Newsome, J.A., Jenson, A.B., and Schlegel, R. (1995). Systemic immunization with papillomavirus L1 protein completely prevents the development of viral mucosal papillomas. *Proc. Natl. Acad. Sci. USA* 92:11553–11557.

Um, S.J., Kim, E.J., Hwang, E.S., Kim, S.J., Namkoong, S.E., and Park, J.S. (2000). Antiproliferative effects of retinoic acid/interferon in cervical carcinoma cell lines: Cooperative growth suppression of IRF-1 and p53. *Int. J. Cancer* 85:416–423.

Wells, S.I., Francis, D.A., Karpova, A.Y., Dowhanick, J.J., Benson, J.D., and Howley, P.M. (2000). Papillomavirus E2 induces senescence in HPV-positive cells via pRB- and p21(CIP)-dependent pathways. *Embo. J.* 19:5762–5771.

Werness, B.A., Levine, A.J., and Howley, P.M. (1990). Association of human papillomavirus types 16 and 18 E6 proteins with p53. *Science* 248:76–79.

Wu, X., Xiao, W., and Brandsma, J.L. (1994). Papilloma formation by cottontail rabbit papillomavirus requires E1 and E2 regulatory genes in addition to E6 and E7 transforming genes. *J. Virol.* 68:6097–6102.

Yee, C., Krishnan, H.I., Baker, C.C., Schlegel, R., and Howley, P.M. (1985). Presence and expression of human papillomavirus sequences in human cervical carcinoma cell lines. *American Journal of Pathology* 119:361–366.

Young, M., Farrell, L., Lambert, P.F., Awasthi, P., and Colburn, N. (2002). Protection against human papillomavirus type 16-E7 oncogene-induced tumorigenesis by *in vivo* expression of dominant negative c-jun. *Mol. Carcinogen* 34:72–77.

Yuan, F., Chen, D.Z., Liu, K., Sepkovic, D.W., Bradlow, H.L., and Auborn, K. (1999). Anti-estrogenic activities of indole-3-carbinol in cervical cells: Implication for prevention of cervical cancer. *Anticancer Res.* 19:1673–1680.

Yuan, H., Estes, P.A., Chen, Y., Newsome, J., Olcese, V.A., Garcea, R.L., and Schlegel, R. (2001). Immunization with a pentameric L1 fusion protein protects against papillomavirus infection. *J. Virol.* 75:7848–7853.

Zeltner, R., Borenstein, L.A., Wettstein, F.O., and Iftner, T. (1994). Changes in RNA expression pattern during the malignant progression of cottontail rabbit papillomavirus-induced tumors in rabbits. *J. Virol.* 68:3620–3630.

12
The Humoral Immune Response to Human Papillomavirus

Erin M. Egelkrout and Denise A. Galloway
Program in Cancer Biology, Fred Hutchinson Cancer Research Center, Seattle, WA 98109-1024

12.1. Introduction

HPV has evolved many mechanisms to evade the host immune system. One of these is its replication in differentiated epithelial cells that have a relatively short natural life span. In addition, the virus does not cause cell lysis that usually triggers an immune response. Perhaps for these reasons there is a lack of a robust inflammatory response. Nevertheless, most HPV infections are cleared naturally and reinfection with the same HPV type is rare. Thus, HPV does cause an immune response that has been extensively studied. This review will focus on the humoral immune response caused by HPV infection.

12.2. Early Work on Identification of Humoral Immune Response

Much of the early work on the humoral immune response to HPV was done using a variety of antigens, including viruses, viral proteins, and peptides. In the case of some of the low-risk types, e.g., 1, 6, and 11, which cause warts, viruses could be obtained from either wart tissue or from xenografts (Pfister and zur Hausen, 1978; Kienzler et al., 1983; Pfister, 1984; Steele and Gallimore, 1990; Bonnez et al., 1992, 1991, 1993). Some early work focused on attempts to detect antibodies reactive with bovine papillomavirus or cottontail rabbit papillomavirus in human sera (Baird, 1983; Beiss et al., 1991; Dillner et al., 1990), but most studies have not detected reactivity to these papillomavirus types in human sera (Jenson et al., 1980; Baird, 1983; Nakai et al., 1986; Jenison et al., 1989; Dillner et al., 1990; Steele and Gallimore, 1990; Beiss et al., 1991; Christensen et al., 1992; Kirnbauer et al., 1994). However, neither warts nor mouse xenograft systems could produce adequate numbers of virions for large-scale studies. Thus many studies used viral proteins, parts of the proteins expressed in bacteria or yeast, or peptides. Because many HPV infections are asymptomatic, it was difficult to identify individuals who were clearly uninfected, and additionally the diversity of HPV types was not recognized. Experimental exposure to HPV suggested that intact virions provoke a type-specific antibody response, while denatured virions induce antibodies able to cross-react with L1 proteins from BPV or HPV (Jenson et al., 1980; Nakai et al., 1986).

12.3. Serological Assays Using Proteins and Synthetic Peptides

A variety of HPV-derived peptides and proteins have been used in studies defining the humoral immune response to HPV. Human serum antibodies that react with HPV fusion proteins or synthetic peptides have been found in individuals in a number of studies (Dillner et al., 1990; Jenison et al., 1990; Kochel et al., 1991b). The major antigen targets appeared to be the capsid proteins, however reactivity to the minor capsid protein L2 was observed more commonly than to L1. Less frequently antibodies to E2 and E7 were observed, and occasionally to E4. Some studies found interesting correlations with seropositivity and HPV disease or HPV DNA detection (Van Doornum et al., 1994; Wikstrom et al., 1995b). However, in other studies the prevalence of HPV antibodies was not strongly associated with other parameters of HPV infection (Jenison et al., 1990; Kochel et al., 1991b). More recent studies described in a later section identified cross-reacting epitopes on the L2 protein, and it is possible that these assays identified that reactivity. Importantly, as will also be discussed in detail in a later section, the early antigen targets did not detect antibodies that correlated with responses to conformational epitopes found on the virion. Some studies defining epitopes on L1 have used panels of overlapping peptides covering the entire protein (Heino et al., 1995). Rocha-Zavaleta et al. used a peptide derived from HPV-16 L1 to test for reactivity in patients with low-grade squamous intraepithelial lesions (LSIL) associated with low and high-risk HPVs or with cervical cancer (Rocha-Zavaleta et al., 2004). An IgG and IgA response was detected in 90 percent of patients with high-risk, but not low-risk, HPVs, suggesting that this peptide contains an epitope common to many high-risk HPV types. Urquiza et al. used two synthetic peptides from L1-surface-exposed regions and found that they were recognized by antibodies in 91–96 percent of sera from CIN or cervical cancer patients compared to 3.6 percent in control women with normal cytology (Urquiza et al., 2005). While these studies may have identified potential epitopes, most recent studies have used antigen targets in which the viral capsid proteins are correctly folded.

A number of assays have also been developed to detect antibodies to the early HPV proteins E6 and E7 or other early proteins. Some of the earliest tests developed used fragments of the proteins (Stacey et al., 1992, 1993; Viscidi et al., 1993; Nindl et al., 1994, 1996; Sun et al., 1994b; Chee et al., 1995). Others have used full-length proteins expressed in yeast (Meschede et al., 1998; Zumbach et al., 2000a,b; Herrero et al., 2003), and full-length GST-fusion proteins expressed in bacteria (Sehr et al., 2001). As will be described in more detail below, these assays have shown some correlation between the presence of antibodies to E6 and E7 and late stage cervical cancer; however, the sensitivity has usually been too low to make these assays useful as a diagnostic test. Tests for other early proteins such as E2 have also been used (Dillner et al., 1990; Lehtinen et al., 1992b,a; Heino et al., 1993; Veress et al., 1994; Kim et al., 1999; Hamsikova et al., 2000; Davidson et al., 2003).

In spite of their limitations, useful information has been obtained using these methods. Nonnenmacher et al. examined the humoral response to HPV-16 E6 and E7 and found that positivity did not correlate with the response to capsid proteins (Nonnenmacher et al., 1995). A number of studies used HPV early protein peptides to attempt to correlate the presence of antibodies to E6 or E7 with progression to cancer or survival (Jochmus-Kudielka et al., 1989; Mann et al., 1990; Mandelson et al., 1992; Muller et al., 1992; Hamsikova et al., 1994; Sun et al., 1994a,b, 1996; Fisher et al., 1996; Dillner et al., 1997). These studies have in general found a correlation between the presence of antibodies to E7 and late-stage cervical cancer. Vonka et al. found that antibodies directed against peptides from E2, E4, and E7 could be detected at a higher rate in women with cervical neoplasia than in matched controls (Vonka et al., 1999). Malcolm et al. examined the reactivity of sera from cervical cancer patients to HPV-16 E7 in a number of assays (Malcolm et al., 2000). A higher proportion of cervical cancer sera relative to controls were positive for E7 antibodies. Waterboer et al. found a correlation between cervical cancer and detection of antibodies to HPV-52 and HPV-58 E6 (Waterboer et al., 2005). In contrast to this evidence that antibodies to E7 correlate with late-stage cervical cancer, Viladiu et al. suggested that there is an association with early-stage cervical cancer (Viladiu et al., 1997). Nguyen et al. examined the IgG and IgA response to HPV-16 E7 in women who had had a hysterectomy (Nguyen et al., 2005). There was a selective down-regulation of the IgA response relative to IgG in cervical cancer, suggesting differences in regulation of the various immunoglobulin types.

Other recent studies have shown that the presence of these antibodies cannot be used diagnostically (Park et al., 1998). Silins et al. did not find a significant association between the presence of antibodies to E6 and E7 and disease prognosis (Silins et al., 2002). In another study that followed women for up to 20 years for development of squamous cell carcinoma (SCC), the presence of antibodies to E6 and E7 was also not found to be a good marker for progression to cervical cancer (Lehtinen et al., 2003).

Relatively little work has been done on the other early HPV proteins, but a few studies examined antibodies to E2 and E4 (Dillner et al., 1989, 1990; Jochmus-Kudielka et al., 1989; Jenison et al., 1990; Kochel et al., 1991a,b; Mann et al., 1990; Mandelson et al., 1992; Van Doornum et al., 1994). Wikstrom et al.

assayed sera for reactivity to L1-, L2-, and E2-derived-peptides from HPV-6, -16, and -18 (Wikstrom et al., 1995b). A high IgG response to the E2 peptide was observed at the time of new infection. Lenner et al. examined IgA and IgG levels against several different HPV epitopes, including 245:16 and 245:18 in E2 (Lenner et al., 1995). There was some association between elevated levels of IgA and poor prognosis. Using baculovirus-expressed HPV 16 E2, Rocha-Zavaleta et al. observed that anti-E2 IgA is elevated in CIN patients but not in cervical cancer patients. This is consistent with a lack of E2 expression at late stages of infection (Rocha-Zavaleta et al., 1997).

12.4.   VLPS/Capsomers

12.4.1.   Introduction/Overview of Serological Methods

A number of different methods have been used to detect the presence of antibodies to HPV. For many years the development of serological assays for the detection of HPV antibodies was hampered by the inability to grow HPVs in culture. Thus viral antigens were not available. Most current tests are based on the use of virus-like particles (VLPs) in enzyme-linked immunosorbent assays (ELISA). The subunits of VLPs, capsomers, have also been used as antigen targets. Neutralization assays, based on the ability of test serum antibodies to prevent infection of cells by pseudovirus, have been developed recently. One characteristic of these tests has been a lack of cross-reactivity between different HPV types. This makes efficient detection of infection with multiple HPV types time-consuming. The recent development of new methods making use of large numbers of different HPV proteins bound to beads may make high-throughput detection of multiple types practical.

12.5.   Capsid/Capsomer/VLP Production

Human papillomavirus is encapsidated by two proteins, the major capsid protein L1 and the minor capsid protein L2 (Orth and Favre, 1985; Kirnbauer et al., 1992; Roden et al., 1996a). Five molecules of the L1 protein form an intermediate structure known as a capsomere. These associate into an icosahedral structure containing 72 of these capsomers. Many fewer molecules of L2 are incorporated into virions and their position is not clearly defined. As will be described in more detail in a following section, much of the known humoral immune response to HPV infection is triggered by epitopes on the highly antigenic capsid structure. It should also be noted that a large proportion of these studies have used HPV-16, one of the most common types in cervical cancer (Munoz et al., 2003). VLPs consisting of an empty shell of L1 retain most of the antigenic properties of the native virion (Salunke et al., 1986; Hagensee et al., 1993, 1994; Kirnbauer et al., 1993; Rose et al., 1993). These VLPs have been a very useful tool in the development both of tests for the presence of an immune response to HPV infection and of potential vaccines (Breitburd et al., 1995; Suzich et al., 1995; Kirnbauer, 1996; Koutsky et al., 2002; Harper et al., 2004).

A number of different methods have been used in VLP production. These involve expression of L1 from eukaryotic vectors including vaccinia virus, baculovirus, or yeast. Alternatively, capsomers produced in bacteria have also been used (Li et al., 1997; Onda et al., 2003; Aires et al., 2006). Early efforts at VLP production were hampered by low expression but modifying L1 codon usage improves expression (Zhou et al., 1999; Leder et al., 2001; Liu et al., 2001; Baud et al., 2004). Recently, the process of VLP maturation and the role of disulfide bonds in stabilization have been studied (Buck et al., 2005b). The stabilization of VLPs by non-ionic surfactants has also been described (Shi et al., 2005). Expressed VLPs are usually purified by methods based on cesium chloride density gradient purification. Interestingly, although the L2 protein is not required for VLP assembly, one recent report suggested that the presence of L2 may aid assembly independent of disulfide bonds (Ishii et al., 2005). Although empty VLPs are highly immunogenic, some studies suggest that immunogenicity can be further enhanced with encapsidation of genetic adjuvants such as IL2 (Oh et al., 2004).

12.5.1. Serological Assays

12.5.1.1. ELISA

The use of VLPs has allowed the development of a number of different forms of enzyme-linked immunosorbent assay (ELISA) for the detection of antibodies to HPV. In the most basic form of this method, the direct ELISA, VLPs are used to coat microtiter plates. Antibodies in positive test serum will bind to the VLPs and are detected using labeled secondary antibody. This approach has been used for the high risk HPV types 16, 18, 31, 33, 35,and 45 and the low-risk types 6 and 11 (Rose et al., 1993; Hines et al., 1994a,b; Kirnbauer et al., 1994; Rose et al., 1994; Sapp et al., 1994; Eklund and Dillner, 1995; Le Cann et al., 1994, 1995; Volpers et al., 1995; Peng et al., 1999; Marais et al., 2000a,b; Giroglou et al., 2001; Combita et al., 2002a). It has also been used for cutaneous HPV types 5, 8, 15, 20, 24 and 38 (Favre et al., 1998; Stark et al., 1998; Wieland et al., 2000; Feltkamp et al., 2003).

In another version of this assay, the antigen capture ELISA, the wells are coated with a monoclonal antibody to a conformational epitope that "captures" the VLP (Carter et al., 1994, 1995; Hagensee et al., 1993). A third version is the competitive ELISA. In this case a labeled monoclonal antibody is added to the assay along with the sample to be tested. Antibodies in the test serum compete with this antibody, actually leading to a reduction in signal when sera containing HPV-specific antibodies are used (Palker et al., 2001; Opalka et al., 2003). In a fourth approach, HPV antibodies in test sera interfere with VLP-induced hemagglutination of erythrocytes (Roden et al., 1996b).

Assays have also been developed using capsomers expressed in bacteria (Li et al., 1997; Rose et al., 1998; Yuan et al., 2001; Onda et al., 2003). In the original report it was found that full-length HPV-11 L1 produced in bacteria and cleaved with trypsin at R415 could assemble into a pentameric structure known as a capsomer. ELISAs showed that these capsomers have the same reactivity as VLPs (Li et al., 1997). It was later shown that the monoclonal antibodies H11.F1 and H11.H3 recognize these capsomers in an ELISA (Rose et al., 1998). In the same study capsomers were used to induce production of type-specific polyclonal immune sera that could block the infectivity of HPV-11 virions. Immunization with capsomers protected dogs from viral infection and induced an antibody response detectable by ELISA (Yuan et al., 2001). Finally, Onda et al. developed a capsomer-based ELISA to HPV-16 that was used in a study of IgA levels in women with incident HPV-16 infection (Onda et al., 2003).

A number of approaches have been taken to address problems with sensitivity and specificity of these assays. For example, a common problem with these tests is that improperly folded VLPs can lead to spurious signal, often from cross-reactive linear epitopes (Wang et al., 2005). Recently two papers have described the use of heparin to circumvent this problem (Wang et al., 2005; Rommel et al., 2005). In another version of the capture ELISA, a coating of heparin on the well of a microtiter plate will selectively capture correctly folded VLPs. It has been reported that VLPs require heparan sulfates to bind to dendritic cells (Bousarghin et al., 2005). Various polymers have also been used to increase sensitivity and specificity (Studentsov et al., 2002, 2003).

The sensitivity of serologic assays is usually addressed by comparison with HPV DNA detected by PCR-based methods. Levels of sensitivity have generally been shown to be about 50 percent (Kirnbauer et al., 1994; Carter et al., 1996; Wideroff et al., 1995, 1996; Andersson-Ellstrom et al., 1994, 1996), although sensitivity as high as 75 percent has been reported (Kjellberg et al., 1999). Furthermore, there is often a correlation between the level of DNA detected and the likelihood of detection of HPV antibodies (Viscidi et al., 1997).

A large body of evidence suggests that VLP ELISAs are specific for different HPV types (Christensen et al., 1990; Hines et al., 1994a,b; Kirnbauer et al., 1994; Sapp et al., 1994; Wikstrom et al., 1995b; Christensen et al., 1996b; Nonnenmacher et al., 1996; Roden et al., 1996a; Touze et al., 1998; Carter et al., 2000; Combita et al., 2002b). A general approach of many of these studies assessing type specificity is to use ELISAs for a particular HPV type to compare serum samples from uninfected women, women infected with that HPV type as determined by DNA testing, and women infected with different HPV types. It must be taken into consideration in such studies that any reactivity detected may reflect a previous infection with a HPV type other than the one being assayed for. The fact that very low rates of infection of 2–7 percent are observed in monogamous women, sexually inexperienced young women, and children also indicates there is little cross-reactivity between the genital and other HPV types, although the possibility of chance false positives should always be considered on detection of high-risk types in children (Andersson-Ellstrom et al., 1994, 1996; Carter et al., 1996; Dillner et al., 1996; Wideroff et al., 1996; Mund et al., 1997; Viscidi et al., 1997; af Geijersstam et al., 1999; Kjellberg et al., 1999; Dunne et al., 2005). An important finding from these studies is that while most surface-exposed epitopes on intact capsids are type-specific (Cowsert et al., 1987; Christensen et al., 1996a), denaturation of capsids often exposes some cross-reactive epitopes (Jenson et al., 1980; Dillner et al., 1991). It should be noted, however, that HPV-6 and -11 appear to share some surface-exposed conformation-dependent epitopes (Christensen et al., 1994, 1996b). It would not be surprising if other very closely related types, e.g., HPV-18 and -45 also have some cross-reactive epitopes.

12.5.1.2.   Neutralization Assays

Another type of assay, the neutralization assay, depends on the ability of antibodies in a test serum to prevent infection of cells by infectious HPV particles or by pseudovirions (McLaughlin-Drubin et al., 2004; Pastrana et al., 2004; Sapp and Selinka, 2005; Buck et al., 2005a). Pseudovirions are produced by transfection of cells with plasmids coding for codon-optimized L1 and L2 and a plasmid that has a SV40 T antigen origin of replication and expresses a reporter protein such as secreted alkaline phosphatase (SEAP). Pseudovirions are purified on density gradients and are used to infect a cell line such as 293TT (Pastrana et al., 2004). Thus, the presence of antibodies to HPV in a test serum blocks infection of pseudovirions and reduces signal from the reporter gene. Beta-lactamase has also been used as a reporter (Yeager et al., 2000). There is some evidence that these assays are more type-specific than the ELISAs. For example, in a set of serum samples from a vaccine trial testing HPV-16 VLPs, Pastrana et al. observed some cross-reactivity between HPV-16 and HPV-18 using a standard ELISA, but not using HPV-16 and HPV-18 neutralization assays (Pastrana et al., 2004).

12.5.1.3.   Luminex Bead Method

As most antibodies to HPV are specific for the HPV type of the antigen and will not recognize the corresponding protein of another HPV type, it has been cumbersome to test for multiple infections with a number of different HPV types. Recently, new methods involving the binding of antigens to color-coded beads have been described. In one method VLPs were bound to beads and mixed with phycoerythrin-conjugated antibodies (Opalka et al., 2003). In this competitive assay the presence of antibody in a test serum reduces the signal. Optimization of this method and comparison to RIA has been described. The multiplex method compared well to previously used RIAs in this study of levels of antibodies to neutralizing epitopes on VLPs for HPV types 6, 11, 16, and 18 (Dias et al., 2005).

In another approach many different HPV proteins are expressed as fusions with glutathione-S-transferase (GST). This builds on earlier work showing that a GST-L1 fusion expressed in bacteria retains most of the properties of VLPs (Rose et al., 1998; Yuan et al., 2001; Sehr et al., 2002). In the multiplex method each GST fusion protein is bound to color coded-beads that are glutathione-conjugated (Waterboer et al., 2005). The beads can be separated and analyzed by FACS. This method compared well with ELISAs, although there was enhanced detection of some weak reactivities. Both of these methods can potentially be used to test for antibodies directed against up to 100 viral proteins at once. They also have the advantage of using very small amounts of serum.

12.5.1.4. Summary and Standardization

Much of the early discrepancy among serological assays to detect HPV antibodies was eliminated by using VLP-based ELISAs. However, variability between laboratories remains in assay performance and in defining cut-off points for positive and negative results. A few reports have attempted to address this issue. Strickler et al. reported the testing of the same samples in three different laboratories and found that the ELISAs used in the three labs performed similarly, but there were significant interlaboratory differences (Strickler et al., 1997). There were variation coefficients of 0.61 to 0.81 between labs. The authors suggest more sharing of positive and negative control samples and also treating results falling near a cut-off point as indeterminate. In an attempt to assess the variability between laboratories as a first step to defining more widely used standards, the World Health Organization recently sponsored a collaborative study in which ten different laboratories in various locations tested the same samples from women with natural HPV infection or from vaccine trials (Ferguson et al., 2005). Although there was substantial variation between labs, the samples were generally ranked in the same order of relative reactivity.

12.5.2. Natural History and Serological Response to HPV Capsid Proteins

12.5.2.1. Introduction

The prevalence of HPV infections is difficult to determine. Many HPV infections are clinically asymptomatic and self-limiting. Most studies on the prevalence of genital HPV infection have relied on detecting HPV DNA in the genital tract, however it is usually uncertain whether the DNA is the result of newly acquired (incident) infection, persistent infection, or a previous infection that had resolved and then recurred (the latter two situations being considered as prevalent infection). Additionally, as most HPV infections will resolve spontaneously, the prevalence is dependent on how often the genital tract is examined. Thus, determining the frequency at which individuals seroconvert is difficult to establish in most studies.

HPV-16 infects roughly 20 percent of adults (Koutsky et al., 2002). Studies in the Nordic countries over the past several decades suggest that overall levels of seroprevalence have increased over time (Dillner, 2000). Infection in young children or the sexually inactive is much rarer (af Geijersstam et al., 1999). In one study of newly sexually active women an infection rate of 32.3 percent was found (Winer et al., 2003). A number of risk factors have been studied for potential association with HPV infection (Baseman and Koutsky, 2005). Genital HPVs are sexually transmitted diseases; thus, infection is associated with the number of sex partners, and new infections are associated with new partners. Smoking and oral contraceptive use have also been reported to be associated with HPV infection in some studies but not others.

Many studies have examined the levels of persistence of HPV infection and immune response. A high percentage of HPV infections are cleared without progression to cancer. It has been estimated that roughly 70 percent of infections clear naturally within a year (Ho et al., 1998). In one 5-year study, a clearance rate of 92 percent was found (Elfgren et al., 2000). Studies too numerous to list entirely have attempted to correlate the appearance of HPV capsid antibodies with infection detected by other methods such as cytology and PCR-based DNA detection methods and with progression to cervical cancer (Wikstrom et al., 1995b; Wideroff et al., 1995; Nonnenmacher et al., 1995; Heim et al., 1995; Dillner et al., 1997; Carter et al., 2000; Lehtinen et al., 2001; Sigstad et al., 2002; Wang, S.S. et al., 2003; Ho et al., 2004). Many infections may clear before causing a detectable immune response. Thus, not all infected women develop a detectable humoral immune response. It is also not surprising that women in whom DNA was detected on more than one occasion had a greater likelihood of seroconversion (Wideroff et al., 1995; Carter et al., 2000). Different studies show seroconversion levels of approximately 50–60 percent (Kirnbauer et al., 1994). Some studies report as high as 88 percent (Nakagawa et al., 2002). In corresponding HPV-DNA negative control women a level of 6 percent was found. This immune response usually requires anywhere from 6 to 18 months to appear (Carter et al., 1996, 2000; Ho et al., 2004). Seroreactivity has also been found to be lower in men than women (Svare et al., 1997; Slavinsky et al., 2001).

Many studies also have examined the effects of HPV infection on specific immunoglobulin classes. The humoral immune response normally consists of an initial response mostly consisting of IgA, followed by a secondary response of primarily IgG antibodies. There is some indication that the response to HPV infection does not always follow this pattern. For instance, Ho et al. found that the IgG and IgA responses occurred at the same time (Ho et al., 2004). The duration of the IgG and IgA responses may also differ. It is thought that IgG seroprevalence reflects lifetime number of sex partners. IgG levels have also been suggested to reflect previous infections, while IgA levels are more indicative of recent infections (Wang et al., 2000).

The following section will summarize recent serological studies on HPV humoral immune response. First, those that have focused on the low-risk types will be discussed, followed by a discussion of those focusing on HPV-16, and finally studies examining multiple HPV types.

12.5.2.2. HPV-1/6/11 Seroreactivity Studies

A number of studies suggest that there is an association between seropositivity for HPV-6 or HPV-11 and the detection of the corresponding DNA or of genital warts (Heim et al., 1995; Eisemann et al., 1996; Wikstrom et al., 1997). It has been observed that seroconversion for HPV-6 occurs concurrently with detection of DNA more often than for other HPV types (Wikstrom et al., 1995a; Carter et al., 2000). Heim et al. examined IgG, IgM, and IgA reactivity to HPV-6 and HPV-11 VLPs (Heim et al., 1995). In this study IgG reactivity was observed in 30–46 percent of condyloma or CIN patients, and IgM reactivity in 64–67 percent. Little IgA reactivity was detected. In another study examining HPV-6 and -11, IgG and IgA response to HPV-6 was observed, but little IgM (Wikstrom et al., 1995a). Emeny et al. examined the humoral immune response to vaccination of 18–25 year-old female university students with an HPV-11 VLP (Emeny et al., 2002). 96.7 percent of vaccinees showed IgG response and 83 percent showed IgA response. IgM was detected in 16.7 percent of vaccinees and no IgE was detected. Another interesting result of these studies has also been a lack of correlation between seropositivity and genital warts in men (Carter et al., 1995).

Fewer studies have examined seroreactivity to HPV-1, which causes foot warts. In one early study an overall seroprevalence of 50 percent was found in children (Pfister and zur Hausen, 1978; Pfister, 1984). In more recent studies an overall prevalence of 60–77 percent was found (Kienzler et al., 1983; Steele and Gallimore, 1990; Carter et al., 1994). An even higher prevalence was associated with a history of foot warts.

12.5.2.3. HPV-16 Seroreactivity Studies

Numerous studies have examined in detail the response to HPV-16 infection with respect to the time to seroconversion, percentage of women who seroconvert, the relative levels of the different immunoglobulin isotypes, and the duration of the immune response. In one large population-based study in the United States conducted by the National Center for Health Statistics, overall HPV-16 seropositivity in women 12–59 years was found to be 13.0 percent (Stone et al., 2002). An overall seroprevalence level of 12 percent for IgG and 6 percent for IgA was found in one cohort of 292 female university students (Hagensee et al., 2000). Thompson et al. found an overall seroprevalence of 24.5 percent in patients attending United States sexually transmitted disease clinics (Thompson et al., 2004). As mentioned above, typically over half of women in whom HPV DNA is detected seroconvert. Kirnbauer et al. found a HPV-16 seroconversion level of 59 percent among a group of 122 women attending women's health clinics (Kirnbauer et al., 1994). Studentsov et al. found an IgG seroconversion level of 55 percent in another study of college women with detectable HPV-16 DNA (Studentsov et al., 2003). One study examined the IgA response on incident infection with HPV-16 in cervical secretions and serum in comparison to IgG (Onda et al., 2003). IgA detection in serum (19.1 months) was delayed relative to that in cervical secretions (10.5 months). It was also found that the IgA antibody response reverted with a median time of 12–13 months. This was a much shorter time than observed for IgG. Another recent study looked at anti-HPV-16 IgG and IgA levels in a cohort of university women. IgG was observed in 56.7 percent, and IgA was observed in 37.0 percent, of women with incident infection. The response appeared after 8.3 months for IgG and after 14 months for IgA. The duration of antibody persistence was roughly 36 months (Ho et al., 2004). In one study IgG and IgA levels were compared in HPV-16 DNA-positive women with and without apparent pathology. Both isotypes were detected in women without pathology at rates lower than in women with pathology (6.5 percent vs. 27.5 percent for mucosal IgG and 13.1 percent vs. 27.0 percent for secretory IgA) (Rocha-Zavaleta et al., 2003).

Matusmoto et al. found that the relative ratio of IgG1 and IgG2 is associated with the rate of regression of CIN (Matsumoto et al., 1999). Wang et al. found IgG1 and IgA to be the most common isotypes, while IgG3 and IgM were rare (Wang et al., 2000). Other studies examined systemic relative to local IgA. When they compared levels of systemic and local IgA to HPV-16, Bontkes et al. found that only the systemic IgA correlated with virus clearance (Bontkes et al., 1999). Bard et al. studied levels of IgG, IgA, and IgM (Bard et al., 2004). They suggest that HPV infection may cause an increase in local IgA concentration.

Some studies have also linked the HPV immune response to particular HLA alleles. For example, deGruijl examined IgG response in a cohort of women with cervical intraepithelial neoplasia (de Gruijl et al., 1999). The IgG response was linked to the presence of the HLA DRB1*0101/DQB1*0501 in these women.

It is clear from these studies that the humoral immune response to HPV is complex and diverse. A consistent finding is that not all infected women develop a detectable antibody response. In general, it appears that IgG seroconversion is detected at a higher rate than for the other immunoglobulin types. IgA seroconversion is detected at a lower rate and is often, but not always, delayed and of shorter duration than IgG seroconversion. The timing of the immune response may also differ in serum and cervical secretions. Sasagawa et al. found mucosal IgA response preceded mucosal IgG response, though Onda et al. found the appearance of cervical IgA and serum IgG to be similar (Onda et al., 2003; Sasagawa et al., 2003).

Finally, there is still some conflicting evidence as to how detection of capsid antibodies correlates with disease progression. Some studies have found seroconversion using capsid VLPs is more likely to be associated with earlier stages of pathology, in contrast to detection of antibodies to HPV early proteins in late stage cervical cancer (Nonnenmacher et al., 1995). One recent study found an inverse correlation between mucosal anti-HPV-16 IgA and IgG and cervical disease (Bierl et al., 2005). This would be consistent with the loss of expression of the capsid proteins over the course of infection. However, other studies have found increasing seropositivity with disease progression (Heim et al., 2002;

Wang, S.S. et al., 2003). More standardization of study methods will be needed to fully address this issue.

12.5.2.4.   Studies Addressing Multiple HPV Types

Infection with multiple HPV types is common. A number of studies have examined multiple HPV types. In one study of women referred for colposcopy, detection of antibodies to HPV-16, -18, and -33 correlated with the detection of DNA of the corresponding type (Wang et al., 1996). There are also contradictory reports as to whether infection with one HPV type is protective against subsequent infection with another type or reinfection with the same type. In one study, Ho et al. found that IgG to HPV-16 was protective (Ho et al., 2002). Others have not found a protective effect (Viscidi et al., 2005). Several recent studies examined the prevalence of HPV types 16, 18, 31, and 45 in Costa Rica (Wang, S.S. et al., 2003; Viscidi et al., 2004; Wang et al., 2004). The authors examined seroreactivity to these types at baseline and 5–7 years later. Initial seroreactivity was not associated with a decreased risk of later infection, suggesting that infection may not protect against subsequent infection, or more likely that antibodies do not mediate clearance. It was also found in the same study that seroconversion decreased with age (Wang et al., 2004). It should be noted that in these studies it is difficult to distinguish between reinfection and reactivation of a latent infection. Van Doornum et al. examined IgA, IgM, and IgG levels in response to HPV types 6, 11, 16, 18, and 33 (van Doornum et al., 1998). While they did not see a well-defined pattern of reactivity, seropositivity was generally higher in women than in men. Nonnenmacher et al. examined serum IgG antibodies to VLPs for HPV types 6, 11, 16, and 18 in Brazilian women considered to be at low risk for developing cervical cancer on the basis of factors including age, marital status, number of lifetime sex partners, and lack of previous abnormal Pap smear (Nonnenmacher et al., 2003). In this study seropositivity was correlated with lifetime number of sexual partners and with multiparity but appeared to be a better marker of past infection than current infection. Wideroff et al. examined HPV types 16, 18, 31 and 45. While the seroreactivity was type-specific, reactivity to one type increased the likelihood of reactivity to the other types (Wideroff et al., 1999). Luostarinen et al. examined levels of antibodies to HPV-16, and to HPV-6 and -11 (Luostarinen et al., 1999). Co-infection with more than one HPV type did not increase the incidence of cervical carcinoma.

In general, HPV-16 is the most commonly detected HPV type. However, one interesting result from these studies is that the relative levels of seropositivity for different types do not always correlate with the relative levels of different types as detected by DNA. Wang et al. found the seroprevalence rates for HPV-16 and HPV-18 to be the same while the overall DNA prevalence was 3.5 percent for HPV-16 and 1.3 percent for HPV-18 (Wang, S.S. et al., 2003). These studies also are largely consistent with work to be described later suggesting that many antibodies to HPV are specific for a given type and that vaccines to HPV will need to target more than one HPV type.

12.5.2.5.   Other Sites of Infection and Other Types of Cancer

Other studies have examined the humoral response to HPV at sites other than the cervix and in nongenital cancers (Bjorge et al., 1997a; Schwartz et al., 1998; Strickler et al., 1998b). Carter et al. examined HPV-16 and HPV-18 antibody prevalence in cases with cancer at a variety of anogenital sites (Carter et al., 2001). A positive association was detected for in situ and invasive vulvar, vaginal, and anal (both male and female) cancer, as well as for invasive penile cancer. Only in situ penile cancer did not show an association with the detection of HPV-16 antibodies. Van Doornum et al. studied the association between seropositivity for antibodies to HPV-16 L1 and E7 and carcinoma of the oropharynx, the oesophagus, penis, and vagina (Van Doornum et al., 2003). An association with penile and oropharyngeal cancer was found. Bjorge et al. assayed for the presence of HPV types 16, 18, 33, and 73 antibodies in patients with anal or perianal skin cancer and controls and found an association with detection of antibodies to HPV-16 and -18 (Bjorge et al., 2002). Kreimer et al. studied HPV-16 in oropharyngeal SCC (Kreimer et al., 2005). While they were primarily concerned with correlating HPV-16 viral load with SCC, they found an increasing seropositivity with increasing viral load. Chen et al. found HPV-16 DNA in 6.3 percent of tonsillitis cases, but not other types (Chen et al., 2005). However, this did not correlate with detection of antibodies by the new luminex method. In adult-onset laryngeal papillomatosis, no correlation was found between disease and the presence of antibodies to HPV-16 or HPV-6 (Aaltonen et al., 2001). There have been some contradictory reports of an association with prostate cancer (Dillner et al., 1998; Strickler et al., 1998a; Adami et al., 2003; Rosenblatt et al., 2003; Korodi et al., 2005). Similarly, seropositivity for HPV has also been reported to be associated with an increased risk of esophageal cancer in some studies (Bjorge et al., 1997b) but not others (Lagergren et al., 1999).

Other studies have addressed the presence of HPV antibodies in patients with skin cancer (Majewski and Jablonska, 2002). Bouwes et al. developed an ELISA using HPV-8 L1 VLPs (Bouwes Bavinck et al., 2000). In a relatively small number of patients, a correlation between seropositivity for HPV-8 and SCC was found. In another study serologic detection of HPV-8 was also found to be strongly associated with cutaneous SCC (Masini et al., 2003), while in the same study the presence of HPV-15 antibodies was not associated with SCC. Pfister et al. used an HPV-8 VLP ELISA to compare high and low grade actinic keratosis but found no differences in seropositivity (Pfister et al., 2003).

12.5.2.6.   Geographical Distribution of Seroprevalence

A number of studies have examined geographical differences in HPV seroprevalence, particularly between Spain and Latin America (Nonnenmacher et al., 1995). A higher level of capsid antibody detection correlated with a higher incidence of cervical cancer in Columbia. Similarly, a higher level of capsid antibody detection was also detected in Jamaica relative to the United States (Strickler et al., 1999). Another study compared Greenland and Denmark

(Nonnenmacher et al., 1996). The HPV-16 IgG seroprevalence rates among women attending sexually transmitted disease clinics were 56 percent and 41 percent in Greenland and Denmark, respectively. Interestingly, the corresponding rates for detection of DNA were 24 percent and 36 percent, respectively.

12.5.2.7. Co-Infection with HIV

A number of studies have examined the effects of co-infection with HIV on HPV seroprevalence and persistence (Sun et al., 1997; Svare et al., 1997). Petter et al. examined levels of IgG, IgA, and IgM to HPV types 6, 11, 16, 18, and 31 in women with HIV (Petter et al., 2000). A seropositivity rate of 58 percent for HPV types 16, 18, and 31 combined was found in HIV–positive women, compared to 19 percent in HIV-negative women. The seroprevalence in HIV-positive women was higher than in HIV-negative CIN patients (48 percent vs. 31 percent for HPV-16). In another study comparing HIV-positive and HIV-negative women, HIV-positive women had a higher prevalence of HPV-16 DNA, a lower rate of seropositivity for HPV-16 IgA, but a higher rate of HPV-16 IgG (Marais et al., 2000c). In a study of HIV-positive and negative men, however, higher seroprevalence to the high-risk HPV types, but not to the low-risk types, was found (Hopfl et al., 2003). Viscidi et al. looked at rates of HPV-16 IgG and IgA seropositivity in HIV-positive and negative women (Viscidi et al., 2003a,b, 2005). The correlation between HIV infection and increased seropositivity for HPV was less clear in these studies. Hagensee et al. assayed for HPV-6 and HPV-16 antibodies in HIV-infected and uninfected homosexual men (Hagensee et al., 1997). The HIV-infected men did not show an increase in HPV antibody levels. Finally, Cameron et al. examined levels of antibodies to HPV types 6, 11, and 16 in serum and oral fluids of HIV-positive patients and found a seroprevalence of 55 percent for HPV-6 and HPV-11 and of 37 percent for HPV-16, with lower levels in the oral fluids (Cameron et al., 2003). While there remains some conflicting data, it appears from these studies that overall infection with HIV increases the likelihood of HPV seropositivity.

12.6.   Studies Defining HPV Virion Epitopes

12.6.1.   Details of Known L1 Epitopes

The crystal structure of the T = 1 HPV-16 L1 VLP has been described (Chen et al., 2000). The basic structure of the VLP consists of five hypervariable loops on the capsid surface. Some nonconserved residues near the C-terminus are also exposed on the capsid surface (Modis et al., 2002), due to the C-terminus of one capsomer extending into a nearby capsomer. The structure has been useful in understanding the nature of HPV epitopes. Epitopes were identified using monoclonal antibodies or using sera produced during the course of natural infection. Chimeric VLPs in which part of the L1 sequence, often within a hypervariable loop, for one HPV type is replaced by the sequence of another type have been useful in mapping epitopes (Christensen et al., 2001; Orozco et al., 2005; Carter et al., 2006). Changing as few as one or two amino acids may change type specificity (Ludmerer et al., 1996). The chimeric L1 VLPs can retain conformational epitopes and induce neutralizing antibodies to both of the parent HPV types (Christensen et al., 2001). A general result of these studies is that most of the epitopes are specific for a single HPV type and depend on conformation, although a few epitopes that are cross-reactive between different HPV types have been identified. (Christensen et al., 1990; Kirnbauer et al., 1994; Hines et al., 1994a,b; Sapp et al., 1994; Wikstrom et al., 1995b; Nonnenmacher et al., 1996; Roden et al., 1996a,b; Touze et al., 1998; Carter et al., 2000; Combita et al., 2002b). The epitopes that have been characterized as cross-reactive tend to be the linear epitopes dependent on sequence alone as opposed to conformation (Firzlaff et al., 1988; Jin et al., 1990). Denaturation of the capsid will usually disrupt conformational type-specific epitopes but not cross-reactive epitopes (Steele and Gallimore, 1990; Bonnez et al., 1991; Carter et al., 1993; Kirnbauer et al., 1994; Dillner et al., 1995). It has also been found that the type-specific antibodies to conformational epitopes tend to show higher titers than the non-specific antibodies to linear epitopes. Finally, many conformational epitopes are retained on a single capsomer (Rose et al., 1998; Yuan et al., 2001).

The epitopes for many antibodies are complex and composed of multiple regions, often on more than one hypervariable loop. Mapping of the epitope for the 16A monoclonal antibody suggested two different loop regions in the N-terminus are involved (Olcese et al., 2004). Two particularly well-studied monoclonal antibodies to HPV are H16.V5 and H16.E70. Several groups have contributed to mapping the epitopes to these antibodies (Roden et al., 1997; Wang et al., 1997; White et al., 1999; Chen et al., 2000; Carter et al., 2003). The FG and HI loops are both important for H16.V5 binding. Carpentier et al. also examined mutations of the FG loop on the surface of HPV-16 L1 (Carpentier et al., 2005). These mutations affected recognition by type-specific neutralizing antibodies and cross-reactive antibodies, suggesting that this loop contributes to multiple epitopes. HPV-16 VLPs in which HPV-16 sequence was exchanged with sequence from HPV-31 or -52 were prepared by (Carter et al., 2003). VLPs with mutations between residues 260–273 and residues 285–290, at different ends of the FG loop, were found to have reduced reactivity to H16.V5 and H16.E70. The presence of a polar amino acid at residue 270 is also important and suggests that formation of a hydrogen bond is required. Both the FG and DE loops were important for H16.E70 binding. This is consistent with the findings of Ludmerer et al. (Ludmerer et al., 1996, 1997) that the DE loop is important for binding of antibodies to HPV-11, but not with other studies showing that binding of antibody H16.E70 can be transferred from one type VLP to another without changing the DE loop (Christensen et al., 2001). The role of residue F50 and its potential effect on the conformation of the adjacent BC loop was also examined. Intact HPV-16 VLPs with an F50L mutation could not be obtained, but binding of all antibodies tested was largely unaffected by exchanging the BC loop. Finally, the C-terminal arm was found to contain the epitope for antibody H16.U4 (Carter et al., 2003).

McClements et al. studied epitopes for several HPV-6 neutralizing monoclonal antibodies by exchanging regions of the HPV-6 and HPV-11 L1 capsid protein (McClements et al., 2001). Two different regions, one near amino acid 53 (BC loop) and another between residues 169 and 178 (EF loop), were found to be important. These regions were also areas not well conserved between different HPV types. It has also been found that HPV-6 epitopes that are type-specific and that cross-react with HPV-11 are distinct but partially overlap (Wang et al., 2003b).

Orozco et al. performed an extensive study to map the epitopes on the surface of HPV-6 capsomers recognized by human sera as opposed to monoclonal antibodies (Orozco et al., 2005). This was done by mutating HPV-6 sequences in the hypervariable loops and the C-terminus to the corresponding HPV-11 sequence. Twelve out of thirty-six sera from a study of incident infection showed specific reactivity for HPV-6. The sera detected a variety of different epitopes, suggesting that there may not be a single immunodominant epitope. This is in contrast to earlier work in which monoclonal antibodies against HPV types 6, 16, 18, and 33 were used to block the reactivity of human sera with capsids of the corresponding HPV type. The reactivity of the majority of sera was blocked by a single monoclonal antibody for each type, which was interpreted to mean that one epitope plays a dominant role in the immune response (Wang et al., 1997). It should be noted that a single IgG molecule is about the same size as three L1 proteins. Thus an antibody to a given epitope could block binding to other adjacent epitopes by steric hindrance.

To define the regions containing dominant L1 epitopes further, a new study has used a HPV-16 neutralization assay to examine the ability of hybrid VLPs to block the neutralizing activity of sera from women enrolled in a natural history study (Carter et al., 2006). This study tested a panel of chimeric VLPs consisting of an HPV-16 or 31 L1 backbone with substitutions from the other type. Some chimeric VLPs retained wild-type HPV-16 ability to block neutralization, others were like HPV-31 in being unable to block neutralization, and some had intermediate activity. Interestingly, a single region substitution in the FG loop abrogated the ability of the HPV-16 VLPs to block neutralization, which was unexpected from previous studies. However, transfer of neutralizing activity required sequences from more than one loop. The DE, FG, and HI loops were most often required, while the BC loop and variable sequences between residues 400 and 450 were less important. This is further evidence that no single loop contains an immunodominant epitope.

Finally, some cross-reactivity between HPV-31 and HPV-33 and between HPV-45 and HPV-18 has been reported. Surprisingly, the cross-reactivity did not always directly correlate with the level of sequence identity or phylogenetic relatedness of different HPV types (Giroglou et al., 2001). Cross-reactivity between HPV-16 and HPV-33 has also been reported (White et al., 1998). In another recent study hybrid capsids with different parts of HPV-11 and HPV-16 were examined (Wang et al., 2003a). Specific reactivity depended on the presence of the C-terminus for each type. Slupetzky et al. studied capsids that were a hybrid of HPV-16 and bovine papillomavirus 1 (Slupetzky et al., 2001). The residues 282–286 and 351–355 were important for recognition by neutralizing antibodies.

All of these studies suggest that the majority of epitopes require multiple regions of the capsid protein. In general they have highlighted the importance of the FG, DE, and HI loops to many epitopes. The complexity of many of these epitopes also suggests that it will be difficult to design a single hybrid L1 protein for vaccines that will be effective against multiple HPV types.

12.6.2. Antibodies to the L2 Protein

While the L2 protein is not necessary for the formation of VLPs, it is required for viral infectivity (Roden et al., 1996a). Clearly the predominant antibody response is to the major capsid protein L1, and a minor component of the antibody response may be directed against the L2 protein (Christensen et al., 1991; Roden et al., 2000). Neutralizing antibodies directed against a small surface-exposed portion of L2 have been identified though they tend not to be as effective as those directed at L1 (Christensen and Kreider, 1991; White et al., 1999; Roden et al., 2000). Other buried epitopes of L2 may be exposed on infection (Carter et al., 1993). Kawana et al. have identified a neutralizing epitope in the N-terminal region of L2 consisting of amino acids 108–111 (Kawana et al., 1999). Pastrana et al., found that BPV L2 amino acids 1–88 induced neutralizing antibody to HPV (Pastrana et al., 2005). Roden et al. assessed the efficacy of immunization of sheep with L2 from HPV6, 16, and 18 (Roden et al., 2000). They found a relatively high degree of cross-neutralization in the resultant sera. Short peptides of L2 have been used in immunization of rabbits and protected the animals from viral challenge (Embers et al., 2002). Recently, cross-neutralization of authentic HPV with antisera from immunization with L2 peptides has been described (Embers et al., 2004). Thus the L2 protein has some role in the immune response to HPV.

12.7.   Conclusions and Future Perspectives

Early work using peptides and proteins suggested an association between the presence of antibodies to the early HPV proteins and development of cervical cancer. The poor growth of HPV in culture limited study of the humoral immune response to HPV until the development of methods to produce virus-like particles incorporating the HPV capsid proteins. These VLPs have allowed a much more detailed analysis of the humoral immune response to HPV infection. This has included extensive study of the natural history of the immune response, showing that not all infections as detected by HPV DNA lead to a detectable antibody response. Much progress has been made in characterization of type-specific and cross-reactive epitopes for the many different HPV types, although more study is needed to determine the extent to which one immunodominant epitope triggers the immune response. The study of the immune response to HPV described here has also provided an extensive body of technical and basic knowledge towards the production of a prophylactic vaccine. Virus-like particles have been used in many candidate vaccine trials and much progress has been made towards the development of prophylactic vaccines in recent years, with up to 100 percent efficacy in some trials (Harro et al., 2001; Koutsky et al., 2002; Brown et al., 2004; Poland et al., 2005; Villa et al., 2005).

References

Aaltonen, L.M., Auvinen, E., Dillner, J., Lehtinen, M., Paavonen, J., Rihkanen, H., and Vaheri, A. (2001). Poor antibody response against human papillomavirus in adult-onset laryngeal papillomatosis. *J. Med. Microbiol.* 50:468–471.

Adami, H.O., Kuper, H., Andersson, S.O., Bergstrom, R., and Dillner, J. (2003). Prostate cancer risk and serologic evidence of human papilloma virus infection: A population-based case-control study. *Cancer Epidemiol. Biomarkers Prev.* 12:872–875.

Af Geijersstam, V., Eklund, C., Wang, Z., Sapp, M., Schiller, J.T., Dillner, J., and Dillner, L. (1999). A survey of seroprevalence of human papillomavirus types 16, 18 and 33 among children. *Int. J. Cancer.* 80:489–493.

Aires, K.A., Cianciarullo, A.M., Carneiro, S.M., Villa, L.L., Boccardo, E., Perez-Martinez, G., Perez-Arellano, I., Oliveira, M.L., and Ho, P.L. (2006). Production of Human Papillomavirus Type 16 L1 Virus-Like Particles by Recombinant Lactobacillus casei Cells. *Appl. Environ. Microbiol.* 72:745–752.

Andersson-Ellstrom, A., Dillner, J., Hagmar, B., Schiller, J., and Forssman, L. (1994). No serological evidence for non-sexual spread of HPV16. *Lancet.* 344:1435.

Andersson-Ellstrom, A., Dillner, J., Hagmar, B., Schiller, J., Sapp, M., Forssman, L., and Milsom, I. (1996). Comparison of development of serum antibodies to HPV16 and HPV33 and acquisition of cervical HPV DNA among sexually experienced and virginal young girls. A longitudinal cohort study. *Sex. Transm. Dis.* 23:234–238.

Baird, P.J. (1983). Serological evidence for the association of papillomavirus and cervical neoplasia. *Lancet.* 2:17–18.

Bard, E., Riethmuller, D., Meillet, D., Pretet, J.L., Schaal, J.P., Mougin, C., and Seilles, E. (2004). High-risk papillomavirus infection is associated with altered antibody responses in genital tract: non-specific responses in HPV infection. *Viral. Immunol.* 17:381–389.

Baseman, J.G. and Koutsky, L.A. (2005). The epidemiology of human papillomavirus infections. *J. Clin. Virol.* 32:S16–S24.

Baud, D., Ponci, F., Bobst, M., De Grandi, P., and Nardelli-Haefliger, D. (2004). Improved efficiency of a Salmonella-based vaccine against human papillomavirus type 16 virus-like particles achieved by using a codon-optimized version of L1. *J. Virol.* 78:12901–12909.

Beiss, B.K., Heimer, E., Felix, A., Burk, R.D., Ritter, D.B., Mallon, R.G., and Kadish, A.S. (1991). Type-specific and cross-reactive epitopes in human papillomavirus type 16 capsid proteins. *Virology.* 184:460–464.

Bierl, C., Karem, K., Poon, A.C., Swan, D., Tortolero-Luna, G., Follen, M., Wideroff, L., Unger, E.R., and Reeves, W.C. (2005). Correlates of cervical mucosal antibodies to human papillomavirus 16: Results from a case control study. *Gynecol. Oncol.* 99:S262–S268.

Bjorge, T., Dillner, J., Anttila, T., Engeland, A., Hakulinen, T., Jellum, E., Lehtinen, M., Luostarinen, T., Paavonen, J., Pukkala, E., Sapp, M., Schiller, J., Youngman, L., and Thoresen, S. (1997a). Prospective seroepidemiological study of role of human papillomavirus in non-cervical anogenital cancers. *BMJ.* 315:646–649.

Bjorge, T., Hakulinen, T., Engeland, A., Jellum, E., Koskela, P., Lehtinen, M., Luostarinen, T., Paavonen, J., Sapp, M., Schiller, J., Thoresen, S., Wang, Z., Youngman, L., and Dillner, J. (1997b). A prospective, seroepidemiological study of the role of human papillomavirus in esophageal cancer in Norway. *Cancer Res.* 57:3989–3992.

Bjorge, T., Engeland, A., Luostarinen, T., Mork, J., Gislefoss, R.E., Jellum, E., Koskela, P., Lehtinen, M., Pukkala, E., Thoresen, S.O., and Dillner, J. (2002). Human papillomavirus infection as a risk factor for anal and perianal skin cancer in a prospective study. *Br. J. Cancer.* 87:61–64.

Bonnez, W., Da Rin, C., Rose, R.C., and Reichman, R.C. (1991). Use of human papillomavirus type 11 virions in an ELISA to detect specific antibodies in humans with condylomata acuminata. *J. Gen. Virol.* 72:1343–1347.

Bonnez, W., Kashima, H.K., Leventhal, B., Mounts, P., Rose, R.C., Reichman, R.C., and Shah, K.V. (1992). Antibody response to human papillomavirus (HPV) type 11 in children with juvenile-onset recurrent respiratory papillomatosis (RRP). *Virology.* 188:384–387.

Bonnez, W., Rose, R.C., Da Rin, C., Borkhuis, C., Mesy Jensen, K.L., and Reichman, R.C. (1993). Propagation of human papillomavirus type 11 in human xenografts using the severe combined immunodeficiency (SCID) mouse and comparison to the nude mouse model. *Virology.* 197:455–458.

Bontkes, H.J., de Gruijl, T.D., Walboomers, J.M., Schiller, J.T., Dillner, J., Helmerhorst, T.J., Verheijen, R.H., Scheper, R.J., and Meijer, C.J. (1999). Immune responses against human papillomavirus (HPV) type 16 virus-like particles in a cohort study of women with cervical intraepithelial neoplasia. II. Systemic but not local IgA responses correlate with clearance of HPV-16. *J. Gen. Virol.* 80:409–417.

Bousarghin, L., Hubert, P., Franzen, E., Jacobs, N., Boniver, J., and Delvenne, P. (2005). Human papillomavirus 16 virus-like particles use heparan sulfates to bind dendritic cells and colocalize with langerin in Langerhans cells. *J. Gen. Virol.* 86:1297–1305.

Bouwes Bavinck, J.N., Stark, S., Petridis, A.K., Marugg, M.E., ter Schegget, J., Westendorp, R.G., Fuchs, P.G., Vermeer, B.J., and Pfister, H. (2000). The presence of antibodies against virus-like particles of epidermodysplasia verruciformis-associated humanpapillomavirus type 8 in patients with actinic keratoses. *Br. J. Dermatol.* 142:103–109.

Breitburd, F., Kirnbauer, R., Hubbert, N.L., Nonnenmacher, B., Trin-Dinh-Desmarquet, C., Orth, G., Schiller, J.T., and Lowy, D.R. (1995). Immunization with viruslike particles from cottontail rabbit papillomavirus (CRPV) can protect against experimental CRPV infection. *J. Virol.* 69:3959–3963.

Brown, D.R., Fife, K.H., Wheeler, C.M., Koutsky, L.A., Lupinacci, L.M., Railkar, R., Suhr, G., Barr, E., DiCello, A., Li, W., Smith, J.F., Tadesse, A., and Jansen, K.U. (2004). Early assessment of the efficacy of a human papillomavirus type 16 L1 virus-like particle vaccine. *Vaccine* 22:2936–2942.

Buck, C.B., Pastrana, D.V., Lowy, D.R., and Schiller, J.T. (2005a). Generation of HPV pseudovirions using transfection and their use in neutralization assays. *Methods Mol. Med.* 119:445–462.

Buck, C.B., Thompson, C.D., Pang, Y.Y., Lowy, D.R., and Schiller, J.T. (2005b). Maturation of papillomavirus capsids. *J. Virol.* 79:2839–2846.

Cameron, J.E., Snowhite, I.V., Chaturvedi, A.K., and Hagensee, M.E. (2003). Human papillomavirus-specific antibody status in oral fluids modestly reflects serum status in human immunodeficiency virus-positive individuals. *Clin. Diagn. Lab. Immunol.* 10:431–438.

Carpentier, G.S., Fleury, M.J., Touze, A., Sadeyen, J.R., Tourne, S., Sizaret, P. Y., and Coursaget, P. (2005). Mutations on the FG surface loop of human papillomavirus type 16 major capsid protein affect recognition by both type-specific neutralizing antibodies and cross-reactive antibodies. *J. Med. Virol.* 77:558–565.

Carter, J.J., Hagensee, M., Taflin, M.C., Lee, S.K., Koutsky, L.A., and Galloway, D.A. (1993). HPV-1 capsids expressed in vitro detect human serum antibodies associated with foot warts. *Virology.* 195:456–462.

Carter, J.J., Hagensee, M.B., Lee, S.K., McKnight, B., Koutsky, L.A., and Galloway, D.A. (1994). Use of HPV 1 capsids produced by recombinant vaccinia viruses in an ELISA to detect serum antibodies in people with foot warts. *Virology.* 199:284–291.

Carter, J.J., Wipf, G.C., Hagensee, M.E., McKnight, B., Habel, L.A., Lee, S.K., Kuypers, J., Kiviat, N., Daling, J.R., Koutsky, L.A., and Galloway, D.A. (1995). Use of human papillomavirus type 6 capsids to detect antibodies in people with genital warts. *J. Infect. Dis.* 172:11–18.

Carter, J.J., Koutsky, L.A., Wipf, G.C., Christensen, N.D., Lee, S.K., Kuypers, J., Kiviat, N., and Galloway, D.A. (1996). The natural history of human papillomavirus type 16 capsid antibodies among a cohort of university women. *J. Infect. Dis.* 174:927–936.

Carter, J.J., Koutsky, L.A., Hughes, J.P., Lee, S.K., Kuypers, J., Kiviat, N., and Galloway, D.A. (2000). Comparison of human papillomavirus types 16, 18, and 6 capsid antibody responses following incident infection. *J. Infect. Dis.* 181:1911–1919.

Carter, J.J., Madeleine, M.M., Shera, K., Schwartz, S.M., Cushing-Haugen, K.L., Wipf, G.C., Porter, P., Daling, J.R., McDougall, J.K., and Galloway, D.A. (2001). Human papillomavirus 16 and 18 L1 serology compared across anogenital cancer sites. *Cancer Res.* 61:1934–1940.

Carter, J.J., Wipf, G.C., Benki, S.F., Christensen, N.D., and Galloway, D.A. (2003). Identification of a human papillomavirus type 16-specific epitope on the C-terminal arm of the major capsid protein L1. *J. Virol.* 77:11625–11632.

Carter, J.J., Wipf, G.C., Madeleine, M.M., Schwartz, S.M., Koutsky, L.A., and Galloway, D.A. (2006). Identification of HPV 16 L1 surface loops required for neutralization by human sera. *J. Virol.* 80:4664–4672.

Chee, Y.H., NamKoong, S.E., Kim, D.H., Kim, S.J., and Park, J.S. (1995). Immunologic diagnosis and monitoring of cervical cancers using in vitro translated HPV proteins. *Gynecol. Oncol.* 57:226–231.

Chen, R., Sehr, P., Waterboer, T., Leivo, I., Pawlita, M., Vaheri, A., and Aaltonen, L.M. (2005). Presence of DNA of human papillomavirus 16 but no other types in tumor-free tonsillar tissue. *J. Clin. Microbiol.* 43:1408–1410.

Chen, X.S., Garcea, R.L., Goldberg, I., Casini, G., and Harrison, S.C. (2000). Structure of small virus-like particles assembled from the L1 protein of human papillomavirus 16. *Mol. Cell.* 5:557–567.

Christensen, N.D. and Kreider, J.W. (1991). Neutralization of CRPV infectivity by monoclonal antibodies that identify conformational epitopes on intact virions. *Virus Res.* 21:169–179.

Christensen, N.D., Kreider, J.W., Cladel, N.M., and Galloway, D.A. (1990). Immuno-logical cross-reactivity to laboratory-produced HPV-11 virions of polysera raised against bacterially derived fusion proteins and synthetic peptides of HPV-6b and HPV-16 capsid proteins. *Virology.* 175:1–9.

Christensen, N.D., Kreider, J.W., Kan, N.C., and DiAngelo, S.L. (1991). The open reading frame L2 of cottontail rabbit papillomavirus contains antibody-inducing neutralizing epitopes. *Virology.* 181:572–579.

Christensen, N.D., Kreider, J.W., Shah, K.V., and Rando, R.F. (1992). Detection of human serum antibodies that neutralize infectious human papillomavirus type 11 virions. *J. Gen. Virol.* 73:1261–1267.

Christensen, N.D., Kirnbauer, R., Schiller, J.T., Ghim, S.J., Schlegel, R., Jenson, A.B., and Kreider, J.W. (1994). Human papillomavirus types 6 and 11 have antigenically distinct strongly immunogenic conformationally dependent neutralizing epitopes. *Virology.* 205:329–335.

Christensen, N.D., Dillner, J., Eklund, C., Carter, J.J., Wipf, G.C., Reed, C.A., Cladel, N.M., and Galloway, D.A. (1996a). Surface conformational and linear epitopes on HPV-16 and HPV-18 L1 virus-like particles as defined by monoclonal antibodies. *Virology.* 223:174–184.

Christensen, N.D., Reed, C.A., Cladel, N.M., Hall, K., and Leiserowitz, G.S. (1996b). Monoclonal antibodies to HPV-6 L1 virus-like particles identify conformational and linear neutralizing epitopes on HPV-11 in addition to type-specific epitopes on HPV-6. *Virology.* 224:477–486.

Christensen, N.D., Cladel, N.M., Reed, C.A., Budgeon, L.R., Embers, M.E., Skulsky, D.M., McClements, W.L., Ludmerer, S.W., and Jansen, K.U. (2001). Hybrid papillomavirus L1 molecules assemble into virus-like particles that reconstitute confor-mational epitopes and induce neutralizing antibodies to distinct HPV types. *Virology.* 291:324–334.

Combita, A.L., Bravo, M.M., Touze, A., Orozco, O., and Coursaget, P. (2002a). Serologic response to human oncogenic papillomavirus types 16, 18, 31, 33, 39, 58 and 59 virus-like particles in Colombian women with invasive cervical cancer. *Int. J. Cancer.* 97:796–803.

Combita, A.L., Touze, A., Bousarghin, L., Christensen, N.D., and Coursaget, P. (2002b). Identification of two cross-neutralizing linear epitopes within the L1 major capsid protein of human papillomaviruses. *J. Virol.* 76:6480–6486.

Cowsert, L.M., Lake, P., and Jenson, A.B. (1987). Topographical and conformational epitopes of bovine papillomavirus type 1 defined by monoclonal antibodies. *J. Natl. Cancer Inst.* 79:1053–1057.

Davidson, E.J., Sehr, P., Faulkner, R.L., Parish, J.L., Gaston, K., Moore, R.A., Pawlita, M., Kitchener, H.C., and Stern, P.L. (2003). Human papillomavirus type 16 E2- and L1-specific serological and T-cell responses in women with vulval intraepithelial neoplasia. *J. Gen. Virol.* 84:2089–2097.

de Gruijl, T.D., Bontkes, H.J., Walboomers, J.M., Coursaget, P., Stukart, M.J., Dupuy, C., Kueter, E., Verheijen, R.H., Helmerhorst, T.J., Duggan-Keen, M.F., Stern, P.L., Meijer, C.J., and Scheper, R.J. (1999). Immune responses against human papillo-mavirus (HPV) type 16 virus-like particles in a cohort study of women with cervical intraepithelial neoplasia. I. Differential T-helper and IgG responses in relation to HPV infection and disease outcome. *J. Gen. Virol.* 80:399–408.

Dias, D., Van Doren, J., Schlottmann, S., Kelly, S., Puchalski, D., Ruiz, W., Boerckel, P., Kessler, J., Antonello, J.M., Green, T., Brown, M., Smith, J., Chirmule, N., Barr, E.,

Jansen, K.U., and Esser, M.T. (2005). Optimization and validation of a multiplexed luminex assay to quantify antibodies to neutralizing epitopes on human papillomaviruses 6, 11, 16, and 18. *Clin. Diagn. Lab. Immunol.* 12:959–969.

Dillner, J. (1990). Mapping of linear epitopes of human papillomavirus type 16: the E1, E2, E4, E5, E6 and E7 open reading frames. *Int. J. Cancer.* 46:703–711.

Dillner, J. (2000). Trends over time in the incidence of cervical neoplasia in comparison to trends over time in human papillomavirus infection. *J. Clin. Virol.* 19:7–23.

Dillner, J., Dillner, L., Robb, J., Willems, J., Jones, I., Lancaster, W., Smith, R., and Lerner, R. (1989). A synthetic peptide defines a serologic IgA response to a human papillomavirus-encoded nuclear antigen expressed in virus-carrying cervical neoplasia. *Proc. Natl. Acad. Sci. U. S. A.* 86:3838–3841.

Dillner, L., Moreno-Lopez, J., and Dillner, J. (1990). Serological responses to papillomavirus group-specific antigens in women with neoplasia of the cervix uteri. *J. Clin. Microbiol.* 28:624–627.

Dillner, L., Heino, P., Moreno-Lopez, J., and Dillner, J. (1991). Antigenic and immunogenic epitopes shared by human papillomavirus type 16 and bovine, canine, and avian papillomaviruses. *J. Virol.* 65:6862–6871.

Dillner, J., Wiklund, F., Lenner, P., Eklund, C., Frederiksson-Shanazarian, V., Schiller, J.T., Hibma, M., Hallmans, G., and Stendahl, U. (1995). Antibodies against linear and conformational epitopes of human papillomavirus type 16 that independently associate with incident cervical cancer. *Int. J. Cancer.* 60:377–382.

Dillner, J., Kallings, I., Brihmer, C., Sikstrom, B., Koskela, P., Lehtinen, M., Schiller, J.T., Sapp, M., and Mardh, P.A. (1996). Seropositivities to human papillomavirus types 16, 18, or 33 capsids and to Chlamydia trachomatis are markers of sexual behavior. *J. Infect. Dis.* 173:1394–1398.

Dillner, J., Lehtinen, M., Bjorge, T., Luostarinen, T., Youngman, L., Jellum, E., Koskela, P., Gislefoss, R.E., Hallmans, G., Paavonen, J., Sapp, M., Schiller, J.T., Hakulinen, T., Thoresen, S., and Hakama, M. (1997). Prospective seroepidemiologic study of human papillomavirus infection as a risk factor for invasive cervical cancer. *J. Natl. Cancer Inst.* 89:1293–1299.

Dillner, J., Knekt, P., Boman, J., Lehtinen, M., af Geijersstam, V., Sapp, M., Schiller, J., Maatela, J., and Aromaa, A. (1998). Sero-epidemiological association between human-papillomavirus infection and risk of prostate cancer. *Int. J. Cancer.* 75:564–567.

Dunne, E.F., Karem, K.L., Sternberg, M.R., Stone, K.M., Unger, E.R., Reeves, W.C., and Markowitz, L.E. (2005). Seroprevalence of human papillomavirus type 16 in children. *J. Infect. Dis.* 191:1817–1819.

Eisemann, C., Fisher, S.G., Gross, G., Muller, M., and Gissmann, L. (1996). Antibodies to human papillomavirus type 11 virus-like particles in sera of patients with genital warts and in control groups. *J. Gen. Virol.* 77:1799–1803.

Eklund, C. and Dillner, J. (1995). A two-site enzyme immunoassay for quantitation of human papillomavirus type 16 particles. *J. Virol. Methods.* 53:11–23.

Elfgren, K., Kalantari, M., Moberger, B., Hagmar, B., and Dillner, J. (2000). A population-based five-year follow-up study of cervical human papillomavirus infection. *Am. J. Obstet. Gynecol.* 183:561–567.

Embers, M.E., Budgeon, L.R., Culp, T.D., Reed, C.A., Pickel, M.D., and Christensen, N.D. (2004). Differential antibody responses to a distinct region of human papillomavirus minor capsid proteins. *Vaccine* 22:670–680.

Embers, M.E., Budgeon, L.R., Pickel, M., and Christensen, N.D. (2002). Protective immunity to rabbit oral and cutaneous papillomaviruses by immunization with short peptides of L2, the minor capsid protein. *J. Virol.* 76:9798–9805.

Emeny, R.T., Wheeler, C.M., Jansen, K.U., Hunt, W.C., Fu, T.M., Smith, J.F., MacMullen, S., Esser, M.T., and Paliard, X. (2002). Priming of human papillomavirus type 11-specific humoral and cellular immune responses in college-aged women with a virus-like particle vaccine. *J. Virol.* 76:7832–7842.

Favre, M., Orth, G., Majewski, S., Baloul, S., Pura, A., and Jablonska, S. (1998). Psoriasis: A possible reservoir for human papillomavirus type 5, the virus associated with skin carcinomas of epidermodysplasia verruciformis. *J. Invest. Dermatol.* 110:311–317.

Feltkamp, M.C., Broer, R., di Summa, F.M., Struijk, L., van der, M.E., Verlaan, B.P., Westendorp, R.G., ter Schegget, J., Spaan, W.J., and Bouwes Bavinck, J.N. (2003). Seroreactivity to epidermodysplasia verruciformis-related human papillomavirus types is associated with nonmelanoma skin cancer. *Cancer Res.* 63:2695–2700.

Ferguson, M., Heath, A., Johnes, S., Pagliusi, S., and Dillner, J. (2005). Results of the first WHO international collaborative study on the standardization of the detection of antibodies to human papillomaviruses. *Int. J. Cancer.* Epub ahead of print. http://www3.interscience.wiley.com/cgi-bin/fulltext/112094836/HTMLSTART.

Firzlaff, J.M., Kiviat, N.B., Beckmann, A.M., Jenison, S.A., and Galloway, D.A. (1988). Detection of human papillomavirus capsid antigens in various squamous epithelial lesions using antibodies directed against the L1 and L2 open reading frames. *Virology.* 164:467–477.

Fisher, S.G., Benitez-Bribiesca, L., Nindl, I., Stockfleth, E., Muller, M., Wolf, H., Perez-Garcia, F., Guzman-Gaona, J., Gutierrez-Delgado, F., Irvin, W., and Gissmann, L. (1996). The association of human papillomavirus type 16 E6 and E7 antibodies with stage of cervical cancer. *Gynecol. Oncol.* 61:73–78.

Giroglou, T., Sapp, M., Lane, C., Fligge, C., Christensen, N.D., Streeck, R.E., and Rose, R.C. (2001). Immunological analyses of human papillomavirus capsids. *Vaccine.* 19:1783–1793.

Hagensee, M.E., Yaegashi, N., and Galloway, D.A. (1993). Self-assembly of human papillomavirus type 1 capsids by expression of the L1 protein alone or by coexpression of the L1 and L2 capsid proteins. *J. Virol.* 67:315–322.

Hagensee, M.E., Olson, N.H., Baker, T.S., and Galloway, D.A. (1994). Three-dimensional structure of vaccinia virus-produced human papillomavirus type 1 capsids. *J. Virol.* 68:4503–4505.

Hagensee, M.E., Kiviat, N., Critchlow, C.W., Hawes, S.E., Kuypers, J., Holte, S., and Galloway, D.A. (1997). Seroprevalence of human papillomavirus types 6 and 16 capsid antibodies in homosexual men. *J. Infect. Dis.* 176:625–631.

Hagensee, M.E., Koutsky, L.A., Lee, S.K., Grubert, T., Kuypers, J., Kiviat, N.B., and Galloway, D.A. (2000). Detection of cervical antibodies to human papillomavirus type 16 (HPV-16) capsid antigens in relation to detection of HPV-16 DNA and cervical lesions. *J. Infect. Dis.* 181:1234–1239.

Hamsikova, E., Novak, J., Hofmannova, V., Munoz, N., Bosch, F.X., De Sanjose, S., Shah, K., Roth, Z., and Vonka, V. (1994). Presence of antibodies to seven human papillomavirus type 16-derived peptides in cervical cancer patients and healthy controls. *J. Infect. Dis.* 170:1424–1431.

Hamsikova, E., Ludvikova, V., Tachezy, R., Kovarik, J., Brouskova, L., and Vonka, V. (2000). Longitudinal follow-up of antibody response to selected antigens of human papillomaviruses and herpesviruses in patients with invasive cervical carcinoma. *Int. J. Cancer.* 86:351–355.

Harper, D.M., Franco, E.L., Wheeler, C., Ferris, D.G., Jenkins, D., Schuind, A., Zahaf, T., Innis, B., Naud, P., De Carvalho, N.S., Roteli-Martins, C.M., Teixeira, J., Blatter, M.M., Korn, A.P., Quint, W., and Dubin, G. (2004). Efficacy of a bivalent L1 virus-like particle vaccine in prevention of infection with human papillomavirus types 16 and 18 in young women: a randomised controlled trial. *Lancet.* 364:1757–1765.

Harro, C.D., Pang, Y.Y., Roden, R.B., Hildesheim, A., Wang, Z., Reynolds, M.J., Mast, T.C., Robinson, R., Murphy, B.R., Karron, R.A., Dillner, J., Schiller, J.T., and Lowy, D.R. (2001). Safety and immunogenicity trial in adult volunteers of a human papillomavirus 16 L1 virus-like particle vaccine. *J. Natl. Cancer Inst.* 93:284–292.

Heim, K., Christensen, N.D., Hoepfl, R., Wartusch, B., Pinzger, G., Zeimet, A., Baumgartner, P., Kreider, J.W., and Dapunt, O. (1995). Serum IgG, IgM, and IgA reactivity to human papillomavirus types 11 and 6 virus-like particles in different gynecologic patient groups. *J. Infect. Dis.* 172:395–402.

Heim, K., Widschwendter, A., Pirschner, G., Wieland, U., Awerkiew, S., Christensen, N.D., Bergant, A., Marth, C., and Hopfl, R. (2002). Antibodies to human papillomavirus 16 L1 virus-like particles as an independent prognostic marker in cervical cancer. *Am. J. Obstet. Gynecol.* 186:705–711.

Heino, P., Goldman, S., Lagerstedt, U., and Dillner, J. (1993). Molecular and serological studies of human papillomavirus among patients with anal epidermoid carcinoma. *Int. J. Cancer.* 53:377–381.

Heino, P., Skyldberg, B., Lehtinen, M., Rantala, I., Hagmar, B., Kreider, J.W., Kirnbauer, R., and Dillner, J. (1995). Human papillomavirus type 16 capsids expose multiple type-restricted and type-common antigenic epitopes. *J. Gen. Virol.* 76:1141–1153.

Herrero, R., Castellsague, X., Pawlita, M., Lissowska, J., Kee, F., Balaram, P., Rajkumar, T., Sridhar, H., Rose, B., Pintos, J., Fernandez, L., Idris, A., Sanchez, M.J., Nieto, A., Talamini, R., Tavani, A., Bosch, F.X., Reidel, U., Snijders, P.J., Meijer, C.J., Viscidi, R., Munoz, N., and Franceschi, S. (2003). Human papillomavirus and oral cancer: the International Agency for Research on Cancer multicenter study. *J. Natl. Cancer Inst.* 95:1772–1783.

Hines, J.F., Ghim, S.J., Christensen, N.D., Kreider, J.W., Barnes, W.A., Schlegel, R., and Jenson, A.B. (1994a). Role of conformational epitopes expressed by human papillomavirus major capsid proteins in the serologic detection of infection and prophylactic vaccination. *Gynecol. Oncol.* 55:13–20.

Hines, J.F., Ghim, S.J., Christensen, N.D., Kreider, J.W., Barnes, W.A., Schlegel, R., and Jenson, A.B. (1994b). The expressed L1 proteins of HPV-1, HPV-6, and HPV-11 display type-specific epitopes with native conformation and reactivity with neutralizing and nonneutralizing antibodies. *Pathobiology.* 62:165–171.

Ho, G.Y., Bierman, R., Beardsley, L., Chang, C.J., and Burk, R.D. (1998). Natural history of cervicovaginal papillomavirus infection in young women. *N. Engl. J. Med.* 338:423–428.

Ho, G.Y., Studentsov,Y., Hall, C.B., Bierman, R., Beardsley, L., Lempa, M., and Burk, R.D. (2002). Risk factors for subsequent cervicovaginal human papillomavirus (HPV) infection and the protective role of antibodies to HPV-16 virus-like particles. *J. Infect. Dis.* 186:737–742.

Ho, G.Y., Studentsov, Y.Y., Bierman, R., and Burk, R.D. (2004). Natural history of human papillomavirus type 16 virus-like particle antibodies in young women. Cancer Epidemiol. *Biomarkers Prev.* 13:110–116.

Hopfl, R., Petter, A., Thaler, P., Sarcletti, M., Widschwendter, A., and Zangerle, R. (2003). High prevalence of high risk human papillomavirus-capsid antibodies in human immunodeficiency virus-seropositive men: a serological study. *BMC. Infect. Dis.* 3:6

Ishii, Y., Ozaki, S., Tanaka, K., and Kanda, T. (2005). Human papillomavirus 16 minor capsid protein L2 helps capsomeres assemble independently of intercapsomeric disulfide bonding. *Virus Genes.* 31:321–328.

Jenison, S.A., Yu, X.P., Valentine, J.M., and Galloway, D.A. (1989). Human antibodies react with an epitope of the human papillomavirus type 6b L1 open reading frame which is distinct from the type-common epitope. *J. Virol.* 63:809–818.

Jenison, S.A., Yu, X.P., Valentine, J.M., Koutsky, L.A., Christiansen, A.E., Beckmann, A.M., and Galloway, D.A. (1990). Evidence of prevalent genital-type human papillomavirus infections in adults and children. *J. Infect. Dis.* 162:60–69.

Jenson, A.B., Rosenthal, J.D., Olson, C., Pass, F., Lancaster, W.D., and Shah, K. (1980). Immunologic relatedness of papillomaviruses from different species. *J. Natl. Cancer Inst.* 64:495–500.

Jin, X.W., Cowsert, L., Marshall, D., Reed, D., Pilacinski, W., Lim, L.Y., and Jenson, A.B. (1990). Bovine serological response to a recombinant BPV-1 major capsid protein vaccine. *Intervirology.* 31:345–354.

Jochmus-Kudielka, I., Schneider, A., Braun, R., Kimmig, R., Koldovsky, U., Schneweis, K.E., Seedorf, K., and Gissmann, L. (1989). Antibodies against the human papillomavirus type 16 early proteins in human sera: correlation of anti-E7 reactivity with cervical cancer. *J. Natl. Cancer Inst.* 81:1698–1704.

Kawana, K., Yoshikawa, H., Taketani, Y., Yoshiike, K., and Kanda, T. (1999). Common neutralization epitope in minor capsid protein L2 of human papillomavirus types 16 and 6. *J. Virol.* 73:6188–6190.

Kienzler, J.L., Lemoine, M.T., Orth, G., Jibard, N., Blanc, D., Laurent, R., and Agache, P. (1983). Humoral and cell-mediated immunity to human papillomavirus type 1 (HPV-1) in human warts. *Br. J. Dermatol.* 108:665–672.

Kim, C.J., Um, S.J., Hwang, E.S., Park, S.N., Kim, S.J., NamKoong, S.E., and Park, J.S. (1999). The antibody response to HPV proteins and the genomic state of HPVs in patients with cervical cancer. *Int. J. Gynecol. Cancer.* 9:1–11.

Kirnbauer, R. (1996). Papillomavirus-like particles for serology and vaccine development. *Intervirology.* 39:54–61.

Kirnbauer, R., Booy, F., Cheng, N., Lowy, D.R., and Schiller, J.T. (1992). Papillomavirus L1 major capsid protein self-assembles into virus-like particles that are highly immunogenic. *Proc. Natl. Acad. Sci. U. S. A.* 89:12180–12184.

Kirnbauer, R., Taub, J., Greenstone, H., Roden, R., Durst, M., Gissmann, L., Lowy, D.R., and Schiller, J.T. (1993). Efficient self-assembly of human papillomavirus type 16 L1 and L1-L2 into virus-like particles. *J. Virol.* 67:6929–6936.

Kirnbauer, R., Hubbert, N.L., Wheeler, C.M., Becker, T.M., Lowy, D.R., and Schiller, J.T. (1994). A virus-like particle enzyme-linked immunosorbent assay detects serum antibodies in a majority of women infected with human papillomavirus type 16. *J. Natl. Cancer Inst.* 86:494–499.

Kjellberg, L., Wang, Z., Wiklund, F., Edlund, K., Angstrom, T., Lenner, P., Sjoberg, I., Hallmans, G., Wallin, K.L., Sapp, M., Schiller, J., Wadell, G., Mahlck, C.G., and Dillner, J. (1999). Sexual behaviour and papillomavirus exposure in cervical intraepithelial neoplasia: a population-based case-control study. *J. Gen. Virol.* 80, 391–398.

Kochel, H.G., Monazahian, M., Sievert, K., Hohne, M., Thomssen, C., Teichmann, A., Arendt, P., and Thomssen, R. (1991a). Occurrence of antibodies to L1, L2, E4 and E7 gene products of human papillomavirus types 6b, 16 and 18 among cervical cancer patients and controls. *Int. J. Cancer.* 48:682–688.

Kochel, H.G., Sievert, K., Monazahian, M., Mittelstadt-Deterding, A., Teichmann, A., and Thomssen, R. (1991b). Antibodies to human papillomavirus type-16 in human sera as revealed by the use of prokaryotically expressed viral gene products. *Virology.* 182:644–654.

Korodi, Z., Dillner, J., Jellum, E., Lumme, S., Hallmans, G., Thoresen, S., Hakulinen, T., Stattin, P., Luostarinen, T., Lehtinen, M., and Hakama, M. (2005). Human papillomavirus 16, 18, and 33 infections and risk of prostate cancer: a Nordic nested case-control study. *Cancer Epidemiol. Biomarkers Prev.* 14:2952–2955.

Koutsky, L.A., Ault, K.A., Wheeler, C.M., Brown, D.R., Barr, E., Alvarez, F.B., Chiacchierini, L.M., and Jansen, K.U. (2002). A controlled trial of a human papillomavirus type 16 vaccine. *N. Engl. J. Med.* 347:1645–1651.

Kreimer, A.R., Clifford, G.M., Snijders, P.J., Castellsague, X., Meijer, C.J., Pawlita, M., Viscidi, R., Herrero, R., and Franceschi, S. (2005). HPV16 semiquantitative viral load and serologic biomarkers in oral and oropharyngeal squamous cell carcinomas. *Int. J. Cancer.* 115:329–332.

Lagergren, J., Wang, Z., Bergstrom, R., Dillner, J., and Nyren, O. (1999). Human papillomavirus infection and esophageal cancer: a nationwide seroepidemiologic case-control study in Sweden. *J. Natl. Cancer Inst.* 91:156–162.

Le Cann, P., Coursaget, P., Iochmann, S., and Touze, A. (1994). Self-assembly of human papillomavirus type 16 capsids by expression of the L1 protein in insect cells. *FEMS. Microbiol. Lett.* 117:269–274.

Le Cann, P., Touze, A., Enogat, N., Leboulleux, D., Mougin, C., Legrand, M.C., Calvet, C., Afoutou, J.M., and Coursaget, P. (1995). Detection of antibodies against human papillomavirus (HPV) type 16 virions by enzyme-linked immunosorbent assay using recombinant HPV 16 L1 capsids produced by recombinant baculovirus. *J. Clin. Microbiol.* 33:1380–1382.

Leder, C., Kleinschmidt, J.A., Wiethe, C., and Muller, M. (2001). Enhancement of capsid gene expression: preparing the human papillomavirus type 16 major structural gene L1 for DNA vaccination purposes. *J. Virol.* 75:9201–9209.

Lehtinen, M., Leminen, A., Kuoppala, T., Tiikkainen, M., Lehtinen, T., Lehtovirta, P., Punnonen, R., Vesterinen, E., and Paavonen, J. (1992a). Pre- and posttreatment serum antibody responses to HPV 16 E2 and HSV 2 ICP8 proteins in women with cervical carcinoma. *J. Med. Virol.* 37:180–186.

Lehtinen, M., Leminen, A., Paavonen, J., Lehtovirta, P., Hyoty, H., Vesterinen, E., and Dillner, J. (1992b). Predominance of serum antibodies to synthetic peptide stemming from HPV 18 open reading frame E2 in cervical adenocarcinoma. *J. Clin. Pathol.* 45:494–497.

Lehtinen, M., Luukkaala, T., Wallin, K.L., Paavonen, J., Thoresen, S., Dillner, J., and Hakama, M. (2001). Human papillomavirus infection, risk for subsequent development of cervical neoplasia and associated population attributable fraction. *J. Clin.Virol.* 22:117–124.

Lehtinen, M., Pawlita, M., Zumbach, K., Lie, K., Hakama, M., Jellum, E., Koskela, P., Luostarinen, T., Paavonen, J., Pukkala, E., Sigstad, E., Thoresen, S., and Dillner, J. (2003). Evaluation of antibody response to human papillomavirus early proteins in women in whom cervical cancer developed 1 to 20 years later. *Am. J. Obstet. Gynecol.* 188:49–55.

Lerner, P., Dillner, J., Wiklund, F., Hallmans, G., and Stendahl, U. (1995). Serum antibody responses against human papillomavirus in relation to tumor characteristics, response to treatment, and survival in carcinoma of the uterine cervix. *Cancer Immunol. Immunother.* 40:201–205.

Li, M., Cripe, T.P., Estes, P.A., Lyon, M.K., Rose, R.C., and Garcea, R.L. (1997). Expression of the human papillomavirus type 11 L1 capsid protein in Escherichia coli: characterization of protein domains involved in DNA binding and capsid assembly. *J. Virol.* 71:2988–2995.

Liu, W.J., Zhao, K.N., Gao, F.G., Leggatt, G.R., Fernando, G.J., and Frazer, I.H. (2001). Polynucleotide viral vaccines: codon optimisation and ubiquitin conjugation enhances prophylactic and therapeutic efficacy. *Vaccine.* 20:862–869.

Ludmerer, S.W., Benincasa, D., and Mark, G.E., III (1996). Two amino acid residues confer type specificity to a neutralizing, conformationally dependent epitope on human papillomavirus type 11. *J. Virol.* 70:4791–4794.

Ludmerer, S.W., Benincasa, D., Mark, G.E., III, and Christensen, N.D. (1997). A neutralizing epitope of human papillomavirus type 11 is principally described by a continuous set of residues which overlap a distinct linear, surface-exposed epitope. *J. Virol.* 71:3834–3839.

Luostarinen, T., af Geijersstam, V., Bjorge, T., Eklund, C., Hakama, M., Hakulinen, T., Jellum, E., Koskela, P., Paavonen, J., Pukkala, E., Schiller, J.T., Thoresen, S., Youngman, L.D., Dillner, J., and Lehtinen, M. (1999). No excess risk of cervical carcinoma among women seropositive for both HPV16 and HPV6/11. *Int. J. Cancer.* 80:818–822.

Majewski, S. and Jablonska, S. (2002). Do epidermodysplasia verruciformis human papillomaviruses contribute to malignant and benign epidermal proliferations? *Arch. Dermatol.* 138:649–654.

Malcolm, K., Meschede, W., Pawlita, M., Koutsky, L.A., and Frazer, I.H. (2000). Multiple conformational epitopes are recognized by natural and induced immunity to the E7 protein of human papilloma virus type 16 in man. *Intervirology.* 43:165–173.

Mandelson, M.T., Jenison, S.A., Sherman, K.J., Valentine, J.M., McKnight, B., Daling, J.R., and Galloway, D.A. (1992). The association of human papillomavirus antibodies with cervical cancer risk. *Cancer Epidemiol. Biomarkers Prev.* 1:281–286.

Mann, V.M., de Lao, S.L., Brenes, M., Brinton, L.A., Rawls, J.A., Green, M., Reeves, W.C., and Rawls, W.E. (1990). Occurrence of IgA and IgG antibodies to select peptides representing human papillomavirus type 16 among cervical cancer cases and controls. *Cancer Res.* 50:7815–7819.

Marais, D., Rose, R.C., Lane, C., Aspinall, S., Bos, P., and Williamson, A.L. (2000a). Seroresponses to virus-like particles of human papillomavirus types 16, 18, 31, 33, and 45 in San people of Southern Africa. *J. Med. Virol.* 60:331–336.

Marais, D.J., Rose, R.C., Lane, C., Kay, P., Nevin, J., Denny, L., Soeters, R., Dehaeck, C.M., and Williamson, A.L. (2000b). Seroresponses to human papillomavirus types 16, 18, 31, 33, and 45 virus-like particles in South African women with cervical cancer and cervical intraepithelial neoplasia. *J. Med. Virol.* 60;403–410.

Marais, D.J., Vardas, E., Ramjee, G., Allan, B., Kay, P., Rose, R.C., and Williamson, A.L. (2000c). The impact of human immunodeficiency virus type 1 status on human papillomavirus (HPV) prevalence and HPV antibodies in serum and cervical secretions. *J. Infect. Dis.* 182:1239–1242.

Masini, C., Fuchs, P.G., Gabrielli, F., Stark, S., Sera, F., Ploner, M., Melchi, C.F., Primavera, G., Pirchio, G., Picconi, O., Petasecca, P., Cattaruzza, M.S., Pfister, H.J., and Abeni, D. (2003). Evidence for the association of human papillomavirus infection and cutaneous squamous cell carcinoma in immunocompetent individuals. *Arch. Dermatol.* 139:890–894.

Matsumoto, K., Yoshikawa, H., Yasugi, T., Nakagawa, S., Kawana, K., Nozawa, S., Hoshiai, H., Shiromizu, K., Kanda, T., and Taketani, Y. (1999). Balance of IgG subclasses toward human papillomavirus type 16 (HPV16) L1-capsids is a possible predictor for the regression of HPV16-positive cervical intraepithelial neoplasia. *Biochem. Biophys. Res. Commun.* 258:128–131.

McClements, W.L., Wang, X.M., Ling, J.C., Skulsky, D.M., Christensen, N.D., Jansen, K.U., and Ludmerer, S.W. (2001). A novel human papillomavirus type 6 neutralizing domain comprising two discrete regions of the major capsid protein L1. *Virology.* 289:262–268.

McLaughlin-Drubin, M.E., Christensen, N.D., and Meyers, C. (2004). Propagation, infection, and neutralization of authentic HPV16 virus. *Virology.* 322:213–219.

Meschede, W., Zumbach, K., Braspenning, J., Scheffner, M., Benitez-Bribiesca, L., Luande, J., Gissmann, L., and Pawlita, M. (1998). Antibodies against early proteins of human papillomaviruses as diagnostic markers for invasive cervical cancer. *J Clin. Microbiol.* 36:475–480.

Modis, Y., Trus, B.L., and Harrison, S.C. (2002). Atomic model of the papillomavirus capsid. *EMBO J.* 21:4754–4762.

Muller, M., Viscidi, R.P., Sun, Y., Guerrero, E., Hill, P.M., Shah, F., Bosch, F.X., Munoz, N., Gissmann, L., and Shah, K.V. (1992). Antibodies to HPV-16 E6 and E7 proteins as markers for HPV-16-associated invasive cervical cancer. *Virology.* 187:508–514.

Mund, K., Han, C., Daum, R., Helfrich, S., Muller, M., Fisher, S.G., Schiller, J.T., and Gissmann, L. (1997). Detection of human papillomavirus type 16 DNA and of antibodies to human papillomavirus type 16 proteins in children. *Intervirology.* 40:232–237.

Munoz, N., Bosch, F.X., De Sanjose, S., Herrero, R., Castellsague, X., Shah, K.V., Snijders, P.J., and Meijer, C.J. (2003). Epidemiologic classification of human papillomavirus types associated with cervical cancer. *N. Engl. J. Med.* 348:518–527.

Nakagawa, M., Viscidi, R., Deshmukh, I., Costa, M.D., Palefsky, J.M., Farhat, S., and Moscicki, A.B. (2002). Time course of humoral and cell-mediated immune responses to human papillomavirus type 16 in infected women. *Clin. Diagn. Lab. Immunol.* 9:877–882.

Nakai, Y., Lancaster, W.D., Lim, L.Y., and Jenson, A.B. (1986). Monoclonal antibodies to genus- and type-specific papillomavirus structural antigens. *Intervirology.* 25:30–37.

Nguyen, H.H., Broker, T.R., Chow, L.T., Alvarez, R.D., Vu, H.L., Andrasi, J., Brewer, L.R., Jin, G., and Mestecky, J. (2005). Immune responses to human papillomavirus in genital tract of women with cervical cancer. *Gynecol. Oncol* 96:452–461.

Nindl, I., Benitez-Bribiesca, L., Berumen, J., Farmanara, N., Fisher, S., Gross, G., Lopez-Carillo, L., Muller, M., Tommasino, M., Vazquez-Curiel, A., and Gissman, L. (1994). Antibodies against linear and conformational epitopes of the human papillomavirus (HPV) type 16 E6 and E7 oncoproteins in sera of cervical cancer patients. *Arch. Virol.* 137:341–353.

Nindl, I., Gissmann, L., Fisher, S.G., Bribiesca, L.B., Berumen, J., and Muller, M. (1996). The E7 protein of human papillomavirus (HPV) type 16 expressed by recombinant vaccinia virus can be used for detection of antibodies in sera from cervical cancer patients. *J. Virol. Methods.* 62:81–85.

Nonnenmacher, B., Hubbert, N.L., Kirnbauer, R., Shah, K.V., Munoz, N., Bosch, F.X., De Sanjose, S., Viscidi, R., Lowy, D.R., and Schiller, J.T. (1995). Serologic response to human papillomavirus type 16 (HPV-16) virus-like particles in HPV-16 DNA-positive invasive cervical cancer and cervical intraepithelial neoplasia grade III patients and controls from Colombia and Spain. *J. Infect. Dis.* 172:19–24.

Nonnenmacher, B., Kruger, K.S., Svare, E.I., Scott, J.D., Hubbert, N.L., van den Brule, A.J., Kirnbauer, R., Walboomers, J.M., Lowy, D.R., and Schiller, J.T. (1996). Seroreactivity to HPV16 virus-like particles as a marker for cervical cancer risk in high-risk populations. *Int. J. Cancer.* 68:704–709.

Nonnenmacher, B., Pintos, J., Bozzetti, M.C., Mielzinska-Lohnas, I., Lorincz, A.T., Ikuta, N., Schwartsmann, G., Villa, L.L., Schiller, J.T., and Franco, E. (2003). Epidemiologic correlates of antibody response to human papillomavirus among women at low risk of cervical cancer. *Int. J. STD. AIDS.* 14:258–265.

Oh, Y.K., Sohn, T., Park, J.S., Kang, M.J., Choi, H.G., Kim, J.A., Kim, W.K., Ko, J.J., and Kim, C.K. (2004). Enhanced mucosal and systemic immunogenicity of human papillomavirus-like particles encapsidating interleukin-2 gene adjuvant. *Virology.* 328:266–273.

Olcese, V.A., Chen, Y., Schlegel, R., and Yuan, H. (2004). Characterization of HPV16 L1 loop domains in the formation of a type-specific, conformational epitope. *BMC. Microbiol.* 4:29.

Onda, T., Carter, J.J., Koutsky, L.A., Hughes, J.P., Lee, S.K., Kuypers, J., Kiviat, N., and Galloway, D.A. (2003). Characterization of IgA response among women with incident HPV 16 infection. *Virology.* 312:213–221.

Opalka, D., Lachman, C.E., MacMullen, S.A., Jansen, K.U., Smith, J.F., Chirmule, N., and Esser, M.T. (2003). Simultaneous quantitation of antibodies to neutralizing epitopes on virus-like particles for human papillomavirus types 6, 11, 16, and 18 by a multiplexed luminex assay. *Clin. Diagn. Lab. Immunol.* 10:108–115.

Orozco, J.J., Carter, J.J., Koutsky, L.A., and Galloway, D.A. (2005). Humoral immune response recognizes a complex set of epitopes on human papillomavirus type 6 11 capsomers. *J. Virol.* 79:9503–9514.

Orth, G. and Favre, M. (1985). Human papillomaviruses. Biochemical and biologic properties. *Clin. Dermatol.* 3:27–42.

Palker, T.J., Monteiro, J.M., Martin, M.M., Kakareka, C., Smith, J.F., Cook, J.C., Joyce, J.G., and Jansen, K.U. (2001). Antibody, cytokine and cytotoxic T lymphocyte responses in chimpanzees immunized with human papillomavirus virus-like particles. *Vaccine.* 19:3733–3743.

Park, J.S., Park, D.C., Kim, C.J., Ahn, H.K., Um, S.J., Park, S.N., Kim, S.J., and NamKoong, S.E. (1998). HPV-16-related proteins as the serologic markers in cervical neoplasia. *Gynecol. Oncol.* 69:47–55.

Pastrana, D.V., Buck, C.B., Pang, Y.Y., Thompson, C.D., Castle, P.E., FitzGerald, P.C., Kruger, K.S., Lowy, D.R., and Schiller, J.T. (2004). Reactivity of human sera in a sensitive, high-throughput pseudovirus-based papillomavirus neutralization assay for HPV16 and HPV18. *Virology.* 321:205–216.

Pastrana, D.V., Gambhira, R., Buck, C.B., Pang, Y.Y., Thompson, C.D., Culp, T.D., Christensen, N.D., Lowy, D.R., Schiller, J.T., and Roden, R.B. (2005). Cross-neutralization of cutaneous and mucosal Papillomavirus types with anti-sera to the amino terminus of L2. *Virology.* 337:365–372.

Peng, S., Qi, Y., Christensen, N., Hengst, K., Kennedy, L., Frazer, I.H., and Tindle, R.W. (1999). Capture ElISA and in vitro cell binding assay for the detection of antibodies to human papillomavirus type 6b virus-like particles in patients with anogenital warts. *Pathology.* 31:418–422.

Petter, A., Heim, K., Guger, M., Ciresa-Ko, N.A., Christensen, N., Sarcletti, M., Wieland, U., Pfister, H., Zangerle, R., and Hopfl, R. (2000). Specific serum IgG, IgM and IgA antibodies to human papillomavirus types 6, 11, 16, 18 and 31 virus-like particles in human immunodeficiency virus-seropositive women. *J. Gen. Virol.* 81:701–708.

Pfister, H. (1984). Biology and biochemistry of papillomaviruses. *Rev. Physiol. Biochem. Pharmacol.* 99:111–181.

Pfister, H., Fuchs, P.G., Majewski, S., Jablonska, S., Pniewska, I., Malejczyk, M. (2003). High prevalence of epidermodysplasia verruciformis-associated human papillomavirus DNA in actinic keratoses of the immunocompetent population. Arch. Dermatol. Res. 295:273–279.

Pfister, H. and zur Hausen, H. (1978). Seroepidemiological studies of human papilloma virus (HPV-1) infections. *Int. J. Cancer.* 21:161–165.

Poland, G.A., Jacobson, R.M., Koutsky, L.A., Tamms, G.M., Railkar, R., Smith, J.F., Bryan, J.T., Cavanaugh, P.F., Jr., Jansen, K.U., and Barr, E. (2005). Immunogenicity and reactogenicity of a novel vaccine for human papillomavirus 16: a 2-year randomized controlled clinical trial. *Mayo Clin. Proc.* 80:601–610.

Rocha-Zavaleta, L., Jordan, D., Pepper, S., Corbitt, G., Clarke, F., Maitland, N.J., Sanders, C.M., Arrand, J.R., Stern, P.L., and Stacey, S.N. (1997). Differences in serological IgA responses to recombinant baculovirus-derived human papillomavirus E2 protein in the natural history of cervical neoplasia. *Br. J. Cancer.* 75:1144–1150.

Rocha-Zavaleta, L., Pereira-Suarez, A.L., Yescas, G., Cruz-Mimiaga, R.M., Garcia-Carranca, A., and Cruz-Talonia, F. (2003). Mucosal IgG and IgA responses to human papillomavirus type 16 capsid proteins in HPV16-infected women without visible pathology. *Viral. Immunol.* 16:159–168.

Rocha-Zavaleta, L., Ambrosio, J.P., Mora-Garcia, M.L., Cruz-Talonia, F., Hernandez-Montes, J., Weiss-Steider, B., Ortiz-Navarrete, V., and Monroy-Garcia, A. (2004). Detection of antibodies against a human papillomavirus (HPV) type 16 peptide that differentiate high-risk from low-risk HPV-associated low-grade squamous intraepithelial lesions. *J. Gen. Virol.* 85:2643–2650.

Roden, R.B., Greenstone, H.L., Kirnbauer, R., Booy, F.P., Jessie, J., Lowy, D.R., and Schiller, J.T. (1996a). In vitro generation and type-specific neutralization of a human papillomavirus type 16 virion pseudotype. *J. Virol.* 70:5875–5883.

Roden, R.B., Hubbert, N.L., Kirnbauer, R., Christensen, N.D., Lowy, D.R., and Schiller, J.T. (1996b). Assessment of the serological relatedness of genital human papillomaviruses by hemagglutination inhibition. *J. Virol.* 70:3298–3301.

Roden, R.B., Armstrong, A., Haderer, P., Christensen, N.D., Hubbert, N.L., Lowy, D.R., Schiller, J.T., and Kirnbauer, R. (1997). Characterization of a human papillomavirus type 16 variant-dependent neutralizing epitope. *J. Virol.* 71:6247–6252.

Roden, R.B., Yutzy, W.H., Fallon, R., Inglis, S., Lowy, D.R., and Schiller, J.T. (2000). Minor capsid protein of human genital papillomaviruses contains subdominant, cross-neutralizing epitopes. *Virology.* 270:254–257.

Rommel, O., Dillner, J., Fligge, C., Bergsdorf, C., Wang, X., Selinka, H.C., and Sapp, M. (2005). Heparan sulfate proteoglycans interact exclusively with conformationally intact HPV L1 assemblies: basis for a virus-like particle ELISA. *J. Med. Virol.* 75:114–121.

Rose, R.C., Bonnez, W., Reichman, R.C., and Garcea, R.L. (1993). Expression of human papillomavirus type 11 L1 protein in insect cells: in vivo and in vitro assembly of viruslike particles. *J. Virol.* 67:1936–1944.

Rose, R.C., Reichman, R.C., and Bonnez, W. (1994). Human papillomavirus (HPV) type 11 recombinant virus-like particles induce the formation of neutralizing antibodies and detect HPV-specific antibodies in human sera. *J. Gen. Virol.* 75:2075–2079.

Rose, R.C., White, W.I., Li, M., Suzich, J.A., Lane, C., and Garcea, R.L. (1998). Human papillomavirus type 11 recombinant L1 capsomeres induce virus-neutralizing antibodies. *J. Virol.* 72:6151–6154.

Rosenblatt, K.A., Carter, J.J., Iwasaki, L.M., Galloway, D.A., and Stanford, J.L. (2003). Serologic evidence of human papillomavirus 16 and 18 infections and risk of prostate cancer. *Cancer Epidemiol. Biomarkers Prev.* 12:763–768.

Salunke, D.M., Caspar, D.L., and Garcea, R.L. (1986). Self-assembly of purified polyomavirus capsid protein VP1. *Cell.* 46:895–904.

Sapp, M., Kraus, U., Volpers, C., Snijders, P.J., Walboomers, J.M., and Streeck, R.E. (1994). Analysis of type-restricted and cross-reactive epitopes on virus-like particles of human papillomavirus type 33 and in infected tissues using monoclonal antibodies to the major capsid protein. *J. Gen. Virol.* 75:3375–3383.

Sapp, M. and Selinka, H.C. (2005). Pseudovirions as specific tools for investigation of virus interactions with cells. *Methods Mol. Biol.* 292:197–212.

Sasagawa, T., Rose, R.C., Azar, K.K., Sakai, A., and Inoue, M. (2003). Mucosal immunoglobulin-A and -G responses to oncogenic human papilloma virus capsids. *Int. J. Cancer.* 104:328–335.

Schwartz, S.M., Daling, J.R., Doody, D.R., Wipf, G.C., Carter, J.J., Madeleine, M.M., Mao, E.J., Fitzgibbons, E.D., Huang, S., Beckmann, A.M., McDougall, J.K., and Galloway, D.A. (1998). Oral cancer risk in relation to sexual history and evidence of human papillomavirus infection. *J. Natl. Cancer Inst.* 90:1626–1636.

Sehr, P., Muller, M., Hopfl, R., Widschwendter, A., and Pawlita, M. (2002). HPV antibody detection by ELISA with capsid protein L1 fused to glutathione S-transferase. *J. Virol. Methods.* 106:61–70.

Sehr, P., Zumbach, K., and Pawlita, M. (2001). A generic capture ELISA for recombinant proteins fused to glutathione S-transferase: validation for HPV serology. *J. Immunol. Methods.* 253:153–162.

Shi, L., Sanyal, G., Ni, A., Luo, Z., Doshna, S., Wang, B., Graham, T.L., Wang, N., and Volkin, D.B. (2005). Stabilization of human papillomavirus virus-like particles by non-ionic surfactants. *J. Pharm. Sci.* 94:1538–1551.

Sigstad, E., Lie, A.K., Luostarinen, T., Dillner, J., Jellum, E., Lehtinen, M., Thoresen, S., and Abeler, V. (2002). A prospective study of the relationship between prediagnostic human papillomavirus seropositivity and HPV DNA in subsequent cervical carcinomas. *Br. J. Cancer.* 87:175–180.

Silins, I., Avall-Lundqvist, E., Tadesse, A., Jansen, K.U., Stendahl, U., Lenner, P., Zumbach, K., Pawlita, M., Dillner, J., and Frankendal, B. (2002). Evaluation of antibodies to human papillomavirus as prognostic markers in cervical cancer patients. *Gynecol. Oncol.* 85:333–338.

Slavinsky, J., III, Kissinger, P., Burger, L., Boley, A., DiCarlo, R.P., and Hagensee, M.E. (2001). Seroepidemiology of low and high oncogenic risk types of human papillomavirus in a predominantly male cohort of STD clinic patients. *Int. J. STD. AIDS.* 12:516–523.

Slupetzky, K., Shafti-Keramat, S., Lenz, P., Brandt, S., Grassauer, A., Sara, M., and Kirnbauer, R. (2001). Chimeric papillomavirus-like particles expressing a foreign epitope on capsid surface loops. *J. Gen. Virol.* 82:2799–2804.

Stacey, S.N., Bartholomew, J.S., Ghosh, A., Stern, P.L., Mackett, M., and Arrand, J.R. (1992). Expression of human papillomavirus type 16 E6 protein by recombinant baculovirus and use for detection of anti-E6 antibodies in human sera. *J. Gen. Virol.* 73:2337–2345.

Stacey, S.N., Ghosh, A., Bartholomew, J.S., Tindle, R.W., Stern, P.L., Mackett, M., and Arrand, J.R. (1993). Expression of human papillomavirus type 16 E7 protein by recombinant baculovirus and use for the detection of E7 antibodies in sera from cervical carcinoma patients. *J. Med. Virol.* 40:14–21.

Stark, S., Petridis, A.K., Ghim, S.J., Jenson, A.B., Bouwes Bavinck, J.N., Gross, G., Stockfleth, E., Fuchs, P.G., and Pfister, H. (1998). Prevalence of antibodies against virus-like particles of Epidermodysplasia verruciformis-associated HPV8 in patients at risk of skin cancer. *J. Invest. Dermatol.* 111:696–701.

Steele, J.C. and Gallimore, P.H. (1990). Humoral assays of human sera to disrupted and nondisrupted epitopes of human papillomavirus type 1. *Virology.* 174:388–398.

Stone, K.M., Karem, K.L., Sternberg, M.R., McQuillan, G.M., Poon, A.D., Unger, E.R., and Reeves, W.C. (2002). Seroprevalence of human papillomavirus type 16 infection in the United States. *J. Infect. Dis.* 186:1396–1402.

Strickler, H.D., Hildesheim, A., Viscidi, R.P., Shah, K.V., Goebel, B., Drummond, J., Waters, D., Sun, Y., Hubbert, N.L., Wacholder, S., Brinton, L.A., Han, C.L., Nasca, P.C., McClimens, R., Turk, K., Devairakkam, V., Leitman, S., Martin, C., and Schiller, J.T. (1997). Interlaboratory agreement among results of human papillomavirus type 16 enzyme-linked immunosorbent assays. *J. Clin. Microbiol.* 35:1751–1756.

Strickler, H.D., Burk, R., Shah, K., Viscidi, R., Jackson, A., Pizza, G., Bertoni, F., Schiller, J.T., Manns, A., Metcalf, R., Qu, W., and Goedert, J.J. (1998a). A multifaceted study of human papillomavirus and prostate carcinoma. *Cancer.* 82:1118–1125.

Strickler, H.D., Schiffman, M.H., Shah, K.V., Rabkin, C.S., Schiller, J.T., Wacholder, S., Clayman, B., and Viscidi, R.P. (1998b). A survey of human papillomavirus 16 antibodies in patients with epithelial cancers. *Eur. J. Cancer Prev.* 7:305–313.

Strickler, H.D., Kirk, G.D., Figueroa, J.P., Ward, E., Braithwaite, A.R., Escoffery, C., Drummond, J., Goebel, B., Waters, D., McClimens, R., and Manns, A. (1999). HPV 16 antibody prevalence in Jamaica and the United States reflects differences in cervical cancer rates. *Int. J. Cancer.* 80:339–344.

Studentsov, Y.Y., Schiffman, M., Strickler, H.D., Ho, G.Y., Pang, Y.Y., Schiller, J., Herrero, R., and Burk, R.D. (2002). Enhanced enzyme-linked immunosorbent assay for detection of antibodies to virus-like particles of human papillomavirus. *J. Clin. Microbiol.* 40:1755–1760.

Studentsov, Y.Y., Ho, G.Y., Marks, M.A., Bierman, R., and Burk, R.D. (2003). Polymer-based enzyme-linked immunosorbent assay using human papillomavirus type 16 (HPV16) virus-like particles detects HPV16 clade-specific serologic responses. *J. Clin. Microbiol.* 41:2827–2834.

Sun, Y., Eluf-Neto, J., Bosch, F.X., Munoz, N., Booth, M., Walboomers, J.M., Shah, K.V., and Viscidi, R.P. (1994a). Human papillomavirus-related serological markers of invasive cervical carcinoma in Brazil. *Cancer Epidemiol. Biomarkers Prev.* 3:341–347.

Sun, Y., Shah, K.V., Muller, M., Munoz, N., Bosch, X.F., and Viscidi, R.P. (1994b). Comparison of peptide enzyme-linked immunosorbent assay and radioimmunoprecipitation assay with in vitro-translated proteins for detection of serum antibodies to human papillomavirus type 16 E6 and E7 proteins. *J. Clin. Microbiol.* 32:2216–2220.

Sun, Y., Hildesheim, A., Brinton, L.A., Nasca, P.C., Trimble, C.L., Kurman, R.J., Viscidi, R.P., and Shah, K.V. (1996). Human papillomavirus-specific serologic response in vulvar neoplasia. *Gynecol. Oncol.* 63:200–203.

Sun, X.W., Kuhn, L., Ellerbrock, T.V., Chiasson, M.A., Bush, T.J., and Wright, T.C., Jr. (1997). Human papillomavirus infection in women infected with the human immunodeficiency virus. *N. Engl. J. Med.* 337:1343–1349.

Suzich, J.A., Ghim, S.J., Palmer-Hill, F.J., White, W.I., Tamura, J.K., Bell, J.A., Newsome, J.A., Jenson, A.B., and Schlegel, R. (1995). Systemic immunization with papillomavirus L1 protein completely prevents the development of viral mucosal papillomas. *Proc. Natl. Acad. Sci. U. S. A.* 92:11553–11557.

Svare, E.I., Kjaer, S.K., Nonnenmacher, B., Worm, A.M., Moi, H., Christensen, R.B., van den Brule, A.J., Walboomers, J.M., Meijer, C.J., Hubbert, N.L., Lowy, D.R., and Schiller, J.T. (1997). Seroreactivity to human papillomavirus type 16 virus-like particles is lower in high-risk men than in high-risk women. *J. Infect. Dis.* 176:876–883.

Thompson, D.L., Douglas, J.M., Jr., Foster, M., Hagensee, M.E., Diguiseppi, C., Baron, A.E., Cameron, J.E., Spencer, T.C., Zenilman, J., Malotte, C.K., Bolan, G., Kamb, M.L., and Peterman, T.A. (2004). Seroepidemiology of infection with human papillomavirus 16, in men and women attending sexually transmitted disease clinics in the United States. *J. Infect. Dis.* 190:1563–1574.

Touze, A., Dupuy, C., Mahe, D., Sizaret, P.Y., and Coursaget, P. (1998). Production of recombinant virus-like particles from human papillomavirus types 6 and 11, and study of serological reactivities between HPV 6, 11, 16 and 45 by ELISA: implications for papillomavirus prevention and detection. *FEMS. Microbiol. Lett.* 160:111–118.

Urquiza, M., Guevara, T., Espejo, F., Bravo, M.M., Rivera, Z., and Patarroyo, M.E. (2005). Two L1-peptides are excellent tools for serological detection of HPV-associated cervical carcinoma lesions. *Biochem. Biophys. Res. Commun.* 332:224–232.

Van Doornum, G.J., Prins, M., Pronk, L., Coutinho, R.A., and Dillner, J. (1994). A prospective study of antibody responses to defined epitopes of human papillomavirus (HPV) type 16 in relationship to genital and anorectal presence of HPV DNA. *Clin. Diagn. Lab. Immunol.* 1:633–639.

van Doornum, G., Prins, M., Andersson-Ellstrom, A., and Dillner, J. (1998). Immunoglobulin A, G, and M responses to L1 and L2 capsids of human papillomavirus types 6, 11, 16, 18, and 33 L1 after newly acquired infection. *Sex. Transm. Infect.* 74:354–360.

Van Doornum, G.J., Korse, C.M., Buning-Kager, J.C., Bonfrer, J.M., Horenblas, S., Taal, B.G., and Dillner, J. (2003). Reactivity to human papillomavirus type 16 L1 virus-like particles in sera from patients with genital cancer and patients with carcinomas at five different extragenital sites. *Br. J. Cancer.* 88:1095–1100.

Veress, G., Konya, J., Csiky-Meszaros, T., Czegledy, J., and Gergely, L. (1994). Human papillomavirus DNA and anti-HPV secretory IgA antibodies in cytologically normal cervical specimens. *J. Med. Virol.* 43:201–207.

Viladiu, P., Bosch, F.X., Castellsague, X., Munoz, N., Escriba, J.M., Hamsikova, E., Hofmannova, V., Guerrero, E., Izquierdo, A., Navarro, C., Moreo, P., Izarzugaza, I., Ascunce, N., Gili, M., Munoz, M.T., Tafur, L., Shah, K.V., and Vonka, V. (1997). Human papillomavirus DNA and antibodies to human papillomaviruses 16 E2, L2, and E7 peptides as predictors of survival in patients with squamous cell cervical cancer. *J. Clin. Oncol.* 15:610–619.

Villa, L.L., Costa, R.L., Petta, C.A., Andrade, R.P., Ault, K.A., Giuliano, A.R., Wheeler, C.M., Koutsky, L.A., Malm, C., Lehtinen, M., Skjeldestad, F.E., Olsson, S.E.,

Steinwall, M., Brown, D.R., Kurman, R.J., Ronnett, B.M., Stoler, M.H., Ferenczy, A., Harper, D.M., Tamms, G.M., Yu, J., Lupinacci, L., Railkar, R., Taddeo, F.J., Jansen, K.U., Esser, M.T., Sings, H.L., Saah, A.J., and Barr, E. (2005). Prophylactic quadrivalent human papillomavirus (types 6, 11, 16, and 18) L1 virus-like particle vaccine in young women: a randomised double-blind placebo-controlled multicentre phase II efficacy trial. *Lancet Oncol.* 6:271–278.

Viscidi, R.P., Sun, Y., Tsuzaki, B., Bosch, F.X., Munoz, N., and Shah, K.V. (1993). Serologic response in human papillomavirus-associated invasive cervical cancer. *Int. J. Cancer.* 55:780–784.

Viscidi, R.P., Kotloff, K.L., Clayman, B., Russ, K., Shapiro, S., and Shah, K.V. (1997). Prevalence of antibodies to human papillomavirus (HPV) type 16 virus-like particles in relation to cervical HPV infection among college women. *Clin. Diagn. Lab. Immunol.* 4:122–126.

Viscidi, R.P., Ahdieh-Grant, L., Clayman, B., Fox, K., Massad, L.S., Cu-Uvin, S., Shah, K.V., Anastos, K.M., Squires, K.E., Duerr, A., Jamieson, D.J., Burk, R.D., Klein, R.S., Minkoff, H., Palefsky, J., Strickler, H., Schuman, P., Piessens, E., and Miotti, P. (2003a). Serum immunoglobulin G response to human papillomavirus type 16 virus-like particles in human immunodeficiency virus (HIV)-positive and risk-matched HIV-negative women. *J. Infect. Dis.* 187:194–205.

Viscidi, R.P., Ahdieh-Grant, L., Schneider, M.F., Clayman, B., Massad, L.S., Anastos, K.M., Burk, R.D., Minkoff, H., Palefsky, J., Levine, A., and Strickler, H. (2003b). Serum immunoglobulin A response to human papillomavirus type 16 virus-like particles in human immunodeficiency virus (HIV)-positive and high-risk HIV-negative women. *J. Infect. Dis.* 188:1834–1844.

Viscidi, R.P., Schiffman, M., Hildesheim, A., Herrero, R., Castle, P.E., Bratti, M.C., Rodriguez, A.C., Sherman, M.E., Wang, S., Clayman, B., and Burk, R.D. (2004). Seroreactivity to human papillomavirus (HPV) types 16, 18, or 31 and risk of subsequent HPV infection: results from a population-based study in Costa Rica. *Cancer Epidemiol. Biomarkers Prev.* 13:324–327.

Viscidi, R.P., Snyder, B., Cu-Uvin, S., Hogan, J.W., Clayman, B., Klein, R.S., Sobel, J., and Shah, K.V. (2005). Human papillomavirus capsid antibody response to natural infection and risk of subsequent HPV infection in HIV-positive and HIV-negative women. *Cancer Epidemiol. Biomarkers Prev.* 14:283–288.

Volpers, C., Sapp, M., Snijders, P.J., Walboomers, J.M., and Streeck, R.E. (1995). Conformational and linear epitopes on virus-like particles of human papillomavirus type 33 identified by monoclonal antibodies to the minor capsid protein L2. *J. Gen. Virol.* 76:2661–2667.

Vonka, V., Hamsikova, E., Kanka, J., Ludvikova, V., Sapp, M., and Smahel, M. (1999). Prospective study on cervical neoplasia IV. Presence of HPV antibodies. *Int. J. Cancer.* 80:365–368.

Wang, S.S., Schiffman, M., Shields, T.S., Herrero, R., Hildesheim, A., Bratti, M.C., Sherman, M.E., Rodriguez, A.C., Castle, P.E., Morales, J., Alfaro, M., Wright, T., Chen, S., Clayman, B., Burk, R.D., and Viscidi, R.P. (2003). Seroprevalence of human papillomavirus-16, -18, -31, and -45 in a population-based cohort of 10000 women in Costa Rica. *Br. J. Cancer.* 89:1248–1254.

Wang, S.S., Schiffman, M., Herrero, R., Carreon, J., Hildesheim, A., Rodriguez, A.C., Bratti, M.C., Sherman, M.E., Morales, J., Guillen, D., Alfaro, M., Clayman, B., Burk, R.D., and Viscidi, R.P. (2004). Determinants of human papillomavirus 16 serological conversion and persistence in a population-based cohort of 10 000 women in Costa Rica. *Br. J. Cancer.* 91:1269–1274.

Wang, X., Wang, Z., Christensen, N.D., and Dillner, J. (2003a). Mapping of human serum-reactive epitopes in virus-like particles of human papillomavirus types 16 and 11. *Virology.* 311:213–221.

Wang, X.M., Cook, J.C., Lee, J.C., Jansen, K.U., Christensen, N.D., Ludmerer, S.W., and McClements, W.L. (2003b). Human papillomavirus type 6 virus-like particles present overlapping yet distinct conformational epitopes. *J. Gen. Virol.* 84:1493–1497.

Wang, X., Sapp, M., Christensen, N.D., and Dillner, J. (2005). Heparin-based ELISA reduces background reactivity in virus-like particle-based papillomavirus serology. *J. Gen. Virol.* 86:65–73.

Wang, Z., Hansson, B.G., Forslund, O., Dillner, L., Sapp, M., Schiller, J.T., Bjerre, B., and Dillner, J. (1996). Cervical mucus antibodies against human papillomavirus type 16, 18, and 33 capsids in relation to presence of viral DNA. *J. Clin. Microbiol.* 34:3056–3062.

Wang, Z., Christensen, N., Schiller, J.T., and Dillner, J. (1997). A monoclonal antibody against intact human papillomavirus type 16 capsids blocks the serological reactivity of most human sera. *J. Gen. Virol.* 78:2209–2215.

Wang, Z.H., Kjellberg, L., Abdalla, H., Wiklund, F., Eklund, C., Knekt, P., Lehtinen, M., Kallings, I., Lenner, P., Hallmans, G., Mahlck, C.G., Wadell, G., Schiller, J., and Dillner, J. (2000). Type specificity and significance of different isotypes of serum antibodies to human papillomavirus capsids. *J. Infect. Dis.* 181:456–462.

Waterboer, T., Sehr, P., Michael, K.M., Franceschi, S., Nieland, J.D., Joos, T.O., Templin, M.F., and Pawlita, M. (2005). Multiplex human papillomavirus serology based on in situ-purified glutathione s-transferase fusion proteins. *Clin. Chem.* 51:1845–1853.

White, W.I., Wilson, S.D., Palmer-Hill, F.J., Woods, R.M., Ghim, S.J., Hewitt, L.A., Goldman, D.M., Burke, S.J., Jenson, A.B., Koenig, S., and Suzich, J.A. (1999). Characterization of a major neutralizing epitope on human papillomavirus type 16 L1. *J. Virol.* 73:4882–4889.

White, W.I., Wilson, S.D., Bonnez, W., Rose, R.C., Koenig, S., and Suzich, J.A. (1998). In vitro infection and type-restricted antibody-mediated neutralization of authentic human papillomavirus type 16. *J. Virol.* 72:959–964.

Wideroff, L., Schiffman, M.H., Nonnenmacher, B., Hubbert, N., Kirnbauer, R., Greer, C.E., Lowy, D., Lorincz, A.T., Manos, M.M., Glass, A.G., Scott, D.R., Sherman, M.E., Buckland, J., Lowy, D., and Schiller, J. (1995). Evaluation of seroreactivity to human papillomavirus type 16 virus-like particles in an incident case-control study of cervical neoplasia. *J. Infect. Dis.* 172:1425–1430.

Wideroff, L., Schiffman, M.H., Hoover, R., Tarone, R.E., Nonnenmacher, B., Hubbert, N., Kirnbauer, R., Greer, C.E., Lorincz, A.T., Manos, M.M., Glass, A.G., Scott, D.R., Sherman, M.E., Buckland, J., Lowy, D., and Schiller, J. (1996). Epidemiologic determinants of seroreactivity to human papillomavirus (HPV) type 16 virus-like particles in cervical HPV-16 DNA-positive and-negative women. *J. Infect. Dis.* 174:937–943.

Wideroff, L., Schiffman, M., Haderer, P., Armstrong, A., Greer, C.E., Manos, M.M., Burk, R.D., Scott, D.R., Sherman, M.E., Schiller, J.T., Hoover, R.N., Tarone, R.E., and Kirnbauer, R. (1999). Seroreactivity to human papillomavirus types 16, 18, 31, and 45 virus-like particles in a case-control study of cervical squamous intraepithelial lesions. *J. Infect. Dis.* 180:1424–1428.

Wieland, U., Ritzkowsky, A., Stoltidis, M., Weissenborn, S., Stark, S., Ploner, M., Majewski, S., Jablonska, S., Pfister, H.J., and Fuchs, P.G. (2000). Communication: papillomavirus DNA in basal cell carcinomas of immunocompetent patients: an accidental association? *J. Invest. Dermatol.* 115:124–128.

Wikstrom, A., Van Doornum, G.J., Kirnbauer, R., Quint, W.G., and Dillner, J. (1995a). Prospective study on the development of antibodies against human papillomavirus type 6 among patients with condyloma acuminata or new asymptomatic infection. *J. Med. Virol.* 46:368–374.

Wikstrom, A., Van Doornum, G.J., Quint, W.G., Schiller, J.T., and Dillner, J. (1995b). Identification of human papillomavirus seroconversions. *J. Gen. Virol.* 76:529–539.

Wikstrom, A., Eklund, C., von Krogh, G., Lidbrink, P., and Dillner, J. (1997). Antibodies against human papillomavirus type 6 capsids are elevated in men with previous condylomas. *APMIS.* 105:884–888.

Winer, R.L., Lee, S.K., Hughes, J.P., Adam, D.E., Kiviat, N.B., and Koutsky, L.A. (2003). Genital human papillomavirus infection: incidence and risk factors in a cohort of female university students. *Am. J. Epidemiol.* 157:218–226.

Yeager, M.D., Aste-Amezaga, M., Brown, D.R., Martin, M.M., Shah, M.J., Cook, J.C., Christensen, N.D., Ackerson, C., Lowe, R.S., Smith, J.F., Keller, P., and Jansen, K.U. (2000). Neutralization of human papillomavirus (HPV) pseudovirions: a novel and efficient approach to detect and characterize HPV neutralizing antibodies. *Virology.* 278:570–577.

Yuan, H., Estes, P.A., Chen, Y., Newsome, J., Olcese, V.A., Garcea, R.L., and Schlegel, R. (2001). Immunization with a pentameric L1 fusion protein protects against papillomavirus infection. *J. Virol.* 75:7848–7853.

Zhou, J., Liu, W.J., Peng, S.W., Sun, X.Y., and Frazer, I. (1999). Papillomavirus capsid protein expression level depends on the match between codon usage and tRNA availability. *J. Virol.* 73:4972–4982.

Zumbach, K., Hoffmann, M., Kahn, T., Bosch, F., Gottschlich, S., Gorogh, T., Rudert, H., and Pawlita, M. (2000a). Antibodies against oncoproteins E6 and E7 of human papillomavirus types 16 and 18 in patients with head-and-neck squamous-cell carcinoma. *Int. J. Cancer.* 85:815–818.

Zumbach, K., Kisseljov, F., Sacharova, O., Shaichaev, G., Semjonova, L., Pavlova, L., and Pawlita, M. (2000b). Antibodies against oncoproteins E6 and E7 of human papillomavirus types 16 and 18 in cervical-carcinoma patients from Russia. *Int. J. Cancer.* 85:313–318.

13
Cell-Mediated Immune Responses to Human Papillomavirus

Gretchen Eiben Lyons[1,2], Michael I. Nishimura[2], and W. Martin Kast[1]

[1] *Norris Comprehensive Cancer Center and Department of Molecular Microbiology & Immunology, University of Southern California, Los Angeles, CA 90033*
[2] *Department of Surgery, University of Chicago, Chicago IL 60637*

13.1. Introduction

Genital human papillomavirus (HPV) infection is one of the most prevalent sexually transmitted diseases, with over 5 million new cases per year in the United States alone. Fortunately, more than 95 percent of HPV infections resolve over 3 to 5 years, and cancer develops in less than 0.1 percent of HPV-infected individuals over their lifetime. The small number of HPV infections that eventually will lead to cervical cancer emphasizes the importance of our immune system's ability to control HPV infections and prevent malignant progression. Cell-mediated immunity is likely to play an important role in protecting against tumor progression, as shown by the increased frequency of HPV-associated tumors in individuals treated with immunosuppressive drugs or suffering from AIDS. In addition, malignant lesions are characterized by an infiltration of CD8[+] cells. Therefore, the development of HPV-induced carcinomas can be seen as an escape from cell-mediated immune surveillance. There is increasing evidence that HPV encodes proteins that can directly subvert the host immune response, thereby preventing resolution of infection. This chapter focuses on the cell-mediated immune response to HPV infections and the ability of HPV to evade the immune system.

13.2. Virally Induced Tumors

The list of human viruses presently known to cause or contribute to tumor development comprises four DNA viruses—namely, Epstein–Barr virus, certain human papilloma virus types, hepatitis B virus, and Kaposi's sarcoma herpesvirus (HHV-8)—and two RNA viruses—human T-cell leukemia virus and hepatitis C virus. Together these viruses contribute significantly to the total incidence of cancer worldwide, and there are certainly other human tumor viruses that have yet to be identified (Epstein, 2001). Of course none of the human tumor viruses are directly carcinogenic; rather each is necessary but not on its own sufficient to cause malignancy. In the absence of these infections, the incidence of these cancers would be reduced by 95 percent with a significant reduction in morbidity and mortality, especially in developing countries. Current work in each of these virus systems seeks to understand the mechanisms of viral action and identify intervention strategies to combat viral infection and subsequent carcinogenic progression. It is thought that oncogenic proliferation is instigated by the presence and expression of viral oncogenes which may be integrated into the host genome. Critical viral genes may also activate cellular proto-oncogenes and/or inactivate anti-oncogenes and their products. Mere infection with any of these viruses, however, does not mean that cancer will result, since many more people are infected than will develop a malignancy, indicating that there are additional factors involved in the progression from an infected cell to a transformed cell with invasive potential. Thus, cancer can be seen as a rare outcome of a common infection (Klein, 2002). This situation is particularly true in the case of HPV infections. While there are approximately 630 million HPV infected individuals worldwide, only approximately 510,000 of these will proceed on to invasive cervical cancer each year (World Health Organization, 2004). This proves to be one of the major problems in understanding the molecular causality and cellular control of HPV induced human cancers.

To date, over 100 different HPV types have been identified, and about one-third of these infect epithelial cells in the genital tract. HPV types infecting the genital tract have been classified as either "high risk" or "low risk" depending on the outcome and prognosis of the lesion which they cause. Low-risk HPV types 6, 11, 42, 43, and 44 are frequently associated with low-grade cervical intraepithelial neoplasia (CIN) and benign genital warts (Stone, 1995). The high-risk types, 16, 18, 31, 33, and 45 are associated with premalignant dysplasia and cancer (Munoz et al., 2003; Remmink et al., 1995; Schiffman et al., 1993). Specifically, HPV-16 and HPV-18 infections represent the major causal factors for cervical cancer (Walboomers et al., 1999). Integration of high-risk viral DNA into the genome of the host cell generally results in the loss of transcriptional control of viral gene expression. As a consequence the transformed cells overexpress the HPV E6 and E7 oncoproteins, initiating the malignant transformation process (Pei, 1996). HPV E6 and E7 expression delays keratinocyte differentiation and stimulates cell cycle progression, allowing the virus to utilize host DNA polymerases to replicate its genome. The primary targets of the E6 and E7 proteins are the p53 and retinoblastoma tumor suppressor proteins, respectively, but there are also additional cellular targets (Mantovani and Banks, 2001) (Munger et al., 2001). The HPV E6 and E7 genes are expressed at low levels in proliferating basal cells, but are highly expressed in HPV-associated genital cancers (Crish et al., 2000; Durst et al., 1992). As a consequence, the E6 and E7 proteins can serve as major targets for the cell-mediated immune response and are therefore, attractive targets for specific immunotherapies.

13.3. Immunology to Viral Assault

Knowledge of T-cell activation is crucial to comprehend fully the potential of the immune system to reject virus-induced tumors and to understand why the natural immune response sometimes fails to do so. Cellular immune responses, especially antigen-specific T lymphocytes, are most likely the critical defense mechanism against HPV-infected cells. The immunological recognition of viral antigens by T cells is restricted by the polymorphic class I and class II human leukocyte antigens (HLA) of the major histocompatibility complex (MHC). HLA class I antigens (HLA-A, -B, or -C) are expressed on virtually all nucleated cells and present intracellularly processed peptides to $CD8^+$ T cells. HLA class II antigens (HLA-DR, -DQ or -DP) are present on some cells of the immune system and present antigen to $CD4^+$ T-helper cells. The extensive polymorphisms of HLA molecules are concentrated within the peptide-binding groove, and the different HLA haplotypes provide a broad repertoire of peptide-binding capacity. Viral immunity is initiated by T-cell receptor engagement of specific peptides, derived from the virus or tumor, presented by the cell-surface MHC class I molecule. An efficient $CD8^+$ T-cell response may therefore be characterized as the generation and clonal expansion of effector cytotoxic T lymphocytes (CTLs) that are capable of recognizing and eliminating cells bearing a specific antigen. Through their cytolytic activity and cytokine production, $CD8^+$ T cells are key components of the cellular immune response against infections and tumors.

The differentiation of naïve CD8 T cells into CTLs occurs in response to two signals; antigenic peptide on MHC class I and a costimulatory signal. Only professional antigen-presenting cells, such as dendritic cells (DCs) express the required costimulatory molecules to provide the second signal. Numerous studies have also demonstrated a requirement for $CD4^+$ T-cell help during the generation of primary $CD8^+$ T-cell responses in vivo (Matloubian et al., 1994; Stohlman et al., 1998; von Herrath et al., 1996). A major pathway of T-cell help (Th) for CTL priming is mediated through CD40–CD40 ligand interactions (Schoenberger et al., 1998). The recognition of a Th epitope on a DC allows the simultaneous activation of both cell types through the interaction of the CD40 ligand on the Th cell with CD40 on the DC, thereby activating the DC to express additional costimulatory molecules and adhesion molecules such intercellular cell adhesion molecule-1 (ICAM-1). Activated DCs then migrate to secondary lymphoid organs to select and stimulate antigen-specific T cells. Through cytokine secretion and activation of DCs, Th cells enhance the differentiation of $CD8^+$ T cells into CTLs that can eradicate virus-infected cells or tumor cells.

13.4. The Importance of Cell-Mediated Immunity

The critical role of cell-mediated immunity is directly implied by the increased incidence of HPV lesions and carcinomas in individuals with impaired cellular immune function, including HIV-infected individuals and renal transplant patients receiving immunosuppressive therapy (Halpert et al., 1986; Petry et al., 1994). On the basis of the largest population based data set available for the study of cancer in immunosuppressed individuals, HIV-infected individuals demonstrated a considerably increased risk for HPV associated cervical cancer (Frisch et al., 2000).

The natural CTL responses against HPV-16 and -18 displayed by immunocompetent patients with cervical lesions but not by uninfected individuals provides additional evidence that cellular immunity is important (Evans et al., 1997). Several studies have detected E7 specific responses at low levels in the peripheral blood of patients with HPV-associated lesions or cancers (Bontkes et al., 2000; Kadish et al., 1997, 2002; van der Burg et al., 2001; Youde et al., 2000). A study of CTL responses in women with HPV-16 infection and squamous intraepithelial lesions has implicated differential E6 responses in viral persistence (Nakagawa et al., 2000). These responses are absent in carcinoma patients (Bontkes et al., 2000). Such HPV-16 E6 specific immunity is often detected in healthy individuals, suggesting that it plays a role in protection against persistent HPV infection and the development of malignancies. (Welters et al., 2003) In addition, HPV-16 E6 and E7 specific CD4$^+$ T-helper cells have been identified in patients with abnormal cytology that were not detectable in individuals without viral infection (de Gruijl et al., 1996, 1998; Kadish et al., 1997; Luxton et al., 1996). The significance of MHC class I restricted HPV specific CD8$^+$ CTLs in HLA-A*0201 positive individuals has been examined in tumor infiltrating lymphocytes as well (Evans et al., 1997). CTLs are specific to particular HPV epitopes based on two lines of evidence: 1) Using individual HPV-16 E7 peptides that bind HLA-A*0201, memory CTLs could be detected in patients with cervical carcinomas or CIN but not in controls (Evans et al., 1997; Ressing et al., 1996). 2) CTLs from cancer patients lysed HLA-A*201+, HPV-16 positive cervical carcinoma-derived Caski cells but not HLA-A*201+ MS751 cervical carcinoma cells harboring HPV45 (Geisbill et al., 1997; Ressing et al., 1996). Furthermore, women who were capable of mounting a cell-mediated immune response to HPV-16 E6 and E7 proteins in vitro are likely to undergo regression of CIN and resolution of HPV infection, whereas patients who fail to exhibit these responses are likely to have persistent disease (Kadish et al., 1997). The presence of infiltrating lymphocytes and HPV specific T cells in spontaneously regressing papillomas also strongly implies that the cellular immune response does in fact control HPV infection and associated disease (Kadish et al., 2002; Coleman et al., 1994; Evans et al., 1997; Ghosh and Moore, 1992). Taken together, these data underscore the importance of CTLs in effectively controlling HPV infection and eliminating HPV transformed cells and suggest that the lack of such a response leads to further progression of disease.

In contrast to the CTL response, antibodies to HPV-16 E6 or E7 oncoproteins are rarely detected in patients with premalignant cervical lesions and found in only 50 percent of patients with late stage cervical cancers (Stern, 1996; Viscidi et al., 1993). Furthermore, humoral immunodeficiency, characterized by a failure to produce antibodies, does not increase susceptibility to the development of

HPV lesions (Benton et al., 1999). The intracellular location of the HPV E6 and E7 proteins also strongly implies that a cellular immune response is more efficacious than a humoral response in clearing HPV lesions. All of these data suggest that while neutralizing antibodies may be an effective way of preventing viral infection and spread, cell-mediated immune responses are required for the ultimate resolution of established infection or disease.

13.5.   Immune Activation Against HPV Infection

More than 95 percent of HPV-positive lesions resolve spontaneously, suggesting that a natural immune response against HPV though not capable of eradicating the virus is capable of clearing most viral associated lesions (Ho et al., 1998; Schlecht et al., 2003). Two observations suggest that host immune factors are involved in the prevention of malignant progression: 1) The small number of infected individuals that eventually develop cancer of the cervix, and 2) the long latency period between primary infection and cancer emergence. In this context, the local immune state within the transformation zone of the cervix, where the majority of intraepithelial and invasive neoplasms develop, might be expected to play a key role in the host defense against HPV infection and associated precancerous lesions. There are two main arms to the immune response that may play a role in the host's natural clearance of HPV infection, innate and adaptive immunity. Innate immunity consists of a rapidly induced, nonspecific response, which does not result in memory. The innate immune system is important at epithelial borders, where most viral infections take place, and is delivered via cytokines and cellular effectors. Activation of the innate immune system then induces a milieu of effector cytokines, which are capable of activating the adaptive immune response. There is considerable evidence to suggest that the effective activation of both the innate and adaptive immune system may be crucial for the eradication of HPV infection.

The presence of infiltrating macrophages and NK cells in regressing genital warts suggests that the innate immune system is most likely the first line of defense against HPV infection (Coleman et al., 1994). NK cells are an important component of the innate cellular immune system and anti-viral host defense. These cells respond to aberrant expression of cell surface molecules in an antigen nonspecific manner. Virally infected cells often lose or express lower levels of surface MHC class I. Since all nucleated cells should express MHC class I molecules, virally infected cells become susceptible to recognition by NK cells, via killing inhibitory receptors (KIR) on their cell surface. If the KIR does not bind to an MHC class I molecule, the NK cell is triggered to destroy the cell. Because HPV infection leads to down-regulation of surface MHC class I molecules (Cromme et al., 1993, 1994; Keating et al., 1995; Vambutas et al., 2000), NK cells may play an important role in innate immune surveillance against HPV.

The regression of HPV lesions is also controlled by cytokines, most importantly IL-2 and interferon gamma (IFN-γ or type II interferon), that boost the adaptive immune response by activating CTLs (Stellato et al., 1997). IL-2, which is secreted primarily by Th-cells, is the principle cytokine required for proliferation and differentiation of activated precursor CTLs into effector CTLs. IFN-γ has also been shown to repress HPV-16 gene expression in HPV-16-immortalized cell lines and inhibit cell growth (Woodworth et al., 1992). Another key factor of immunosurveillance against HPV may be supplied by type I interferons (IFN-α or IFN-β) released from keratinocytes. Interferons are proteins that interfere with viral replication in eukaryotic cells and form an early cytokine barrier against viral disease, including HPV infection, and also inhibit the differentiation of HPV-infected cells (Rockley and Tyring, 1995). Keratinocytes also release TNF-α upon infection with HPV (Arany et al., 1993). This cytokine exerts an inhibitory effect on HPV replication as well, and attracts T cells via the up-regulation of ICAM-1 and HLA-DR on keratinocytes, which encourages the presentation of antigens to infiltrating CD4+ T cells and can result in the resolution of infection (Al Saleh et al., 1998; Coleman et al., 1994). Furthermore, TNFα enhances the migration of Langerhans cells (LCs) into the regional lymph nodes, where they present antigen to CD8+ T cells (Cumberbatch and Kimber, 1992). TNFα is also able to recruit NK cells to the tumor, providing a valuable mechanism to eliminate tumor cells (Glas et al., 2000; Kashii et al., 1999).

Taken together, it is evident that combinations of innate and adaptive immune responses to HPV are capable of clearing most infections and can protect an individual from the development of tumors. It can then be postulated that in the case of cervical malignancy the host immune response to HPV has failed. Boosting adaptive immune responses to HPV by vaccination and boosting innate immunity by drugs like Imiquimod and Polyphenon E are therefore appealing new therapeutic approaches.

13.6. Immune Evasion by HPV

Viral persistence, which is required for malignancy, requires avoidance of immune attack by the host. Therefore, defining the relationship between HPV and the immune system is paramount in understanding the progression from infection to tumorigenesis. HPVs are expert at subverting host immune responses with a multitude of mechanisms to minimize or prevent exposure of the virus to the immune system. By infecting epithelial cells, HPV causes little tissue damage and avoids a strong anti-inflammatory response. Because papillomaviruses are not lytic, there is little opportunity for APCs to engulf virions or present HPV derived antigens to the immune system. HPV also encodes mainly nonsecreted nucleoproteins, which are not accessible for extracellular immune recognition. Furthermore, most HPV nonstructural proteins are expressed at low levels compared to proteins expressed by more immunogenic viruses. Therefore, it appears that HPVs seem to evade the innate immune response by avoiding the main triggers that should initiate an immune response to a viral infection.

Numerous studies have focused on the specific mechanisms whereby HPV can modify the immune response to interfere with antigen presentation in order to hide from immune recognition and establish a persistent infection.

13.6.1. MHC Regulation

Comparison of invasive carcinoma to regressing lesions sheds light on the immune escape mechanisms of HPV. Cervical neoplasias exhibit downregulation of MHC class I antigens and upregulation of class II molecules (Al Saleh et al., 1998; Coleman et al., 1994; Cromme et al., 1993; Glew et al., 1992; Keating et al., 1995). MHC class I antigens are typically expressed on the surface of most cells, including tumor cells, and are essential for presenting epitopes of the rejection antigens to T cells. It has been suggested that HPV may evade T cell recognition of infected cells by decreasing the surface MHC class I complex through modulation of transporter associated with antigen presentation (TAP-1) (Vambutas et al., 2000). TAP-1 is essential for assembling MHC class I proteins in the endoplasmic reticulum, and its downregulation occurs in carcinomas of the cervix (Cromme et al., 1994). The bovine papillomavirus E5 protein has also been implicated for its ability to down-regulate class I molecules in trans-formed cells (Ashrafi et al., 2002). Although the absence of MHC class I would render the infected cells more sensitive to NK cell attack, HPV infections are also characterized by a down-regulation of type I IFNs which will result in reduced NK cytotoxicity (Barnard and McMillan, 1999; Li et al., 1999; Park et al., 2000; Ronco et al., 1998). Therefore, the ability of HPV gene products to modulate the IFN response pathways of infected cells would compromise any protection from interferon induced NK immunity. MHC class I down-regulation would undoubtedly hamper tumor presentation of antigen by APCs, leading to a decrease in immunological recognition. Furthermore, loss of class I molecules on the surface of tumor cells will reduce their potential as targets for CTLs. In contrast, the significance of the up-regulation of class II molecules on tumor cells, namely, HLA-DR, is unclear. As opposed to malignant tissue, HLA-DR is not detectable in normal cervical keratinocytes. It has been postulated that expression of HLA-DR on keratinocytes after HPV infection may contribute to the host's immunity by presenting antigen to CD4+ T cells (Coleman et al., 1994; Konya and Dillner, 2001). However, this assumption is contradicted by the observation that an increase in HLA-DR expression on cervical lesions is positively correlated with the grade of the lesion (Al Saleh et al., 1998; Cromme et al., 1993; Glew et al., 1992). This result suggests that HLA-DR expression has a suppressive effect on the immune function. One such mechanism may be the activation of CD4+ CD25+ regulatory T cells. Upon antigen-specific stimulation, these regulatory CD4+ CD25+ T cells contribute to tumor growth by inhibiting the activation of normally responsive T cells (Onizuka et al., 1999; Thornton and Shevach, 1998). Thus, the increased expression of class II molecules may lead to an increased activation of these suppressive T cells, which in turn could inhibit both CTLs and potentially cytotoxic CD4+ T cells.

13.6.2. Modulation of Antigen Presentation

Another immune evasion mechanism that has recently been reported after HPV infection involves cross-presentation by immature LCs. Because HPV infects basal cells of the mucosa, LCs are the only APCs that HPV will interact with during its life cycle. LCs lack costimulatory signals necessary for efficient T-cell activation but are very efficient in capturing antigens (Fausch et al., 2003). After contact with an antigen, LCs normally up-regulate costimulatory molecules, down-regulate endocytosis of other antigens, and migrate toward a lymph node for MHC class I presentation to T cells. In the case of an HPV infection, however, it has been reported that although LCs are able to bind and internalize HPV virus like particles (VLPs), they do not up-regulate activation markers necessary for the generation of a CTL response (Fausch et al., 2002). Furthermore, LCs did not migrate out of the epidermis in an in vivo skin explant assay in response to HPV VLP stimulation (Fausch et al., 2002). The mechanism responsible for these HPV-induced effects in LCs has recently been identified as an up-regulation of PI_3K in LCs, and indeed these HPV related immune phenomena are reversed by PI_3K inhibitors (Fausch et al., 2005). Inhibition of LC function in this manner is a novel immune escape mechanism used by HPV to evade host immunity and prolong HPV infection. Furthermore, the inability of LCs and DCs to mature may protect the infected tissue from the attention of the host adaptive immunity. The inefficient activation of T cells by these immature LCs may also result in immunological ignorance or tolerance to the virally infected cells. Other reports have observed an overall loss of LCs in cervical malignancies. A decline of LCs in cervical lesions is accompanied by a marked decrease in CD1a expression in correlation with the severity of the lesion (Al Saleh et al., 1998; Tay et al., 1987). The E6 protein has been shown to inhibit epithelial cell-dendritic cell interactions (Matthews et al., 2003), perhaps accounting for the depletion of LCs observed in infected cervical epithelium. These results suggest that interference with local APCs of the cervix may protect the infected tissue from the attention of the host adaptive immunity.

13.6.3. Loss of T-Cell-Receptor ζ-Chain Expression

Another immune escape mechanism observed during the progression toward cervical cancer is derived from the T cell itself. Expression of the ζ-chain of the T-cell receptor (TCR) is required for T-cell activation, and the reduced levels or absence of ζ-chain expression is indicative of defects in T-cell activation (Chan et al., 1992). Reduced expression of the ζ–chain in T cells occurs in large percentages of cancer patients for many different tumor types (Baniyash, 2004; Whiteside, 1999). Data indicate that the tumor microenvironment has negative effects on the infiltrating T cells (de Gruijl et al., 1999). The pro-inflammatory cytokine, TNF-α has been shown to directly down-regulate ζ-chain expression (Cope, 2002, 2003; Isomaki et al., 2001). Consequently, the TCR–CD3 complex in these cells showed impaired assembly and stability at the cell surface, thereby uncoupling the TCR signal transduction pathway. A decrease in ζ-chain expression also occurs in response to exposure to hydrogen peroxide produced by activated granulocytes and macrophages (Kono et al., 1996b; Schmielau, 2001). The downregulation of ζ-chains in T cells is not limited to tumor infiltrating lymphocytes (TILs), but can also be measured in peripheral blood lymphocytes (PBLs) and NK cells of tumor-bearing patients (Baniyash, 2004). Furthermore, studies on ζ-chain expression in PBLs of patients with CIN or cervical cancer indicated that reduced ζ-chain expression may correlate with disease progression (Kono et al., 1996a; Nieland et al., 1998). These data indicate that reduced levels of ζ-chain expression may contribute to the inability of the host to mount an effective immune response against HPV-infected cells. Furthermore, ζ-chain expression may be a prognostic marker for survival.

In conclusion, there appear to be several cellular mechanisms working in concert to disrupt both antigen presentation and T-cell activation against HPV-infected cells during cervical carcinogenesis. Additionally, E7 and other HPV early gene products might act indirectly by affecting different cytokines, which may inhibit the innate immune response and the activation of adaptive immunity.

13.6.4.   Subversion of HPV by Cytokines

Production of type I interferons (IFN-α/β) by virally infected cells is one of the first lines of host defense against viral infection. IFN gene transcription is initiated when viral double stranded RNA is recognized by Toll-like receptor 3 (TLR-3) on DCs (Alexopoulou et al., 2001; Matsumoto et al., 2002). IFNs then induce resistance to viral infection by inhibiting expression of viral proteins, activating NK cells and inducing MHC class I on uninfected cells, thus making virally infected cells more susceptible to NK activity. These mechanisms allow IFNs to both limit viral spread and activate virus specific CTLs. Therefore, it seems logical that HPV would have evolved ways to disrupt this pathway. The E7 proteins interact with IFN regulatory factor (IRF-1) which inhibits activation of the IFN-β promoter (Park et al., 2000), while the E6 protein appears to target multiple IFN-responsive genes, including IFN-α, IFN-β, and STAT-1 (Nees et al., 2001). Furthermore, since HPV does not have a known dsRNA intermediate, it would not signal via TLR-3 on DCs to release type I interferons and thus may not activate NK activity or downstream IFN-inducible genes (Matsumoto et al., 2004).

Interleukin-18 (IL-18) is another cytokine that plays a crucial role in the generation of immune responses against viral infection. IL-18 released from macrophages stimulates the production of IFN-γ, which is required for the activation of CTLs. Both E6 and E7 inhibit IL-18 induced IFN-γ production in PBMC and NK cells, by reducing the binding of IL-18 to the α chain of its receptor (Lee et al., 2001). Loss of IL-18 signaling is critical because IL-18 helps to prime the T-cell compartment to launch a CD8[+] -mediated response (Le Buanec et al., 1999). Thus, the ability of HPV to interfere with both the IFN pathway and the expression of IL-18 provides an effective means for HPV to hide from the immune system and establish a persistent infection that could lead to malignancy.

Another critical immunodulatory cytokine is IL-2. As opposed to the high levels of IL-2 observed in regressing papillomas, high-grade squamous intraepthelial lesions (SIL) exhibit a dramatic decrease in IL-2 expression (Al Saleh et al., 1998). Because IL-2 production is primarily mediated by lymphocytes, such a loss suggests a decrease in the activation of lymphocytes in cervical cancer. In addition, the production of IL-10 in cervical carcinomas may provide another mechanism to weaken T-cell-mediated immune surveillance (El Sherif et al., 2001; Kim et al., 1995; Sheu et al., 2001a). IL-10 down-regulates costimulatory molecules on DCs and inhibits their migration to regional lymph nodes, thereby, preventing antigen presentation to $CD8^+$ CTLs. IL-10 treated human DCs can induce anergy in tumor-specific CTLs, resulting in a failure to lyse tumor cells (Steinbrink et al., 1999).

TGF-β is another cytokine capable of inhibiting CTL generation and diminishing the production of immunostimulatory cytokines, such as TNF-α and IFN-γ. Recently, the role of TGF-β in cervical cancer was investigated by examining cytokine profiles from ten different human cervical cancer lines, all of which displayed an increase in TGF-β expression (Hazelbag et al., 2001). TGF-β has been directly implicated in the escape of tumor recognition in immunocompetent hosts (Chang et al., 1993; Torre-Amione et al., 1990). Furthermore, TGF-β is reported to be secreted by CD4+ CD25+ suppressive T cells and is necessary for the immune suppressive function of these cells (Read et al., 2000). TGF-β also enhances IL-10 production by macrophages (Kitani et al., 2000). Taken together, these data directly imply that an excess of TGF-β secreted in precancerous lesions may prevent the development of an efficient cell-mediated response to HPV. Therefore, depleting TGF-β may be an effective strategy to boost cellular immunity. This approach was recently implemented by examining the antibody-mediated depletion of TGF-β in HPV-16$^+$ tumor bearing mice (Gunn et al., 2001). Addition of anti-TGF-β treatment to HPV vaccination dramatically increased the regression of HPV established tumors (Gunn et al., 2001). In addition, a sharp decline in TNF-α and IFN-γ expression was observed in the cervical cancer lines (El Sherif et al., 2001; Hazelbag et al., 2001). The reduced transcription of these cytokines in the higher grades of HPV associated precancerous lesions may contribute to a local environment which favors progression of these lesions by downregulating MHC class I expression (Maudsley, 1991).

The tumor suppressor function of the immune system is critically dependent on the actions of IFN-γ as well, which are directed at regulating tumor cell immunogenicity (Boursnell et al., 1996). IFN-γ expression is required for the formation and the catalytic activity of the immunoproteasome (Morel et al., 2000). In the absence of the immunoproteasome, antigen processing takes place via the standard proteasome (Macagno et al., 1999), which can result in differentially processed epitopes and therefore the lack of T-cell recognition (Boes et al.,

1994; Morel et al., 2000). Therefore, in the absence of IFN-γ, tumor cells could switch from the immunoproteasome to the standard proteasome and thereby escape an immune response directed against improperly processed epitopes. It remains to be determined whether there is a difference in the HPV epitopes processed by the standard and the immunoproteasome.

Taken together, these data indicate that HPV-infected cells, via the production of TGF-β, up-regulate IL-10 while down-regulating TNFα and IFN-γ. This situation can result in a reduced level of antigen presentation and decreased recognition by cytotoxic T cells. Such immune deficiency might result in prolonged HPV infection and thus increase the chance of malignancy.

13.7. The Importance of the CD4 Helper Response

A positive role for HPV-specific Th immunity was suggested by the predominance of $CD4^+$ T cells in regressing genital warts, as well as by the detection of delayed type hypersensitivity responses to HPV-16 E7 in the majority of subjects with spontaneous regressing CIN lesions (Coleman et al., 1994; Hopfl et al., 2000). It is now well accepted that the induction of a strong $CD4^+$ T-helper response is advantageous for the activation and perpetuation of a strong cell-mediated immune response (Velders et al., 2003). Upon recognition of antigenic peptides presented by APCs in MHC class II molecules, $CD4^+$ helper T cells release cytokines that regulate the immune response to the antigen. Cytokines are also important immunological mediators of cell-mediated defenses against tumors.

CD4 T-helper cells are divided into two subsets on the basis of the immunomodulatory cytokines they produce and their effector functions. Th1 cytokines (IL-2, IFN-γ, TNFα) are pro-inflammatory or immunostimulatory cytokines that boost the cellular immune response by promoting the outgrowth of CTLs. Th2 cytokines (IL-4, IL-5, IL-10, and IL-13) are assumed to have the opposite effect, impairing the cell-mediated immune response. A shift toward cytokines produced by Th2 cells has been associated with a less effective immune response in a number of cancer types (Huang et al., 1995; Nakagomi et al., 1995). Naïve $CD8^+$ T cells are able to differentiate into a T-cytotoxic 1 (Tc1) subset that produces IL-2 and IFN-γ, or a Tc2 subset that produces IL-4, IL-5, and IL-10 (Carter and Dutton, 1996; Mosmann et al., 1997; Sad et al., 1995; Seder et al., 1992). Although both subsets are cytolytic in humans, Th1/Tc1 cells seem to enhance cell-mediated immunity. In contrast, Th2/Tc2 cells predominate during parasite infiltrations and are less protective than Tc1 cells against viral infection. Cervical cancer cells can directly drive the tumor encountered T cells toward the Th2/Tc2 phenotype through an IL-10- and TGF-β-mediated pathway (Sheu et al., 2001b). Thus, a type 2 cytokine release from Th2 cells drives the production of Tc2 cells and further dampens the immune response. This inappropriate response to viral infection may lead to the persistence of HPV, preventing infected cells from being eliminated by cell-mediated immune responses. In vitro studies have also revealed an inverse association between the degree of CIN and IL-2 production by peripheral blood mononuclear cells in response to HPV-16 E6 and E7 peptides (Hildesheim et al., 1997; Tsukui et al., 1996). Women with CIN3 or cancer appear to have a decreased ability to mount a Th1-mediated immune response to HPV E6/E7, compared with women with CIN1 or HPV-infected women without lesions. It is possible that a Th1 -mediated immune response may play a crucial role in the host immunological control of HPV infection and that the lack of such a response may lead to the advancement of disease.

Several T-helper epitopes have been identified in HPV-16 L1, E6 and E7 proteins by their ability to induce T-cell proliferative responses from PBMCs (Strang et al., 1990; van der Burg et al., 2001; Warrino et al., 2004). T-cell-proliferative responses against HPV-16 E7 proteins or peptides were found in healthy individuals and more frequently in patients with CIN or carcinomas (Kadish et al., 1997). Immunization protocols that generate strong effector CTL responses in the absence of $CD4^+$ T cells are unable to induce long-term CTL memory, illustrating the importance of this helper population (Clarke, 2000). The observations that regression of genital warts is associated with increased numbers of CD4+ cells and that women with invasive cancer are at a higher risk of developing progressive disease if they exhibit impaired or decreased $CD4^+$ T cells provides indirect evidence that $CD4^+$ T cells may indeed be important in mediating or augmenting antitumor immune responses (Gemignani et al., 1995). Recently, it has been observed that TILs from a cervical cancer patient can recognize the autologous tumor in an MHC class II restricted fashion (Hohn et al., 1999). Thus, MHC class II molecules expressed on cervical cancer cells may serve as restricting molecules to present HPV derived epitopes to $CD4^+$ T cells. MHC class II epitopes may then be implemented into vaccination strategies to augment T cells responses against tumor cells. The recent identification of such epitopes, as mentioned above, will provide new opportunities for developing effective cancer vaccines and improve our understanding of the mechanisms by which CD4+ T cells regulate the host immune system.

13.8. Genetic Susceptibility to HPV Associated Carcinogenesis

Associations between certain HLA alleles and susceptibility to or protection against CIN lesions and cervical carcinoma probably reflect T-cell responses directed against HPV oncoproteins. The fact that some women develop cervical cancer argues that peptides derived from the oncoproteins do not bind as well to any of the HLA molecules in that persons' repertoire or might not be processed well by proteasomal cleavage of the oncoproteins and thus are not presented to T cells. On the other hand, protection indicates that there are certain HLA molecules that bind well to peptides derived from HPV oncoproteins. Given the pivotal role of HLA molecules in the recognition of tumor peptides by CTLs, several studies have been performed to examine the association of specific HLA alleles with HPV infection status and development of cervical cancer. Since individual HLA molecules can present different viral peptides, the various combinations of inherited HLA alleles should result in different immune responsiveness to HPV encoded antigens among individuals. Several groups have reported positive or negative associations of particular HLA class II alleles with CIN or invasive cervical carcinomas (Apple et al., 1994; David et al., 1992; Duggan-Keen et al., 1996; Glew et al., 1993; Helland et al., 1994; Vandenvelde et al., 1993; Wank and Thomssen, 1991). As a consequence, carriers of these alleles may be at a higher or lower risk of developing cervical cancer. Unfortunately, results from different groups evaluating HLA types and HPV induced disease have not been in agreement. This disparity is in part due to the heterogeneity in HLA frequencies among different populations and to differences in the prevalence of infections with specific HPV types. The most frequently reported positive association was the presence of the HLA class II, DQ3 antigen or DQB1*03 allele (Odunsi et al., 1995; Wang, 2001; Wank, 1991). Associations with persistent HPV-16 infection or CIN III were also described for DRB1*07, DRB1*1501/DQB1*0602 and DRB1*07/DQB1*0201 haplotypes (Apple et al., 1995). In contrast, it has been observed that DRB1*1301 protects against HPV infection and cervical disease (Bontkes et al., 1998; Sastre-Garau et al., 1996). These findings support the hypothesis that multiple risk alleles are required to confer an increased risk that is detectable at the population level, possibly through inadequate presentation of viral antigens to the immune system. In contrast, the presence of a single protective allele may be sufficient to confer protection.

Studies have revealed no HLA class I associations with HPV-16 infection but have demonstrated an HLA-A2, HLA-B44, and HLA-B7 correlation with disease progression (Bontkes et al., 1998; Duggan-Keen et al., 1996; Montoya et al., 1998). HLA-A2 and B44 are the predominant haplotypes in patients with late-stage carcinomas compared with patients with early stage carcinomas and are among the alleles most commonly down-regulated in carcinoma cells (Keating et al., 1995). Furthermore, down-regulation of HLA-B7 on cervical cancer cells is associated with a worse survival compared to normal expression of this antigen (Duggan-Keen et al., 1996). Taken together, these findings suggest that certain individuals may or may not be more prone to mount a robust cellular immune response against HPV-infected cells. However, no correlation between disease progression and the presence or absence of cellular immunity has been established. Therefore, studies are required to determine whether certain HLA types may present more or less immunodominant epitopes for CTL recognition that may contribute to the progression of premalignant lesions of the cervix.

13.9.   Conclusions

Evidence has clearly shown that cell-mediated immunity is crucial for the eradication of HPV-infected cells. Furthermore, the failure to induce or maintain a T-cell response leads to persistent infection and the development of malignancy.

There are several mechanisms whereby the activity of T cells is compromised upon HPV infection. Suppressive Th2 cytokines released within the tumor environment have been shown to inhibit CTL activation and antigen presentation. The down-regulation of class I molecules on the surface of tumor cells prevents both antigen presentation and recognition of target antigens by effector CTLs, while the up-regulation of class II molecules may activate CD4+ CD25+ regulatory T cells to inhibit CTLs. Additionally, there are other mechanisms, such as modulation of antigen presentation by LCs and the downregulation of the TCR ζ-chain, which can inhibit T-cell activation and contribute to immune escape. Since more than 99 percent of all HPV infections and low-grade lesions will eventually regress, it appears that in the majority of cases, the host immune response eventually overcomes the immunosuppressive effects of the virus. The development of cervical cancer can therefore be seen as a rare event whereby the host–virus relationship is disturbed and HPV continues to persist and subvert the immune response toward malignancy. The knowledge of how HPV interacts with and escapes T-cell immunity also allows the identification of strategies to counteract the T-cell immune escape mechanisms and induce HPV specific T-cell immunity. Major efforts by a large number of academic institutions, biotech companies and pharmaceutical companies in the development of therapeutic vaccines that induce HPV specific T-cell immunity will likely materialize in the coming years and provide new treatment options for patients with HPV-associated diseases.

Acknowledgments. W.M. Kast holds the Walter A. Richter Cancer Research Chair and is supported by the V and Whittier Foundations and by NIH grants RO1 CA74397 and PO1 CA97296.

References

Al Saleh, W., Giannini, S.L., Jacobs, N., Moutschen, M., Doyen, J., Boniver, J., and Delvenne, P. (1998). Correlation of T-helper secretory differentiation and types of antigen-presenting cells in squamous intraepithelial lesions of the uterine cervix. *J. Pathol.* 184:283–290.

Alexopoulou, L., Holt, A.C., Medzhitov, R., and Flavell, R.A. (2001). Recognition of double-stranded RNA and activation of NF-kappaB by Toll-like receptor 3. *Nature* 413:732–738.

Apple, R.J., Becker, T.M., Wheeler, C.M., and Erlich, H.A. (1995). Comparison of human leukocyte antigen DR-DQ disease associations found with cervical dysplasia and invasive cervical carcinoma. *J. Natl. Cancer Inst.* 87:427–436.

Apple, R.J., Erlich, H.A., Klitz, W., Manos, M.M., Becker, T.M., and Wheeler, C.M. (1994). HLA DR-DQ associations with cervical carcinoma show papillomavirus-type specificity. *Nat. Genet.* 6:157–162.

Arany, I., Rady, P., and Tyring, S.K. (1993). Alterations in cytokine/antioncogene expression in skin lesions caused by "low-risk" types of human papillomaviruses. *Viral Immunol.* 6:255–265.

Ashrafi, G.H., Tsirimonaki, E., Marchetti, B., O'Brien, P.M., Sibbet, G.J., Andrew, L., and Campo, M.S. (2002). Down-regulation of MHC class I by bovine papillomavirus E5 oncoproteins. *Oncogene* 21:248–259.

Baniyash, M. (2004). TCR zeta-chain downregulation: curtailing an excessive inflammatory immune response. *Nat. Rev. Immunol.* 4:675–687.

Barnard, P. and McMillan, N.A. (1999). The human papillomavirus E7 oncoprotein abrogates signaling mediated by interferon-alpha. *Virology* 259:305–313.

Benton, C., Shyahidullah H., Hunter J. A. M. (1999). Human papillomavirus in the immunocompromised. *Papillomavirus Report* 3:23–26.

Boes, B., Hengel, H., Ruppert, T., Multhaup, G., Koszinowski, U.H., and Kloetzel, P.M. (1994). Interferon gamma stimulation modulates the proteolytic activity and cleavage site preference of 20S mouse proteasomes. *J. Exp. Med.* 179:901–909.

Bontkes, H.J., de Gruijl, T.D., van den Muysenberg, A.J., Verheijen, R.H., Stukart, M.J., Meijer, C.J., Scheper, R.J., Stacey, S.N., Duggan-Keen, M.F., Stern, P.L., Man, S., Borysiewicz, L.K., and Walboomers, J.M. (2000). Human papillomavirus type 16 E6/E7-specific cytotoxic T lymphocytes in women with cervical neoplasia. *Int. J. Cancer* 88:92–98.

Bontkes, H.J., van Duin, M., de Gruijl, T.D., Duggan-Keen, M.F., Walboomers, J.M., Stukart, M.J., Verheijen, R.H., Helmerhorst, T.J., Meijer, C.J., Scheper, R.J., Stevens, F.R., Dyer, P.A., Sinnott, P., and Stern, P.L. (1998). HPV 16 infection and progression of cervical intra-epithelial neoplasia: analysis of HLA polymorphism and HPV 16 E6 sequence variants. *Int. J. Cancer* 78:166–171.

Boursnell, M.E., Rutherford, E., Hickling, J.K., Rollinson, E.A., Munro, A.J., Rolley, N., McLean, C.S., Borysiewicz, L.K., Vousden, K., and Inglis, S.C. (1996). Construction and characterisation of a recombinant vaccinia virus expressing human papillomavirus proteins for immunotherapy of cervical cancer. *Vaccine* 14:1485–1494.

Carter, L.L., and Dutton, R.W. (1996). Type 1 and type 2: a fundamental dichotomy for all T-cell subsets. *Curr. Opin. Immunol.* 8:336–342.

Chan, A.C., Irving, B.A., and Weiss, A. (1992). New insights into T-cell antigen receptor structure and signal transduction. *Curr. Opin. Immunol.* 4:246–251.

Chang, H.L., Gillett, N., Figari, I., Lopez, A.R., Palladino, M.A., and Derynck, R. (1993). Increased transforming growth factor beta expression inhibits cell proliferation in vitro, yet increases tumorigenicity and tumor growth of Meth A sarcoma cells. *Cancer Res.* 53:4391–4398.

Clarke, S.R. (2000). The critical role of CD40/CD40L in the CD4-dependent generation of CD8+ T cell immunity. *J. Leukoc. Biol.* 67:607–614.

Coleman, N., Birley, H.D., Renton, A.M., Hanna, N.F., Ryait, B.K., Byrne, M., Taylor-Robinson, D., and Stanley, M.A.(1994). Immunological events in regressing genital warts. *Am. J. Clin. Pathol.* 102:768–774.

Cope, A.P. (2002). Studies of T-cell activation in chronic inflammation. *Arthritis Res.* 4(Suppl. 3):S197–S211.

Cope, A.P. (2003). Exploring the reciprocal relationship between immunity and inflammation in chronic inflammatory arthritis. *Rheumatology (Oxford)* 42:716–731.

Crish, J.F., Bone, F., Balasubramanian, S., Zaim, T.M., Wagner, T., Yun, J., Rorke, E.A., and Eckert, R.L. (2000). Suprabasal expression of the human papillomavirus type 16 oncoproteins in mouse epidermis alters expression of cell cycle regulatory proteins. *Carcinogenesis* 21:1031–1037.

Cromme, F.V., Airey, J., Heemels, M.T., Ploegh, H.L., Keating, P.J., Stern, P.L., Meijer, C.J., and Walboomers, J.M. (1994). Loss of transporter protein, encoded by the

TAP-1 gene, is highly correlated with loss of HLA expression in cervical carcinomas. *J. Exp. Med.* 179:335–340.

Cromme, F.V., Meijer, C.J., Snijders, P.J., Uyterlinde, A., Kenemans, P., Helmerhorst, T., Stern, P.L., van den Brule, A.J., and Walboomers, J.M. (1993). Analysis of MHC class I and II expression in relation to presence of HPV genotypes in premalignant and malignant cervical lesions. *Br. J. Cancer* 67:1372–1380.

Cumberbatch, M. and Kimber, I. (1992). Dermal tumour necrosis factor-alpha induces dendritic cell migration to draining lymph nodes, and possibly provides one stimulus for Langerhans' cell migration. *Immunology* 75:257–263.

David, A.L., Taylor, G.M., Gokhale, D., Aplin, J.D., Seif, M.W., and Tindall, V.R. (1992). HLA-DQB1*03 and cervical intraepithelial neoplasia type III. *Lancet* 340:52.

de Gruijl, T.D., Bontkes, H.J., Peccatori, F., Gallee, M.P., Helmerhorst, T.J., Verheijen, R.H., Aarbiou, J., Mulder, W.M., Walboomers, J.M., Meijer, C.J., van de Vange, N., and Scheper, R.J. (1999). Expression of CD3-zeta on T-cells in primary cervical carcinoma and in metastasis-positive and -negative pelvic lymph nodes. *Br. J. Cancer* 79:1127–1132.

de Gruijl, T.D., Bontkes, H.J., Stukart, M.J., Walboomers, J.M., Remmink, A.J., Verheijen, R.H., Helmerhorst, T.J., Meijer, C.J., and Scheper, R.J. (1996). T cell proliferative responses against human papillomavirus type 16 E7 oncoprotein are most prominent in cervical intraepithelial neoplasia patients with a persistent viral infection. *J. Gen. Virol.* 77:2183–2191.

de Gruijl, T.D., Bontkes, H.J., Walboomers, J.M., Stukart, M.J., Doekhie, F.S., Remmink, A.J., Helmerhorst, T.J., Verheijen, R.H., Duggan-Keen, M.F., Stern, P.L., Meijer, C.J., and Scheper, R.J. (1998). Differential T helper cell responses to human papillomavirus type 16 E7 related to viral clearance or persistence in patients with cervical neoplasia: a longitudinal study. *Cancer Res.* 58:1700–1706.

Duggan-Keen, M.F., Keating, P.J., Stevens, F.R., Sinnott, P., Snijders, P.J., Walboomers, J.M., Davidson, S., Hunter, R.D., Dyer, P.A., and Stern, P.L. (1996). Immunogenetic factors in HPV-associated cervical cancer: influence on disease progression. *Eur. J. Immunogenet.* 23:275–284.

Durst, M., Glitz, D., Schneider, A., and zur, H.H. (1992). Human papillomavirus type 16 (HPV 16) gene expression and DNA replication in cervical neoplasia: analysis by *in situ* hybridization. *Virology* 189:132–140.

El Sherif, A.M., Seth, R., Tighe, P.J., and Jenkins, D. (2001). Quantitative analysis of IL-10 and IFN-gamma mRNA levels in normal cervix and human papillomavirus type 16 associated cervical precancer. *J. Pathol.* 195:179–185.

Epstein, M.A. (2001). Reflections on Epstein-Barr virus: some recently resolved old uncertainties. *J. Infect.* 43:111–115.

Evans, E.M., Man, S., Evans, A.S., and Borysiewicz, L.K. (1997). Infiltration of cervical cancer tissue with human papillomavirus-specific cytotoxic T-lymphocytes. *Cancer Res.* 57:2943–2950.

Fausch, S.C., Fahey, L.M., DaSilva D.M., and Kast, W.M. (2005). HPV can escape immune recognition through Langerhans cell PI3-kinase activation, *J. Immunol.* 174:7172–7178.

Fausch, S.C., da Silva, D.M., and Kast, W.M. (2003). Differential uptake and cross-presentation of human papillomavirus virus-like particles by dendritic cells and Langerhans cells. *Cancer Res.* 63:3478–3482.

Fausch, S.C., da Silva, D.M., Rudolf, M.P., and Kast, W.M. (2002). Human papillomavirus virus-like particles do not activate Langerhans cells: a possible immune escape mechanism used by human papillomaviruses. *J. Immunol.* 169:3242–3249.

Frazer, I.H. (1996). Immunology of papillomavirus infection. *Curr. Opin. Immunol.* 8:484–491.

Frisch, M., Biggar, R.J., and Goedert, J.J. (2000). Human papillomavirus-associated cancers in patients with human immuno deficiency virus infection and acquired immuno deficiency syndrome. *J. Natl. Cancer Inst.* 92:1500–1510.

Geisbill, J., Osmers, U., and Durst, M. (1997). Detection and characterization of human papillomavirus type 45 DNA in the cervical carcinoma cell line MS751. *J. Gen. Virol.* 78:655–658.

Gemignani, M., Maiman, M., Fruchter, R.G., Arrastia, C.D., Gibbon, D., and Ellison, T. (1995). CD4 lymphocytes in women with invasive and preinvasive cervical neoplasia. *Gynecol. Oncol.* 59:364–369.

Ghosh, A.K. and Moore, M. (1992). Tumour-infiltrating lymphocytes in cervical carcinoma. *Eur. J. Cancer* 28A:1910–1916.

Glas, R., Franksson, L., Une, C., Eloranta, M.L., Ohlen, C., Orn, A., and Karre, K. (2000). Recruitment and activation of natural killer (NK) cells *in vivo* determined by the target cell phenotype. An adaptive component of NK cell-mediated responses. *J. Exp. Med.* 191:129–138.

Glew, S.S., Duggan-Keen, M., Cabrera, T., and Stern, P.L. (1992). HLA class II antigen expression in human papillomavirus-associated cervical cancer. *Cancer Res.* 52:4009–4016.

Glew, S.S., Duggan-Keen, M., Ghosh, A.K., Ivinson, A., Sinnott, P., Davidson, J., Dyer, P.A., and Stern, P.L. (1993). Lack of association of HLA polymorphisms with human papillomavirus-related cervical cancer. *Hum. Immunol.* 37:157–164.

Gunn, G.R., Zubair, A., Peters, C., Pan, Z.K., Wu, T.C., and Paterson, Y. (2001). Two Listeria monocytogenes vaccine vectors that express different molecular forms of human papilloma virus-16 (HPV-16) E7 induce qualitatively different T cell immunity that correlates with their ability to induce regression of established tumors immortalized by HPV-16. *J. Immunol.* 167:6471–6479.

Halpert, R., Fruchter, R.G., Sedlis, A., Butt, K., Boyce, J.G., and Sillman, F.H. (1986). Human papillomavirus and lower genital neoplasia in renal transplant patients. *Obstet. Gynecol.* 68:251–258.

Hazelbag, S., Fleuren, G.J., Baelde, J.J., Schuuring, E., Kenter, G.G., and Gorter, A. (2001). Cytokine profile of cervical cancer cells. *Gynecol. Oncol.* 83:235–243.

Helland, A., Borresen, A.L., Kristensen, G., and Ronningen, K.S. (1994). DQA1 and DQB1 genes in patients with squamous cell carcinoma of the cervix: relationship to human papillomavirus infection and prognosis. *Cancer Epidemiol. Biomarkers Prev.* 3:479–486.

Hildesheim, A., Schiffman, M.H., Tsukui, T., Swanson, C.A., Lucci, J., III, Scott, D.R., Glass, A.G., Rush, B.B., Lorincz, A.T., Corrigan, A., Burk, R.D., Helgesen, K., Houghten, R.A., Sherman, M.E., Kurman, R.J., Berzofsky, J.A., and Kramer, T.R. (1997). Immune activation in cervical neoplasia: cross-sectional association between plasma soluble interleukin 2 receptor levels and disease. *Cancer Epidemiol. Biomarkers Prev.* 6:807–813.

Ho, G.Y., Bierman, R., Beardsley, L., Chang, C.J., and Burk, R.D. (1998). Natural history of cervicovaginal papillomavirus infection in young women. *N. Engl. J. Med.* 338:423–428.

Hohn, H., Pilch, H., Gunzel, S., Neukirch, C., ilmes, C., Kaufmann, A., Seliger, B., and Maeurer, M.J. (1999). CD4+ tumor-infiltrating lymphocytes in cervical cancer recognize HLA-DR-restricted peptides provided by human papillomavirus-E7. *J. Immunol.* 163:5715–5722.

Hopfl, R., Heim, K., Christensen, N., Zumbach, K., Wieland, U., Volgger, B., Widschwendter, A., Haimbuchner, S., Muller-Holzner, E., Pawlita, M., Pfister, H., and Fritsch, P. (2000). Spontaneous regression of CIN and delayed-type hypersensitivity to HPV-16 oncoprotein E7. *Lancet* 356:1985–1986.

Huang, M., Wang, J., Lee, P., Sharma, S., Mao, J.T., Meissner, H., Uyemura, K., Modlin, R., Wollman, J., and Dubinett, S.M. (1995). Human non-small cell lung cancer cells express a type 2 cytokine pattern. *Cancer Res.* 55:3847–3853.

Isomaki, P., Panesar, M., Annenkov, A., Clark, J.M., Foxwell, B.M., Chernajovsky, Y., and Cope, A.P. (2001). Prolonged exposure of T cells to TNF down-regulates TCR zeta and expression of the TCR/CD3 complex at the cell surface. *J. Immunol.* 166:5495–5507.

Kadish, A.S., Ho, G.Y., Burk, R.D., Wang, Y., Romney, S.L., Ledwidge, R., and Angeletti, R.H. (1997). Lymphoproliferative responses to human papillomavirus (HPV) type 16 proteins E6 and E7: outcome of HPV infection and associated neoplasia. *J. Natl. Cancer Inst.* 89:1285–1293.

Kadish, A.S., Timmins, P., Wang, Y., Ho, G.Y., Burk, R.D., Ketz, J., He, W., Romney, S.L., Johnson, A., Angeletti, R., and Abadi, M. (2002). Regression of cervical intraepithelial neoplasia and loss of human papillomavirus (HPV) infection is associated with cell-mediated immune responses to an HPV type 16 E7 peptide. *Cancer Epidemiol. Biomarkers Prev.* 11:483–488.

Kashii, Y., Giorda, R., Herberman, R.B., Whiteside, T.L., and Vujanovic, N.L. (1999). Constitutive expression and role of the TNF family ligands in apoptotic killing of tumor cells by human NK cells. *J. Immunol.* 163:5358–5366.

Keating, P.J., Cromme, F.V., Duggan-Keen, M., Snijders, P.J., Walboomers, J.M., Hunter, R.D., Dyer, P.A., and Stern, P.L. (1995). Frequency of down-regulation of individual HLA-A and -B alleles in cervical carcinomas in relation to TAP-1 expression. *Br. J. Cancer* 72:405–411.

Kim, J., Modlin, R.L., Moy, R.L., Dubinett, S.M., McHugh, T., Nickoloff, B.J., and Uyemura, K. (1995). IL-10 production in cutaneous basal and squamous cell carcinomas. A mechanism for evading the local T cell immune response. *J. Immunol.* 155:2240–2247.

Kitani, A., Fuss, I.J., Nakamura, K., Schwartz, O.M., Usui, T., Strober, W., (2000) Treatment of experimental (Trinitrobenzene sulfonic acid) colitis by intranasal administration of transforming growth factor (TGF)-betal plasmid: TGF-betal-mediated suppression of T helper cell type 1 response occurs by interleukin (IL)-10 induction and IL-12 receptor beta2 chain downregulation. *J Exp Med.* 192:41–52.

Klein, G. (2002). Perspectives in studies of human tumor viruses. *Front Biosci.* 7:d268–d274.

Kono, K., Ressing, M.E., Brandt, R.M., Melief, C.J., Potkul, R.K., Andersson, B., Petersson, M., Kast, W.M., and Kiessling, R. (1996a). Decreased expression of signal-transducing zeta chain in peripheral T cells and natural killer cells in patients with cervical cancer. *Clin. Cancer Res.* 2:1825–1828.

Kono, K., Salazar-Onfray, F., Petersson, M., Hansson, J., Masucci, G., Wasserman, K., Nakazawa, T., Anderson, P., and Kiessling, R. (1996b). Hydrogen peroxide secreted by tumor-derived macrophages down-modulates signal-transducing zeta molecules and inhibits tumor-specific T cell-and natural killer cell-mediated cytotoxicity. *Eur. J. Immunol.* 26:1308–1313.

Konya, J., and Dillner, J. (2001). Immunity to oncogenic human papillomaviruses. *Adv. Cancer Res.* 82:205–238.

Le Buanec, H., Lachgar, A., D'Anna, R., Zagury, J.F., Bizzini, B., Bernard, J., Ittele, D., Hallez, S., Giannouli, C., Burny, A., and Zagury, D. (1999). Induction of cellular immunosuppression by the human papillomavirus type 16 E7 oncogenic protein. *Biomed. Pharmacother.* 53:323–328.

Lee, S.J., Cho, Y.S., Cho, M.C., Shim, J.H., Lee, K.A., Ko, K.K., Choe, Y.K., Park, S.N., Hoshino, T., Kim, S., Dinarello, C.A., and Yoon, D.Y. (2001). Both E6 and E7 oncoproteins of human papillomavirus 16 inhibit IL-18-induced IFN-gamma production in human peripheral blood mononuclear and NK cells. *J. Immunol.* 167:497–504.

Li, S., Labrecque, S., Gauzzi, M.C., Cuddihy, A.R., Wong, A.H., Pellegrini, S., Matlashewski, G.J., and Koromilas, A.E. (1999). The human papilloma virus (HPV)-18 E6 oncoprotein physically associates with Tyk2 and impairs Jak-STAT activation by interferon-alpha. *Oncogene* 18:5727–5737.

Luxton, J.C., Rowe, A.J., Cridland, J.C., Coletart, T., Wilson, P., and Shepherd, P.S. (1996). Proliferative T cell responses to the human papillomavirus type 16 E7 protein in women with cervical dysplasia and cervical carcinoma and in healthy individuals. *J. Gen. Virol.* 77:1585–1593.

Macagno, A., Gilliet, M., Sallusto, F., Lanzavecchia, A., Nestle, F.O., and Groettrup, M. (1999). Dendritic cells up-regulate immunoproteasomes and the proteasome regulator PA28 during maturation. *Eur. J. Immunol.* 29:4037–4042.

Mantovani, F. and Banks, L. (2001). The human papillomavirus E6 protein and its contribution to malignant progression. *Oncogene* 20:7874–7887.

Matloubian, M., Concepcion, R.J., and Ahmed, R. (1994). CD4+ T cells are required to sustain CD8+ cytotoxic T-cell responses during chronic viral infection. *J. Virol.* 68:8056–8063.

Matsumoto, M., Funami, K., Oshiumi, H, and Seya, T. (2004). Toll-like receptor 3: a link between toll-like receptor, interferon and viruses. *Microbiol. Immunol.* 48:147–154.

Matsumoto, M., Kikkawa, S., Kohase, M., Miyake, K., and Seya, T., (2002). Establishment of a monoclonal antibody against human Toll-like receptor 3 that blocks double-stranded RNA-mediated signaling. *Biochem. Biophys. Res. Commun.* 293:1364–1369.

Matthews, K., Leong, C.M., Baxter, L., Inglis, E., Yun, K., Backstrom, B.T., Doorbar, J., and Hibma, M., (2003). Depletion of Langerhans cells in human papillomavirus type 16-infected skin is associated with E6-mediated down regulation of E-cadherin. *J. Virol.* 77:8378–8385.

Maudsley, D.J. (1991). Role of oncogenes in the regulation of MHC antigen expression. *Biochem. Soc. Trans.* 19:291–296.

Montoya, L., Saiz, I., Rey, G., Vela, F., and Clerici-Larradet, N., (1998). Cervical carcinoma: human papillomavirus infection and HLA-associated risk factors in the Spanish population. *Eur. J. Immunogenet.* 25:329–337.

Morel, S., Levy, F., Burlet-Schiltz, O., Brasseur, F., Probst-Kepper, M., Peitrequin, A.L., Monsarrat, B., Van Velthoven, R., Cerottini, J.C., Boon, T., Gairin, J.E., and Van den Eynde, B.J. (2000). Processing of some antigens by the standard proteasome but not by the immunoproteasome results in poor presentation by dendritic cells. *Immunity* 12:107–117.

Mosmann, T.R., Li, L., and Sad, S., (1997). Functions of CD8 T-cell subsets secreting different cytokine patterns. *Semin. Immunol.* 9:87–92.

Munger, K., Basile, J.R., Duensing, S., Eichten, A., Gonzalez, S.L., Grace, M., and Zacny, V.L. (2001). Biological activities and molecular targets of the human papillomavirus E7 oncoprotein. *Oncogene* 20:7888–7898.

Munoz, N., Bosch, F.X., de Sanjose, S., Herrero, R., Castellsague, X., Shah, K.V., Snijders, P.J., and Meijer, C.J. (2003). Epidemiologic classification of human papillomavirus types associated with cervical cancer. *N. Engl. J. Med.* 348:518–527.

Nakagawa, M., Stites, D.P., Patel, S., Farhat, S., Scott, M., Hills, N.K., Palefsky, J.M., and Moscicki, A.B. (2000). Persistence of human papillomavirus type 16 infection is associated with lack of cytotoxic T lymphocyte response to the E6 antigens. *J. Infect. Dis.* 182:595–598.

Nakagomi, H., Pisa, P., Pisa, E.K., Yamamoto, Y., Halapi, E., Backlin, K., Juhlin, C., and Kiessling. R., (1995). Lack of interleukin-2 (IL-2) expression and selective expression of IL-10 mRNA in human renal cell carcinoma. *Int. J. Cancer* 63:366–371.

Nees, M., Geoghegan, J.M., Hyman, T., Frank, S., Miller, L., and Woodworth, C.D. (2001). Papillomavirus type 16 oncogenes downregulate expression of interferon-responsive genes and upregulate proliferation-associated and NF-kappaB-responsive genes in cervical keratinocytes. *J. Virol.* 75:4283–4296.

Nieland, J.D., Loviscek, K., Kono, K., Albain, K.S., McCall, A.R., Potkul, R.K., Fisher, S.G., Velders, M.P., Petersson, M., Kiessling, R., and Kast, W.M. (1998). PBLs of early breast carcinoma patients with a high nuclear grade tumor unlike PBLs of cervical carcinoma patients do not show a decreased TCR zeta expression but are functionally impaired. *J. Immunother.* 21:317–322.

Odunsi, K., Terry, G., Ho, L., Bell, J., Cuzick, J., and Ganesan, T.S. (1995). Association between HLA DQB1 * 03 and cervical intra-epithelial neoplasia. *Mol. Med.* 1:161–171.

Onizuka, S., Tawara, I., Shimizu, J., Sakaguchi, S., Fujita, T., and Nakayama, E., (1999). Tumor rejection by *in vivo* administration of anti-CD25 (interleukin-2 receptor alpha) monoclonal antibody. *Cancer Res.* 59:3128–3133.

Park, J.S., Kim, E.J., Kwon, H.J., Hwang, E.S., Namkoong, S.E., and Um, S.J. (2000). Inactivation of interferon regulatory factor-1 tumor suppressor protein by HPV E7 oncoprotein. Implication for the E7-mediated immune evasion mechanism in cervical carcinogenesis. *J. Biol. Chem.* 275:6764–6769.

Pei, X.F. (1996). The human papillomavirus E6/E7 genes induce discordant changes in the expression of cell growth regulatory proteins. *Carcinogenesis* 17:1395–1401.

Petry, K.U., Scheffel, D., Bode, U., Gabrysiak, T., Kochel, H., Kupsch, E., Glaubitz, M., Niesert, S., Kuhnle, H., and Schedel, I. (1994). Cellular immunodeficiency enhances the progression of human papillomavirus-associated cervical lesions. *Int. J. Cancer* 57:836–840.

Read, S., Malmstrom, V., and Powrie, F. (2000). Cytotoxic T lymphocyte-associated antigen 4 plays an essential role in the function of CD25(+)CD4(+) regulatory cells that control intestinal inflammation. *J. Exp. Med.* 192:295–302.

Remmink, A.J., Walboomers, J.M., Helmerhorst, T.J., Voorhorst, F.J., Rozendaal, L., Risse, E.K., Meijer, C.J., and Kenemans, P. (1995). The presence of persistent high-risk HPV genotypes in dysplastic cervical lesions is associated with progressive disease: natural history up to 36 months. *Int. J. Cancer* 61:306–311.

Ressing, M.E., van Driel, W.J., Celis, E., Sette, A., Brandt, M.P., Hartman, M., Anholts, J.D., Schreuder, G.M., ter Harmsel, W.B., Fleuren, G.J., Trimbos, B.J., Kast, W.M., and Melief, C.J. (1996). Occasional memory cytotoxic T-cell responses of patients with human papillomavirus type 16-positive cervical lesions against a human leukocyte antigen-A *0201-restricted E7-encoded epitope. *Cancer Res.* 56:582–588.

Rockley, P.F., and Tyring, S.K. (1995). Interferons alpha, beta and gamma therapy of anogenital human papillomavirus infections. *Pharmacol. Ther.* 65:265–287.

Ronco, L.V., Karpova, A.Y., Vidal, M., and Howley, P.M. (1998). Human papillomavirus 16 E6 oncoprotein binds to interferon regulatory factor-3 and inhibits its transcriptional activity. *Genes Dev.* 12:2061–2072.

Sad, S., Marcotte, R., and Mosmann, T.R. (1995). Cytokine-induced differentiation of precursor mouse CD8+ T cells into cytotoxic CD8+ T cells secreting Th1 or Th2 cytokines. *Immunity* 2:271–279.

Sastre-Garau, X., Loste, M.N., Vincent-Salomon, A., Favre, M., Mouret, E., de la, R.A., Durand, J.C., Tartour, E., Lepage, V., and Charron, D. (1996). Decreased frequency of HLA-DRB1 13 alleles in Frenchwomen with HPV-positive carcinoma of the cervix. *Int. J. Cancer* 69:159–164.

Schiffman, M.H., Bauer, H.M., Hoover, R.N., Glass, A.G., Cadell, D.M., Rush, B.B., Scott, D.R., Sherman, M.E., Kurman, R.J., and Wacholder, S., (1993). Epidemiologic evidence showing that human papillomavirus infection causes most cervical intraepithelial neoplasia. *J. Natl. Cancer Inst.* 85:958–964.

Schlecht, N.F., Platt, R.W., Negassa, A., Duarte-Franco, E., Rohan, T.E., Ferenczy, A., Villa, L.L., and Franco, E.L. (2003). Modeling the time dependence of the association between human papillomavirus infection and cervical cancer precursor lesions. *Am. J. Epidemiol.* 158:878–886.

Schmielau, J., and Finn, O.J. (2001). Activated granulocytes and granulocyte-derived hydrogen peroxide are the underlying mechanism of suppression of t-cell function in advanced cancer patients. *Cancer Res.* 61:4756–4760.

Schoenberger, S.P., Toes, R.E., van der Voort, E.I., Offringa, R., and Melief, C.J. (1998). T-cell help for cytotoxic T lymphocytes is mediated by CD40–CD40L interactions. *Nature* 393:480–483.

Seder, R.A., Boulay, J.L., Finkelman, F., Barbier, S., Ben Sasson, S.Z., Le Gros, G., and Paul, W.E. (1992). CD8+ T cells can be primed in vitro to produce IL-4. *J. Immunol.* 148:1652–1656.

Sheu, B.C., Lin, R.H., Lien, H.C., Ho, H.N., Hsu, S.M., and Huang, S.C. (2001b). Predominant Th2/Tc2 polarity of tumor-infiltrating lymphocytes in human cervical cancer. *J. Immunol.* 167:2972–2978.

Sheu, B.C., Lin, R.H., Lien, H.C., Ho, H.N., Hsu, S.M., and Huang, S.C. (2001a). Predominant Th2/Tc2 polarity of tumor-infiltrating lymphocytes in human cervical cancer. *J. Immunol.* 167:2972–2978.

Steinbrink, K., Jonuleit, H., Muller, G., Schuler, G., Knop, J., and Enk, A.H. (1999). Interleukin-10-treated human dendritic cells induce a melanoma-antigen-specific anergy in CD8(+) T cells resulting in a failure to lyse tumor cells. *Blood* 93:1634–1642.

Stellato, G., Paavonen, J., Nieminen, P., Hibma, M., Vilja, P., and Lehtinen. M. (1997). Diagnostic phase antibody response to the human papillomavirus type 16 E2 protein is associated with successful treatment of genital HPV lesions with systemic interferon alpha-2b. *Clin. Diagn. Virol.* 7:167–172.

Stern, P.L. (1996). Immunity to human papillomavirus-associated cervical neoplasia. *Adv. Cancer Res.* 69:175–211.

Stohlman, S.A., Bergmann, C.C., Lin, M.T., Cua, D.J., and Hinton, D.R. (1998). CTL effector function within the central nervous system requires CD4+ T cells. *J. Immunol.* 160:2896–2904.

Stone, K.M. (1995). Human papillomavirus infection and genital warts: update on epidemiology and treatment. *Clin. Infect. Dis.* 20(Suppl. 1):S91–S97.

Strang, G., Hickling, J.K., McIndoe, G.A., Howland, K., Wilkinson, D., Ikeda, H., and Rothbard, J.B. (1990). Human T cell responses to human papillomavirus type 16 L1 and E6 synthetic peptides: identification of T cell determinants, HLA-DR restriction and virus type specificity. *J. Gen. Virol.* 71:423–431.

Tay, S.K., Jenkins, D., Maddox, P., Campion, M., and Singer, A. (1987). Subpopulations of Langerhans' cells in cervical neoplasia. *Br. J. Obstet. Gynaecol.* 94:10–15.

Thornton, A.M. and Shevach, E.M. (1998). CD4 + CD25+ immunoregulatory T cells suppress polyclonal T cell activation *in vitro* by inhibiting interleukin 2 production. *J. Exp. Med.* 188:287–296.

Torre-Amione, G., Beauchamp, R.D., Koeppen, H., Park, B.H., Schreiber, H., Moses, H.L., and Rowley, D.A. (1990). A highly immunogenic tumor transfected with a murine transforming growth factor type beta 1 cDNA escapes immune surveillance. *Proc. Natl. Acad. Sci. U S A* 87:1486–1490.

Tsukui,T., Hildesheim, A., Schiffman, M.H., Lucci, J., III, Contois, D., Lawler, P., Rush, B.B., Lorincz, A.T., Corrigan, A., Burk, R.D., Qu, W., Marshall, M.A., Mann, D., Carrington, M., Clerici, M., Shearer, G.M., Carbone, D.P., Scott, D.R., Houghten, R.A., and Berzofsky, J.A. (1996). Interleukin 2 production in vitro by peripheral lymphocytes in response to human papillomavirus-derived peptides: correlation with cervical pathology. *Cancer Res.* 56:3967–3974.

Vambutas, A., Bonagura, V.R., and Steinberg, B.M. (2000). Altered expression of TAP-1 and major histocompatibility complex class I in laryngeal papillomatosis: correlation of TAP-1 with disease. *Clin. Diagn. Lab Immunol.* 7:79–85.

van der Burg, S.H., Ressing, M.E., Kwappenberg, K.M., de Jong, A., Straathof, K., de Jong, J., Geluk, A., van Meijgaarden, K.E., Franken, K.L., Ottenhoff, T.H., Fleuren, G.J., Kenter, G., Melief, C.J., and Offringa, R. (2001). Natural T-helper immunity against human papillomavirus type 16 (HPV16) E7-derived peptide epitopes in patients with HPV16-positive cervical lesions: identification of 3 human leukocyte antigen class II-restricted epitopes. *Int. J. Cancer* 91:612–618.

Vandenvelde, C., De Foor, M., and Van Beers, D. (1993). Precision about the association between cervical carcinoma and HLA-DQB1*03 alleles. *Lancet* 342:553.

Velders, M.P., Markiewicz, M.A., Eiben, G.L., and Kast, W.M. (2003). CD4+ T cell matters in tumor immunity. *Int. Rev. Immunol.* 22:113–140.

Viscidi, R.P., Sun, Y., Tsuzaki, B., Bosch, F.X., Munoz, N., and Shah, K.V. (1993). Serologic response in human papillomavirus-associated invasive cervical cancer. *Int. J. Cancer* 55:780–784.

von Herrath, M.G., Yokoyama, M., Dockter, J., Oldstone, M.B., and Whitton, J.L. (1996). CD4-deficient mice have reduced levels of memory cytotoxic T lymphocytes after immunization and show diminished resistance to subsequent virus challenge. *J. Virol.* 70:1072–1079.

Walboomers, J.M., Jacobs, M.V., Manos, M.M., Bosch, F.X., Kummer, J.A., Shah, K.V., Snijders, P.J., Peto, J., Meijer, C.J., and Munoz, N. (1999). Human papillomavirus is a necessary cause of invasive cervical cancer worldwide. *J. Pathol.* 189:12–19.

Wang, R.F. (2001). The role of MHC class II-restricted tumor antigens and CD4+ T cells in antitumor immunity. *Trends Immunol.* 22:269–276.

Wank, R. and Thomssen, C. (1991). High risk of squamous cell carcinoma of the cervix for women with HLA-DQw3. *Nature* 352:723–725.

Warrino, D.E., Olson, W.C., Knapp, W.T., Scarrow, M.I., Brennan, L.J., Guido, R.S., Edwards, R.P., Kast, W.M. and Storkus, W.J. (2004). Disease-stage variance in functional CD4+ T cell responses against novel Pan-HLA-DR presented HPV-16 E7 epitopes. *Clin. Cancer Res.* 10:3301–3308.

Welters, M.J., de Jong, A., van den Eeden, S.J., van der Hulst, J.M., Kwappenberg, K.M., Hassane, S., Franken, K.L., Drijfhout, J.W., Fleuren, G.J., Kenter, G., Melief, C.J., Offringa, R., and van der Burg, S.H. (2003). Frequent display of human papillomavirus type 16 E6-specific memory t-Helper cells in the healthy population as witness of previous viral encounter. *Cancer Res.* 63:636–641.

Whiteside, T.L. (1999). Signaling defects in T lymphocytes of patients with malignancy. *Cancer Immunol. Immunother.* 48:346–352.

Woodworth, C.D., Lichti, U., Simpson, S., Evans, C.H., and DiPaolo, J.A. (1992). Leukoregulin and gamma-interferon inhibit human papillomavirus type 16 gene transcription in human papillomavirus-immortalized human cervical cells. *Cancer Res.* 52:456–463.

World Health Organization (2004). Incidence of HPV 2004. who.int/vaccines Immunization, Vaccine and Biologicals.

Youde, S.J., Dunbar, P.R., Evans, E.M., Fiander, A.N., Borysiewicz, L.K., Cerundolo, V., and Man, S. (2000). Use of fluorogenic histocompatibility leukocyte antigen-A*0201/HPV 16 E7 peptide complexes to isolate rare human cytotoxic T-lymphocyte-recognizing endogenous human papillomavirus antigens. *Cancer Res.* 60:365–371.

14
Papillomavirus Vaccines

John Schiller

National Cancer Institute

14.1.   Introduction

Research on papillomavirus vaccines has steadily intensified over the last 10–15 years. This activity is a direct outgrowth of the enormous gains in our understanding of papillomavirus biology and the increased appreciation of HPVs as human tumor viruses, coupled with major advances in basic immunology and molecular biology. Most significantly, vaccination studies have increasingly moved from animal models to clinic trials. Vaccines to prevent HPV infections (prophylactic vaccines) and treat HPV-induced neoplasia (therapeutic vaccines) are actively being pursued. Like successful prophylactic vaccines for other virus infections (Ehreth, 2003; Zinkernagel, 2003), the lead prophylactic HPV candidates function primarily by inducing virion-neutralizing antibodies. In contrast, HPV therapeutic vaccines are expected to function primarily through T-cell-mediated effector mechanisms directed against HPV nonstructural proteins expressed in pre-cancer or cancer cells (Stern et al., 2000). Therapeutic vaccines against other viral infections and cancers have had limited success, despite extensive efforts (Mocellin et al., 2004; Moingeon, 2003). However, major advances in the understanding of the activation and regulation of cell-mediated immunity (CMI) suggest that we may be on the verge of important breakthroughs in therapeutic vaccine development. This chapter reviews the virologic and immunologic basis for the development of therapeutic and prophylactic HPV vaccines, with particular emphasis on clinical studies, as well as strengths and weaknesses of the current animal models for evaluating papillomavirus vaccination strategies. Finally, important public-health-related issues raised by the vaccines will be discussed.

14.2.   Therapeutic Vaccines

14.2.1.   Introduction

Therapeutic HPV vaccines have mostly targeted cervical neoplasia, in large measure because cervical cancer accounts for most cases of HPV-associated cancers (Bosch and de Sanjose, 2003; Gillison and Shah, 2003). There are several features of premalignant cervical neoplasia that make it an attractive disease for therapeutic vaccine intervention. First, cervical cancers progress from a series of well-defined precursors, and the effects of immunotherapeutic intervention can be routinely monitored by noninvasive methods such as colposcopy, cytology, and HPV DNA testing. Thus, immunotherapeutic approaches can be evaluated at multiple stages of carcinogenic progression (Schiffman and Kjaer, 2003). Second, dysplasias and early-stage cancers are routinely identified in Pap screening programs, providing a ready pool of potential candidates for clinical trials and, ultimately, identifying at-risk individuals for therapeutic vaccination. Third, HPV infections typically take many years to progress to malignant cancer, providing a long window of opportunity for therapeutic intervention. Fourth, participants in dysplasia trials need not be placed at undue risk of progression to cancer after the trials are completed, because effective treatment of premalignant lesions exists. Finally, the E6 and E7 viral oncogenes are uniformly retained and expressed during carcinogenic progression and are required for maintenance of the transformed phenotype (zur Hausen, 2002). Therefore, lesions invariably express viral neo-antigens and may not be able to evade immune responses by down regulation or loss of these proteins. Because E6 and E7 are nonself, they may be less subject to mechanisms of immune tolerance than cell-derived tumor antigens, and so are attractive target antigens for immunotherapeutic vaccines. Based on these features, cervical neoplasia may well be the most amenable clinical model for establishing general principles of tumor immunotherapy, in addition to being an important disease in its own right.

What stage of neoplastic disease should be targeted by therapeutic HPV vaccination? The answer to this question is complex, requiring public health, commercial, and likelihood-of-success considerations. The most important public health goal is to prevent deaths due to HPV-induced cancers. However, as discussed below, cancers will almost surely be the most difficult stage of the HPV disease spectrum for which to develop effective immunotherapy. Several of the early clinical trials of HPV immunotherapies were safety and immunogenicity studies in cervical cancer patients. Clinical responses were not observed, and T-cell immune responses to vaccination were not always detected (Borysiewicz et al., 1996; Kaufmann et al., 2002; Steller et al., 1998). These results are consistent with the limited success of therapeutic vaccines for other human solid tumors. There are also nonbiological barriers to development of therapeutic vaccines against cervical cancer. In comparison to other diseases that have attracted intense interest from the pharmaceutical industry in recent years, cervical and other HPV-associated cancers are not particularly common in the developed world. For example, there are approximately 10,000 cervical cancer cases annually in the United States, compared with an estimated 300,000 high-grade cervical dysplasias, 2,000,000 low-grade dysplasias, and 10,000,000 subclinical genital HPV infections (ACOG Bulletin, 2005). Based on numbers alone, premalignant HPV diseases would be more attractive targets for commercialization than cancer. However, unless they targeted cancer, therapeutic vaccines would likely have little effect in developing countries, where

80 percent of the cervical cancer cases occur (Parkin et al., 2001), because few women in these countries have access to quality Pap screening programs to identify premalignant lesions for treatment. Dysplasia-specific vaccines could only be effective in these settings if they were widely distributed to the general public and provided long-term protection.

Most recent therapeutic HPV vaccines trials have targeted the relatively common intermediate stages of cervical disease, CIN2 and CIN3. Unlike primary infections, CIN 3 is a treatable disease, with ablative therapies being quite effective and well tolerated (Cox, 2002b). Thus, the bar for development of alternative treatments is rather high for CIN3. Because decisions on whether to follow or treat CIN2 can be difficult, especially in young women (Cox, 2002a), it would be useful to have a safe and effective vaccine as an alternative management approach to surgical intervention. Many vaccine trials have included both CIN2 and CIN3 patients, which can complicate interpretation of the results since these lesions encompass a broad biological spectrum of HPV-induced neoplasia, from relatively recent low-grade infections to long-standing aneuploid carcinomas in situ (Schiffman and Kjaer, 2003). One might expect better responses for more benign lesions. In most of these trials involving CIN3 or CIN2/3 patients, the potential for progression to cancer has limited the observation period following vaccination to 6 months or less followed by ablative therapy, so recurrence rates or other longer term effects of vaccination on cervical disease have not been investigated.

In addition to having the highest incidence, primary HPV infection and its cytological and histological manifestations LSIL and CIN 1, respectively, are attractive vaccine targets for three reasons. First, they express the full spectrum of HPV genes (Longworth and Laimins, 2004), and therefore provide the most targets for immunotherapy. For example, in the basal layers of the epithelium where the viral infection is thought to be maintained, E1 and E2 are expressed, and so provide additional therapeutic vaccine targets. Second, these lesions are genetically stable and therefore are least likely to have evolved to evade immune recognition (Koopman et al., 2000). Third, these lesions have the shortest duration, so they are least likely to have induced tolerance to the viral antigens (Matsumoto et al., 2004). For these reasons, immunotherapy of primary HPV infections is the most likely to be successful. However, primary infections also have significant liabilities as targets for immunotherapies. Most problematically, they usually regress spontaneously over a period of months and are not considered a treatable disease (Winer et al., 2005). To be attractive to the public, vaccines targeting these lesions would have to have high efficacy and very modest side effects. The expected increase in the use of HPV DNA testing as a screen for cervical cancer risk could lead to increasing numbers of women with knowledge of their genital HPV infection status, which could very well increase demand for a nonsurgical treatment that safely and rapidly eliminates these infections. Recognition of this demand might in turn increase industry efforts to develop vaccines to treat primary infections. The public health benefit of this type of vaccine would be debatable, since most treated women would likely undergo spontaneous regression in the absence of treatment. However, there might be some benefit by preventing cancer in the minority of women who would develop persistent infection and be lost to follow-up. There might also be some inhibition of viral transmission, if the vaccine rapidly induced a virologic cure.

There are several potential noncervical targets for HPV vaccines. It has been estimated that high-grade anal dysplasia (AIN 3) is as prevalent in men who have sex with men as CIN 3 is in heterosexual women, and may have similar propensity for progression to cancer (Klencke and Palefsky, 2003). However, compared to CIN 3, ablative therapy has more potential for adverse side effects. Thus, AIN is a very attractive target for immunotherapy, and AIN vaccine trials have been initiated (Klencke et al., 2002). Given the biologic and virologic similarities of AIN and CIN, trials in one disease may well inform trials in the other, and a successful vaccine developed for one disease may be effective for the other. Persistent vulvar intraepithelial neoplasia (VIN), although rarely progressing to cancer, can cause considerable morbidity, and has also been targeted in at least two therapeutic vaccine trials (Davidson et al., 2003; Muderspach et al., 2000). A substantial number of oral cancers, particularly those of the tonsils and oropharynx, express high-risk HPV oncogenes (Syrjanen, 2005). However, the precursor lesions for these cancers have not been unambiguously identified and consequently there is no screening technology. Therefore immunotherapy of HPV-associated oral lesions would be limited to cancers.

Anogenital warts present a quite different disease for immunotherapeutic intervention. They are found on both cornyefied and non-cornyefied epidermis, are not subject to carcinogenic progression and are predominately caused by distinct viral types, predominantly HPV6 and HPV11. Anogenital warts have a rather high incidence, cause considerable morbidity and psychosocial distress, and often recur following currently available treatments (Wiley et al., 2002). Thus they are attractive targets for therapeutic vaccines, and clinical vaccine trials for anogenital warts have been initiated. Recurrent respiratory papillomatosis (RRP) is also primarily an HPV6/11-induced benign neoplasia and can be a debilitating disease with high recurrence rates after standard therapy, requiring frequent surgeries to prevent airway obstruction (Silverman and Pitman, 2004). However, RRP is a rare disease and is unlikely to be a prime focus of commercial vaccine development.

Common cutaneous and plantar warts, which are caused by a distinct set of HPVs, have received almost no attention as targets for immunotherapy (Rivera and Tyring, 2004). Although they spontaneously resolve and are not cancer-associated, these lesions are very common and cause significant morbidity in a subset of individuals. Successful development of immunotherapies against genital warts would hopefully spur interest in developing similar vaccines against cutaneous warts. Unlike the case for anogenital disease, there are reasonable animal models for evaluating therapeutic vaccination strategies against cutaneous papillomas (Brandsma, 1994). However, cutaneous warts may be caused by a variety of HPV types (Rubben et al., 1997), so the valency of a cutaneous wart vaccine may need to be high, unless cross protective.

14.2.2.   Lessons from Natural History Studies

It is clear that T-cell responses are important for control of HPV infections, since individuals with T-cell defects, such as AIDS and renal transplant patients, have increased rates of HPV persistence, premalignant anogenital lesions, and cervical cancer (Palefsky and Holly, 2003). It is less clear which viral antigens are critical for immune control of infection or elimination of virally induced lesions. Two recent studies found that T-cell responses to E6 and E2 are preferentially detected in women who have undergone regression of cervical lesions, or are simply currently uninfected, in comparison to women with persistent cervical lesions (de Jong et al., 2004; Nakagawa et al., 2000). However, responses to E7 did not differ significantly between the two groups. These findings suggest that immune responses to E6 and E2 may more often be involved in control of cervical infections, than are responses to E7. This conclusion is interesting considering that E7 has been the target antigen for the majority of therapeutic vaccination studies. E7 was initially chosen for several reasons unrelated to the natural immunity to infection. First, E7 appeared to be present at the highest steady-state levels in early studies of viral gene expression in cell lines derived from cervical cancer (Androphy et al., 1987). Second, E7 contains high-affinity peptides for the MHC class I and class II alleles of the mouse strain most commonly used in immunogenicity studies, C57Bl-6, as well as a high-affinity peptide for HLA-A2, a common human class I allele (Kast et al., 1993).

The immune effector mechanisms that are primarily responsible to viral clearance (or control) and lesion regression remain undefined. Both CD8- and CD4-positive tumor infiltrating lymphocytes have been detected in regressing lesions, but their relative contribution to regression is unknown (Stanley, 2001). Thus, the emphasis on induction of CD8+ cytotoxic T cells (CTLs) in most therapeutic vaccine studies is based more on findings in mouse immunotherapy models than on natural history data in humans.

14.2.3.   Animal Models for Therapeutic HPV Vaccines

Animal models for immunotherapy of HPV-induced neoplasias are less than satisfactory because HPV do not cause hyperproliferative diseases in animals and none of the established animal papillomavirus infection models involves anogenital neoplasia (Brandsma, 1994). Most animal studies of HPV therapeutic vaccines involve transplantable murine tumors that express HPV genes. Such studies are relatively easy to perform, and the cell inoculum can be titrated such that protection from tumor challenge or regression of pre-established tumors can be induced through vaccination. Subcutaneous injection of the murine TC-1 tumor line, which expresses HPV16 E6E7 in a H-2Db background, is the most commonly used model (Lin et al., 1996). An alternative C57Bl/6 tumor line is C3, which responds to HPV16 L1-specific as well as E6/E7 specific T cells (De Bruijn et al., 1998). Eiben et al. (2002) recently described another line, which presents E6/E7 peptides in the context of human HLA-A.2 class I molecules and forms tumors in A.2 transgenic mice (Eiben et al., 2002). Protection in these assays, in both tumor challenge after vaccination and vaccination after tumor establishment protocols, is primarily CD8 CTL-mediated. However, these models have little biological similarity to the ultimate vaccine target, i.e., HPV-infected cells proliferating slowly in situ as focal neoplasia in an intact epithelium. In addition, the models can have relatively low stringency. Many vaccination strategies have been shown to be effective, especially in the TC-1 model, implying that the ability of this assay to discriminate between the effectiveness of different vaccine strategies, beyond a minimal standard, may be low. Greater stringency can be achieved by increasing the tumor cell dose or size of the established tumor prior to vaccination. However, unlike human tumors, these transplanted tumors develop rapidly, so the models are not useful for evaluating strategies involving boosting over relatively long intervals.

A somewhat more biologically relevant model for evaluating HPV immunotherapies involves transplantation of the skin of transgenic mice expressing E6 and/or E7 in their epidermis onto syngeneic mice lacking the transgenes (Dunn et al., 1997). This model is particularly stringent in that grafted E7-transgenic skin is not rejected by mice vaccinated with E7 protein in adjuvant, even though the mice rejected E7-expressing transplantable tumors. Interestingly, the grafts are rejected after systemic administration of endotoxin or killed Listeria (Frazer et al., 2001). Thus it appears that induction of strong innate immune responses can induce regression. However, regression may ultimately involve adaptive, presumably E7-specific, immune responses, since mice that reject the grafts will also reject a second E7-expressing graft, even without further Listeria or endotoxin exposure. Adoptive transfer of E7-specific, T-cell-receptor transgenic CD8+ T cells combined with E7 protein immunization also led to graft rejection (Matsumoto et al., 2004), but this approach will clearly be difficult to translate to human use.

Cervical neoplasias arising in K14-E6E7 transgenic mice after estrogen treatment would seemingly provide an even more relevant model for assessing immunotherapies against HPV-induced genital tract disease (Arbeit et al., 1996), but this model has not been utilized as yet in therapeutic vaccine studies. It is likely that it would be a very stringent model for assessing tumor rejection, since the tumors arise in a situation in which the entire tissue is expressing the vaccine target. This expression profile might not only increase tolerance but also exhaust effector cells before a clinically apparent response to the tumor could be generated.

The two most widely employed animal models for examining immunotherapies for warts are cottontail rabbit papillomavirus (CRPV) infection of domestic rabbits and canine oral papillomavirus (COPV) infection of dogs. The COPV model is attractive because it involves mucosal lesions, but it has been difficult to study immunotherapies in this model because of the rapid rate of spontaneous regression (Nicholls et al., 2001; Peh et al., 2002). However, the studies that have been conducted suggest that E2 may be the most attractive viral protein target for inducing immune-mediated regression of canine warts (Moore et al., 2003). Most immunotherapeutic approaches have not consistently induced regression of established CRPV warts (Han et al., 2000a). Several studies have found E2 to be the most effective single target (Brandsma et al., 2004; Han et al., 2000b), but a principal lesson from these studies appears to be that inclusion of more antigen targets in the vaccine increases the frequency of regression.

14.2.4. Therapeutic Clinical Trials

Two broad classes of therapeutic vaccines have been tested in human trials: peptide/protein-based and genetic immunization based on transfer of viral nucleic acid sequences. Almost all the trials have targeted E7 or E6 plus E7, of HPV16 or HPV16 plus HPV18.

Peptide-based vaccines are attractive because of their ease of manufacture and the ease of monitoring the immune response, since it can be focused on a single or small number of determinants (Salit et al., 2002). However, peptide-based vaccines are HLA restricted, which would limit their utility in human populations. Because peptides usually are not inherently immunogenic, they require an inducer of innate immune responses, i.e., an adjuvant. This requirement can lead to nonspecific local and/or systemic inflammatory reactions. The acceptability of these types of side effects may depend upon the severity of the lesion targeted, and serious side effects will almost certainly be more acceptable in a cancer vaccine than in one that targets low grade dysplasias/productive infections. One of the earliest therapeutic trials employed a lipodated HPV16 HLA-A.2-restricted peptide vaccine in 12 A2-positive patients with refractory cervical cancer. Peptide-specific cytotoxic lymphocytes were induced in the majority of vaccinees, but there were no clinical responses (Steller et al., 1998). In another single arm study, women with CIN or VIN 2/3 were vaccinated with a different HPV16 E7 A.2-restricted peptide in combination with incomplete Freund's adjuvant (Muderspach et al., 2000). The majority of vaccinees mounted peptide-specific T-cell responses. Three of eighteen patients cleared their dysplasias and another six displayed partial regression.

Whole protein-based vaccines have the advantage of containing more peptides that may bind to diverse HLA haplotypes. One the first HPV protein-based vaccines to be developed was TA-CIN by Cantab (now Zenova). It consists of full-length HPV16 L2, E6, and E7 as a single fusion protein, which was administered to healthy volunteers without adjuvant (de Jong et al., 2002). Substantial T-cell responses were detected only in the group receiving three injections of the highest dose of antigen. Phase 1 clinical trials of two additional protein-based therapeutic vaccines were recently published. CSL tested an HPV16 E6E7 fusion protein mixed with their ISCOMATRIX adjuvant in 31 CIN patients (Frazer et al., 2004), and GSK tested a HPV16 E7-*Haemophilus influenzae* protein D fusion protein mixed with their AS02B adjuvant in seven CIN patients (Hallez et al., 2004). Both vaccines were safe and induced antigen-specific T-cell responses.

Protein vaccines involving fusions with certain proteins, such as heat shock proteins or papillomavirus-like particles (VLPs, discussed below), can have the intrinsic ability to activate innate immune responses (Barton and Medzhitov, 2002). Stressgen has tested a mycobacterial HSP65-HPV16 E7 protein for safety and immunogenicity in a number of uncontrolled phase 1 trials (Hunt, 2001). This vaccine was well tolerated after parenteral injection without adjuvant and, surprisingly, there was a preliminary suggestion that this vaccine may have induced regression of genital warts, which are usually caused by HPV6 or HPV11, not HPV16. HSP65 alone or HSP65 plus unlinked E7 did not induce regression of a transplantable mouse tumor, suggesting an E7-specific response was required in this model (Chu et al., 2000). However, the effector mechanisms that clear in situ lesions in people may be different. An uncontrolled phase 2 HspE7 trial demonstrated regression in 70 percent of CIN patients over a 2-month observation period, and a controlled phase 3 trial has begun (Maciag and Paterson, 2005). A Medigene sponsored trial of a chimeric HPV16 L1-E7 VLP vaccine induced good responses to L1 but relatively weak CMI responses to E7 after parenteral injection without adjuvant in CIN 2/3 patients. In this blinded and placebo-controlled trial, there was a trend toward increased clinical responses in the vaccinees compared to controls, but the difference was not significant (John Nieland, personal communication).

Genetic immunization involving transfer of wild-type E6 and E7 genes to humans is unacceptable because they are oncogenes with the potential to induce tumors if they were included in a vaccine. Three solutions to this problem have been tested, at least in animal models: 1) minigene constructs of MHC-specific epitopes (Klencke et al., 2002), 2) mutated E6 and E7 genes that lack p53 and pRb binding activities, respectively (Borysiewicz et al., 1996), and 3) gene shuffling strategies, in which the entire gene is rearranged as segments of overlapping peptides (Osen et al., 2001).

Genetic immunization strategies fall into two subclasses, direct DNA transfer and viral-vector-mediated gene transfer. Simple parenteral injection of naked DNA induces CMI reasonably well in rodents but appears to be considerably less effective in humans (Liu, 2003). Therefore, studies in humans utilize facilitated DNA transfer. A recent Zycos trial in CIN 2/3 involved DNA encapsulated in a biodegradable polymer microparticles, which expressed HPV16 and 18 E6 and E7 epitopes (Garcia et al., 2004). A clinical response was demonstrated in this placebo-controlled trial, although the response was significant only in the subgroup of younger women enrolled in the trial.

Several clinical trials of HPV vaccines involving vaccinia-based viral vectors have been conducted. A Cantab (now Zenova) sponsored trial involved expression of HPV16 and HPV18 E6/E7 from wild-type vaccinia is underway (Baldwin et al., 2003), and an attenuated Modifed Vaccinia Ankara (MVA) expressing mutated HPV16 E6/E7 plus IL-2 is undergoing clinical development by Transgene (Liu et al., 2004). The MVA vaccine demonstrated a good safety profile and perhaps some partial efficacy after subcutaneous injection in trials of CIN2/3. Intravaginal application of an HPV16 E2 expressing MVA to women with CIN 1–3 caused the lesions in most women to regress, but the trial was not placebo controlled (Corona Gutierrez et al., 2004). It is unclear whether the effect of this vaccine has an E2 antigen-specific component or is simply the result of the inflammatory response to infection by this nonreplicating pox virus.

Finally, there has been one trial employing a heterologous prime/boost strategy (Smyth et al., 2004). In some situations, this strategy may be more effective at boosting T-cell responses than homologous prime/boost, but the increase in manufacturing complexity is clearly a drawback to this approach. This trial involved three intramuscular infections of the HPV16 L2/E6/E7 fusion protein described above, followed by a single dermal scarification with the 16/18 E6/E7 recombinant vaccinia virus. While this vaccination strategy was immunogenic, it is unclear whether it is superior to others discussed above.

In summary, evidence of antigen-specific CD8 and/or CD4 responses to the target viral antigen was detected in the majority of vaccinees in most recent trials of vaccines for premalignant neoplasia. However, unequivocal evidence for consistent clinical regression of HPV induced neoplasia and virus clearance has not been produced, in part because most studies were early phase trials involving small numbers of vaccinees and were not placebo controlled. However, based on the results to date, it would be rather surprising if any of the current therapeutic vaccine candidates prove to be highly effective in larger and better controlled efficacy trials, at least in trials of CIN 3 or cancer patients. Because there was no clear correlation between clinical response and immune response in a specific CMI assay, the CMI response most predictive of a clinical response remains unknown.

14.2.5. Type Specificity

Since more than a dozen HPV DNA types are detected in cervical cancers, it is important to consider whether therapeutic vaccines have the potential to induce cross-protective immunity. Most therapeutic vaccines tested to date are based on E6 and E7 antigens, which are not as highly conserved as some other viral proteins, e.g., E2 and L1. Therefore broad protection across HPV types might not be predicted, although cross protection of closely related types such as HPV16/31, -18/45, and -6/11 would not be unexpected. However, there is no clear evidence that the immunity that arises after natural infection is cross-protective, even for closely related types (Baseman and Koutsky, 2005). Nevertheless, one could argue that a vaccine could induce a type or magnitude of cross-protecting immunity that is not elicited by natural infection. There has been little evaluation of cross protection of therapeutic vaccines in the animal models, but some recent clinical studies employing HPV16/18 specific immunogens have reported evidence suggesting activity against non-HPV16/18 cervical disease. For example, the clinical responses seen in the Zycos microparticle DNA trial were not type-specific (Garcia et al., 2004). Also, as noted above, the Stressgen HSP-HPV16 E7 immunogen may induce clinical responses against genital warts, which are predominantly caused by HPV6 and HPV 11. However, there is no evidence to date that these cross-type clinical responses are viral antigen specific, and regression could be due to the nonspecific induction of innate immune responses, as seen in the mouse skin graft model. It will be important to follow up these observations in additional blinded and placebo controlled trials, ideally with control arms that involve preparations that induce similar innate responses as the vaccine arm. It would also be interesting to examine the extent to which CMI measured in in vitro assays is cross-reactive.

14.2.6. Why Haven't Therapeutic HPV Vaccines Been More Successful?

There are at least five reasons why therapeutic HPV vaccines display limited success in clinical trials. First, the T-cell responses measured after vaccination are generally rather modest, at least in comparison to those measured after systemic viral infections. There is a need to develop more effective strategies to prime and multiply boost CMI to specific antigens, without inducing unacceptable autoimmune and/or pro-inflammatory reactions. It is encouraging to note that the CMI that appears to induce regression of natural HPV lesions is effective without eliciting substantial systemic or local side effects. However, most of the natural immune response is probably directed to the lesion, in contrast to the situation after systemic vaccination. Second, persistent HPV infections may induce specific immune tolerance to the viral antigens due to long-term exposure of the immune system to the antigen in a noninflammatory epithelial setting coupled with immune evasion mechanisms of the virus (Woodworth, 2005). There is evidence for peripheral tolerance in the E6E7 transgenic mouse models (Azoury-Ziadeh et al., 2001). The extent to which specific tolerance prevents effective therapeutic vaccination against HPV in humans is largely unknown. Third, cervical cancers and many high-grade dysplasias are genetically unstable. As a consequence, they are able to evolve to evade immune induction and recognition (Koopman et al., 2000). The majority of cervical cancers and many CIN3 lesions have defects in MHC class I presentation, often through down-regulation of specific class I alleles or defects in the processing of peptides for class I presentation (Bontkes et al., 1998). The failure to detect a significant clinical response to the Zycos vaccine in older CIN2/3 patients may in part be due to longer mean duration of infection in the older women, and hence greater likelihood of immune tolerance to E6 and E7 and evolution of immune evasion by their lesions. Fourth, systemic immune responses are primarily generated and evaluated in therapeutic vaccine trials, but local CMI responses are likely to be critical for clearing of HPV infection, since precancerous HPV-induced neoplasia are local mucosal infections (Revaz and Nardelli-Haefliger, 2005). Some evidence suggests that T cells generated in response to systemic antigens do not efficiently migrate to anogenital mucosal tissues (Belyakov and Berzofsky, 2004). Generation of T-cell responses in mucosal-associated lymphoid tissues and/or better recruitment of T cells to the sites of infection may be needed to produce effective therapeutic vaccines. Finally, as noted above, the current animal models of evaluating therapeutic HPV vaccine are not predictive of efficacy against anogenital disease in humans. It may be necessary to develop animal models that more closely resemble human disease to gain better insight into the immune mechanisms involved in elimination of superficial mucosal cells expressing HPV proteins and into the best vaccination strategies to induce these effectors.

14.2.7.   Conclusions: Therapeutic Vaccines

The number of therapeutic HPV vaccines that have been tested in clinical trials in the last few years is very encouraging. However, not all the vaccine candidates that have shown activity in animal models can be tested in clinical trials (recently reviewed in (Roden et al., 2004)). It is difficult to prioritize these candidates on a rational basis. This dilemma arises because comparative immunogenicity studies are rarely conducted, and because we do not understand what type of CMI is important for regression of anogenital lesions. Therefore, the vaccines that have been and will be tested are often selected by companies based on access to intellectual property rights as much as on a scientific basis. Vaccines are selected by academic scientists primarily on the basis of the availability of clinical grade vaccines from companies. Nevertheless, there are reasons to be optimistic that recent advances in basic immunology and vaccinology can be translated into effective HPV immunotherapies. The many attractive features of HPV-induced neoplasia as a clinical model of tumor immunotherapy, coupled with the public health significance of the lesions, argues that this research should receive increased public, private, and commercial support, despite the impending introduction of the prophylactic HPV vaccines discussed below. Furthermore, the lack of cross protection by the prophylactic vaccines and the likelihood that these vaccines will have minimal benefit to women who are already infected at the time of vaccination indicates that therapeutic vaccination may well be an important tool in the management of cervical neoplasia, even after prophylactic vaccination becomes established.

14.3.   Prophylactic Vaccines

14.3.1.   Introduction

A vaccine strategy based on papillomavirus virus-like particles (VLPs) is clearly the leading prophylactic vaccination approach. L1, the major capsid protein, self-assembles into VLPs that are morphologically indistinguishable from the ordered array of 72 L1 pentamers that are the principal component of the outer shell of authentic virions. However, because none of the other viral genes are involved in VLP production and the VLPs do not contain DNA, they are inherently noninfectious and nononcogenic. Most importantly, they resemble authentic virions in their ability to generate high titers of virus-neutralizing antibodies (Kirnbauer et al., 1992), the prerequisite for effective prophylactic vaccines against most other viruses (Zinkernagel, 2003). Vaccination studies in mice and rabbits confirmed that purified HPV VLP preparations could generate high titers of antibodies that neutralized HPV virions or pseudovirions in in vitro assays (Roden et al., 1996; Rose et al., 1994). The in vitro neutralizing activity of HPV VLP antibodies was largely type specific, although there was some low titer cross-neutralization detected for closely related types, such as HPV6 and -11 or -16 and -31 (Christensen et al., 1994; Roden et al., 1996; Pastrana et al., 2004, Schiller, unpublished).

Proper L1 conformation is critical for generating neutralizing antibodies to L1, since neither denatured L1 nor shorter L1 peptides induce such antibodies in appreciable amounts. Early attempts to generate L1-based vaccines were unsuccessful because the L1 was not in its native conformation (Pilacinski et al., 1986). L1 VLPs can be generated using a variety of production systems. VLPs were first observed using recombinant baculovirus and vaccinia viruses (Hagensee et al., 1993; Kirnbauer et al., 1992; Rose et al., 1993; Zhou et al., 1993), but have since been generated after transfection of cultured human cells (Zhou et al., 1999), in yeast (Hofmann et al., 1995), or even in bacteria (Nardelli-Haefliger et al., 1997).

14.3.2.   Preclinical Studies

Papillomavirus infections are remarkably species restricted; hence, there are no animal models for productive HPV infection or the generation of hyperproliferative disease after application of infectious HPVs. Therefore, proof-of-concept studies to examine the ability of VLP vaccination to prevent papillomavirus infection have relied on animal papillomavirus types, specifically cottontail rabbit papillomavirus (CRPV) in domestic rabbits (Breitburd et al., 1995), canine oral papillomavirus (COPV) in dogs (Suzich et al., 1995), and bovine papillomavirus type 4 (BPV4) in cattle (Kirnbauer et al., 1996). In each of these models, low-dose intramuscular injection of VLPs, even without adjuvant, generated excellent protection from experimental challenge with high-dose virus. Protection from virus challenge could be passively transferred using serum from vaccinated animals, indicating that antibodies were sufficient to confer protection (Breitburd et al., 1995; Suzich et al., 1995). Unfortunately there is no animal model for venereal transmission of a papillomavirus, the route of infection leading to cervical cancer. Therefore, preclinical studies cannot assess the levels of cervicovaginal antibodies needed to prevent sexual transmission of oncogenic HPVs. Although the animal studies were encouraging in demonstrating the ability of the VLPs to induce high systemic titers of neutralizing antibodies, they could not predict the success of VLP-based vaccines in preventing genital mucosal infections in women.

L1 VLPs did not induce regression of preexisting warts, i.e., they were not effective as a therapeutic vaccine, even though L1 can induce specific CMI responses that lead to rejection of L1-expressing transplantable syngeneic tumors in mice (De Bruijn et al., 1998). The inability of VLPs to induce wart regression may be related to the restriction of L1 expression to the superficial layers in the infected stratified squamous epithelium, while virus infection is likely maintained in the underlying basal layers (Doorbar, 2005). CMI responses directed against an antigen expressed only in the more superficial layers would not be expected to eliminate infected basal cells.

The minor capsid protein, L2, can be incorporated into VLPs when co-expressed with L1. However, L2 incorporation increases VLP yields only marginally, at best (Kirnbauer et al., 1993). Also, type-specific and cross-neutralizing titers were not greater after vaccination of animals with L1/L2 VLPs than with L1-only VLPs (Roden et al., 2000). In the absence of a clear advantage of L1/L2 VLPs in terms of vaccine yield or immunogenicity, all clinical trials of VLP vaccines have used L1 only VLPs, simply due to manufacturing complexity considerations.

14.3.3. Clinical Trials

HPV VLPs for clinical trials have been produced in two systems, cultured insect cell lines infected with L1-recombinant baculoviruses and *Saccharomyces cerevisiae* carrying L1 expression plasmids. In initial phase I clinical trials, HPV6, HPV11, and HPV16 VLPs proved to be safe and highly immunogenic after two to three intramuscular injections of 10–50-μg doses, even without adjuvant (Evans et al., 2001; Harro et al., 2001; Zhang et al., 2000). In one trial directly comparing VLP vaccination with and without adjuvant, the titers of VLP antibodies were no higher if 50-μg doses of HPV 16 VLPs were administered with alum or MF-59 (an oil-in-water emulsion adjuvant) versus without adjuvant, although adjuvant did increase antibody responses to 10-μg doses of VLPs (Harro et al., 2001). Alum-based adjuvants are included in the two vaccines undergoing commercial development. Presumably their inclusion not only increases immunogenicity at the 20–40-μg VLP doses used in the vaccines, but also may help stabilize VLPs.

Two large vaccine manufactures, GlaxoSmithKline and Merck, have published phase 2B proof-of-concept efficacy trials in healthy young women. The GSK trial involved the vaccine that they hope to commercialize (Harper et al., 2004), which contains HPV16 and HPV18 VLPs produced in insect cells combined with their proprietary ASO4 adjuvant. A large Merck trial was conducted with an HPV16 VLP vaccine (Koutsky et al., 2002), and a smaller trial used the tetravalent HPV6, -11, -16, -18 vaccine that they hope to commercialize (Villa et al., 2005). The VLPs for both Merck trials were produced in yeast and combined with an alum adjuvant. Despite the differences in formulation, the published results of the trials were very similar (Table 14.1). The vaccines were safe and well tolerated, with no serious vaccine-related adverse events. Seroconversion following VLP vaccination was virtually 100 percent. Mean serum VLP antibody titers after the third vaccination were 50–100-fold higher than those measured in women with natural infections. There has been no indication that increasing the valency (i.e., number of L1 types) of the vaccine decreased the antibody responses to the individual VLP types.

Strikingly, efficacy against biopsy confirmed vaccine type-specific cervical dysplasias was 100 percent in all three trials with relatively short follow-up (Table 14.1). It is very encouraging that 100 percent efficacy against HPV16 DNA positive CIN2-3 after 3.5 years of follow-up in the HPV16 VLP trial was recently reported (Laura Koutsky, personal communication). Efficacy in preventing persistent type-specific HPV infection (primarily defined as detection of cervical HPV DNA at two consecutive visits at least six months apart) was 90–100 percent. Protection against type-specific transient infection was somewhat less than against persistent infection. This finding does not necessarily imply that the vaccine induces clearance of newly infected cells. Rather it could indicate that neutralizing antibodies prevent successive rounds of auto-inoculation that might often be required to establish persistent infection. For instance, initial infection may often be in keratinocytes with limited capacity to replicate. It may usually take several rounds of infection of this cell type before a stem cell with unlimited capacity is infected. Furthermore, it remains unclear what proportion of the single-time detections of cervical HPV DNA represent true incident infections rather than intermittent detection of latent infections or contaminating HPV DNA from a partner. Regardless of the biological explanation for the cases of transient HPV DNA detection in the vaccines, it is important to note that protection from persistent infection is the critical virologic outcome, since persistent, but not transient, HPV infection (particularly by types 16 and 18) is the critical risk factor for carcinogenic progression (Khan et al., 2005).

TABLE 14.1. Proof of Concept VLP Prophylactic Efficacy Trials

Study	Koutsky et al., 2002	Harper et al., 2004	Villa et al., 2005
VLP types	16	16, 18	6, 11, 16, 18
Adjuvant	Alum	AS04	Alum
Sponsor	Merck	GSK	Merck
Trial sites	US	US/CA/BR	US/EU/BR
Age	16–23	15–25	16–23
N (ATP)	1533	721	468
Vaccination schedule (Mos)	0, 2, 6	0, 1, 6	0, 2, 6
Follow-up (yr)	1.5	1.5	2.5
No. of cervical infections:			
Control/vaccine	68/6	23/2	Not reported
% efficacy	91%	92%	
No. of persistent			
Infections: Cont/Vac	42/0	7/0	35/4*
% Efficacy	100%	100%	89%
No. of Clinical outcomes:			
Cont/Vac	9/0 (CIN)	6/0 (CIN)	3/0 (CIN), 3/0 (warts)
% Efficacy	100%	100%	100%

All data for according to protocol (ATP) groups. Outcomes reported only for HPV types with corresponding VLPs in the vaccine.
* 3 of 4 outcomes were first HPV DNA detections at the last study visit.

Analysis of cross-protection by VLP vaccines against other genital HPV types has been incompletely reported for the above trials. Cross-protection would be very desirable, since many cervical cancer and CIN3s, and the majority of low-grade cytological abnormalities detected in Pap screens, are caused by HPV types not in the vaccines (Munoz et al., 2004; Schiffman et al., 2005). No protection against CIN containing DNA of other HPV types was seen in the Merck HPV16 trial (22 cases in both vaccine and placebo arms), but protection against infection by non-vaccine types was not reported. In contrast, data from GSK (Gary Dubin, personal communication) indicate that the GSK bivalent vaccine provided significant protection against nonvaccine types. The protection against incident infection by other high-risk types was over 40 percent and protection against cervical cytological abnormalities caused by other high-risk types was almost 70 percent. These results are rather unexpected, because previous in vitro neutralizing assays using VLP sera from animals suggested that cross-neutralizing titers were generally low or undetectable for closely related types, such as HPV16 and HPV 33, and HPV18 and HPV 45 (Roden et al., 1996; White et al., 1998), although these assays were rather insensitive measures of neutralizing activity (Pastrana et al., 2005). An attractive explanation for the GSK results is that the neutralizing antibody titers produced after HPV16/18 VLP vaccination are so high that they afford protection against heterologous types, even if only a small fraction of the induced antibodies is cross-neutralizing. If so, long-term protection against homologous types may occur, while protection against heterologous types might be more short-lived. An alternative explanation is that cross-protection results from L1-specific cell-mediated immunity. Much of L1 is highly conserved across high-risk types, raising the possibility that cross-reactive T-cell epitopes exist. Also L1 VLPs are known to induce potent T-cell responses in vaccinees, some cross-reactive, after vaccination (Pinto et al., 2003, and Ligia Pinto, personal communication). However, as noted above, L1 is not detectably expressed in basal keratinocytes, where the infection is thought to be maintained (Doorbar, 2005), so it is unclear how a CMI response to L1 would prevent establishment of HPV infection.

What could account for the apparent discrepancy in the Merck and GSK results with respect to cross-protection? One explanation is a difference in end points. Merck only reported cross-protection results for CIN, while GSK presented both HPV DNA and cytology data. This explanation seems unlikely, since protection from persistent DNA and cytologic abnormalities would be expected to translate into protection against CIN. Alternatively, there may be subtle differences in the surface conformation of the two vaccines and hence differences in the presentation of cross-neutralizing epitopes. A third possibility is that AS04 adjuvant used in the GSK trial induces more cross-protective immunity. In preliminary GSK studies, VLPs in AS04 did induce modestly higher homologous VLP antibody titers than did VLPs in alum. Further, AS04 may predominately induce a Th1-biased CMI response. Th1 responses would typically be expected to better promote clearance of virally infected cells than the Th2 bias seen after vaccination with VLPs plus alum (Palker et al., 2001). (To distinguish between these possibilities it will be interesting to determine the relative induction of cross-neutralizing antibodies by the two vaccines and to determine whether the extent of cross-protection against specific types correlates with the in vitro cross-neutralizing antibody titers of the vaccinated women against those types. If it does not, then a CMI mechanism of cross-protection would be favored). Clearly, it will be of great interest to determine the degree of cross protection induced by the two vaccines in the large-scale efficacy trials discussed below.

The solid protection against persistent cervical infection observed in the above trials may seem surprising, if antibodies are the primary, if not the only, mediators of protection. Intramuscular injection of protein antigens is generally considered effective at inducing systemic IgG responses but not mucosal IgA, the main antibody at most mucosal surfaces (Revaz and Nardelli-Haefliger, 2005). However generation of a systemic IgG response could protect against cervical infection in two ways. First, the cervix is unusual for a mucosal tissue in that there is extensive transudation of serum IgG, much more than, for example, in the gut (Mestecky and Russell, 2000). In most women, approximately half of the immunoglobulin in cervical mucus is IgG. Second, infection in many instances may require trauma to expose the basal cells of the squamous epithilium to infection. This trauma could directly expose the site of infection to serum exudates. At present, the relative importance of transudated and exudated serum antibodies in preventing infection at the cervix is unclear. However, substantial levels of VLP antibodies, mostly IgG, can be detected at the cervical os after intramuscular VLP vaccination. In a pilot study examining this phenomenon, women taking oral contraceptives, and therefore not undergoing normal menstrual cycling, maintained relatively constant levels of cervical VLP antibodies, approximately 10 percent of that detected in serum (Nardelli-Haefliger et al., 2003). In women with normal menstrual cycles, mucosal titers of both VLP specific and total IgG decreased an additional tenfold around the time of ovulation. This observation raises the possibility that protection in women not taking oral contraceptives may be less complete and/or of shorter duration.

The encouraging results of the phase 2b trials prompted large-scale phase 3 efficacy trials. There are three ongoing phase 3 trials sponsored by Merck, GSK, and the U.S. National Cancer Institute (NCI), respectively. The Merck and GSK trials are multicentric, involving more than 20,000 young women each at a large number of different sites in North America, Latin America, Europe and Asia. The NCI-sponsored trial is the only population-based study, enrolling 7,500 women in the province of Guanacaste, Costa Rica. The Merck study is testing their tetravalent (HPV6, -11, -16, -18) vaccine, while both the GSK and NCI trials are testing the bivalent (HPV16, -18) vaccine manufactured by GSK. The primary clinical end point for these trials is type-specific protection from CIN2/3, and the primary virologic end point is type-specific protection against persistent cervical infection. An advisory group to the U.S. Food and Drug Administration (FDA) suggested that protection from CIN2/3 was the most desirable end point for licensure (Pagliusi and Teresa Aguado, 2004). The Merck vaccine, Gardasil, has received regulatory approval for sale to 9-26 year old females in the U.S., European Union, Brazil, and several other countries. GSK appears to be about one-year behind in its phase 3 trial and will likely seek regulatory approval for its vaccine in the last quarter of 2006.

It will be important to establish plans for long-term follow-up of vaccinees in the phase 3 trials to evaluate the duration of protection afforded by these vaccines. Important questions include—will protection be more durable for homologous than heterologous types? Will it be more durable in women taking oral contraceptives? Will the type of adjuvant influence duration of protection? A substantial increase in the number of breakthough infections in vaccinees over time would encourage secondary trials of booster vaccination. At present it is difficult to predict whether protection will be life-long or require periodic booster vaccination. However, protection may not need to be life-long to have a large impact on cervical cancer rates. Both the chance of acquisition of a high-risk HPV infection and its progression to cervical cancer before death by another cause are expected to diminish with age, i.e., a vaccinated woman who only becomes susceptible to oncogenic HPV infection after age 50 would have a small probability of dying of cervical cancer.

14.3.4. Remaining Scientific Questions

The phase 2 VLP trials have established the vaccines' safety, remarkably consistent immunogenicity, and high efficacy at protecting young women against persistent type-specific HPV infection and type-specific CIN. It is highly likely that the vaccines would protect against cervical cancer, given the critical importance of persistent HPV infection in the development of cervical cancer (Bosch et al., 2002). In contrast, the answers to other questions are more difficult to predict. As noted above, the duration of protection is unknown, and there is no animal model for estimating the levels of cervical and/or serum antibodies required for preventing infection. The degree of cross protection and its duration relative to protection from homologous virus infection is also uncertain, especially given the contradictory results from the Merck and GSK studies. The greater the cross protection, the less pressure to include additional VLP types in the vaccines or develop more cross-protective vaccines, such as the L2-based vaccines described below.

It is important to determine if correlates of protection can be established in order to facilitate the development of next generation vaccines. Whether neutralizing antibodies are the primary mediators of protection will need to be assessed in the efficacy trials. VLP-based ELISA measurements may not be adequate for this purpose since they measure both neutralizing and non-neutralizing antibodies. For example, cross-type reactivity in VLP ELISAs is greater than that measured using in vitro neutralization assays, when sera of VLP vaccinated women are tested (Pastrana et al., 2004). It may be preferable to use a high throughput in vitro neutralization assay that is highly type specific, at least for more distantly related mucosal types such as HPV16 and HPV18.

An interesting question is whether serum antibody (neutralization) titers can provide a good correlate of protection. This situation may well be the case if exudated serum antibodies are principally responsible for protection. However, they may not if transudated and secreted antibodies in cervicovaginal mucus are primarily responsible. If the ratio of vaccine-induced cervical mucosal to serum antibodies varies across vaccinees (Nardelli-Haefliger et al., 2003), then neutralizing antibody titers in mucosal samples might correlate better with protection.

14.3.5.   Target Populations

Who should be vaccinated with an HPV VLP vaccine? Who will be vaccinated? The answers to these questions go to the heart of the implementation issues surrounding the vaccines and may vary depending on a number of factors, most significantly on health-care resources. Cervical cancer accounts for approximately 90 percent of the cases of HPV-associated cancers (Parkin et al., 2001). This statistic provides a strong rationale to target women in settings with limited health-care resources devoted to Pap screening programs. Since many women acquire genital HPV infections soon after becoming sexually active (Baseman and Koutsky, 2005), and there is no indication that the VLP vaccines will act therapeutically, the vaccine would almost certainly have the most impact if young women are vaccinated before they become sexually active. Therefore the primary target group for this vaccine should be preadolescent girls aged 9–13. Safety and immunogenicity studies of the GSK and Merck vaccines in this age group have generated encouraging results.

It is likely that initial demand for the vaccine will be greatest in sexually active young women (Kahn et al., 2003). Several reasons suggest they should be vaccinated: 1) the vaccine appears safe, 2) they will be protected from primary infection by the vaccine HPV types, 3) cross-protection against the less prevalent nonvaccine types may occur, 4) the vaccine could reduce the chances of the establishment of persistent infection in women with prevalent infection if persistent cervical infection requires multiple rounds of auto-inoculation. Some data suggest that many genital HPV infections can ascend anatomically, starting as a vulvar/lower vaginal infection and only later reaching the cervix (Winer et al., 2003). Vaccine-induced neutralizing antibodies could interrupt this process. Whether vaccination of women with prevalent infection reduced the rate of establishment of persistent infection can be evaluated in the on-going efficacy trials, since a substantial number of women with prevalent infection were enrolled.

Should boys and/or men be vaccinated? Since genital HPV infections are primarily sexually transmitted, it would seem at first glance that it would be desirable to vaccine both sexes. However, there are several reasons why vaccination of males is likely to have less public health impact, than vaccinating females: 1) Males have a lower incidence of HPV-associated cancer, although HPV-induced anal and penile cancers are certainly a concern (Parkin et al.,

2001). 2) There is as yet no indication that the vaccines will be effective in preventing anogenital infections in men, since the sites of penile infection are not bathed in antibody-containing mucus. In this regard, the evidence of prophylactic efficacy of a herpes simplex gD vaccine in HSV seronegative women, but not in men, provides a cautionary note (Stanberry et al., 2002). This question should be answered in the next few years by the recently initiated Merck and GSK trials in men. 3) It is unclear how much herd immunity would be increased by vaccinating men. In some models, vaccination of men would have relatively little impact relative to cost, provided that coverage in women is high (Garnett, 2005). Vaccinated women with prevalent or breakthrough infections may be less likely to transmit the virus than vaccinated infected men, since virus shed from vaccinated women is expected to contact the neutralizing antibodies in her genital mucosal secretions. Based on this argument, one would predict that mass vaccination of women would have more impact on herd immunity than would vaccination of men. If the VLP vaccines are shown to prevent genital warts in men, inclusion of HPV6 and HPV11 VLPs in the Merck vaccine will provide an added incentive for men to be vaccinated, since men and women have similar incidences of this disease (Wiley et al., 2002), but only if the vaccine is shown to prevent disease in men.

14.3.6.   Implementation Issues

Interesting implementation issues are associated with prophylactic HPV vaccines, and the public health impact of the vaccines may well depend largely on how well these issues are addressed. One issue is how well the public, particularly the parents of preadolescent girls, will accept a vaccine targeting a sexually transmitted disease. Relatively few studies have been published on this topic, although current data suggest that the vaccines would be acceptable to a substantial majority of women and parents of adolescent girls, and that acceptance would increase with knowledge about the disease and the vaccine. In a recent survey in Georgia (USA) 55 percent of parents of 10–15-year-old children initially responded that they would have their children vaccinated (Davis et al., 2004). The number supporting vaccination increased to 75 percent after reading a short educational pamphlet on HPV and cervical cancer. However, groups supporting "abstinence only" solutions to STI control have already voiced their opposition to the vaccines, and it is likely that there will be a minority, at least in the United States, that will not accept the HPV vaccines for their children (see *JNCI*, "News" 97:1030–31, 2005). Stressing cancer prevention may well aid in promoting vaccination. However, the relation of sexual activity to HPV/cervical cancer cannot and should not be ignored. The vaccination of preadolescents needs to be justified, as would vaccination of males. Also vaccine promotion programs would provide an exceptional opportunity to increase the public's awareness of cervical cancer etiology and prevention strategies, necessitating a discussion of their sexual transmission.

Establishing preadolescent vaccination programs will likely prove challenging since there are no current health surveillance guidelines that require this age group to visit health clinics three times within six months. A possible long-term solution would be deliver the initial vaccinations as part of the infant vaccination program and then deliver a single booster dose during preadolescence, if necessary. However, demonstrating the effectiveness of this strategy would take at least 15 years, since the trial endpoints would have to include protection from acquisition of sexually transmitted HPV infections. Noninjectable delivery of the vaccine outside of a clinic setting might also increase compliance in this age group. There are several such approaches that might be used, but, as discussed below, none have elicited consistent high-level antibody responses in humans.

Another major concern is the effect of widespread vaccination on cervical screening programs. The current vaccines should not replace Pap screening in countries with effective programs (Schiller and Davies, 2004). Pap screening has reduced cervical cancer rates by at least 80 percent in the United States. A vaccine that prevents 90 percent of cervical cancer caused by HPV16 and HPV18 could potentially prevent 63 percent of cervical cancer, but the 30 percent of cancer caused by other HPV types would not be prevented. However, models suggest that vaccination coupled with later onset and/or increased intervals of Pap and/or HPV DNA screening would both cost less than existing screening programs and reduce the number of cancer deaths (Goldie et al., 2004; Kulasingam and Myers, 2003). The cost per life saved by a screening program will increase if the vaccines are effective, since there will be fewer women with high-grade disease to diagnose and treat. Thus, the success of vaccination could diminish enthusiasm of governments to support screening programs and decrease compliance rates in such programs. It is worrisome that a substantial number of women may abandon their screening programs because of misconceptions concerning the decree of protection against all high-risk types and the effectiveness of the vaccines on prevalent disease. Well-designed and implemented public health information programs will be needed to counteract this tendency (Schiller and Davies, 2004). Alternatively women may believe that the vaccines are not working because an appreciable number of them, or their acquaintances, will be diagnosed with a Pap abnormality despite vaccination. This perception could be quite common if the vaccines are type specific, since only about one third of LSIL diagnoses are due to HPV16 or HPV18 infections (ACOG Bulletin, 2005). It will be important to manage expectation in this regard.

Finally, a major implementation concern is delivering a vaccine to the economically disadvantaged women of the world. Unfortunately, the women who can least afford the vaccine are also the women who are least likely to have access to quality screening programs which would reduce their risk of cervical cancer. A costly vaccine will thus be available to women who are least likely to develop cervical cancer because they are regularly screened. One solution is for the governments of developing countries or private benefactors to buy the vaccines for distribution to poorer women, although it is questionable how extensive the coverage or sustainable such a program could be. It is also unclear when substantial quantities of the Merck and GSK vaccines will be available at low cost to women in developing countries. It seems unlikely that corporations will initially provide a large number of vaccine doses at low cost to developing countries at the same time they are developing manufacturing capacity to meet the needs of "steady state" vaccination programs of an adolescent/preadolescent cohort, plus the large initial demand for catch-up vaccination of older women in developed counties. It is more likely that the manufacturing capacity initially used for lucrative catch-up vaccination would gradually shift to underdeveloped countries as the demand from older women in developed countries wanes. The time frame is difficult to estimate, given the uncertainties over the volume of vaccine sales after initial introduction. Production of the VLP vaccines by manufacturers in second-tier industrializing countries, e.g. in India, China, or Brazil, might result in substantially more vaccine being delivered to poor women sooner and at lower costs. This supply stream could be accomplished by licensing agreements and technology transfer from Merck or GSK. Alternatively, independent vaccine production programs could be established, at least in countries where the patents covering the VLP technologies were not filed. There certainly is precedence for this approach. For instance, the GSK Hepatitis B VLP vaccine recently sold in India for 500 Rs. It is now is produced and sold in India by Shantha Biotechnics at 18 Rs (US $0.40) (see *Nature*, "News" 436: 480–3, 2005), and a similar low-cost vaccine is produced in Korea. Overall, the cost of the hepatitis B vaccine has dropped more than 100-fold, in large measure due to its production in Asia.

14.3.7.   Second-Generation Vaccines

A second possible solution to problem of limited access to the HPV vaccines is to develop and implement second-generation prophylactic vaccines that are less expensive to manufacture and distribute. In the long term, this approach might greatly increase the number of women vaccinated. However, this approach is less certain and may require longer development time than an approach involving regional manufacturing of the existing vaccine. There are a number of second-generation vaccine strategies that have been developed and many have shown promising results in animal studies. Few have been tested in clinical trials, and none of the trials have provided information on prophylactic efficacy.

These vaccine strategies address a number of limitations of the current VLP vaccine candidates. Each approach has its potential strengths and weaknesses (Table 14.2). Ultimately, the cost of implementing a vaccine program, rather than the cost of purchasing the vaccine, is likely to be the most critical factor limiting poor women's access to the vaccine. It would be surprising if the cost of the current HPV VLP vaccines would not eventually approach that of the HBV vaccines discussed above, especially if they were manufactured in second tier industrializing countries. This cost would still be formidable in many settings, but not as formidable as the cost of delivering three injections of a heat-labile vaccine over the course of 6 months to preadolescents. Cost of

TABLE 14.2. Second-Generation HPV Vaccines

Vaccine	Potential Advantages	Potential Limitations	Reference
Additional VLP types (HPV 31, 45, 33, 52, etc)	Established technology	Increased cost, modest increase in protection from cervical cancer	Munoz et al., 2004
Heat stabilization of VLPs	Decreased implementation costs	Unproven technology for HPV VLPs	Brandau et al., 2003
Slow-release formulation	Lower cost of administration, if fewer doses required.	Unproven technology for HPV VLPs	Sturesson et al., 1999
Upper respiratory tract delivery of purified VLPs	Needle free delivery; induction of sIgA; lower cost of implementation?	Consistency of immune response? Safety?	Nardelli-Haefligar et al., 2005
Oral delivery of VLP in crude plant or yeast extract	Low-cost production and administration; Induction of sIgA	Very low level expression in plants; Low immunogenicity in animal models	Warzecha et al., 2003
L1 DNA	Lower cost of production	Less immunogenic than VLPs? Difficult to estimate oncogenic potential of injected vectors	Tobery et al., 2003
L1 pentameric subunits	Lower cost of production (made in bacteria)	Same distribution costs as VLPs; less immunogenic than VLPs?	Yuan et al., 2001
L1 recombinant bacteria	Low cost of production and administration (i.e., mucosal)	Regulatory issues with GM organisms; safety/immunogenticity uncertain	Baud et al., 2004
L1 recombinant virus	Lower cost of delivery if mucosal; Lower cost of production?	Regulatory issues GM organisms; safety/immunogenticity uncertain	Liu et al., 2005
Chimeric VLPs	Combined prophylactic/therapeutic efficacy; Earlier benefits	Modest therapeutic effect in early trial	Greenstone et al., 1998
VLPs combined with a therapeutic HPV vaccines	Combined prophylactic/therapeutic efficacy; earlier benefits	Efficacy of current therapeutic vaccines limited; interaction with VLPs uncertain	None
L2 protein or peptide	Induction of broadly cross-neutralizing antibodies; lower production costs	Lower titers of neutralizing antibodies than VLPs	Pastrana et al., 2005

delivery is probably the most important factor in the under utilization of the HBV vaccine in many parts of the world (Kao and Chen, 2002). Development of a vaccine that would be heat stable, delivered by a simple needle-free method, or requiring fewer than three doses would likely increase the number of poor women receiving the vaccine. Approaches that address more than one of these limitations would seem particularly attractive. For instance, upper respiratory delivery of a slow release powdered formulation of VLPs could address the heat stability, multiple dose, and needle injection limitations. HPV16 pseudovirions are resistant to desiccation (Roden et al., 1997), and it may be possible to stabilize VLPs in a powder formulation. Also, upper respiratory delivery of VLPs in aqueous solution was recently shown to be immunogenic in women (Nardelli-Haefliger et al., 2005). Oral delivery of a live recombinant bacterial vaccine could address the need for both simple low-cost production and delivery. An HPV16 L1 expressing attenuated *Salmonella typhi* strain (Ty21), e.g., that has been given to millions of people to prevent Salmonella-induced intestinal disease, has produced encouraging results in a mouse model (Baud et al., 2004), and may be an attractive HPV vaccine candidate if it can consistently induce high titer of neutralizing antibodies in people.

L2-based vaccines offer an alternative to L1 VLP-based prophylactic vaccines. They have the advantage of inducing a much broader range of type cross-neutralizing antibodies than do L1 VLPs, and can be generated as simple polypeptides in *E. coli* (Pastrana et al., 2005). In theory, a single polypeptide immunogen might protect against all genital types, and cutaneous types as well. However, to date the titers of type-specific antibodies induced by L2 immunogens have not approached the titers induced by VLP-based vaccines, both in several mouse studies (Kawana et al., 2001; Pastrana et al., 2005; Roden et al., 2000) and in one clinical trial (Kawana et al., 2003). It remains to be determined if this deficit can be overcome by, for example, further refinement of the peptide immunogens, novel adjuvants, or virus-like display.

Combined prophylactic/therapeutic vaccines would be attractive for mass immunization campaigns involving women (and perhaps men) over a broad age range. Mass immunization could promote rapid development of herd immunity and thereby control of viral transmission in a population. Combined vaccines might offer at least some protection to the millions of currently infected women, particularly to those without access to cervical screening, and superior prevention of disease in the next generation. However, to have a substantial public heath impact, these vaccines would probably have to induce regression of persistent CIN 3, if not cervical cancers. As discussed above, this outcome has not been demonstrated as yet by any therapeutic vaccine.

Chimeric VLPs are the most thoroughly studied type of combined vaccine. They involve HPV early-gene polypeptides fused to the C-terminus of either L1 or L2 (Greenstone et al., 1998; Muller et al., 1997). HPV16 E7 chimeric VLPs are the best studied. They can induce high titers of neutralizing antibodies and CMI, including CTLs, to the target polypeptide. After a single injection without adjuvant, they protect from challenge with E7-expressing syngeneic tumors (TC-1). However, CMI responses were not effectively boosted in mice (Da Silva et al., 2001; Liu, 2003). An HPV L1-E7 chimeric VLP consistently induced high antibody titers but did not consistently induce regression of CIN 2/3, in a recent placebo controlled clinical trial (John Nieland, personal communication). Additional clinical trials of this approach may not be warranted unless the problem of boosting with the homologous vaccine (thus generating preexisting antibodies neutralizing the effect of subsequent immunizations) is overcome. As an alternative, it might be possible to combine L1 VLPs with a successful therapeutic vaccine, should one be identified in future trials. VLP binding activates innate immune responses in several types of immunocytes, including dendritic cells, B cells and monocytes (Lenz et al., 2001; Yang et al., 2005). Whether these activities would facilitate or antagonize CMI responses to the therapeutic component of the vaccine remains to be determined.

An important question for second-generation vaccine development is whether there is a realistic possibility for their commercial development. It is unlikely that a VLP-based vaccine would be developed in any of the most industrialized countries over the next two decades, because sale of such vaccines would probably infringe on patents that have been licensed to GSK and/or Merck. These companies themselves will have little incentive to develop alternative vaccines, with the possible exception of simply increasing the number of VLP types, in competition with one another. However, companies in countries in which protection for the dominating VLP patents has not been sought would presumably have the freedom to produce alternative vaccines for country-specific or regional distribution, although they could not sell the vaccines in countries in which the existing patents were filed. This situation provides an interesting opportunity for vaccine manufacturers in middle tier countries to develop alternative vaccines with superior properties with little fear of competition from the major vaccine manufacturers in Europe, North America, Japan, or Australia. A vaccine that is proven safe and effective in developing countries would likely also be attractive for implementation in more developed countries, for example if it required fewer doses or was delivered by a needle-free method. Eventually it might be tested and sold in the major industrialized countries, once the dominating patents expire.

14.3.8. Conclusions: Prophylactic Vaccines

The HPV VLP prophylactic vaccine clinical trials have produced uniformly encouraging results, well beyond even the most optimistic projections. There are high expectations that two independently developed vaccines will be commercially available within the next few years. While some basic scientific questions remain, implementation questions will likely dominate the prophylactic HPV vaccine agenda in the near future, particularly those involved in maximizing acceptance and delivery of the vaccine, especially to the economically disadvantaged women of the world who need it the most.

References

ACOG Practice Bulletin. (April 2005). Clinical management guidelines for obstetrician-gynecologists. Number 61. Human papillomavirus. *Obstet. Gynecol.* **105**(4), 905–918.

Androphy, E.J., Hubbert, N.L., Schiller, J.T., and Lowy, D.R. (1987). Identification of the HPV16 E6 protein from transformed mouse cells and human cervical carcinoma cell lines. *EMBO J.* 6:989–992.

Arbeit, J.M., Howley, P.M., and Hanahan, D. (1996). Chronic estrogen-induced cervical and vaginal squamous carcinogenesis in human papillomavirus type 16 transgenic mice. *Proc. Natl. Acad. Sci. U S A* 93(7):2930–2935.

Azoury-Ziadeh, R., Herd, K., Fernando, G.J., Lambert, P., Frazer, I.H., and Tindle, R.W. (2001). Low level expression of human papillomavirus type 16 (HPV16) E6 in squamous epithelium does not elicit E6 specific B- or T-helper immunological responses, or influence the outcome of immunisation with E6 protein. *Virus Res.* 73(2):189–199.

Baldwin, P.J., van der Burg, S.H., Boswell, C.M., Offringa, R., Hickling, J.K., Dobson, J., Roberts, J.S., Latimer, J.A., Moseley, R.P., Coleman, N., Stanley, M.A., and Sterling, J.C. (2003). Vaccinia-expressed human papillomavirus 16 and 18 E6 and E7 as a therapeutic vaccination for vulval and vaginal intraepithelial neoplasia. *Clin. Cancer Res.* 9(14):5205–5213.

Barton, G.M., and Medzhitov, R. (2002). Toll-like receptors and their ligands. *Curr. Top. Microbiol. Immunol.* 270:81–92.

Baseman, J.G., and Koutsky, L.A. (2005). The epidemiology of human papillomavirus infections. *J. Clin. Virol.* 32(Suppl 1):S16–S24.

Baud, D., Ponci, F., Bobst, M., De Grandi, P., and Nardelli-Haefliger, D. (2004). Improved efficiency of a Salmonella-based vaccine against human papillomavirus type 16 virus-like particles achieved by using a codon-optimized version of L1. *J. Virol.* 78(23):12901–1209.

Belyakov, I.M., and Berzofsky, J.A. (2004). Immunobiology of mucosal HIV infection and the basis for development of a new generation of mucosal AIDS vaccines. *Immunity* 20(3):247–253.

Bontkes, H.J., Walboomers, J.M., Meijer, C.J., Helmerhorst, T.J., and Stern, P.L. (1998). Specific HLA class I down-regulation is an early event in cervical dysplasia associated with clinical progression. *Lancet* 351(9097):187–188.

Borysiewicz, L.K., Fiander, A., Nimako, M., Man, S., Wilkinson, G.W., Westmoreland, D., Evans, A.S., Adams, M., Stacey, S.N., Boursnell, M.E., Rutherford, E., Hickling, J.K., and Inglis, S.C. (1996). A recombinant vaccinia virus encoding human papillomavirus types 16 and 18, E6 and E7 proteins as immunotherapy for cervical cancer. *Lancet* 347(9014):1523–1527.

Bosch, F.X., and de Sanjose, S. (2003). Chapter 1: Human papillomavirus and cervical cancer—burden and assessment of causality. *J. Natl. Cancer Inst. Monogr.* 31:3–13.

Bosch, F.X., Lorincz, A., Munoz, N., Meijer, C.J., and Shah, K.V. (2002). The causal relation between human papillomavirus and cervical cancer. *J. Clin. Pathol.* 55(4):244–265.

Brandau, D.T., Jones, L.S., Wiethoff, C.M., Rexroad, J., and Middaugh, C.R. (2003). Thermal stability of vaccines. *J. Pharm. Sci.* 92(2):218–231.

Brandsma, J.L. (1994). Animal models of Human-Papillomavirus-associated oncogenesis. *Intervirology* 37:189–200.

Brandsma, J.L., Shlyankevich, M., Zhang, L., Slade, M.D., Goodwin, E.C., Peh, W., and Deisseroth, A.B. (2004). Vaccination of rabbits with an adenovirus vector expressing the papillomavirus E2 protein leads to clearance of papillomas and infection. *J. Virol.* 78(1):116–123.

Breitburd, F., Kirnbauer, R., Hubbert, N.L., Nonnenmacher, B., Trin-Dinh-Desmarquet, C., Orth, G., Schiller, J.T., and Lowy, D.R. (1995). Immunization with virus-like particles from cottontail rabbit papillomavirus (CRPV) can protect against experimental CRPV infection. *J. Virol.* 69(6):3959–3963.

Christensen, N.D., Kirnbauer, R., Schiller, J.T., Ghim, S.J., Schlegel, R., Jenson, A.B., and Kreider, J.W. (1994). Human papillomavirus types 6 and 11 have antigenically distinct strongly immunogenic conformationally dependent neutralizing epitopes. *Virology* 205:329–335.

Chu, N.R., Wu, H.B., Wu, T., Boux, L.J., Siegel, M.I., and Mizzen, L.A. (2000). Immunotherapy of a human papillomavirus (HPV) type 16 E7-expressing tumour by administration of fusion protein comprising Mycobacterium bovis bacille Calmette-Guerin (BCG) hsp65 and HPV16 E7. *Clin. Exp. Immunol.* 121(2):216–225.

Corona Gutierrez, C.M., Tinoco, A., Navarro, T., Contreras, M.L., Cortes, R.R., Calzado, P., Reyes, L., Posternak, R., Morosoli, G., Verde, M.L., and Rosales, R. (2004). Therapeutic vaccination with MVA E2 can eliminate precancerous lesions (CIN 1, CIN 2, and CIN 3) associated with infection by oncogenic human papillomavirus. *Hum. Gene Ther.* 15(5):421–431.

Cox, J.T. (2002a). Management of precursor lesions of cervical carcinoma: history, host defense, and a survey of modalities. *Obstet. Gynecol. Clin. North Am.* 29(4):751–785.

Cox, J. T. (2002b). Management of women with cervical cancer precursor lesions. *Obstet. Gynecol. Clin. North Am.* 29(4):787–816.

Da Silva, D.M., Pastrana, D.V., Schiller, J.T., and Kast, W.M. (2001). Effect of preexisting neutralizing antibodies on the anti-tumor immune response induced by chimeric human papillomavirus virus-like particle vaccines. *Virology* 290(2):350–360.

Davidson, E.J., Boswell, C.M., Sehr, P., Pawlita, M., Tomlinson, A.E., McVey, R.J., Dobson, J., Roberts, J.S., Hickling, J., Kitchener, H.C., and Stern, P.L. (2003). Immunological and clinical responses in women with vulval intraepithelial neoplasia vaccinated with a vaccinia virus encoding human papillomavirus 16/18 oncoproteins. *Cancer Res.* 63(18):6032–6041.

Davis, K., Dickman, E.D., Ferris, D., and Dias, J.K. (2004). Human papillomavirus vaccine acceptability among parents of 10- to 15-year-old adolescents. *J. Low Genit. Tract Dis.* 8(3):188–194.

De Bruijn, M.L.H., Greenstone, H.L., Vermeulen, H., Melief, C.J.M., Lowy, D.R., Schiller, J.T., and Kast, W.M. (1998). L1-specific protection from tumor challenge elicited by HPV16 virus-like particles. *Virology* 250:371–376.

de Jong, A., O'Neill, T., Khan, A.Y., Kwappenberg, K.M., Chisholm, S.E., Whittle, N.R., Dobson, J.A., Jack, L.C., St Clair Roberts, J.A., Offringa, R., van der Burg, S.H., and Hickling, J.K. (2002). Enhancement of human papillomavirus (HPV) type 16 E6 and E7-specific T-cell immunity in healthy volunteers through vaccination with TA-CIN, an HPV16 L2E7E6 fusion protein vaccine. *Vaccine* 20(29–30):3456–3464.

de Jong, A., van Poelgeest, M.I., van der Hulst, J.M., Drijfhout, J.W., Fleuren, G.J., Melief, C.J., Kenter, G., Offringa, R., and van der Burg, S.H. (2004). Human papillomavirus type 16-positive cervical cancer is associated with impaired CD4+ T-cell immunity against early antigens E2 and E6. *Cancer Res.* 64(15):5449–5455.

Doorbar, J. (2005). The papillomavirus life cycle. *J. Clin. Virol.* 32(Suppl 1):S7–S15.

Dunn, L.A., Evander, M., Tindle, R.W., Bulloch, A.L., de Kluyver, R.L., Fernando, G.J., Lambert, P.F., and Frazer, I.H. (1997). Presentation of the HPV16E7 protein by skin grafts is insufficient to allow graft rejection in an E7-primed animal. *Virology* 235(1):94–103.

Ehreth, J. (2003). The value of vaccination: a global perspective. *Vaccine* 21(27–28): 4105–4117.

Eiben, G.L., Velders, M.P., Schreiber, H., Cassetti, M.C., Pullen, J.K., Smith, L.R., and Kast, W.M. (2002). Establishment of an HLA-A*0201 human papillomavirus type 16 tumor model to determine the efficacy of vaccination strategies in HLA-A*0201 transgenic mice. *Cancer Res.* 62(20):5792–5799.

Evans, T.G., Bonnez, W., Rose, R.C., Koenig, S., Demeter, L., Suzich, J.A., O'Brien, D., Campbell, M., White, W.I., Balsley, J., and Reichman, R.C. (2001). A Phase 1 Study of a Recombinant Viruslike Particle Vaccine against Human Papillomavirus Type 11 in Healthy Adult Volunteers. *J. Infect. Dis.* 183(10):1485–1493.

Frazer, I.H., Kluyver, R.D., Leggatt, G.R., Yang Guo, H., Dunn, L., White, O., Harris, C., Liem, A., and Lambert, P. (2001). Tolerance or immunity to a tumor antigen expressed in somatic cells can be determined by systemic proinflammatory signals at the time of first antigen exposure. *J. Immunol.* 167(11):6180–6187.

Frazer, I.H., Quinn, M., Nicklin, J.L., Tan, J., Perrin, L.C., Ng, P., O'Connor, V.M., White, O., Wendt, N., Martin, J., Crowley, J.M., Edwards, S.J., McKenzie, A.W., Mitchell, S.V., Maher, D.W., Pearse, M.J., and Basser, R.L. (2004). Phase 1 study of HPV16-specific immunotherapy with E6E7 fusion protein and ISCOMATRIX adjuvant in women with cervical intraepithelial neoplasia. *Vaccine* 23(2):172–181.

Garcia, F., Petry, K.U., Muderspach, L., Gold, M.A., Braly, P., Crum, C.P., Magill, M., Silverman, M., Urban, R.G., Hedley, M.L., and Beach, K.J. (2004). ZYC101a for treatment of high-grade cervical intraepithelial neoplasia: A randomized controlled trial. *Obstet. Gynecol.* 103(2):317–326.

Garnett, G.P. (2005). Role of herd immunity in determining the effect of vaccines against sexually transmitted disease. *J. Infect. Dis.* 191(Suppl 1):S97–106.

Gillison, M.L., and Shah, K.V. (2003). Chapter 9: Role of mucosal human papillomavirus in nongenital cancers. *J. Natl. Cancer Inst. Monogr.* 31:57–65.

Goldie, S.J., Kohli, M., Grima, D., Weinstein, M.C., Wright, T.C., Bosch, F.X., and Franco, E. (2004). Projected clinical benefits and cost-effectiveness of a human papillomavirus 16/18 vaccine. *J. Natl. Cancer Inst.* 96(8):604–615.

Greenstone, H.L., Nieland, J.D., de Visser, K.E., De Bruijn, M.L.H., Kirnbauer, R., Roden, R.B.S., Lowy, D.R., Kast, W.M., and Schiller, J.T. (1998). Chimeric papillomavirus virus-like particles elicit antitumor immunity against the E7 oncoprotein in an HPV16 tumor model. *Proc. Natl. Acad. Sci. U S A* 95(4):1800–1805.

Hagensee, M.E., Yaegashi, N., and Galloway, D.A. (1993). Self-assembly of human papillomavirus type 1 capsids by expression of the L1 protein alone or by coexpression of the L1 and L2 capsid proteins. *J. Virol.* 67(1):315–322.

Hallez, S., Simon, P., Maudoux, F., Doyen, J., Noel, J.C., Beliard, A., Capelle, X., Buxant, F., Fayt, I., Lagrost, A.C., Hubert, P., Gerday, C., Burny, A., Boniver, J., Foidart, J.M., Delvenne, P., and Jacobs, N. (2004). Phase I/II trial of immunogenicity of a human papillomavirus (HPV) type 16 E7 protein-based vaccine in women with oncogenic HPV-positive cervical intraepithelial neoplasia. *Cancer Immunol. Immunother.* 53(7):642–650.

Han, R., Cladel, N.M., Reed, C.A., Peng, X., Budgeon, L.R., Pickel, M., and Christensen, N.D. (2000a). DNA vaccination prevents and/or delays carcinoma development of papillomavirus-induced skin papillomas on rabbits. *J. Virol.* 74(20):9712–9716.

Han, R., Reed, C.A., Cladel, N.M., and Christensen, N.D. (2000b). Immunization of rabbits with cottontail rabbit papillomavirus E1 and E2 genes: Protective immunity induced by gene gun-mediated intracutaneous delivery but not by intramuscular injection. *Vaccine* 18(26):2937–2944.

Harper, D.M., Franco, E.L., Wheeler, C., Ferris, D.G., Jenkins, D., Schuind, A., Zahaf, T., Innis, B., Naud, P., De Carvalho, N.S., Roteli-Martins, C.M., Teixeira, J., Blatter, M.M., Korn, A.P., Quint, W., and Dubin, G. (2004). Efficacy of a bivalent L1 virus-like particle vaccine in prevention of infection with human papillomavirus types 16 and 18 in young women: A randomised controlled trial. *Lancet* 364(9447):1757–1765.

Harro, C.D., Pang, Y.Y., Roden, R.B., Hildesheim, A., Wang, Z., Reynolds, M.J., Mast, T.C., Robinson, R., Murphy, B.R., Karron, R.A., Dillner, J., Schiller, J.T., and Lowy, D.R. (2001). Safety and immunogenicity trial in adult volunteers of a human papillomavirus 16 L1 virus-like particle vaccine. *J. Natl. Cancer Inst.* 93(4):284–292.

Hofmann, K., Cook, J., Joyce, J., Brown, D., Schultz, L., George, H., Rosolowsky, M., Fife, K., and Jansen, K. (1995). Sequence determination of Human Papillomavirus 6a and assembly of virus-like particles in *Saccharomyces cerevisiae*. *Virology* 209:506–518.

Hunt, S. (2001). Technology evaluation: HspE7, StressGen Biotechnologies Corp. *Curr. Opin. Mol. Ther.* 3(4):413–417.

Kahn, J.A., Rosenthal, S.L., Hamann, T., and Bernstein, D.I. (2003). Attitudes about human papillomavirus vaccine in young women. *Int. J. STD AIDS* 14(5):300–306.

Kao, J.H., and Chen, D.S. (2002). Global control of hepatitis B virus infection. *Lancet Infect. Dis.* 2(7):395–403.

Kast, W.M., Brandt, R.M., Drijfhout, J.W., and Melief, C.J. (1993). Human leukocyte antigen-A2.1 restricted candidate cytotoxic T lymphocyte epitopes of human papillomavirus type 16 E6 and E7 proteins identified by using the processing-defective human cell line T2. *J. Immunother.* 14:115–120.

Kaufmann, A.M., Stern, P.L., Rankin, E.M., Sommer, H., Nuessler, V., Schneider, A., Adams, M., Onon, T.S., Bauknecht, T., Wagner, U., Kroon, K., Hickling, J., Boswell, C.M., Stacey, S.N., Kitchener, H.C., Gillard, J., Wanders, J., Roberts, J.S., and Zwierzina, H. (2002). Safety and immunogenicity of TA-HPV, a recombinant vaccinia virus expressing modified human papillomavirus (HPV)-16 and HPV-18 E6 and E7 genes, in women with progressive cervical cancer. *Clin. Cancer Res.* 8(12):3676–3685.

Kawana, K., Kawana, Y., Yoshikawa, H., Taketani, Y., Yoshiike, K., and Kanda, T. (2001). Nasal immunization of mice with peptide having a cross-neutralization epitope on minor capsid protein L2 of human papillomavirus type 16 elicit systemic and mucosal antibodies. *Vaccine* 19(11–12):1496–1502.

Kawana, K., Yasugi, T., Kanda, T., Kino, N., Oda, K., Okada, S., Kawana, Y., Nei, T., Takada, T., Toyoshima, S., Tsuchiya, A., Kondo, K., Yoshikawa, H., Tsutsumi, O., and Taketani, Y. (2003). Safety and immunogenicity of a peptide containing the cross-neutralization epitope of HPV16 L2 administered nasally in healthy volunteers. *Vaccine* 21(27–28):4256–4260.

Khan, M.J., Castle, P.E., Lorincz, A.T., Wacholder, S., Sherman, M., Scott, D.R., Rush, B.B., Glass, A.G., and Schiffman, M. (2005). The elevated 10-year risk of cervical precancer and cancer in women with human papillomavirus (HPV) type 16 or 18 and the possible utility of type-specific HPV testing in clinical practice. *J. Natl. Cancer Inst.* 97(14):1072–1079.

Kirnbauer, R., Booy, F., Cheng, N., Lowy, D.R., and Schiller, J.T. (1992). Papillomavirus L1 major capsid protein self-assembles into virus-like particles that are highly immunogenic. *Proc. Natl. Acad. Sci. U S A* 89(24):12180–12184.

Kirnbauer, R., Taub, J., Greenstone, H., Roden, R.B.S., Durst, M., Gissmann, L., Lowy, D.R., and Schiller, J.T. (1993). Efficient self-assembly of human papillomavirus type 16 L1 and L1-L2 into virus-like particles. *J. Virol.* 67(12):6929–6936.

Kirnbauer, R., Chandrachud, L., O'Neil, B., Wagner, E., Grindlay, G., Armstrong, A., McGarvie, G., Schiller, J., Lowy, D., and Campo, M. (1996). Virus-like particles of Bovine Papillomavirus type 4 in prophylactic and therapeutic immunization. *Virology* 219:37–44.

Klencke, B.J., and Palefsky, J.M. (2003). Anal cancer: An HIV-associated cancer. *Hematol. Oncol. Clin. North Am.* 17(3):859–872.

Klencke, B., Matijevic, M., Urban, R.G., Lathey, J.L., Hedley, M.L., Berry, M., Thatcher, J., Weinberg, V., Wilson, J., Darragh, T., Jay, N., Da Costa, M., and Palefsky, J.M. (2002). Encapsulated plasmid DNA treatment for human papillomavirus 16-associated anal dysplasia: a Phase I study of ZYC101. *Clin. Cancer Res.* 8(5):1028–1037.

Koopman, L.A., Corver, W.E., van der Slik, A.R., Giphart, M.J., and Fleuren, G.J. (2000). Multiple genetic alterations cause frequent and heterogeneous human histocompatibility leukocyte antigen class I loss in cervical cancer. *J. Exp. Med.* 191(6):961–976.

Koutsky, L.A., Ault, K.A., Wheeler, C.M., Brown, D.R., Barr, E., Alvarez, F.B., Chiacchierini, L.M., and Jansen, K.U. (2002). A controlled trial of a human papillomavirus type 16 vaccine. *N. Engl. J. Med.* 347(21):1645–1651.

Kulasingam, S.L., and Myers, E.R. (2003). Potential health and economic impact of adding a human papillomavirus vaccine to screening programs. *JAMA* 290(6):781–789.

Lenz, P., Day, P.M., Pang, Y.S., Frye, S.A., Jensen, P.N., Lowy, D.R., and Schiller, J.T. (2001). Papillomavirus-like particles induce acute activation of dendritic cells. *J. Immunol.* 166:5346–5355.

Lin, K.-Y., Guarnieri, F.G., Staveley-O'Carroll, K.F., Levitsky, H.I., August, J.T., Pardoll, D.M., and Wu, T.-C. (1996). Treatment of established tumors with a novel vaccine that enhances major histocompatibility class II presentation of tumor antigen. *Cancer Res.* 56:21–26.

Liu, M.A. (2003). DNA vaccines: A review. *J. Intern. Med.* 253(4):402–410.

Liu, X.S., Xu, Y., Hardy, L., Khammanivong, V., Zhao, W., Fernando, G.J., Leggatt, G.R., and Frazer, I.H. (2003). IL-10 mediates suppression of the CD8 T cell IFN-gamma response to a novel viral epitope in a primed host. *J. Immunol.* 171(9):4765–4772.

Liu, M., Acres, B., Balloul, J.M., Bizouarne, N., Paul, S., Slos, P., and Squiban, P. (2004). Gene-based vaccines and immunotherapeutics. *Proc. Natl. Acad. Sci. U S A* 101(Suppl 2):14567–14571.

Liu, D.W., Chang, J.L., Tsao, Y.P., Huang, C.W., Kuo, S.W., and Chen, S.L. (2005). Co-vaccination with adeno-associated virus vectors encoding human papillomavirus 16 L1 proteins and adenovirus encoding murine GM-CSF can elicit strong and prolonged neutralizing antibody. *Int. J. Cancer* 113(1):93–100.

Longworth, M.S., and Laimins, L.A. (2004). Pathogenesis of human papillomaviruses in differentiating epithelia. *Microbiol. Mol. Biol. Rev.* 68(2):362–372.

Maciag, P.C., and Paterson, Y. (2005). Technology evaluation: HspE7 (Stressgen). *Curr. Opin. Mol. Ther.* 7(3):256–263.

Matsumoto, K., Leggatt, G.R., Zhong, J., Liu, X., de Kluyver, R.L., Peters, T., Fernando, G.J., Liem, A., Lambert, P.F., and Frazer, I.H. (2004). Impaired antigen presentation and effectiveness of combined active/passive immunotherapy for epithelial tumors. *J. Natl. Cancer Inst.* 96(21):1611–1619.

Mestecky, J., and Russell, M.W. (2000). Induction of mucosal immune responses in the human genital tract. *FEMS Imm. Med. Micro.* 27:351–355.

Mocellin, S., Mandruzzato, S., Bronte, V., Lise, M., and Nitti, D. (2004). Part I: Vaccines for solid tumours. *Lancet Oncol.* 5(11):681–689.

Moingeon, P., Almond, J., and de Wilde, M. (2003). Therapeutic vaccines against infectious diseases. *Curr. Opin. Microbiol.* 6(5):462–471.

Moore, R.A., Walcott, S., White, K.L., Anderson, D.M., Jain, S., Lloyd, A., Topley, P., Thomsen, L., Gough, G.W., and Stanley, M.A. (2003). Therapeutic immunisation with COPV early genes by epithelial DNA delivery. *Virology* 314(2):630–635.

Muderspach, L., Wilczynski, S., Roman, L., Bade, L., Felix, J., Small, L.A., Kast, W.M., Fascio, G., Marty, V., and Weber, J. (2000). A phase I trial of a human papillomavirus (HPV) peptide vaccine for women with high-grade cervical and vulvar intraepithelial neoplasia who are HPV 16 positive. *Clin. Cancer Res.* 6(9):3406–3416.

Muller, M., Zhou, J., Reed, T.D., Rittmuller, C., Burger, A., Gabelsberger, J., Braspenning, J., and Gissmann, L. (1997). Chimeric papillomavirus-like particles. *Virology* 234:93–111.

Munoz, N., Bosch, F.X., Castellsague, X., Diaz, M., de Sanjose, S., Hammouda, D., Shah, K.V., and Meijer, C.J. (2004). Against which human papillomavirus types shall we vaccinate and screen? The international perspective. *Int. J. Cancer* 111(2):278–285.

Nakagawa, M., Stites, D.P., Patel, S., Farhat, S., Scott, M., Hills, N.K., Palefsky, J.M., and Moscicki, A.B. (2000). Persistence of human papillomavirus type 16 infection is associated with lack of cytotoxic T lymphocyte response to the E6 antigens. *J. Infect. Dis.* 182(2):595–598.

Nardelli-Haefliger, D., Lurati, F., Wirthner, D., Spertini, F., Schiller, J.T., Lowy, D.R., Ponci, F., and De Grandi, P. (2005). Immune responses induced by lower airway mucosal immunisation with a human papillomavirus type 16 virus-like particle vaccine. *Vaccine* 23(28):3634–3641.

Nardelli-Haefliger, D., Roden, R.B.S., Benyacoub, J., Sahli, R., Kraehenhuhl, J.P., Schiller, J.T., Lachat, P., Potts, A., and De Grande, P. (1997). Human papillomavirus type 16 virus-like particles expressed in attenuated Salmonella typhimurium elicit mucosal and systemic neutralizing antibodies in mice. *Inf. Immun.* 65:3328–3336.

Nardelli-Haefliger, D., Wirthner, D., Schiller, J.T., Lowy, D.R., Hildesheim, A., Ponci, F., and De Grandi, P. (2003). Specific antibody levels at the cervix during the menstrual cycle of women vaccinated with human papillomavirus 16 virus-like particles. *J. Natl. Cancer Inst.* 95(15):1128–1137.

Nicholls, P.K., Moore, P.F., Anderson, D.M., Moore, R.A., Parry, N.R., Gough, G.W., and Stanley, M.A. (2001). Regression of canine oral papillomas is associated with infiltration of CD4+ and CD8+ lymphocytes. *Virology* 283(1):31–39.

Osen, W., Peiler, T., Ohlschlager, P., Caldeira, S., Faath, S., Michel, N., Muller, M., Tommasino, M., Jochmus, I., and Gissmann, L. (2001). A DNA vaccine based on a shuffled E7 oncogene of the human papillomavirus type 16 (HPV 16) induces E7-specific cytotoxic T cells but lacks transforming activity. *Vaccine* 19(30):4276–4286.

Pagliusi, S.R., and Teresa Aguado, M. (2004). Efficacy and other milestones for human papillomavirus vaccine introduction. *Vaccine* 23(5):569–578.

Palefsky, J.M., and Holly, E.A. (2003). Chapter 6: Immunosuppression and co-infection with HIV. *J. Natl. Cancer Inst. Monogr.* 31:41–46.

Palker, T.J., Monteiro, J.M., Martin, M.M., Kakareka, C., Smith, J.F., Cook, J.C., Joyce, J.G., and Jansen, K.U. (2001). Antibody, cytokine and cytotoxic T lymphocyte responses in chimpanzees immunized with human papillomavirus virus-like particles. *Vaccine* 19(27):3733–3743.

Parkin, D.M., Bray, F., Ferlay, J., and Pisani, P. (2001). Estimating the world cancer burden: Globocan 2000. *Int. J. Cancer* 94(2):153–156.

Pastrana, D.V., Buck, C.B., Pang, Y.Y., Thompson, C.D., Castle, P.E., FitzGerald, P.C., Kruger Kjaer, S., Lowy, D.R., and Schiller, J.T. (2004). Reactivity of human sera in a sensitive, high-throughput pseudovirus-based papillomavirus neutralization assay for HPV16 and HPV18. *Virology* 321(2):205–216.

Pastrana, D.V., Gambhira, R., Buck, C.B., Pang, Y.Y., Thompson, C.D., Culp, T.D., Christensen, N.D., Lowy, D.R., Schiller, J.T., and Roden, R.B. (2005). Cross-neutralization of cutaneous and mucosal Papillomavirus types with anti-sera to the amino terminus of L2. *Virology* 337(2):365–372.

Peh, W.L., Middleton, K., Christensen, N., Nicholls, P., Egawa, K., Sotlar, K., Brandsma, J., Percival, A., Lewis, J., Liu, W.J., and Doorbar, J. (2002). Life cycle heterogeneity in animal models of human papillomavirus-associated disease. *J. Virol.* 76(20):10401–10416.

Pilacinski, W.P., Glassman, D.L., Glassman, K.F., Reed, D.E., Lum, M.A., Marshall, R.F., Muscoplat, C.C., and Faras, A.J. (1986). Immunization against bovine papillomavirus infection. In *Papillomaviruses: Ciba Foundation Symposium 120*. Chichester: Wiley, pp. 136–156.

Pinto, L.A., Edwards, J., Castle, P.E., Harro, C.D., Lowy, D.R., Schiller, J.T., Wallace, D., Kopp, W., Adelsberger, J.W., Baseler, M.W., Berzofsky, J.A., and Hildesheim, A. (2003). Cellular immune responses to human papillomavirus (HPV)-16 L1 in healthy volunteers immunized with recombinant HPV-16 L1 virus-like particles. *J. Infect. Dis.* 188(2):327–338.

Revaz, V., and Nardelli-Haefliger, D. (2005). The importance of mucosal immunity in defense against epithelial cancers. *Curr. Opin. Immunol.* 17(2):175–179.

Rivera, A., and Tyring, S.K. (2004). Therapy of cutaneous human Papillomavirus infections. *Dermatol. Ther.* 17(6):441–448.

Roden, R.B.S., Greenstone, H.L., Kirnbauer, R., Booy, F.P., Jessie, J., Lowy, D.R., and Schiller, J.T. (1996). *In vitro* generation and type-specific neutralization of a human papillomavirus type 16 virion pseudotype. *J. Virol.* 70:5875–5883.

Roden, R.B., Lowy, D.R., and Schiller, J.T. (1997). Papillomavirus is resistant to desiccation. *J. Infect. Dis.* 176(4):1076–1079.

Roden, R.B., Yutzy, W.I., Fallon, R., Inglis, S., Lowy, D.R., and Schiller, J.T. (2000). Minor capsid protein of human genital papillomaviruses contains subdominant, cross-neutralizing epitopes. *Virology* 270(2):254–257.

Roden, R.B., Ling, M., and Wu, T.C. (2004). Vaccination to prevent and treat cervical cancer. *Hum. Pathol.* 35(8):971–982.

Rose, R.C., Bonnez, W., Reichman, R.C., and Garcea, R.L. (1993). Expression of human papillomavirus type 11 L1 protein in insect cells: *In vivo* and *in vitro* assembly of virus-like particles. *J. Virol.* 67(4):1936–1944.

Rose, R.C., Reichman, R.C., and Bonnez, W. (1994). Human papillomavirus (HPV) type 11 recombinant virus-like particles induce the formation of neutralizing antibodies and detect HPV-specific antibodies in human sera. *J. Gen. Virol.* 75(Pt 8):2075–2079.

Rubben, A., Kalka, K., Spelten, B., and Grussendorf-Conen, E.I. (1997). Clinical features and age distribution of patients with HPV 2/27/57-induced common warts. *Arch. Dermatol. Res.* 289(6):337–340.

Salit, R.B., Kast, W.M., and Velders, M.P. (2002). Ins and outs of clinical trials with peptide-based vaccines. *Front. Biosci.* 7:e204-e213.

Schiffman, M., and Kjaer, S.K. (2003). Chapter 2: Natural history of anogenital human papillomavirus infection and neoplasia. *J. Natl. Cancer Inst. Monogr.* 31:14–19.

Schiffman, M., Khan, M.J., Solomon, D., Herrero, R., Wacholder, S., Hildesheim, A., Rodriguez, A.C., Bratti, M.C., Wheeler, C.M., and Burk, R.D. (2005). A study of the impact of adding HPV types to cervical cancer screening and triage tests. *J. Natl. Cancer Inst.* 97(2):147–150.

Schiller, J.T., and Davies, P. (2004). Delivering on the promise: HPV vaccines and cervical cancer. *Nat. Rev. Microbiol.* 2(4):343–347.

Silverman, D.A., and Pitman, M.J. (2004). Current diagnostic and management trends for recurrent respiratory papillomatosis. *Curr. Opin. Otolaryngol. Head Neck Surg.* 12(6):532–537.

Smyth, L.J., Van Poelgeest, M.I., Davidson, E.J., Kwappenberg, K.M., Burt, D., Sehr, P., Pawlita, M., Man, S., Hickling, J.K., Fiander, A.N., Tristram, A., Kitchener, H.C., Offringa, R., Stern, P.L., and Van Der Burg, S.H. (2004). Immunological responses in women with human papillomavirus type 16 (HPV-16)-associated anogenital intraepithelial neoplasia induced by heterologous prime-boost HPV-16 oncogene vaccination. *Clin. Cancer Res.* 10(9):2954–2961.

Stanberry, L.R., Spruance, S.L., Cunningham, A.L., Bernstein, D.I., Mindel, A., Sacks, S., Tyring, S., Aoki, F.Y., Slaoui, M., Denis, M., Vandepapeliere, P., Dubin, G., and Glaxo-Smith-Kline Herpes Vaccine Efficacy Study Group. (2002). Glycoprotein-D-adjuvant vaccine to prevent genital herpes [comment]. *N. Engl. J. Med. Online* 347(21): 1652–1661.

Stanley, M.A. (2001). Immunobiology of papillomavirus infections. *J. Reprod. Immunol.* 52(1–2):45–59.

Steller, M.A., Gurski, K.J., Murakami, M., Daniel, R.W., Shah, K.V., Celis, E., Sette, A., Trimble, E.L., Park, R.C., and Marincola, F.M. (1998). Cell-mediated immunological responses in cervical and vaginal cancer patients immunized with a lipidated epitope of human papillomavirus type 16 E7. *Clin. Cancer Res.* 4(9):2103–2109.

Stern, P.L., Brown, M., Stacey, S.N., Kitchener, H.C., Hampson, I., Abdel-Hady, E.S., and Moore, J.V. (2000). Natural HPV immunity and vaccination strategies. *J. Clin. Virol.* 19(1–2):57–66.

Sturesson, C., Artursson, P., Ghaderi, R., Johansen, K., Mirazimi, A., Uhnoo, I., Svensson, L., Albertsson, A.C., and Carlfors, J. (1999). Encapsulation of rotavirus into poly(lactide-co-glycolide) microspheres. *J. Control Release* 59(3):377–389.

Suzich, J.A., Ghim, S., Palmer-Hill, F.J., White, W.I., Tamura, J.K., Bell, J., Newsome, J.A., Jenson, A.B., and Schlegel, R. (1995). Systemic immunization with papillomavirus L1 protein completely prevents the development of viral mucosal papillomas. *Proc. Natl. Acad. Sci. U S A* 92:11553–11557.

Syrjanen, S. (2005). Human papillomavirus (HPV) in head and neck cancer. *J. Clin. Virol.* 32(Suppl 1):S59–S66.

Tobery, T.W., Smith, J.F., Kuklin, N., Skulsky, D., Ackerson, C., Huang, L., Chen, L., Cook, J.C., McClements, W.L., and Jansen, K.U. (2003). Effect of vaccine delivery system on the induction of HPV16L1-specific humoral and cell-mediated immune responses in immunized rhesus macaques. *Vaccine* 21(13–14):1539–1547.

Villa, L.L., Costa, R.L., Petta, C.A., Andrade, R.P., Ault, K.A., Giuliano, A.R., Wheeler, C.M., Koutsky, L.A., Malm, C., Lehtinen, M., Skjeldestad, F.E., Olsson, S.E., Steinwall, M., Brown, D.R., Kurman, R.J., Ronnett, B.M., Stoler, M.H., Ferenczy, A., Harper, D.M., Tamms, G.M., Yu, J., Lupinacci, L., Railkar, R., Taddeo, F.J., Jansen, K.U., Esser, M.T., Sings, H.L., Saah, A.J., and Barr, E. (2005). Prophylactic quadrivalent human papillomavirus (types 6, 11, 16, and 18) L1 virus-like particle vaccine in young women: a randomised double-blind placebo-controlled multicentre phase II efficacy trial. *Lancet Oncol.* 6(5):271–278.

Warzecha, H., Mason, H.S., Lane, C., Tryggvesson, A., Rybicki, E., Williamson, A.L., Clements, J.D., and Rose, R.C. (2003). Oral immunogenicity of human papillomavirus-like particles expressed in potato. *J. Virol.* 77(16):8702–8711.

White, W.I., Wilson, S.D., Bonnez, W., Rose, R.C., Koenig, S., and Suzich, J.A. (1998). *In vitro* infection and type-restricted antibody-mediated neutralization of authentic human papillomavirus type 16. *J. Virol.* 72(2):959–964.

Wiley, D.J., Douglas, J., Beutner, K., Cox, T., Fife, K., Moscicki, A.B., and Fukumoto, L. (2002). External genital warts: diagnosis, treatment, and prevention. *Clin. Infect. Dis.* 35(Suppl 2):S210–S224.

Winer, R.L., Lee, S.K., Hughes, J.P., Adam, D.E., Kiviat, N.B., and Koutsky, L.A. (2003). Genital human papillomavirus infection: Incidence and risk factors in a cohort of female university students. *Am. J. Epidemiol.* 157(3):218–226.

Winer, R.L., Kiviat, N.B., Hughes, J.P., Adam, D.E., Lee, S.K., Kuypers, J.M., and Koutsky, L.A. (2005). Development and duration of human papillomavirus lesions, after initial infection. *J. Infect. Dis.* 191(5):731–738.

Woodworth, C.D. (2005). HPV innate immunity. *Front Biosci.* 7:2058–2071.

Yang, R., Murillo, F.M., Delannoy, M.J., Blosser, R.L., Yutzy, W.H.T., Uematsu, S., Takeda, K., Akira, S., Viscidi, R.P., and Roden, R.B. (2005). B lymphocyte activation by human papillomavirus-like particles directly induces Ig class switch recombination via TLR4-MyD88. *J. Immunol.* 174(12):7912–7919.

Yuan, H., Estes, .A., Chen, Y., Newsome, J., Olcese, V.A., Garcea, R.L., and Schlegel, R. (2001). Immunization with a pentameric L1 fusion protein protects against papillomavirus infection. *J. Virol.* 75(17):7848–7853.

Zhang, L.F., Zhou, J., Chen, S., Cai, L.L., Bao, Q.Y., Zheng, F.Y., Lu, J.Q., Padmanabha, J., Hengst, K., Malcolm, K., and Frazer, I.H. (2000). HPV6b virus like particles are potent immunogens without adjuvant in man. *Vaccine* 18(11–12): 1051–1058.

Zhou, J., Stenzel, D.J., Sun, X.Y., and Frazer, I.H. (1993). Synthesis and assembly of infectious bovine papillomavirus particles *in vitro. J. Gen. Virol.* 74:763–768.

Zhou, J., Liu, W.J., Peng, S.W., Sun, X.Y., and Frazer, I. (1999). Papillomavirus capsid protein expression level depends on the match between codon usage and tRNA availability. *J. Virol.* 73:4972–4982.

Zinkernagel, R.M. (2003). On natural and artificial vaccinations. *Annu. Rev. Immunol.* 21:515–546.

zur Hausen, H. (2002). Papillomaviruses and cancer: From basic studies to clinical application. *Nat. Rev. Cancer* 2(5):342–350.

15
Clinical Assessment, Therapies, New Tests, and Algorithms

Christopher P. Crum,[1] and Ralph M. Richart[2]

[1] *Division of Women's and Perinatal Pathology, Department of Pathology, Brigham and Women's Hospital and Harvard Medical School, Boston, MA 02115*
[2] *Department of Pathology, Columbia University College of Physicians and Surgeons, New York, NY 10032*

15.1.   Introduction

Human papillomaviruses (HPV) are common in sexually active populations and infect over 70 percent of reproductive-age women at some point in their lives. HPVs are directly linked to cervical preinvasive and invasive neoplasia, an association underlying strategies that employ viral testing in screening for cervical neoplasia and in patient care algorithms. The data supporting such approaches have been supported by the ASCUS/LSIL Triage Study for Cervical Cancer (ALTS) trial and affirmed in consensus conferences, which have led to national management guidelines that have included a significant commitment to HPV DNA testing (Solomon et al., 2001; Wright et al., 2003).

This review will address the recent evolution of HPV DNA testing, including the intellectual basis for its use, potential applications to screening, algorithms for the management of nondiagnostic abnormalities, "problem-oriented" HPV DNA testing, and new therapeutic strategies.

15.2.   Human Papillomaviruses and Risk

15.2.1.   HPV Infection Is the Principal Cause of Cervical Neoplasia

The connection between HPV infection and cervical neoplasia has been well established, and the accumulated experimental, molecular, and clinical evidence establishing this link have left no doubt that HPV causes cervical neoplasia.

15.2.2.   HPV Infection Is Extremely Common in Sexually Active, Reproductive-Age Women

Sexually active teenagers have a positive point prevalence of HPV DNA as high as 40 percent. Most infections are transient, often appearing and disappearing without detectable cytologic abnormality, but persistent infection by the same oncogenic-risk HPV type is strongly associated with an increased risk of a current or subsequent cervical neoplasm (Rosenfeld et al., 1992; Bory et al., 2002).

15.2.3.   There Is a Broad Gradient of Risk Imposed by Cancer-Associated (High-Risk) HPV Types, with HPV Type 16 Conferring the Greatest Risk; Low-Risk HPV Types May Confer Risk as Surrogate Markers of At-Risk Behavior

Over 100 human papillomavirus types have been identified and characterized, and more than 25 of these have been isolated from the anogenital tract. Genital HPVs have been divided into those with low, intermediate, and high association with cervical neoplasia. Those HPVs with any association of cancer are termed high-risk or "oncogenic-risk" types, and the relative strength of the association between certain HPV groups and cancer has been published (Munoz et al., 2003). About 70 percent of low-grade CINs (cervical intraepithelial neoplasia, graded 1–3) contain one or more viruses from the high-risk group, with HPV16 predominating, followed in prevalence by HPV types 18, 45, and 52. Low-risk HPV types are not found in bone fide high-grade CINs or invasive cancers, but infection with a low-risk type may serve as a surrogate risk factor for a simultaneous or subsequent infection by a high-risk HPV type (Koutsky et al., 1992). Interestingly, the distribution of high-risk HPV types in women with normal Papanicolaou (Pap) smears closely parallels the type frequency of those viruses in cervical neoplasia, suggesting that transient infections are the rule and that many of these will not be accompanied by morphological changes. In adenocarcinomas or AIS, HPV types 16 and 18 predominate (Pirog et al., 2000). Low-risk HPV types do not infect the endocervical epithelium.

15.2.4.   A Woman Harboring a Persistent High-Risk HPV Type in Her Genital Tract Is at Risk for Developing a High-Grade Squamous Intraepithelial Lesion (HSIL)

Although the vast majority of HPV infections are transient and the risk of developing HSIL depends upon the HPV type, the duration of infection is of greatest importance, as persistence places the patient at higher proximate risk of developing a high-grade CIN or invasive cancer. The importance of the amount of virus present (viral load) is still being debated.

Remission of transient HPV infections is due to an immunological response, and type-specific immunity follows remission. As would be expected, many women develop serial infections with the different HPV types, particularly in association with HPV16. Although over 90 percent of infected women clear their infection, persistent infection by the same oncogenic-risk HPV type is strongly associated with an increased risk of developing cervical neoplasia. Retrospective studies of patients with cervical cancer have also documented an association between persistent HPV DNA positivity and neoplasia. Hopman et al. and Nobbenhuis et al. showed that over 90 percent of patients who developed Pap smear abnormalities had persistent high-risk HPV infection (Hopman et al., 2000; Nobbenhuis et al., 2001). Although many studies indicate that viral load is strongly associated with the risk of developing a biopsy-proven squamous intraepithelial lesion (SIL), even small amounts of virus may herald the presence of a preinvasive lesion, and setting too high a threshold for HPV DNA positivity will exclude some patients with coincident neoplasia.

15.2.5. Women Infected by HIV Are at Increased Risk of CIN

The risk of HPV DNA positivity is increased in women who are HIV positive, as is the risk of persistent HPV infection, the risk of subsequent CIN, and the risk of persistent CIN (Ahdieh et al., 2001). Evidence linking HIV infection to HSIL and cancer is less compelling. Ellerbrock reported that 91 percent and 75 percent of SILs in the two different groups of HIV positive women, respectively, were low-grade SILs (LSIL), and Franceschi et al. reported a 15-fold greater risk of low-grade SILs in HIV-infected women (Ellerbrock et al., 2000; Franceschi et al., 1998). However, most studies have not found HIV to increase the risk of developing cervical cancer, perhaps because of competitive risk.

15.2.6. Sexual History of the Male Sexual Partner Influences the Risk of Cervical Neoplasia

Thomas linked CIN risk in Thai women with male behavior, specifically with unprotected intercourse with prostitutes (Thomas et al., 2001). Castilla et al. (1996) found that prevalence rates of penile HPV in Spain and Columbia explained the differences in cervical cancer rate in these two countries. Some groups, but not all, have linked sexual behavior of husbands to the risk of cervical cancer in their monogamous wives (Agarwal et al., 1993).

15.3.   HPV DNA Testing

15.3.1.   Screening

Unlike the subjective interpretation of the Papanicolaou smear and colposcopy, HPV DNA detection provides a test of high sensitivity and an objective measurement of cervical cancer or precursor risk; the increase in risk in HPV-DNA-positive women is from 10- to 40-fold. If the HPV DNA test is positive, the risk of a HSIL will vary depending on whether the smear is normal (HSIL risk of approximately 10 percent), abnormal (at least 20 percent), and whether HPV DNA is detected at more than one visit (up to 33 percent). A single positive hybrid capture II (HC II) HPV DNA test, particularly in young women, has less HSIL (CIN2 to 3) predictive value. However, HPV DNA testing using the high-risk probe set provides a negative predictive value of 99–100 percent when both tests are negative. This evaluation not only provides greater reassurance of negativity than any other test or combination of tests, it does so in a cost-effective fashion (Goldie et al., 2005). The principal value of HPV DNA testing in a screening program is to identify invasive cancer in women whose lesions might have been missed by cytologic screening alone. The use of HPV DNA testing in conjunction with a Pap test should substantially reduce the incidence of "interval cancers" and minimize the number of women whose cancers remain undetected despite regular Pap smear screening. It is important to point out, however, that HPV DNA testing is not recommended for women under 30 years of age, as the positive rate in this age group is 15–25%, a prevalence too high to be clinically manageable (Lorincz and Richart, 2003). Protein-based "secondary markers" are being widely studied and may allow HPV DNA testing to be extended to younger women in the future.

Combining HPV DNA testing and cytology may allow for a wider screening interval and may be more efficient in older women. HPV DNA can almost always be detected prior to the development of a cytologic abnormality, and it is clear that patients who are negative by both HPV DNA testing and Pap smear may be rescreened at longer intervals than those screened by Pap smear alone. An increased screening interval (recommended at 3 years in the United States) can be used once a woman has been ascertained to be free of high-risk HPV and related lesions, because the time from a new infection to HSIL or invasive cancer is prolonged in these women. Even if new HPV infections are acquired, they will almost never progress to a clinically important disease during the recommended screening interval (Lorincz and Richart, 2003).

Some studies have shown that a sizable fraction of postmenopausal women will score positive for HPV DNA (Smith et al., 2004). A recent study using a very sensitive assay for HPV DNA found two peaks of HPV DNA prevalence. The first peak of 16.7 percent was found in women under 25 years of age. HPV DNA prevalence then declined to 3.7 percent in the age group 35–44 years, then increased progressively to 23 percent in women 65 years and older (Herrero et al., 2000). However, the second peak of HPV DNA positivity in older women consisted principally of low-risk HPV types. Increased rates of HPV detection in older women have been inconsistently linked to a specific cause, such as hormone replacement therapy (Smith et al., 2004; Ferenczy et al., 1997).

15.3.2. Application to the Management of Atypical Squamous Cells of Undetermined Significance

The 1988 Bethesda Conference created the term "atypical squamous cells of undetermined significance" (ASCUS) to identify the subset of squamous atypias that cannot readily be designated as benign or HPV related by cytological analysis. At the 2001 Bethesda Conference, the ASCUS diagnostic category was simplified and a new rubric, ASC-US, was introduced. Because the diagnosis of ASC-US is an admission of uncertainty, it is implicit that the criteria for this cytological diagnosis will remain vague and, by definition, less reproducible than the criteria for squamous intraepithelial lesions. This uncertainly has been documented by numerous studies in which intra- and interobserver reproducibility are poorer for a diagnosis of squamous atypia than for a diagnosis of SIL. ASC-US is important, nonetheless, because patients with this diagnosis will, in aggregate, have an approximate 10 percent risk of a coexisting high-grade squamous intraepithelial lesion and up to half the patients with an ASC-US diagnosis will have a low-grade CIN lesion (Wright et al., 2004).

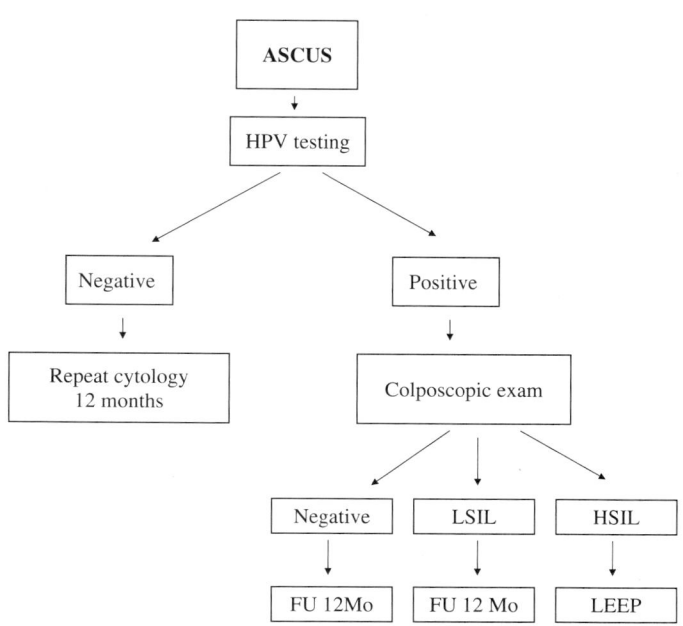

Prior to the development of a sensitive and reproducible HPV DNA test with demonstrated clinical utility, all ASC-US patients were triaged colposcopically at considerable expense. With present generation HPV DNA testing, however, it is possible to triage ASC-US patients cost effectively at the laboratory rather than the clinical level. As liquid-based cytology (LBC) has largely replaced the conventional Pap smear in the United States, it is customary for clinicians to leave a standing order with their laboratory for "reflex testing" for HPV DNA from the residual fluid in all patients with ASC-US. High-risk HPV DNA positive patients are referred for colposcopy, whereas those who are HPV DNA negative are returned to the follow-up pool and receive further testing at a longer interval.

As both LSIL and HSIL cytological diagnoses are, by definition, lesions caused by HPV, further HPV DNA testing for triage or at the time of the first colposcopic follow-up visit is not clinically useful. All such patients must be examined colposcopically to rule out HSIL or invasion and then treated or followed appropriately in accordance with published guidelines. As patients with "ASC favor HSIL" have such a high prevalence of HSIL on colposcopy, they are managed as if they had a HSIL cytological diagnosis and are not triaged with HPV DNA testing.

The 2001 Bethesda Conference and the American Society for Colposcopy and Cervical Pathology (ASCCP) Consensus Conference both endorsed HPV DNA testing in the management of patients with ASC-US Pap smears, and that approach has become the standard of care in the United States. Because HPV DNA testing is so useful as a triage procedure, the older three subsets of ASC-US were combined without further subclassification as ASC-US (except for ASC favor HSIL), enormously simplifying the management of patients with equivocal Pap smear results.

Currently three approaches are recommended for the management of patients with an ASC-US Pap smear, including Papanicolaou smear follow-up, immediate colposcopy, and concurrent (reflex) or subsequent HPV DNA testing. These approaches have been described in a summary of the ASCCP Consensus Conference (www.consensus.asccp.org) and are summarized in Table 1. In short, Papanicolaou smear follow-up is considered effective but less sensitive than HPV DNA testing and requires continued follow up. Colposcopic examination and reflex HPV DNA testing are equivalent in sensitivity.

In general women with ASCUS favor HSIL (ASCUS-H) are referred directly to colposcopy. HPV testing of this group might provide some reassurance in a subset of women whose HPV test and colposcopic exam prove to be negative.

15.3.3. Management of Patients with Atypical Glandular Cells

The patient with a Papanicolaou smear containing atypical glandular cells (AGUS) which may be of endocervical origin poses a greater diagnostic challenge than those with ASC-US because the potential disease state includes

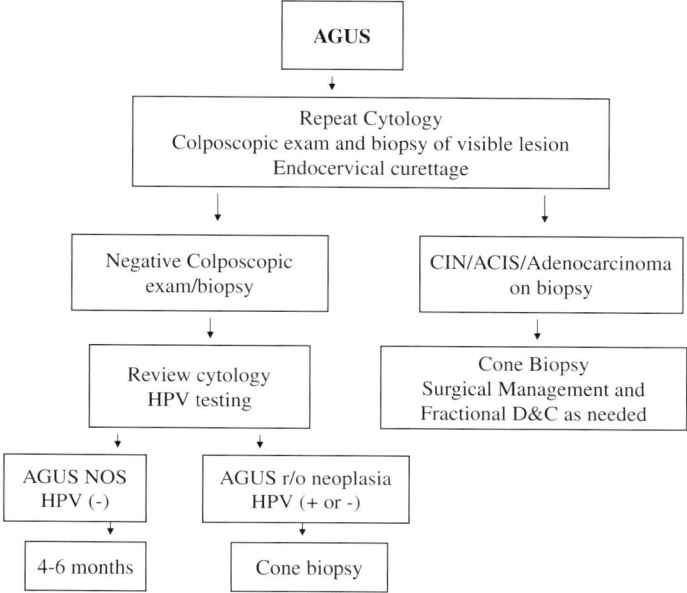

not only adenocarcinoma in situ but HSIL (30 percent) and invasive adeno-carcinoma (approximately 1%). In women over 35, an upper genital tract adenocarcinoma (endometrial, tubal, or ovarian) must also be excluded (Krane et al., 2004). Although the use of HPV DNA testing in AGUS patients has not been agreed upon in consensus reports, nearly all endocervical neoplasms are caused by HPV and can be detected using HPV DNA testing. However, there are fewer reports on HPV DNA testing in AGUS patients than ASC-US patients, and consensus agreements have been more difficult to reach. Like ASCUS-H, HPV testing, if accompanied by a negative exam, might provide a measure of reassurance, inasmuch as preliminary studies suggest a low rate of neoplasia on short-term followup of these patients (Kinney, W., personal communication).

15.3.4. Problem-Oriented HPV DNA Testing

15.3.4.1.   HPV DNA Testing Following CIN Therapy

There is compelling evidence that HPV is usually eliminated following ablative or excisional therapy of CIN. Tate et al. reported that HPV was rarely present in the mucosa adjacent to squamous intraepithelial lesions, implying that if these lesions were removed, the virus would be eliminated (Tate et al., 1996). Post-treatment follow-up studies have uniformly shown that if an HPV DNA test is negative, persistent disease or recurrences are rare. If persistent disease or new infections do occur, they are associated with HPV DNA positivity. Lin et al. and Jain et al. analyzed the relationship between HPV DNA status following cone biopsy and disease detected at hysterectomy (Lin et al., 2001; Jain et al., 2001). Although many of these patients had positive cone margins, if they were HPV DNA negative, the negative predictive value approached 100 percent. In a study of the positive predictive value of HPV DNA testing, Chua et al. reported that 25 of 26 patients with persistent disease scored positive for HPV DNA (Chua and Hjerpe, 1997). Other authors have reported that HPV DNA may be present, at least initially, following successful CIN ablation. The specificity of HPV DNA testing in the studies by Lin et al. and Jain et al. were only 52 and 14 percent, indicating that papillomavirus may persist, despite the absence of morphologically identified disease. In contrast, Chua et al. reported no HPV DNA positive patients in post-cone cytologic specimens from 22 patients with negative follow-up. Strand et al. found that at 6–12 months following cone biopsy or laser vaporization, only 2 of 30 women were HPV DNA positive (Medeiros et al., 2005). It was suggested that HPV DNA testing delayed for 6–12 months post-therapy may provide useful information regarding the risk of persistent disease, particularly in women with abnormal cytology following cone biopsy (Wright et al., 2004).

15.3.4.2.   HPV DNA Testing To Resolve Tumor Origin

Because HPV is so strongly linked to cervical cancer, it may be valuable as a marker to define tumors of uncertain origin. Thus, HPV DNA positivity is *prima facie* evidence that a tumor originates in the lower anogenital tract, although HPV positive head-and-neck and skin cancers also occur.

15.3.4.3.   HPV DNA Testing To Identify Suitable Candidates
for Therapeutics Targeting Related Diseases

Although there are no pharmaceutical agents that are effective in treating CIN or cervical cancer, a number of trials targeting HPV-related tumors in both men and women, including not only the female genital tract but also the anal canal and oropharynx, are underway. If HPV-specific therapeutic agents are successfully developed, HPV DNA testing may provide valuable selection criteria.

15.3.5.   Potential Risks of HPV DNA Testing

Although the HPV DNA test itself is objective, the manner in which the clinician deals with the results of the test may be subjective and not well informed. Patients, similarly, may not fully understand the implications of testing. Although a woman's lifetime risk of acquiring an HPV infection is about 70 percent, less than 5 percent will develop a cervical lesion. Both clinicians and patients must be sufficiently well informed to understand the implications of this fact. Unlike an abnormal Papanicolaou smear which, itself, imposes a considerable psychological burden on the patient, HPV DNA testing may label the individual as afflicted not only with a sexually transmitted disease, but an infection that may not be confirmed on exam and may persist for some time with an uncertain outcome. Although a cancer outcome is highly unlikely in cytologically negative women, the uncertainty of the diagnosis requires careful patient counseling. Conaglen et al. found that considerable psychological impact resulted from a diagnosis of HPV infection, although this reaction was not distinct from anxiety in response to other infections (Conaglen et al., 2001).

15.3.5.1. Limitations in Specificity

Specificity issues fall into two categories. The first is cross-reactivity. The probe cocktails used in the hybrid capture II (HC II) system consists of 13 of the most common high- and intermediate-oncogenic-risk HPV types. The HC II probe set will detect virtually 100 percent of infections by these types. However, to achieve this sensitivity, the test must also detect a proportion of HPV types that do not fall within the high-risk group, including types 6, 11, 53, 54, 66, and others (Poljak et al., 2002; Hughes et al., 2002; Medeiros et al., 2005). Although detection of these HPV types is not a significant disadvantage, physicians must understand these implications and patients must be properly counseled and managed. It is important not to view HPV DNA detection as a "cancer test" but as a test that measures the relative risk of having or developing a high-grade CIN or cancer.

The second issue concerns low positive HC II results. The threshold of the HC II assay is determined by comparing the signals from two sets of standards such that a relative light unit (the metric used by HC II) value of 1.0 is the cut-off point for HC II positivity. A number of studies have established that HPV DNA test results in the low range of positive (RLU from 1.0 to 5.0) have a higher index of variability from test to test than those at higher levels (de Cremoux et al., 2003). Federschneider et al. noted that 40 percent of HC II positive cases in the 0.8 to 1.5 RLU range were not confirmed as HPV DNA positive in a subsequent test. In our experience, RLU values between 1.0 and 5.0 have a lower rate of confirmation by polymerase chain reaction analysis, with only 20 percent containing HPV types in the high-risk probe set and over 40 percent scoring negative (Federschneider et al., 2004). In a recent statistical analysis of signal strengths in the microtiter plates used for HC II, there was a relationship between the adjacent well values that might reflect low level cross-contamination during the assay (Federschneider et al., 2004). This observation has not been confirmed and may be addressed by technical improvements in thes assay.

It is unclear what reporting or clinical recommendations should be made to deal more effectively with these observations. For example, the risk of a HSIL outcome in an HPV DNA-*negative* ASC-US patient is approximately 1%. The clinical significance of this borderline category is not trivial, inasmuch as 7 percent of cytology samples fall within this range. It is not clear whether laboratories should report RLU values and, if positive, whether tests that fall into the low values range should be repeated. Patients with an ASC-US Pap smear and a low positive HC II value are at greater risk of having HSIL than a

HC II negative patient with a normal Papanicolaou smear, and efforts to fine-tune management based on additional testing may or may not prove useful. An alternate approach is to manage such patients expectantly by repeating the HC II in 6 months or by colposcopy (both acceptable under Bethesda 2000 guidelines).

15.3.6. Type-Specific HPV DNA Testing

The preferred management of patients with LSIL Pap smears under the ASCCP guidelines is to rule out HSIL or invasion colposcopically, then follow the patients with HPV DNA testing and Pap smears. If the patient is persistently (over one year) Pap smear or HPV DNA positive, then she should be re-colposcoped and treated appropriately. This strategy is based on the reasoning that the persistent shedding of oncogenic-risk HPV places a patient at risk of having HSIL. As currently practiced, that recommendation implies type-specific persistence. It is known, however, that most HPV infections, irrespective of type, are detected and cleared by the immune system and that many patients with what appears to be persistent HPV shedding are, in fact, experiencing serial infections. At present these considerations are ignored, and the assumption is that when persistent HPV shedding is observed, it is type specific. The substantially increased costs of routine HPV typing has not been judged to be required for adequate clinical management.

15.4. Conventional Approaches to Pre-Invasive HPV-Related Diseases

15.4.1. Cytology- and Pathology-Based Management

The typical approach to a diagnosis of CIN is as follows: If the cytologic diagnosis is ASCUS or LSIL and if the colposcopy reveals either LSIL (on biopsy) or does not confirm an abnormality, recommended management is a follow-up examination in 12 months, including a Papanicolaou smear. One study showed that the follow-up risk for HSIL at 2 years after a negative colposcopic exam, negative biopsy or biopsy of LSIL in the above setting is between 11 and 13 percent (Cox et al., 2003). For this reason, the absolute distinction of LSIL from normal mucosa by the pathologist is of marginal significance in terms of follow-up strategy, provided the findings are consistent with the colposcopic impression. If a diagnosis of LSIL persists at one year, an excisional (Loop electrical excision procedure (LEEP) or cone biopsy) procedure should be considered.

The management of a patient with a biopsy diagnosis of HSIL is typically to perform LEEP or cone biopsy. In the event that the colposcopic exam following an HSIL cytology does not confirm the lesion or the biopsy is negative, colposcopic follow-up is advised. Repeatedly abnormal cytology is grounds for LEEP.

Management of atypical glandular cells should include the following:

If the cytologic findings are diagnostic of ACIS, colposcopic exam and endocervical curettage is warranted. If negative, the cytology should be reviewed. If the diagnosis is confirmed on review, cone biopsy should be performed.

If the findings are AGUS, not otherwise specified, the above protocol should be followed. Alternatively, HPV testing is warranted if the glandular abnormality is inconclusive. A negative HPV testing is most useful if the abnormality is considered inconclusive on review. However, the following concerns must be addressed, including (1) the possibility that the abnormal cells come from the upper genital tract and (2) the possibility of an HPV-negative cervical adenocarcinoma. While the latter are rare, they do occur and impose a responsibility on both pathologist and clinician to never make a therapeutic decision based on HPV data alone. The diagnosis of AGUS is grounds for a thorough investigation, even if cone biopsy is not the initial procedure (Tam et al., 2003).

15.5.   New Therapeutic Approaches

15.5.1.   Aldara

A trial of treating CIN with the immuno-enhancing agent Aldara (Imiquimod) was terminated early because of unacceptable local side effects at the dosages and routes of administration selected for the trials. Buck, however, has reported a high rate of therapeutic efficacy using topical Aldara in the treatment of HPV-related vaginal lesions (Buck et al., 2002).

15.5.2.   Immunotherapeutics

There are a number of on-going studies of immunotherapeutic approaches to treating HPV-related lesions, including a study of the Safety and Immunogenicity of Repeated Vaccination with pNGVL4a-HPV-16 E7(detox)/HSP70 in patients with Stage III or IV HPV16 positive head and neck squamous cell carcinoma (HNSCC); a phase I study of the Safety, Immunogenicity, and Anti-angiogenic Effects of Repeated Vaccination with pNGVL4a-CRT/E7 (detox) DNA in patients with Advanced, Persistent, or Recurrent Cervical Carcinoma; and a phase I/II clinical trial of pNGVL4-Sig/E7(detox)/HSP70 for the treatment of patients with HPV16+ CIN3 (M. Gillison, D. Armstrong, C. Trimble, and T.C. Wu, personal communication).

A recent controlled trial evaluated a DNA vaccine employing 11 immunogenic regions from four HPV target proteins of HPV16 and HPV18. The resulting multivalent immunogen was encoded into a bacterial plasmid DNA expression vector with a strong transcriptional promoter. To enhance potency, the plasmid DNA was encapsulated within biodegradable poly (D,L-lactide-co-glycolide) (PLG) microparticles. Following encapsulation of the plasmid, the 1–2-μm particles were lyophilized. This preparation (ZYC101a) was suspended in saline and administered as an injectable. One hundred fifty subjects with a history of biopsy proven CIN 2 or CIN 3 were randomized into treatment and placebo groups, given three injections over 4 months, and underwent cone biopsy 6 months following entry into the study. Overall, the regression rate was higher in the treatment group (44 percent vs. 25 percent), but the difference was not statistically significant. However, for subjects under age 25, the regression rate (70 percent vs. 30 percent) was statistically significant. Whether this difference reflects a greater susceptibility in less advanced pre-invasive disease (as would be expected for younger women) or other age-related factors is not clear (Garcia et al., 2004).

15.5.3. Preventive Vaccines

With the discovery and subsequent refinements in the laboratory production of papillomavirus-like particles (VLPs), preventive vaccines for both animal and human papillomavirus infections have been developed and tested. They have proven to be highly efficacious in preventing both persistent infection and preinvasive disease (Zhou et al., 1991; Kirnbauer et al., 1993; Ghim et al., 1992; Hagensee et al., 1993; Rose et al., 1993; Koutsky et al., 2002; Harper et al., 2004). Precisely how many HPV types will be combined, the timing of vaccination, and the duration of their protective effects remain to be determined. Nonetheless, a vaccine combining HPV types 6, 11, 16, and 18 has been approved for release. Assuming a lasting level of protection and introduction of the vaccine at puberty, at least two decades will elapse before a noticeable change in cervical cancer rates can be detected. The reduction in preinvasive disease and abnormal Papanicolaou smears will be noticed more rapidly, although at least half of abnormal Papanicolaou smears are due to non-HPV-related causes. This fact will likely place greater emphasis on HPV DNA testing as the proportion of abnormal smears that are associated with HPV decreases.

15.6. Summary

The causal connection between HPV infection and cervical cancer has radically changed the clinical approach to this disease. Combining HPV testing with cytology has resulted in new care algorithms that are less expensive and more sensitive than previous Pap smears. As more sophisticated technologies arise, future care guidelines will continue to change along with new therapeutic options that will be available to patients.

References

Agarwal, S.S., Sehgal, A., Sardana, S., Kumar, A., and Luthra, U.K (1993). Role of male behavior in cervical carcinogenesis among women with one lifetime sexual partner. *Cancer* 72(5):1666–1669.

Ahdieh, L., Klein, R.S., and Burk, R., et al. (2001). Prevalence, incidence, and type-specific persistence of human papillomavirus in human immunodeficiency virus (HIV)-positive and HIV-negative women. *J. Infect. Dis.* 184(6):682–690.

Bory, J.P., Cucherousset, J., Lorenzato, M., Gabriel, R., Quereux, C., Birembaut, P., and Clavel, C. (2002, Dec 10). Recurrent human papillomavirus infection detected with the hybrid capture II assay selects women with normal cervical smears at risk for developing high grade cervical lesions: A longitudinal study of 3,091 women. *Intl. J. Cancer* 102(5):519–525.

Buck, H.W., Fortier, M., Knudsen, J., and Paavonen, J. (2002). Imiquimod 5% cream in the treatment of anogenital warts in female patients. *Intl. J. Gynaecol. Obstet.* 77:231–238.

Castilla, J., Barrio, G., de la Fuente, L., and Belza, M.J. (1998). Sexual behaviour and condom use in the general population of Spain, 1996. *AIDS Care* 10(6):667–676.

Chua, K.L., and Hjerpe, A. (1997). Human papillomavirus analysis as a prognostic marker following conization of the cervix uteri. *Gynecol. Oncol.* 66:108–113.

Conaglen, H.M., Hughes, R., Conaglen, J.V., and Morgan, J. (2001). A prospective study of the psychological impact on patients of first diagnosis of human papillomavirus. *Intl. J. STD & AIDS* 12:651–658.

Cox, J.T., Schiffman, M., and Solomon, D.; ASCUS-LSIL Triage Study (ALTS) Group. (2003). Prospective follow-up suggests similar risk of subsequent cervical intraepithelial neoplasia grade 2 or 3 among women with cervical intraepithelial neoplasia grade 1 or negative colposcopy and directed biopsy. 188:1406–1412.

de Cremoux, P., Coste, J., Sastre-Garau, X., Thioux, M., Bouillac, C., Labbe, S., Cartier, I., Ziol, M., Dosda, A., Le Gales, C., Molinie, V., Vacher-Lavenu, M.C., Cochand-Priollet, B., Vielh, P., and Magdelenat, H.; French Society of Clinical Cytology Study Group. (2003). Efficiency of the hybrid capture 2 HPV DNA test in cervical cancer screening. A study by the French Society of Clinical Cytology. *Am. J. Clin. Pathol.* 120:492–499.

Ellerbrock, T.V., Chiasson, M.A., Bush, T.J., Sun, X.W., Sawo, D., Brudney, K., and Wright, T.C. Jr (2000). Incidence of cervical squamous intraepithelial lesions in HIV-infected women. *JAMA* 283(8):1031–1037.

Federschneider, J.M., Yuan, L., Brodsky, J., Breslin, G., Betensky, R.A., and Crum, C.P. (2004). The borderline or weakly positive Hybrid Capture II HPV test: A statistical and comparative (PCR) analysis. *Am J. Obstet. Gynecol.* 191:757–761.

Ferenczy, A., Gelfand, M.M., Franco, E., and Mansour, N. (1997). Human papillomavirus infection in postmenopausal women with and without hormone therapy. *Obstet. Gynecol.* 90:7–11.

Franceschi, S., Dal Maso, L., Arniani, S., Crosignani, P., Vercelli, M., Simonato, L., Falcini, F., Zanetti, R., Barchielli, A., Serraino, D., and Rezza, G. (1998). Risk of cancer other than Kaposi's sarcoma and non-Hodgkin's lymphoma in persons with AIDS in Italy. Cancer and AIDS Registry Linkage Study. *Br. J. Cancer* 78:966–970.

Garcia, F., Petry, K.U., Muderspach, L., Gold, M.A., Braly, P., Crum, C.P., Magill, M., Silverman, M., Urban, R.G., Hedley, M.L., and Beach, K.J. (2004). ZYC101a for treatment of high-grade cervical intraepithelial neoplasia: A randomized controlled trial. *Obstet. Gynecol.* 103:317–326.

Ghim, S.J., Jenson, A.B., and Schlegel, R. (1992). HPV-1 L1 protein expressed in cos cells displays conformational epitopes found on intact virions. *Virology* 190:548–552.

Goldie, S.J., Gaffikin, L., Goldhaber-Fiebert, J.D., Gordillo-Tobar, A., Levin, C., Mahe, C., and Wright, T.C. (2005). Alliance for Cervical Cancer Prevention Cost Working Group Cost-effectiveness of cervical-cancer screening in five developing countries. *N. Engl. J. Med.* 353:2158–2168.

Hagensee, M.E., Yaegashi, N., and Galloway, D.A. (1993). Self-assembly of human papillomavirus type 1 capsids by expression of the L1 protein alone or by coexpression of the L1 and L2 capsid proteins. *J. Virol.* 67:315–322.

Harper, D.M., Franco, E.L., Wheeler, C., Ferris, D.G., Jenkins, D., Schuind, A., Zahaf, T., Innis, B., Naud, P., De Carvalho, N.S., Roteli-Martins, C.M., Teixeira, J., Blatter, M.M., Korn, A.P., Quint, W., and Dubin, G.; GlaxoSmithKline HPV Vaccine Study Group. (2004). Efficacy of a bivalent L1 virus-like particle vaccine in prevention of infection with human papillomavirus types 16 and 18 in young women: A randomised controlled trial. *Lancet* 364:1757–1765.

Herrero, R., Hildesheim, A., Bratti, C., Sherman, M.E., Hutchinson, M., Morales, J., Balmaceda, I., Greenberg, M.D., Alfaro, M., Burk, R.D., Wacholder, S., Plummer, M., and Schiffman, M. (2000). Population-based study of human papillomavirus infection and cervical neoplasia in rural Costa Rica. *J. Natl. Cancer Inst.* 92:464–474.

Hopman, E.H., Rozendaal, L., Voorhorst, F.J., Walboomers, J.M., Kenemans, P., and Helmerhorst, T.J. (2000). High risk human papillomavirus in women with normal cervical cytology prior to the development of abnormal cytology and colposcopy. *BJOG* 107:600–604.

Hughes, S.A., Sun, D., Gibson, C., Bellerose, B., Rushing, L., Chen, H., Harlow, B.L., Genest, D.R., Sheets, E.E., and Crum, C.P. (2002). Managing atypical squamous cells of undetermined significance (ASCUS): Human papillomavirus testing, ASCUS subtyping, or follow-up cytology? *Am. J. Obstet. Gynecol.* 186:396–403.

Jain, S., Tseng, C.J., Horng, S.G., Soong, Y.K., and Pao, C.C. (2001). Negative predictive value of human papillomavirus test following conization of the cervix uteri. *Gynecol. Oncol.* 82:177–180.

Kirnbauer, R., Taub, J., Greenstone, H., Roden, R., Durst, M., Gissmann, L., Lowy, D.R., and Schiller, J.T. (1993). Efficient self-assembly of human papillomavirus type 16 L1 and L1-L2 into virus-like particles. *J. Virol.* 67(12):6929–6936.

Koutsky, L.A., Holmes, K.K., Critchlow, C.W., Stevens, C.E., Paavonen, J., Beckman, A.M., DeRouen, T.A., Galloway, D.A., Vernon, D., and Kiviat, N.B. (1992). A cohort study of the risk of cervical intraepithelial neoplasia grade 2 or 3 in relation to papillomavirus infection. *N. Engl. J. Med.* 327(18):1272–1278.

Koutsky, L.A., Ault, K.A., Wheeler, C.M., Brown, D.R., Barr, E., Alvarez, F.B., Chiacchierini, L.M., and Jansen, K.U.; Proof of Principle Study Investigators. (2002). A controlled trial of a human papillomavirus type 16 vaccine. *N. Engl. J. Med.* 347:1645–1651.

Krane, J.F., Lee, K.R., Sun, D., Yuan, L., and Crum, C.P. (2004). Atypical glandular cells of undetermined significance. Outcome predictions based on human papillomavirus testing. *Am. J. Clin. Pathol.* 121:87–92.

Lin, C.T., Tseng, C.J., Lai, C.H., Hsueh, S., Huang, K.G., Huang, H.J., and Chao, A. (2001). Related articles, links value of human papillomavirus deoxyribonucleic acid testing after conization in the prediction of residual disease in the subsequent hysterectomy specimen. *Am. J. Obstet. Gynecol.* 184:940–945.

Lorincz, A.T., and Richart, R.M. (2003). Human papillomavirus DNA testing as an adjunct to cytology in cervical screening programs. *Archiv. Pathol. Lab. Med.* 127:959–968.

Medeiros, F., Yuan, L., Breslin, G., Brodsky, J., Cibas, E.S., Feldman, S., Cviko, A., and Crum, C.P. (2005). Type-specific HPV testing identifies low grade squamous intraepithelial lesions with different risks for biopsy-proven HSIL. *J. Lower Genital Tract Dis.* 9:154–159.

Munoz, N., Bosch, F.X., de Sanjose, S., Herrero, R., Castellsague, X., Shah, K.V., Snijders, P.J., and Meijer, C.J.; International Agency for Research on Cancer Multicenter Cervical Cancer Study Group. (2003). Epidemiologic classification of human papillomavirus types associated with cervical cancer. *N. Engl. J. Med.* 348:518–527.

Nobbenhuis, M.A., Meijer, C.J., van den Brule, A.J., Rozendaal, L., Voorhorst, F.J., Risse, E.K., Verheijen, R.H., and Helmerhorst, T.J. (2001). Addition of high-risk HPV testing improves the current guidelines on follow-up after treatment for cervical intraepithelial neoplasia. *Br. J. Cancer* 84:796–801.

Pirog, E.C., Kleter, B., Olgac, S., Bobkiewicz, P., Lindeman, J., Quint, W.G., Richart, R.M., and Isacson, C. (2000). Prevalence of human papillomavirus DNA in different histological subtypes of cervical adenocarcinoma. *Am. J. Pathol.* 157:1055–1062.

Poljak, M., Marin, I.J., Seme, K., and Vince, A. (2002). Hybrid Capture II HPV Test detects at least 15 human papillomavirus genotypes not included in its current high-risk probe cocktail. *J. Clin. Virol.* 25:S89–S97.

Rose, R.C., Bonnez, W., Reichman, R.C., and Garcea, R.L. (1993). Expression of human papillomavirus type 11 L1 protein in insect cells: *in vivo* and *in vitro* assembly of virus like particles. *J. Virol.* 67:1936–1944.

Rosenfeld, W.D., Rose, E., Vermund, S.H., Schreiber, K., and Burk, R.D. (1992). Follow-up evaluation of cervicovaginal human papillomavirus infection in adolescents. *J. Pediatr.* 121:307–311.

Smith, E.M., Johnson, S.R., Ritchie, J.M., Feddersen, D., Wang, D., Turek, L.P., and Haugen, T.H. (2004). Persistent HPV infection in postmenopausal age women. *Intl. J. Gynecol. Obstet.* 87:131–137.

Solomon, D., Schiffman, M., and Tarone, R.; ALTS Study group. (2001). Comparison of three management strategies for patients with atypical squamous cells of undetermined significance: Baseline results from a randomized trial. *J. Natl. Cancer Inst.* 93(4):293–299.

Tam, K.F., Cheung, A.N., Liu, K.L., Ng, T.Y., Pun, T.C., Chan, Y.M., Wong, L.C., Ng, A.W., and Ngan, H.Y. (2003). A retrospective review on atypical glandular cells of undetermined significance (AGUS) using the Bethesda 2001 classification. *Gynecol. Oncol.* 91(3):603–607.

Tate, J.E., Resnick, M., Sheets, E.E., and Crum, C.P. (1996). Absence of papillomavirus DNA in normal tissue adjacent to most cervical intraepithelial neoplasms. *Obstet. Gynecol.* 88:257–260.

Thomas, D.B., Ray, R.M., Kuypers, J., Kiviat, N., Koetsawang, A., Ashley, R.L., Qin, Q., Koetsawang, S. (2001). Human papillomaviruses and cervical cancer in Bangkok. III. The role of husbands and commercial sex workers. *Am. J. Epidemiol.* 153:740–748.

Wright, T.C. Jr., Cox, J.T., Massad, L.S., Carlson, J., Twiggs, L.B., and Wilkinson, E.J.; American Society for Colposcopy and Cervical Pathology. (2003). 2001 consensus guidelines for the management of women with cervical intraepithelial neoplasia. *Am J. Obstet. Gynecol.* 189:295–304.

Wright, T.C. Jr., Schiffman, M., Solomon, D., Cox, J.T., Garcia, F., Goldie, S., Hatch, K., Noller, K.L., Roach, N., Runowicz, C., and Saslow, D. (2004). Interim guidance for the use of human papillomavirus DNA testing as an adjunct to cervical cytology for screening. *Obstet. Gynecol.* 103:304–309.

Zhou, J., Sun, X.Y., Stenzel, D.J., and Frazer, I.H. (1991). Expression of vaccinia recombinant HPV 16 L1 and L2 ORF proteins in epithelial cells is sufficient for assembly of HPV virionlike particles. *Virology* 185(1):251–257.

16
Possible Worldwide Impact of Prevention of Human Papillomavirus Infection

Sonia R. Pagliusi, M. Teresa Aguado, and D. Maxwell Parkin

Initiative for Vaccine Research, Department of Immunization, Vaccines and Biologicals, and Descriptive Epidemiology Group, International Agency for Research on Cancer, World Health Organization, 1211 Geneva, 27 Switzerland

16.1. Introduction

Human papillomavirus (HPV) infections are associated with a wide spectrum of mucocutaneous diseases ranging from benign skin warts through different grades of precancer to invasive cancer (Table 16.1). There are over 120 types of molecularly identified HPVs: all characterized types infect exclusively epithelial cells of the skin, the anogenital tract, and, less commonly, the oropharyngeal mucosa (de Villiers et al., 2004). So far there is no convincing evidence for HPV infections in the gastric, intestinal, or colonic mucosa (zur Hausen, 2002).

16.1.1. HPV Infections Worldwide and Related Human Diseases

Cutaneous types of HPV are found in common skin warts that affect most people, usually in childhood. Certain HPV types cause common flat warts or plantar warts, papillomas of the respiratory tract, and genital warts (Condylomata) (Beutner and Tyring, 1997; Sobhani et al., 2004). Prevalence data of cutaneous HPV in published studies varies from 3 percent to 50 percent (Astori et al., 1998; Pfister, 2003) (Fig. 16.1). In addition, several types of cancer have been associated with HPV infection, including cancers of the cervix uteri, external female genitalia (vulva, vagina), penis, and upper gastrointestinal tract (oral cavity and oropharynx) (IARC, 1995; Herrero and Munoz, 1999). About 40 HPV types infect the genital mucosa, and about 15 types are classified as high-risk types for cancer development because DNA of these types is regularly found in

TABLE 16.1. Sites of Human Papillomavirus Infections and Related Human Diseases.

Skin	Anogenital tract	Upper aero-digestive tract	Others
Common warts	Anogenital warts	Juvenile laryngeal papillomatosis	Conjunctival carcinoma
Plantar warts	Cervical neoplasias and carcinoma	Recurrent respiratory papillomatosis	Kerato carcinomas associated with immune deficiency
Flat warts	Anal carcinoma	Laryngeal squamous cell carcinoma	
Mosaic warts	Penile carcinoma	Oro-pharyngeal cancers (including Tonsil)	
	Vulvar carcinoma		

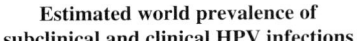

**Estimated world prevalence of
subclinical and clinical HPV infections**

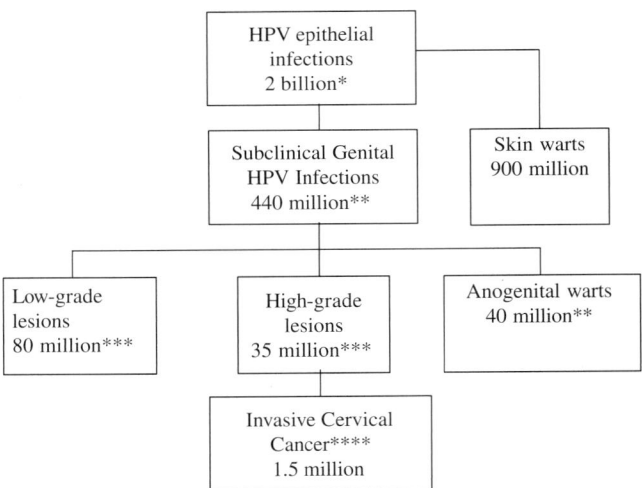

FIGURE 16.1. Estimated prevalence of HPV infections among subjects in reproductive age, based on world population of 6,271,698,000 in 2003 (World Bank) and surveys (15–50 years old) by *Pfister, 2003; **Koutsky, 1997, seroprevalence or DNA prevalence; ***Herrero et al., 2000, 1997;***Sankaranarayanan et al., 2004; ****Parkin et al., 2001.

malignant lesions (zur Hausen, 2000). HPV-16 and -18 are the most commonly found types in cervical cancer cases worldwide.

16.1.2.   Genital HPV Infections and Clinical Outcomes

Genital HPV infections are one of the most common sexually transmitted infections in humans globally. The virus can infect the mucosae of the vagina, cervix, rectum, and urethra. The most commonly seen clinical manifestation of genital HPV infections is warts. Genital HPV is very rare in virgins and is most common in individuals with multiple sexual partners. However, data on genital HPV infection are limited because most infections are subclinical and because sensitive methods to detect the viral DNA have become available only recently (Snijders et al., 2003).

Over 90 percent of genital HPV infections are transient and clear without treatment, causing no clinical symptoms. Estimates based on surveys conducted in developed countries indicate that approximately 1–2 percent of the sexually active population, aged between 15 and 49 years, have clinically recognized genital warts (Koutsky et al., 1988; Koutsky, 1997; Syrjanen, 1990). Furthermore, some HPV-infected cervical cells can develop into abnormal tissue, visualized as flat lesions. These low-grade to moderate or high-grade dysplasias, also characterized as cervical intraepithelial neoplasias 1, 2, or 3 (CIN1, CIN2, CIN3), respectively, are detectable in women by various screening methods (Miller, 1992). Anal and vulvar intraepithelial neoplasias are similarly classified as AINs and VINs.

Natural history studies and surveys in developed and developing countries indicate that low-grade dysplasias are found in 3–4 percent of sexually active adults (Koutsky, 1997), and the majority of them regress spontaneously without treatment. Moderate or high-grade cervical dysplasias are found in about 1–2 percent of women (Herrero et al., 2000; Sankaranarayanan et al., 2004). On a global basis this represents about 35 million women, and, if not treated, as many as one-third of these cases may progress to invasive cervical cancer. Progression from cervical cancer precursor lesions to invasive cancer is usually a slow process, estimated to take on average 5 to 10 years.

16.1.3. *Prevalence and Incidence of HPV Infections*

The prevalence of HPV-related disease appears to correlate with the prevalence of the infection as measured by virus DNA in a given population (Herrero et al., 2000; Sankaranarayanan et al., 2004; Pham et al., 2003). Although it is difficult to estimate the overall prevalence of human papillomavirus infections, it has been suggested that during their lifetime about 60 percent of sexually active adults are infected with HPVs (Koutsky and Kiviat, 1999; Koutsky, 1997; Ho et al., 2002). The prevalence of genital HPV in the general population of reproductive age averages about 10 percent globally, but varies between 2.3 percent in Hanoi, Vietnam (Pham et al., 2003), and 16 percent in Guanacaste, Costa Rica (Herrero et al., 2000). Prevalence is highest (25–50 percent) among sexually active young women (Winer et al., 2003). These figures imply about 440 million people are infected with genital HPV worldwide. In contrast, reported prevalence of other sexually transmitted infections is 62 million infected by gonorrhea and 40 million infected by human immunodeficiency virus (HIV)[1]. Studies using

---

[1] www.niaid.nih.gov/dmid/stds/condomreport.pdf.

direct detection methods for HPV DNA in exfoliated cells from the genital tract suggest that the prevalence of HPV DNA in men is similar to that in women, i.e., between 10 and 20 percent (Hippelainen et al., 1993; Bauer et al., 1991; Martinez et al., 1988; Herrero et al., 2000).

16.2.   HPV Infection as a Cause of Cancer

16.2.1.   Cancer of the Cervix Uteri

Genital infections with oncogenic types of HPV are recognized as the major cause of cervical cancer (IARC, 1995; Walboomers et al., 1999). The causal relationship between HPV infections and cervical cancer was first suggested in the 1970s (zur Hausen, 1976) and subsequently confirmed. This relationship is based on a large body of scientific evidence, fulfilling the criteria for causality proposed by Hill (Hill, 1965; Bosch et al., 2002). Virtually all cervical cancer cases are associated with infection by at least one of the 15 oncogenic types of human papillomavirus, hence the conclusion that HPV infection is a necessary cause of cervical cancer (Walboomers et al., 1999). This conclusion implies that women will not get cervical cancer without an HPV infection. Nevertheless, the majority of HPV infections appear to naturally clear over about 12 months and do not lead to cancer (Ho et al., 1998).

16.2.1.1.   Other Risk Factors for Genital Cancers

HPV is not the only relevant risk factor for cervical cancer. There are very marked differences in risk of cervical cancer according to such demographic variables as social status, religion, occupation, marital status, and ethnicity. Epidemiological studies (mainly case control studies) have shown a consistent association between risk of CIN and early age at initiation of sexual activity, increasing number of sexual partners, and other indicators of sexual activity. Following the recognition of the central role of HPV, it became clear that most of the "demographic" and "behavioral" risk factors were simply related to the probability of infection with HPV. Nevertheless, certain environmental risk factors appear to be associated with increased risk of invasive cancer, independently of their association with probability of infection by HPV. They include an increasing number of pregnancies (Muñoz et al., 2002, 1993; Thomas et al., 2001; Hildesheim et al., 2001), long-term exposure to oral contraceptives (Moreno et al., 2002), smoking (IARC, 2004a), and probably other sexually transmitted infections, including Chlamydia (Smith et al., 2002) and herpes simplex virus type 2 (de Sanjose et al., 1994). In addition, there appears to be some genetic predisposition to cervical cancer, which may be related to the inheritance of alleles affecting immune response or viral DNA replication (Allen et al., 1996; Odunsi et al., 1996; Hildesheim et al., 1998a,b).

Since 1993, cervical cancer has been considered to be an "AIDS-defining" condition, meaning that if it occurs in someone who is positive for HIV, that person is deemed to have AIDS. Studies in the United States, Italy, and France, showed that subjects with HIV/AIDS have increased risk of invasive cancer of the cervix (Goedert et al., 1998; Franceschi et al., 1998; Serraino et al., 1999; Frisch et al., 2000), but a similar study in Australia was negative (Grulich et al., 1999). Some of the excess risk is due to the confounding effect of HPV infection, which is associated with HIV infection because of their common mode of transmission. The prevalence of CIN is clearly higher in HIV-infected women. There appears to be an independent effect of HIV on risk of CIN, and an interaction between the effects of HIV and HPV, as might be expected if the role of HIV on HPV disease progression was indirect, through creation of immune dysfunction (Mandelblatt et al., 1999). HIV infection is also thought to increase the risk of anal carcinoma by 30 times (Sobhani et al., 2004). To date there is no reliable ranking of these risk factors, independent of HPV.

16.2.1.2. Estimated Incidence of Cervical Cancer—An Update

Although invasive cervical cancer is an uncommon consequence of HPV infection, it is the second most common cancer among women worldwide, with an estimated 493,000 new cases and 274,000 deaths in the year 2002 (Parkin et al., 2005). Cervical cancer is much more common in developing countries, where 83 percent of cases occur, accounting for 15 percent of female cancers. In developed countries it accounts for only 3.6 percent of new cancers. The highest incidence rates are observed in sub-Saharan Africa, Melanesia, Latin America and the Caribbean, South-Central Asia, and South East Asia (Fig. 16.2).

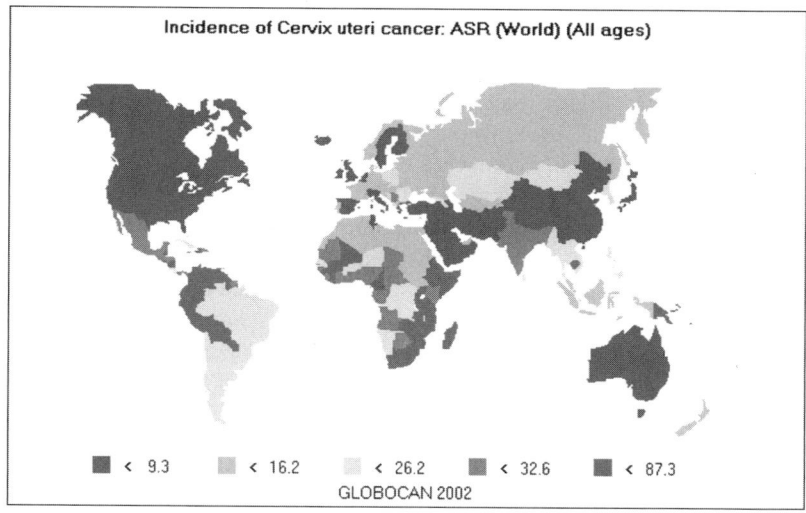

FIGURE 16.2. Estimated age-standardized incidence rates of cervical cancer by country as color-coded from highest (red) to lowest (dark green) per 100,000 women. Available at www.iarc.fr. (See Plate 8)

Incidence rates are generally low in developed countries, with age-standardized rates less than 14.5 per 100,000 (10^5) women. Before the introduction of screening programs in the 1960s and 1970s, the incidence in most of Europe, North America, and Australia/New Zealand was much as we see in developing countries today (38.0 per 10^5) (Gustafsson et al., 1997a; Dorn and Cutler, 1959). Very low rates are also observed in China (6.8 per 10^5), and in Western Asia (5.8 per 10^5); the lowest recorded rate is 0.4 per 10^5 in Ardabil, North West Iran (Sadjadi et al., 2003).

Incidence of cervical cancer begins to rise at age 20–29, and the risk increases rapidly to reach a peak usually around age 45–49 in European populations, but often later in developing countries. Incidence rates then decline somewhat. Screening programs change this pattern profoundly (Gustafsson et al., 1997b).

16.2.1.3. Changes in Incidence Over Time

In economically developed countries, rates of cervical cancer incidence and mortality have declined in the last 40 years; this trend has been ascribed to a combination of a reduction in risk in older generations of women (genital hygiene, parity, etc.), with, more recently, the beneficial effects of population screening programs based on exfoliative cervical cytology. In the Nordic countries, for example, the decline in incidence is related to the coverage and extent of the organized screening programs (Hakama, 1982; Sigurdsson, 1999), and is most marked in the age groups targeted by these programs. Nevertheless, in many of the countries showing declines in overall incidence and/or mortality, increases in risk are seen in young women (Bray et al., 2005). This effect most likely is the result of changes in sexual habits and increased transmission of papillomaviruses in younger generations of women, with the magnitude of the effect on incidence depending upon the countervailing effects of screening. Thus in some countries, e.g., France, there has been no increase in risk of cervical cancer in young women, while the upward trend of cervical disease in England and Wales has been successfully countered by a much improved screening program, implemented in 1988 (Quinn et al., 1999).

A total of 80 to 90 percent of cervical cancers are squamous cell carcinomas, and incidence trends for this histological subtype determine those for cervical cancer as a whole. In several populations where incidence rates of squamous cell carcinomas are declining (presumably as a result of screening), rates of adenocarcinomas are increasing, and the increasing risk appears to affect relatively recent generations of women (Vizcaino et al., 1998). The older methods of obtaining cytological smears (wooden spatulae) were less effective in detecting adenocarcinoma or precursor lesions than for squamous cell tumours (Fu et al., 1987; Sigurdsson, 1995). Increases in exposure to HPV in recent generations may be responsible for increasing the incidence of adenocarcinoma, whereas screening programs have diminished the effect of increased exposure on the incidence of squamous cell tumours. The use of oral contraceptives has also been linked to an increased risk of cervical adenocarcinoma (Ursin et al., 1994).

In developing countries rates of incidence and mortality have, in general, been relatively stable or shown only modest declines. This effect probably reflects the absence of systematic screening programs, or, where programs have been introduced, low population coverage and poor quality cytology (Lazcano-Ponce et al., 1998). For example, in Cuba (Fernandez Garrote et al., 1996) and in Costa Rica (Herrero et al., 1992), the screening programs seem to have had virtually no impact upon incidence of cancer. In contrast, there appear to have been dramatic declines in cervical cancer in China. The age-adjusted incidence of cervical cancer in Shanghai fell from 26.7 per 100,000 to 2.5 between 1972 and 1974 and 1993 and 1994 (Jin et al., 1999); and mortality rates have fallen dramatically, especially in urban populations, although the trend has reversed recently in younger women (Yang et al., 2003). The declines have been attributed to introduction of exfoliative cytology screening testing (i.e., Pap smear screening) treatment programs, and improved genital hygiene, while the increased rates among younger women may reflect changing in economic circumstances and sexual mores, with a greater prevalence of infection with HPV and other agents (Li et al., 2000). The limited data available from Africa do not suggest any decrease in incidence of cervical cancer (Parkin et al., 2003).

16.2.2. Other Ano-Genital Cancers

HPV-16 infection may also cause squamous-cell cancers of the vulva and vagina, carcinoma of the penis, and anal cancer (IARC, 1995). Published studies do not allow quantification of relative risk and infection prevalence, because they are generally small in size, and they do not include measurement of prevalence of infection at these sites in normal subjects. Nevertheless, we can estimate that one third of the cases of cancer of vulva and vagina are attributable to HPV infection, as are one-half of cases of penile cancer and two-thirds of the cases of anal cancer (Herrero and Munoz, 1999).

16.2.2.1.   Upper Aero-Digestive Tract

HPV probably plays a role in the etiology of a fraction of cancers of the oral cavity and pharynx (Shah, 1998), although the major risk factors are tobacco and alcohol. Several studies have investigated prevalence of HPV in cancers of the mouth and pharynx (Franceschi et al., 1996; Snijders et al., 1997; Gillison and Shah, 2001). On average approximately 40 percent of tumours were HPV-positive, but the prevalence varied widely with the population studied, subsites, type of specimen, and detection method. HPV was detected most commonly in oropharynx and tonsil. Based on a large multicentre case-control study in nine countries, Herrero et al. (2003) estimate that the percentage of cases attributable to HPV are 5 percent for mouth cancers, and 16 percent for cancers of the oropharynx, although methodological consideration suggest that these numbers may be slight overestimates (van Houten et al., 2001).

16.2.3. How Much Cancer Globally Is Related to Infection with HPV?

The estimated numbers of new cancer cases in the year 2002, by country, age group, and sex are available for 25 types of cancer in GLOBOCAN 2002 (Ferlay et al., 2004). These estimates do not include certain cancers for which HPV probably plays a causative role: cancers of the external genitalia (vulva, vagina, and penis), cancers of the anus, and cancers of the oropharynx. For cancers of the external female genitalia and penis, numbers of cases were estimated from cancer registry data extracted from *Cancer Incidence in Five Continents, Volume VIII* (Parkin et al., 2002) as the ratio of cases of cancer at these sites, to cases of cervical cancer, by age and region. These estimates are 40,000 annual cases of cancer of the external genitalia in females, and 26,300 cases of penile cancer. A total of 30,400 cases of anal cancers, about equally divided between males and females, was estimated from recorded ratios of colorectal cancer to anal cancer (Fig. 16.3). The estimates for oropharynx cancer have been derived from the numbers of cancers of the pharynx (Globocan, 2002) and the proportion of such cases that are located in the oropharynx, according to registry data in *Cancer Incidence in Five Continents, Volume VIII* (Parkin et al., 2002).

The estimates of HPV-attributable cancers are shown in Table 16.2a (by site; developed vs. developing countries) and Table 16.2b (by site and sex). We assume that HPV is responsible for all of the cervical cancers occurring in the world (492,800), two-thirds of the anal cancers (20,368 cases), one-half of the penile cancers (13,150 cases), and one-third of the cancers of the vulva/vagina (13,200 cases). In addition, we assume that 3 percent of oral cavity cancers, and 12 percent of cancers of the oropharynx, are attributable to HPV, a total of 14,475

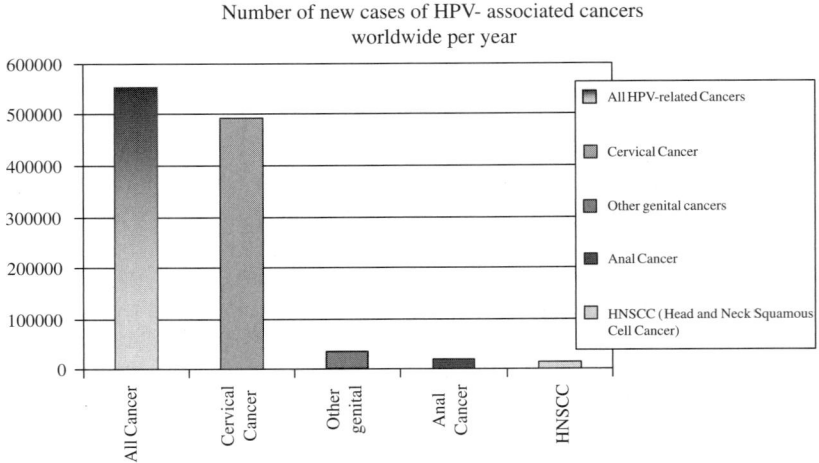

FIGURE 16.3. Incidence of HPV-associated cancers worldwide in 2002. All cancer, all HPV-associated cancer. HNSCC, head-and-neck squamous cell carcinoma.

TABLE 16.2. HPV infection-attributable cancers in 2002 worldwide.

a. DEVELOPED & DEVELOPING COUNTRIES

SITE	DEVELOPED COUNTRIES				DEVELOPING COUNTRIES				WORLD			
	TOTAL cancers	AF%	Attributable cancers	% all cancer	TOTAL cancers	AF%	Attributable cancers	% all cancer	TOTAL cancers	AF%	Attributable cancers	% all cancer
CERVIX	83 400	100	83 400	1.66%	409 400	100	409 400	7.03%	492 800	100	492 800	4.54%
PENIS	5 200	50	2 600	0.05%	21 100	50	10 550	0.18%	26 300	50	13 150	0.12%
VULVA,VAGINA	18 300	33	6 039	0.12%	21 700	33	7 161	0.12%	40 000	33	13 200	0.12%
ANUS	14 500	67	9 715	0.19%	15 900	67	10 653	0.18%	30 400	67	20 368	0.19%
MOUTH	91 100	3	2 733	0.05%	183 000	3	5 490	0.09%	274 100	3	8 223	0.08%
ORO PHARYNX	24 400	12	2 928	0.06%	27 700	12	3 324	0.06%	52 100	12	6 252	0.06%
ALL SITES	5 016 100		107 415	2.14%	5 827 500		446 578	7.66%	10 843 600		553 993	5.11%

b. BY SEX

SITE	MALES				FEMALES				BOTH SEXES			
	TOTAL cancers	AF%	Attributable cancers	% all cancer	TOTAL cancers	AF%	Attributable cancers	% all cancer	TOTAL cancers	AF%	Attributable cancers	% all cancer
CERVIX					492 800	100	492 800	9.74%	492 800	100	492 800	4.54%
PENIS	26 300	50	13 150	0.23%					26 300	50	13 150	0.12%
VULVA,VAGINA					40 000	33	13 200	0.26%	40000	33	13 200	0.12%
ANUS	14 500	67	9 715	0.17%	15 900	67	10 653	0.21%	30 400	67	20 368	0.19%
MOUTH	175 900	3	5 277	0.09%	98 400	3	2 952	0.06%	274 300	3	8 229	0.08%
ORO PHARYNX	42 600	12	5 112	0.09%	9 600	12	1 152	0.02%	52 200	12	6 264	0.06%
ALL SITES	5 801 800		33 254	0.57%	5 060 700		520 757	10.29%	10 862 500		554 011	5.11%

cases (Table 16.2). Thus, HPV is one of the most important infectious agents in cancer causation, responsible for 5.1 percent of the world cancer burden. The distribution is very different between developing and developed countries. The attributable fractions are 2.1 percent in developed countries, and 7.7 percent in developing countries. The distribution is also different by sex: 0.6 percent of cancers in men due to HPV infection, 10.3 percent of cancers in women.

16.3.   Primary and Secondary Prevention Strategies

Cervical cancer is preventable, firstly because it is caused by a preventable infectious agent, and secondly because it develops over many years, offering several opportunities for interventions aimed at preventing progression of infection to disease outcomes. Thus, both primary and secondary prevention strategies are feasible.

Primary prevention strategies are interventions aimed at avoiding or reducing the exposure to the infectious agent(s) and risk factor(s) causing the disease. These are generally based on behavioral, environmental, or biological interventions including immunization programs. Secondary prevention strategies are aimed at early detection and treatment of precursor lesions for invasive cervical cancer. The emphasis has been on screening of asymptomatic women using several different technological approaches. Strategies that combine both primary and secondary prevention are likely to offer the most effective approach in the short term, because the impact of vaccination on cancer will not be immediate (Goldie et al., 2004a). Factors that may accelerate the impact of prevention strategies are discussed below with a particular focus on vaccination.

16.3.1.   Primary Prevention Strategies

Primary prevention requires that a person avoid exposure to the infectious agent, in this case, the sexually transmitted HPV or that exposure is rendered innocuous (e.g., by vaccination). To design primary prevention strategies against a disease, the cause(s) as well as risk factor(s) need to be known and interventions are generally developed and implemented in the context of promoting behavioral, environmental, or biological changes (Fig. 16.4).

16.3.1.1.   Behavioral and Environmental Changes

Primary prevention strategies could comprise approaches to changing the transmission dynamics of HPV, through modification of behaviors that mediate viral transmission, such as delaying age of first sexual contact and of first delivery, as well as decrease in the number of lifetime sexual partners, via counseling in reproductive health and family planning settings. Other environmental/behavioral changes that may reduce viral transmission include improvements in genital hygiene (Bayo et al., 2002), and use of condoms (Manhart and Koutsky, 2002).

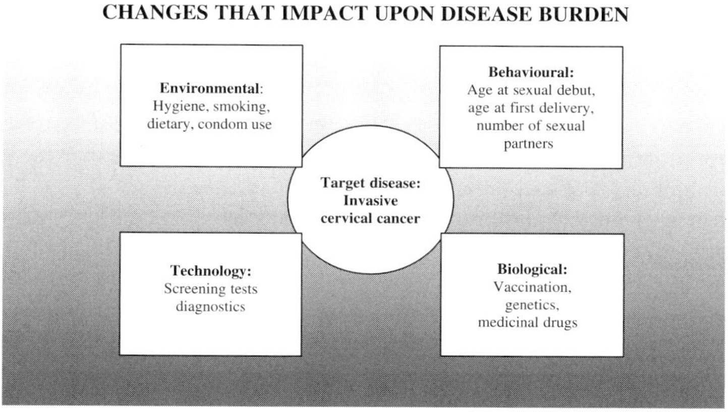

CHANGES THAT IMPACT UPON DISEASE BURDEN

FIGURE 16.4. The four areas of intervention where changes can lead to effective primary and secondary preventive strategies to target a given disease. Specific examples of changes that may possibly offer reduction in cervical cancer morbidity/mortality are listed for each area.

In this respect, while the consistent and correct use of condoms can prevent most sexually transmitted infections, HPVs are abundant in external genital mucosae that are not covered by condoms. Available data on condom use lack consistency to provide precise estimates on their effectiveness. Nevertheless, some data suggest that while condoms may not prevent HPV infection, they may reduce the risk of acquiring genital HPV-related diseases (Manhart and Koutsky, 2002). However, sexual behavior changes are a complex issue and very challenging to achieve.

16.3.1.2.   Vaccination

Vaccination has proven to be one of the most cost-effective life-saving public health interventions in the past century for protecting against infectious diseases such as smallpox and poliomyelitis (Fenner, 1982; Heymann and Aylward, 2004). Advances in HPV vaccine development indicate that a novel prevention tool against cervical cancer will be available in the near future (Galloway, 2003). Briefly, clinical trials showed that specific HPV vaccine candidates are safe, well tolerated and highly immunogenic in men and women (Harro et al., 2002). Antibody responses can be induced in 99.8 percent of subjects, with titers that are 50–100-fold higher than elicited by natural infections. Furthermore, these vaccine candidates conferred high levels of protection against incident HPV-16 and -18 infections, as well as 100 percent protection against persistent HPV-16 and -18 infections and their related cervical lesions, including CIN (Koutsky et al., 2002; Harper et al., 2004). Thus vaccines have the potential to prevent cervical cancer. Large phase III efficacy trials are ongoing and may confirm and extend the preliminary results. Because HPV-16 and -18 are the dominant types involved in cervical cancer etiology in all parts of the world, a vaccine against these two types should, in theory, be able to prevent some 70 percent of cases of cancer (Muñoz et al., 2004). Although it will take many years for a vaccine to affect cancer incidence, effective vaccination will immediately reduce the rate of abnormal pap smears and relieve the health care system of the burden of monitoring and treating these lesions. In addition, vaccination against HPV-16 and -18 may also prevent other HPV-related cancers.

Even if a given vaccine preparation is demonstrated to be highly efficacious in controlled studies, there is still considerable uncertainty about its feasibility, acceptability, and ultimate impact on disease incidence and burden worldwide. Not much is known at present concerning the effectiveness of vaccination among women in areas where other infections, such as HIV, are prevalent. Key factors likely to affect the impact of vaccination need to be considered before the vaccine is introduced. These factors include disease incidence rates, viral epidemiology, and availability of suitable vaccination programs.

(a) *Impact of vaccination on disease incidence rates.* Historically, vaccination to prevent threatening infectious diseases has shown high impact on disease burden, especially if implemented in areas where disease incidence is highest. HPV vaccines can potentially have the biggest impact on the global burden of cervical cancer if made available to countries where cervical cancer incidence is highest and affects a large population (Table 16.3).

The highest age standardized incidence rates reported occur in seven countries of sub-Saharan Africa and three of Latin America (Table 16.3) among the 30 countries displayed. Furthermore, the highest numbers of cervical cancer cases every year are reported in India, Mexico, and Brazil, accounting together for a third of all cases worldwide. Therefore, early implementation of HPV vaccination at high coverage in those countries is likely to accelerate the impact of HPV prevention on related global disease burden rates.

(b) *Impact of local viral-epidemiological profile.* As decisions about vaccination are taken at national level, the local epidemiological profile as well as disease incidence need to be considered. For instance, HPV vaccine candidates are currently based on antigens derived from most common HPV types found in cancer biopsies, namely, 16 and 18, although some vaccine preparations may contain antigens also targeting genital warts. As HPV immune responses are likely to be type specific, it is expected that vaccination will primarily prevent cervical cancer cases caused by these two HPV types. The fractions of cervical cancer attributable to these two types vary slightly between geographical areas (Muñoz et al., 2004). HPV-16/18 vaccines will presumably have higher impact on the disease incidence (Goldie et al., 2004a) in areas where the percentage of cancer cases attributable to these two types is high. Importantly, many countries lack this kind of information and may either rely on assumptions using data from nearby countries, or have to generate local data to inform decisions.

TABLE 16.3. Examples of selected countries with high cervical cancer age-standardized incidence rates (ASR) globally based on cancer registries reporting (*)(Globocan 2002), and high number of child bearing aged women at risk (+). Source: "World Population Prospects: The 2002 Revision", New York, United Nations, 2003. Population between 5 and 15 years old could be considered in school vaccination.

RANK	COUNTRY	Age Standardized Incidence per 100.000*	Total number of cases/ year*	Number of CBA women (15–49 years old) in thousands+	Population between 5 and 15 years in thousands+
1	Haiti	87.3	2774	2118	2076
2	Tanzania	68.6	7515	8692	10454
3	Lesotho	61.5	479	478	471
4	Swaziland	58.9	186	269	303
5	Zambia	53.7	1650	2391	3125
6	Bolivia	55.0	1831	2152	2209
7	Paraguay	53.2	1131	1464	1455
8	Zimbabwe	52.1	1817	3085	3623
9	Guinea	50.9	1444	1951	2257
10	Rwanda	49.4	1087	2109	2283
11	Peru	48.2	5400	7067	5989
12	Nicaragua	47.2	809	1358	1456
13	Malawi	46.6	1766	2698	3369
14	El Salvador	45.6	1213	1716	1476
15	Papua N. G.	40.4	637	1391	1509
16	Ecuador	38.7	1978	3430	2823
17	South Africa	37.5	6742	12332	10036
18	Venezuela	36.0	3845	6747	5525
19	Colombia	36.4	6815	12022	9307
20	Uganda	36.3	2429	5457	7605
21	Ethiopia	35.9	7619	15980	19729
22	Cameroon	35.7	1759	3792	4304
23	Mali	35.2	1336	2840	3830
24	Cambodia	38.7	1768	3497	3758
25	Mozambique	33.6	2058	4590	5137
26	India	30.7	132082	265111	231405
27	Mexico	29.5	12516	28558	22287
28	Nigeria	28.5	9922	8111	8612
29	Rumania	23.9	3448	5850	2561
30	Brazil	23.4	19603	50328	33139

(c) *Impact of potential different vaccination strategies.* Theoretically, every childbearing-aged woman (CBAW) is at risk for genital HPV infections, and consequently for cervical cancer, and the risk may persist throughout life. Yet, most infections appear to be acquired soon after sexual debut, and HPV prevalence declines in older ages up to 65 years old (Winer et al., 2003; Ley et al., 1991; Shin et al., 2004; Herrero et al., 2000). HPV vaccines would primarily be aimed at reducing cervical cancer incidence in adults. Because prophylactic HPV vaccines aim to prevent HPV infections rather than clearing them, it is believed that vaccination is likely to be most effective if given before the beginning of sexual activity, e.g. at adolescence,[2] in order to prevent the peak of HPV prevalence reported between 15 and 25 years of age (Ho et al., 1998; Moscicki et al., 1998; Woodman et al., 2001). Indeed, controlled clinical trials of HPV vaccine candidates have so far reported very high efficacy for protecting previously uninfected young women aged 15 to 25 years against HPV infections (Koutsky et al., 2002; Harper et al., 2004). On the other hand, vaccinating adolescents represents a challenge for public health immunization because there are few formal immunization programs, particularly in developing countries, designed to routinely reach this age group (Clemens et al., 1999).

The administration of prophylactic HPV vaccines to children is not likely to be recommended at the moment, because the protective effect of vaccination against HPV infection and related cervical disease has so far only been documented for a duration of 3 to 4 years. If HPV vaccines are proven to be effective and immunization is demonstrated to confer lasting protection against HPV infections and related diseases, then HPV vaccines could become, together with hepatitis B and other vaccines, part of vaccination programs targeting infants. Because it will take 10 to 20 years to accumulate such evidence, health professionals need to actively search for effective strategies to accelerate disease control and enhance the impact of these HPV vaccines in the near future. Possibly, successful HPV vaccination programs may need to target young women aged 12 to 23, at least at an early rollout phase. As evidence becomes available, strengthening school vaccination programs targeting populations between 5 and 15 years of age (Table 16.3) represents a strategy to be considered in the near future.

Vaccination schedules currently designed to include adolescent age groups, that could potentially be relevant to implementation of HPV vaccination, are tetanus, measles, meningitis, and rubella vaccination campaigns or programs. For instance, some countries, such as Costa Rica, achieve high coverage for a routine second dose of measles-mumps-rubella vaccine, with 95 percent of schoolchildren receiving two or more doses of measles vaccine in the urban area (Calvo et al., 2004). Catch-up campaigns for tetanus elimination involving three-dose vaccination, mostly targeting CBAW, have also achieved high coverage for the initial and booster doses globally (Vandelaer et al., 2003), and may represent synergistic opportunities for HPV vaccine introduction. Taken together, these opportunities highlight the need for coordinated global and national strategies in order to reach adolescents for HPV vaccination. Careful analysis of vaccination strategies and field situations in several regions in order to determine optimal strategies for HPV vaccination is warranted. Immunization programs will need to be tailored to the level of economic and health systems development of a country. Further

---

[2] WHO definition of adolescents is 10–19 years of age (Clements et al., 1999).

research and demonstration projects will be crucial to design successful public sector HPV vaccination strategies.

Although cervical cancer is a women's health problem, HPV is transmitted mostly by a male partner. Prophylactic vaccination of males may therefore have an impact on HPV transmission and consequently women's disease incidence. It has been hypothesized that a female-vaccination-only strategy would be less efficient than those targeting both sexes (Hughes et al., 2002). In contrast, by using a disease transmission model for sexual transmission of HPV, it was speculated that HPV-16/18 vaccination of females only could reduce lifetime cervical cancer cases by 61.8 percent, while inclusion of male participants in a vaccine rollout would further reduce cervical cancer cases by only an additional 2.2 percent (Taira et al., 2004). Still, evidence-based data are lacking in this respect, especially in areas of high disease burden, and further work is needed to resolve this issue. Obviously, vaccination of both sexes implies increased costs than strategies for female vaccination only.

While a higher prevalence of oncogenic HPV types has been demonstrated for HIV-infected patients, the mechanism by which HIV increases the infection risk is unknown. Nevertheless HIV infection may have an impact on HPV vaccination in countries where HIV prevalence is high and should be carefully considered when designing immunization strategies. Clearly, vaccine safety and immunogenicity in HIV-positive subjects needs to be addressed before deployment (Pagliusi and Aguado, 2004).

(d) Impact of vaccine cost on HPV vaccination. Finally, while recommendations to use a vaccine rely on its safety, efficacy, and effectiveness against disease, vaccine cost and ease of distribution and administration are clearly also factors that may considerably influence its large-scale adoption and uptake. In developing countries, where disease incidence is high, an HPV vaccine would need to be as attractive to public health programs as hepatitis B vaccines have been in the last years in terms of cost-effectiveness (Beutels, 2001). Acceptability and use of HPV vaccines will be clearly higher if they are affordable. Although the HPV vaccines most far along in development are administered by needle injection, vaccines that are administered to mucosal surfaces without injection are likely to be more suitable in the areas of the highest need if such vaccines display similar efficacy.

16.3.2. *Secondary Prevention Strategies*

In the context of cervical cancer prevention, secondary prevention strategies include all interventions after a person is infected by HPV, and aim to inhibit or delay the development and progression of HPV infection to invasive cancer. Several levels of prevention are possible. They include—

- Eliminating viral infection or the cancer itself using therapeutic vaccines.
- Antiviral drugs.

- Detection of cellular abnormalities induced by HPV infection (ranging from mild dysplasias to carcinoma in situ), and providing appropriate treatment to prevent progression to invasive cancer. This is the basis of most programs of screening.
- Detection of invasive cancers at an early stage, so that treatment has a high probability of success in preventing disability or death from the disease.

16.3.2.1.   Therapeutic Vaccines

Established HPV infections have been considered as a target for so called therapeutic vaccines, which aim to eliminate viral infection or virally infected cells, and so reverse the oncogenic process induced by the virus. The goal of therapeutic vaccination is to generate a cell-mediated cytolytic T cell response that kills precancer or cancer cells expressing HPV non-structure proteins. In animal models, therapeutic vaccination controls growth of tumours expressing the papillomavirus nonstructural E6 and E7 proteins. Results of phase I clinical trials have been limited to studies on viral load or viral persistence, and have not demonstrated statistically significant regression of HPV-induced lesions (Follen et al., 2003). Because therapeutic vaccination would target HPV-positive women, such women at risk need to be identified by a screening strategy before vaccination can be implemented.

16.3.2.2.   Antiviral Strategies to Inhibit Transmission and Prevent Cancer

Antiviral strategies that inhibit various aspects of the viral life cycle will presumably reduce the concentration of HPV in infected tissue and therefore may reduce the likelihood of carcinogenic progression or transmission of the virus to new hosts. Based on experience with other viral systems, drugs that interfere with the viral DNA replication proteins, E1 and E2, are likely to be of use and are being developed. However, agents that target E1 or E2 are unlikely to be useful in those cases of cervical cancer that contain integrated HPV DNA. As detailed in Chapter 10, continuing expression of the HPV E6 and E7 oncoproteins is also required for continuous proliferation of cervical carcinoma cell lines and HPV-immortalized keratinocytes in culture and in animals. In these experiments, extinction of viral gene expression causes growth arrest/senescence or apoptosis, either of which might be a beneficial therapeutic outcome if it occurs in patients. Antiviral approaches that cause the elimination of the viral genome, repression of viral gene expression, or inhibition of viral protein function may thus prevent the progression of premalignant disease to cancer or even allow the treatment of established cancer. As is the case for therapeutic vaccination, at-risk HPV-positive women need to be identified by a screening strategy before antiviral approaches can be implemented.

16.3.2.3.   Screening for Pre-Invasive Disease

The natural history of cervical cancer is typically of long duration, and opportunities for screening exist at several levels. As noted above, the first event is infection with HPV of the so-called high-risk HPV subtypes (HPV types 16, 18, 31, 33, 45, 51, 52, 58, and 59). Most such infections are transient, but some persist. Higher viral loads are associated with cellular atypia, koilocytosis, and cervical intraepithelial neoplasia (CIN). Most of these mild lesions are transient, like HPV infections. However, a certain number of lesions progress to greater levels of cellular atypia (higher grades of CIN), the probability of which appears to increase with age. Finally, the disease passes into an invasive stage, which is initially visible only microscopically (microinvasive) and then, although more advanced, is subclinical, before being diagnosed.

There are two broad screening approaches:

(a) screening for precursor lesions using cytology (the Pap smear);
(b) screening for presymptomatic cancer (invasive or pre-invasive) by aided visual inspection.

Our longest experience is with screening by the Pap smear. This method depends upon the detection of precursor lesions, and hence the prevention of invasive cancer.

(a) *Screening by cytology: The Pap smear.* It has been known for some time that invasive squamous cell carcinoma of the cervix (accounting for 80–90 percent of cancers) is preceded by recognizable precursor lesions that may be detected by exfoliative cytology (Papanicolaou and Traut, 1943). Several classification systems for the sequence of pathological conditions, which are considered premalignant, or preinvasive, have been introduced. Microscopically, this sequence is characterized by progressive de-differentiation or atypia of epithelial cells, and progressive involvement of the full thickness of the epithelium. Initially, this progression was described in terms of increasing degree of "dysplasia" (mild, moderate, severe), and carcinoma in situ, where the full thickness of the epithelium is involved by undifferentiated cells. Different degrees of dysplasia may coexist at different sites within the same cervix. The "cervical intraepithelial neoplasia" terminology was introduced later, and came progressively into use in the early 1980s. Recently, cytological appearance has been classified by the "Bethesda System," which is sometimes used to describe the cytological changes in terms of high-grade, or low-grade, squamous intraepithelial lesions (SIL). The category "low-grade SIL" also includes the cytological diagnosis of koilocytotic atypia, which corresponds to cytological changes characteristic of infection with human papillomavirus (HPV)-koilocytosis or frank condylomatous change. Prospective studies have shown that cytologically normal women infected with HPV have a high risk of progressing to SIL/CIN, and that infections with the high-risk HPV types predict an elevated risk of progression to high grade SIL/CIN II–III.

There have been literally dozens of observational or retrospective case control studies evaluating the effectiveness of Pap smear screening, although there has never been a true randomized trial. Practically all such studies confirm the protection against invasive cancer provided by Pap smear screening (IARC, 2004b). The success of cytological screening requires training of smear-takers to ensure proper specimen collection, and adequate quality control and quality assurance programs for interpreting the smears. There must be adequate facilities for diagnostic follow-up of abnormal tests, and appropriate treatment of confirmed neoplastic lesions; this scenario implies a carefully designed and agreed referral system. Screening is most effective when delivered as part of an organized program, in which the target population has been identified, individual women are identifiable, and measures are available to guarantee high coverage and attendance. It is usually very difficult to meet these criteria in developing countries. Facilities for taking smears can usually be organized by training appropriate paramedical personnel, although poor sampling technique will reduce the validity of the test and the impact of the screening conferred against cancer. Interpretation of smears demands that trained cytotechnologists are available, but there are rarely suitable facilities for training and employing such staff. Unless rigorous quality-control procedures are in place in the cytology laboratory, the quality of Pap smears can be very poor. In a study in Mexico (Lazcano-Ponce et al., 1994), the level of false negative tests was between 10 and 54 percent. Adequate facilities for the follow-up of suspicious or positive smears and for appropriate treatment of confirmed neoplastic lesions may not be available. Finally, the level of organization of services may be such that it is difficult to coordinate between the woman, the laboratory, and the clinical facility, so that abnormal screening tests are followed up according to an agreed-upon protocol, and that information is provided about negative tests.

(b) *Visual Inspection methods.* Because of the technical difficulties inherent in screening programs based on cytology in developing countries, there has been a renewal of interest in the last decade in programs using aided visual inspection. Two methods have been widely tested: visualization of the cervix after impregnation with 3 to 4 percent acetic acid (VIA) or with Lugol's iodine (VILI). Both are inexpensive, safe, and acceptable methods that require a lower level of infrastructure than laboratory-based tests, and they can be performed by a wide range of personnel after a short period of training. The tests appear to be more sensitive than Pap smear in detecting preinvasive cervical lesions, although specificity is lower (Sankaranarayanan et al., 2004). However, since these methods give an immediate result (positive or negative), it is possible to institute the follow-up examination (by colposcopy) and treatment immediately. An alternative approach that has been advocated is simply to treat all cervices showing abnormalities on visual inspection ("see and treat approach"), accepting the inevitable overtreatment of false positives. This strategy requires the use of simple and complication-free methods of treatment (Sellors and Sankaranarayanan, 2003). Simulation modeling suggests that these methods may have considerable advantages with respect to cost-effectiveness (Goldie et al., 2004b). However, the IARC review (2004) concluded that, in view of uncertainties in the definitions of test results, reproducibility, and quality assurance, long-term results from randomized trials are essential to further evaluate these procedures.

16.3.2.4. Testing for Human Papillomaviruses

There is considerable interest in the use of HPV DNA testing in screening, since it is clear that women with persistent HPV infection have higher risk of developing precursor lesions (CIN) (Koutsky et al., 1992). For mass screening purposes, the Hybrid Capture™ (HC) assay has been the most widely used. Although it is a test to detect viral DNA, the sensitivity of the test is generally reported in terms of its ability to detect CIN, and it is considerably more sensitive (95 percent) than conventional cytology (70 percent). Nevertheless, the place of primary screening by HPV DNA testing has not so far been evaluated. The main problem is that, as already noted, 5 to 15 percent of women are infected with HPV (and an even higher percentage at young ages), so that many more individuals would require a diagnostic follow-up. Because the test is relatively expensive, primary screening does not appear to be cost-effective. In addition, there are questions about the psychological and emotional impact of communicating positive HPV results to women. So far, the main use of HPV testing is in the triage of low grade abnormalities (ASCUS and CIN I) detected by cytology (Lörincz, 1997). Thus, women with abnormal Pap smears containing high-risk HPV DNA present are managed more aggressively than those whose lesions are devoid of HPV DNA. The application of this approach to screening will have to await the development of tests that can be carried out rapidly and relatively cheaply. Since only one test per lifetime would potentially remove the need for up to 80 percent of Pap smears currently being performed (Goldie et al. 2004b) a suitable test may well be available relatively soon.

16.3.2.5. Detection of Early Invasive Cancer

The rationale behind attempts to detect cancer at an early stage is that this enhances the effectiveness of treatment, either in terms of reducing the risk of death from cancer and/or by improving quality of life after cancer detection. Two approaches have been used to bring about diagnoses of invasive cancer at an earlier stage:

(a) education to bring about greater awareness of the population about early symptoms and signs of cancer, and about the disease itself, so that individuals bring themselves to medical attention at the earliest possible time; and
(b) active case finding, through examination of women who are not specifically seeking medical advice.

The latter approach, a form of population screening using simple visual inspection to detect early cancers was proposed as a potential method of early diagnosis in WHO guidelines in the early 1990s (Stjernsward et al., 1987; Miller, 1992). Unfortunately, experience showed that simple visual inspection was unsuitable as a screening test (Sankaranarayanan et al., 2002), and it is now no longer advocated. On the other hand, long before screening was introduced, considerable progress had been made against cervical cancer by a combination of improvements in treatment (particularly the introduction of radium therapy), and by diagnosis of invasive cancers at progressively earlier stages (Ponten et al.,1995). Probably this experience will be replicated in developing countries, and also be accelerated by organized intervention (Jayant et al., 1995).

16.4. Conclusion

The opportunity to reduce HPV-related diseases, particularly cervical cancer, is at hand. With the advances in the understanding of human papillomaviruses, their molecular biology, transforming mechanisms, and etiology of HPV diseases, better primary and secondary prevention strategies can now be designed and implemented. New screening and diagnostic technologies can offer significant advances for secondary prevention in developing countries, while HPV vaccines may offer a relatively simple and cost-effective intervention to accelerate cervical cancer prevention and control further. For the next decades, however, combinations of primary and secondary prevention strategies will be needed, and the modalities of these combinations may change over time, depending on many factors that are likely to be country specific. Defining the best combination for primary and secondary prevention strategies will be the future challenge at the local, regional and global level.

References

Allen, M., Kalantari, M., Ylitalo, N., Pettersson, B., Hagmar, B., Scheibenpflug, L., Johansson, B., Petterson, U., and Gyllensten, U. (1996). HLA DQ-DR haplotype and susceptibility to cervical carcinoma: Indications of increased risk for development of cervical carcinoma in individuals infected with HPV 18. *Tissue Antigens* 48:32–37.

Astori, G., Lavergne, D., Benton, C., Hockmayr, B., Egawa, K., Garbe, C., and de Villiers, E.M. (1998, May). Human papillomaviruses are commonly found in normal skin of immunocompetent hosts. *J. Invest. Dermatol.* 110(5):752–755.

Bayo, S., Bosch, F.X., de Sanjose, S., Munoz, N., Combita, A.L., Coursaget, P., Diaz, M., Dolo, A., van den Brule, A.J., and Meijer, C.J. (2002, Feb). Risk factors of invasive cervical cancer in Mali. *Int. J. Epidemiol.* 31(1):202–209.

Bauer, H.M., Ting, Y., Greer, C.E., Chambers, J.C., Tashiro, C.J., Chimera, J., Reingold, A., and Manos, M.M. (1991, Jan). Genital human papillomavirus infection in female university students as determined by a PCR-based method. *JAMA* 265(4):472–477 (See also pages 23–30).

Beutels, P. (1998, Nov). Economic evaluations applied to HB vaccination: General observations. *Vaccine* 16(Suppl):S84–S92.

Beutner, K.R., and Tyring, S. (1997, May 5). Human papillomavirus and human disease. *Am. J. Med.* 102(5A):9–15. (Review).

Bosch, F.X., Lorincz, A., Muñoz, N., Meijer, C.J.L.M., and Shah, K.V. (2002). The causal relation between human papillomavirus and cervical cancer. *J. Clin. Pathol.* 55:244–265.

Bray, F., Loos, A.H., McCarron, P., Weiderpass, E., Arbyn, M., Møller, H., Hakama, M., and Parkin, D.M. (2005). Trends in cervical squamous cell carcinoma incidence in 13 European countries: Changing risk and the effects of screening . *Cancer Epidemiol. Biomarkers Prev.* 14:677–686.

Calvo, N., Morice, A., Saenz, E., and Navas, L. (2004, Aug) Using surveys of schoolchildren to evaluate coverage with and oppurtunity for vaccination in Costa Rica. *Rev. Panam. Salud Publica.* 16(2):118–124.

Clements, C.J., Chandra-Mouli, V., Byass, P., Ferguson, B.J. (1999). Global strategies, policies and practices for immunization of adolescents. http://www.who.int/vaccines-documents/DocsPDF/www9866.pdf

de Villiers, E.M., Fauquet, C., Broker, T.R., Bernard, H.U., and zur Hausen, H. (2004, Jun 20). Classification of papillomaviruses. *Virology* 324(1):17–27.

Dorn, H.F., and Cutler, S.J. (1959). Morbidity from cancer in the United States: Parts I and II. *Publ. Health Monogr No. 56.* US Dept of Health, Education and Welfare, Washington, D.C.

Fenner, F. (1982, Sep–Oct). A successful eradication campaign. Global eradication of smallpox. *Rev Infect Dis.* 4(5):916–930.

Ferlay, J., Bray, F., Pisani, P., and Parkin, D.M. (2004). *GLOBOCAN 2002: Cancer Incidence, Mortality and Prevalence Worldwide.* Lyon: IARC Press.

Fernandez Garrote, L., Lence Anta, J.J., Cabezas Cruz, E., Romero, T., and Camacho, R. (1996). Evaluation of the cervical cancer control program in Cuba. *Bull. Pan Am. Health Organ.* 30:387–391.

Follen, M., Meyskens, F.L. Jr., Alvarez, R.D., Walker, J.L., Bell, M.C., Storthz, K.A., Sastry, J., Roy, K., Richards-Kortum, R., and Cornelison, T.L. (2003, Nov 1). Cervical cancer chemoprevention, vaccines, and surrogate endpoint biomarkers. *Cancer* 98(9 Suppl):2044–2051.

Franceschi, S., Dal Maso, L., Arniani, S., Crosignani, P., Vercelli, M., Simonato, L., Falcini, F., Zanetti, R., Barchielli, A., Serraino, D., and Rezza, G. (1998). Risk of cancer other than Kaposi's sarcoma and non-Hodgkin's lymphoma in persons with AIDS in Italy. Cancer and AIDS Registry Linkage Study. *Br. J. Cancer* 78:966–970.

Franceschi, S., Muñoz, N., Snijders, P.J., and Walboomers, W.W. (1996). Human papillomavirus and cancers of the upper aerodigestive tract: A review of epidemiological and experimental evidence. *Cancer Epidemiol. Biomarkers Prev.* 5:567–575.

Frisch, M., Biggar, R.J., and Goedert, J.J. (2000). Human papillomavirus-associated cancers in patients with human immunodeficiency virus infection and acquired immunodeficiency syndrome. *J. Natl. Cancer Inst.* 92:1500–1510.

Fu, Y.S., Berek, J.S., and Hilborne, L.H. (1987). Diagnostic problems of in situ and invasive adenocarcinomas of the uterine cervix. *Appl. Pathol.* 5:47–56.

Galloway, D.A. (2003, Aug). Papillomavirus vaccines in clinical trials. *Lancet Infect Dis.* 3(8):469–475.

Gillison, M.L., and Shah, K.V. (2001). Human papillomavirus-associated head and neck squamous cell carcinoma: Mounting evidence for an etiologic role for HPV in a subset of head and neck cancers. *Curr. Opin. Oncol.* 13:183–188.

Goedert, J.J., Cote, T.R., Virgo, P., Scoppa, S.M., Kingma, D.W., Gail, M.H., Jaffe, E.S., and Biggar, R.J. (1998). Spectrum of AIDS-associated malignant disorders. *Lancet* 351:1833–1839.

Goldie, S.J., Kim, J.J., and Wright, T.C. (2004b, Apr). Cost-effectiveness of human papillomavirus DNA testing for cervical cancer screening in women aged 30 years or more. *Obstet. Gynecol.* 103(4):619–631.

Goldie, S.J., Kohli, M., Grima, D., Weinstein, M.C., Wright, T.C., Bosch, F.X., and Franco, E. (2004a, Apr 21). Projected clinical benefits and cost-effectiveness of a human papillomavirus 16/18 vaccine. *J. Natl. Cancer Inst.* 96(8):604–615.

Grulich, A.E., Wan, X., Law, M.G., Coates, M., and Kaldor, J.M. (1999). Risk of cancer in people with AIDS. *AIDS.* 13:839–843.

Gustafsson, L., Ponten, J., Bergstrom, R., and Adami, H.-O. (1997a). International incidence rates of invasive cervical cancer before cytological screening. *Int. J. Cancer* 71:159–165.

Gustafsson, L., Ponten, J., Zack, M., and Adami, H.-O. (1997b). International incidence rates of invasive cervical cancer after introduction of cytological screening. *Cancer Causes Control* 8:755–763.

Hakama, M. (1982). Trends in the incidence of cervical cancer in the Nordic countries. In K. Magnus (ed.): *Trends in Cancer Incidence*, pp. 279–292. New York: Hemisphere Press.

Harper, D.M., Franco, E.L., Wheeler, C., Ferris, D.G., Jenkins, D., Schuind, A., Zahaf, T., Innis, B., Naud, P., De Carvalho, N.S., Roteli-Martins, C.M., Teixeira, J., Blatter, M.M., Korn, A.P., Quint, W., Dubin, G., and the GlaxoSmithKline HPV Vaccine Study Group. (2004, Nov 13). Efficacy of a bivalent L1 virus-like particle vaccine in prevention of infection with human papillomavirus types 16 and 18 in young women: A randomised controlled trial. *Lancet* 364(9447):1757–1765.

Harro, C.D., Pang, Y.Y., Roden, R.B., Hildesheim, A., Wang, Z., Reynolds, M.J., Mast, T.C., Robinson, R., Murphy, B.R., Karron, R.A., Dillner, J., Schiller, J.T., and Lowy, D.R. (2001, Feb 21). Safety and immunogenicity trial in adult volunteers of a human papillomavirus 16 L1 virus-like particle vaccine. *J. Natl. Cancer Inst.* 93(4):284–292

Heymann, D.L., and Aylward, R.B. (2004, Sep 23). Eradicating polio. *N. Engl. J. Med.* 351(13):1275–1277.

Herrero, R., Brinton, L.A., Reeves, W.C., Brenes, M.M., de Britton, R.C., Gaitan, E., and Tenorio, F. (1992). Screening for cervical cancer in Latin America: A case-control study. *Int. J. Epidemiol.* 21:1050–1056.

Herrero, R., Muñoz, N., Lazcano, E., Poss, H., Sukvirach, S., de Sanjose, S., Meijer, C.J.L.M., Coursaget, P., and Walboomers, J.M.M. (2000). HPV International prevalence surveys in general populations. *Proc. 18th Int. Papillomavirus Conf.*, Barcelona, O54, p. 126.

Herrero, R., and Munoz, N. (1999). Human papillomavirus and cancer. In R. Newton, V. Beral & R. Weiss (eds.): *Infections & Human Cancer, Cancer Surveys*, vol. 33. pp. 75–98.

Herrero, R., Hildesheim, A., Bratti, C., Sherman, M.E., Hutchinson, M., Morales, J., Balmaceda, I., Greenberg, M.D., Alfaro, M., Burk, R.D., Wacholder, S., Plummer, M., and Schiffman, M. (2000a). Population-based study of human papillomavirus infection and cervical neoplasia in rural Costa Rica. *J. Natl. Cancer Inst.* 92:464–474.

Herrero, R., Castellsagué, X., Pawlita, M., Lissowska, J., Kee, F., Balaram, P., Rajkumar, T., Sridhar, H., Rose, B., Pintos, J., Fernandez, L., Idris, A., Nieto, M.J., Nieto, A., Talamini, R., Tavani, A., Bosch, X., Riedel, U., Snijders, P.J.F., Meijer, C.J.M., Viscidi, R., Muñoz, N., and Franceschi, S. for the IARC Multi-centric Oral Cancer Study Group. (2003). Human papillomavirus and oral cancer: The International Agency for Research on Cancer multicenter study. *J. Natl. Cancer Inst.* 95:1772–1783.

Hildesheim, A., Schiffman, M., Scott, D.R., Marti, D., Kissner, T., Sherman, M.E., Glass, A.G., Manos, M.M., Lorincz, A.T., Kurman, R.J., Buckland, J., Rush, B.B., and Carrington, M. (1998a, Nov). Human leukocyte antigen class I/II alleles and development of human papillomavirus-related cervical neoplasia: Results from a case-control study conducted in the United States. *Cancer Epidemiol. Biomarkers Prev.* 7(11):1035–1041.

Hildesheim, A., Schiffman, M., Brinton, L.A., Fraumeni, J.F. Jr., Herrero, R., Bratti, M.C., Schwartz, P., Mortel, R., Barnes, W., Greenberg, M., McGowan, L., Scott, D.R., Martin, M., Herrera, J.E., and Carrington, M. (1998b, Dec 10). p53 polymorphism and risk of cervical cancer. *Nature* 396(6711). 531–532.

Hildesheim, A., Herrero, R., Castle, P.E., Wacholder, S., Bratti, M.C., Sherman, M.E., Lorincz, A.T., Burk, R.D., Morales, J., Rodriguez, A.C., Helgesen, K., Alfaro, M., Hutchinson, M., Balmaceda, I., Greenberg, M., and Schiffman, M. (2001). HPV co-factors related to the development of cervical cancer: Results from a population-based study in Costa Rica. *Br. J. Cancer* 84:1219–1226.

Hill, A.B. (1965). The environment and disease: Association or causation?. *Proc. R. Soc. Med.* 58, 295–300.

Hippelainen, M.I., Syrjanen, S., Hippelainen, M.J., Saarikoski, S., and Syrjanen, K. (1993, Oct). Diagnosis of genital human papillomavirus (HPV) lesions in the male: Correlation of peniscopy, histology and in situ hybridisation. *Genitourin Med.* 69(5):346–351.

Ho, G.Y., Studentsov, Y., Hall, C.B., Bierman, R., Beardsley, L., Lempa, M., and Burk, R.D. (2002, Sep 15). Risk factors for subsequent cervicovaginal human papillomavirus (HPV) infection and the protective role of antibodies to HPV-16 virus-like particles. *J. Infect. Dis.* 186(6):737–742.

Ho, G.Y., Bierman, R., Beardsley, L., Chang, C.J., and Burk, R.D. (1998). Natural history of cervicovaginal papillomavirus infection in young women. *N. Engl. J. Med.* 338:423–428.

Hughes, J.P., Garnett, G.P., and Koutsky, L. (2002). The theoretical population-level impact of a prophylactic human papillomavirus vaccines. *Epidemiology.* 13(6):631–639.

IARC. (1995). Monographs on the Evaluation of Carcinogenic Risks to Humans. *Human Papillomaviruses*, vol. 64. Lyon: IARC Press.

IARC. (2004a). Monographs on the Evaluation of Carcinogenic Risks to Humans. *Tobacco Smoke and Involuntary Smoking*, vol. 83. Lyon: IARC Press.

IARC. (2004b). *Handbooks of Cancer Prevention Vol.10. Cervical Cancer Screening.* Lyon: IARC Press.

Jayant, K., Rao, R.S., Nene, B.M., and Dale, P.S. (1995). Improved stage at diagnosis of cervical cancer with increased cancer awareness in a rural Indian population. *Int. J. Cancer* 63, 161–163.

Jin, F., Devesa, S.S., Chow, W.H., Zheng, W., Ji, B.T., Fraumeni, J.F. Jr., and Gao, Y.T. (1999). Cancer incidence trends in urban Shanghai, 1972–1994: An update. *Int. J. Cancer* 83:435–440.

Koutsky, L.A., Holmes, K.K., Critchlow, C.W., Stevens, C.E., Paavonen, J., Beckmann, A.M., DeRouen, T.A., Galloway, D.A., Vernon, D., and Kiviat, N.B. (1992, Oct 29). A cohort study of the risk of cervical intraepithelial neoplasia grade 2 or 3 in relation to papillomavirus infection. *N. Engl. J. Med.* 327(18):1272–1278.

Koutsky, L.A., Ault, K.A., Wheeler, C.M., Brown, D.R., Barr, E., Alvarez, F.B., Chiacchierini, L.M., and Jansen, K.U. (2002). A controlled trial of a human papillomavirus type 16 vaccine. *N. Engl. J. Med.* 347:1645–1651.

Koutsky, L.A., and Kiviat, N.B. (1999). Human papillomavirus infections. In K. K. Holmes, P. F. Sparling P. A. Mardh, S. M. Lemon, W. E. Stamm, P. Piot, and J. N. Wasserheit (eds.): *Sexually transmitted Diseases*. 3rd ed. New York: McGraw-Hill, pp. 347–360.

Koutsky, L.A., Galloway, D.A., and Holmes, K.K. (1988). Epidemiology of genital human papillomavirus infection. *Epidemiol. Rev.*10:122–163 (Review).

Koutsky, L.A. (1997, May 5). Epidemiology of genital human papillomavirus infection. *Am. J. Med.*102(5A):3–8 (Review).

Lazcano-Ponce, E., Alonso, P., Lopez, L., and Hernandez, M. (1994). Quality control study on negative gynaecological cytology in Mexico. *Diagn. Cytopathol.* 10:10–14.

Lazcano-Ponce, E.C., Buiatti, E., Najera-Aguilar, P., Alonso-de-Ruiz, P., and Hernandez-Avila, M. (1998). Evaluation model of the Mexican national program for early cervical cancer detection and proposals for a new approach. *Cancer Causes Control* 9:241–251.

Ley, C., Bauer, H.M., Reingold, A., Schiffman, M.H., Chambers, J.C., Tashiro, C.J., and Manos, M.M. (1991). Determinants of genital human papillomavirus infection in young women. *J. Natl. Cancer Inst.* 83:997–1003.

Li, H.Q., Jin, S.Q., Xu, H.X., and Thomas, D.B. (2000). The decline in the mortality rates of cervical cancer and a plausible explanation in Shandong, China. *Int. J. Epidemiol.* 29(3):398–404.

Lörincz, A.T. (1997). Methods of DNA Hybridization and their Clinical Applicability to Human Papillomavirus Detection. In E. Franco and J. Monsonego (eds.): *New Developments in Cervical Cancer Screening and Prevention*. Chapter 37, pp. 325–337. UK: Blackwell Science.

Mandelblatt, J.S., Kanetsky, P., Eggert, L., and Gold, K. (1999). Is HIV infection a cofactor for cervical squamous cell neoplasia?. *Cancer Epidemiol. Biomarkers Prev.* 8:97–106.

Manhart, L.E., and Koutsky, L.A. (2002, Nov). Do condoms prevent genital HPV infection, external genital warts, or cervical neoplasia? A meta-analysis. *Sex Transm. Dis.* 29(11):725–735.

Martinez, J., Smith, R., Farmer, M., Resau, J., Alger, L., Daniel, R., Gupta, J., Shah, K., and Naghashfar, Z. (1988, Oct). High prevalence of genital tract papillomavirus infection in female adolescents. *Pediatrics* 82(4):604–608.

Miller, A.B. (1992). *Cervical Cancer Screening Programs. Managerial Guidelines*. World Health Organization, Geneva, Switzerland.

Moreno, V., Bosch, F.X., Muñoz, N., Meijer, C.J., Shah, K.V., Walboomers, J.M., Herrero, R., and Franceschi, S. (2002). Effect of oral contraceptives on risk of cervical cancer in women with human papillomavirus infection: The IARC multicentric case-control study. *Lancet* 359:1085–1092.

Moscicki, A.B., Shiboski, S., Broering, J., Powell, K., Clayton, L., Jay, N., Darragh, T.M., Brescia, R., Kanowitz, S., Miller, S.B., Stone, J., Hanson, E., and Palefsky, J. (1998). The natural history of human papillomavirus infection as measured by repeated DNA testing in adolescent and young women. *J. Pediatr.* 132:277–284.

Muñoz, N., Bosch, F.X., de Sanjose, S., Vergara, A., del Moral, A., Munoz, M.T., Tafur, L., Gili, M., Izarzugaza, I., Viladiu, P., Navarro, C., Alonso de Ruiz, P., Aristizabal, N., Sanatamaria, M., Orfilla, J., Daniel, R.E., Guerrero, E., and Shah, K.V. (1993). Risk factors for cervical intraepithelial neoplasia grade III/carcinoma in situ in Spain and Colombia. *Cancer Epidemiol. Biomarkers Prev.* 2:423–431.

Muñoz, N., Franceschi, S., Bosetti, C., Moreno, V., Herrero, R., Smith, J.S., Shah, K.V., Meijer, C.J., and Bosch, F.X. (2002). Role of parity and human papillomavirus in cervical cancer: The IARC multicentric case-control study. *Lancet* 359:1093–1101.

Muñoz, N., Bosch, F.X., Castellsague, X., Diaz, M., de Sanjose, S., Hammouda, D., Shah, K.V., and Meijer, C.J. (2004). Against which human papillomavirus types shall we vaccinate and screen? The international perspective. *Int. J. Cancer* 111:278–285.

Odunsi, K., Terry, G., Ho, L., Bell, J., Cuzick, J., and Ganesan, T.S. (1996). Susceptibility to human papillomavirus-associated cervical intra-epithelial neoplasia is determined by specific HLA DR-DQ alleles. *Int. J. Cancer* 67:595–602.

Pagliusi, S.P., and Aguado, M.T. (2004). Efficacy and other milestones for human papillomavirus vaccine introduction. *Vaccine.* 23(5):569–578.

Papanicolaou, G.N., and Traut, H.F. (1943). *Diagnosis of Uterine Cancer by the Vaginal Smear.* New York: Commonwealth Fund.

Parkin, D.M., Ferlay, J., Handi-Cherif, M., Sitas, F., Thomas, J.O., Wabinga, H., and Whelan, S.L. (2003). *Cancer in Africa: Epidemiology and Preventuion.* IARC Scientific publication No. 153, Lyon: IARC Press.

Parkin, D.M., Bray, F., Ferlay, J., and Pisani, P. (2005). Global Cancer Statistics 2002. CA Cancer J Clin. 55:74–108.

Parkin, D.M., Whelan, S.L., Ferlay, J., Teppo, L., and Thomas, D.B. (eds.). (2002). *Cancer Incidence in Five Continents,* vol. VIII, IARC Scientific Publications No. 155, Lyon: IARC Press.

Pham, T.H., Nguyen, T.H., Herrero, R., Vaccarella, S., Smith, J.S., Nguyen Thuy, T.T., Nguyen, H.N., Nguyen, B.D., Ashley, R., Snijders, P.J., Meijer, C.J., Muñoz, N., Parkin, D.M., and Franceschi, S. (2003). Human papillomavirus infection among women in South and North Vietnam. *Int. J. Cancer* 104, 213–220.

Pfister, H. (2003). Human papillomavirus and skin cancer. Chapter 8. *J Natl. Cancer Inst. Monogr.* 31:52–56.

Pontén, J., Adami, H.O., Bergström, R., Dillner, J., Friberg, L.G., Gustafsson, L., Miller, A.B., Parkin, D.M., Sparén, P., and Trichopoulos, D. (1995). Strategies for global control of cervical cancer. *Int. J. Cancer* 60, 1–26.

Quinn, M., Babb, P., Jones, J., and Allen, E. (1999). Effect of screening on incidence of and mortality from cancer of cervix in England: Evaluation based on routinely collected statistics . *Br. Med. J.* 318:904–908.

Sadjadi, A., Malekzadeh, R., Derakhshan, M.H., Sepehr, A., Nouraie, M., Sotoudeh, M., Yazdanbod, A., Shokoohi, B., Mashayekhi, A., Arshi, S., Majidpour, A., Babaei, M., Mosavi, A., Mohagheghi, M.A., Alimohammadian, M., and Mohagheghi Mosavi, M.A. (2003). Cancer occurrence in Ardabil: Results of a population-based cancer registry from Iran. *Int. J. Cancer* 107:113–118.

de Sanjose, S., Muñoz, N., Bosch, F.X., Reimann, K., Pedersen, N.S., Orfila, J., Ascunce, N., Gonzalez, L.C., Tafur, L., Gili, M., Lette, I., Viladiu, P., Tormo, M.J., Moreo, P., Shah, K., and Wahren, B. (1994). Sexually transmitted agents and cervical neoplasia in Colombia and Spain. *Int. J. Cancer* 56:358–363.

Sankaranarayanan, R., Basu, P., Wesley, R.S., Mahé, C., Keita, N., Gombe Mbalawa, C.C., Sharma, R., Dolo, A., Shastri, S.S., Nacoulma, M., Nayama, M., Somanathan, T., Lucas, E., Muwonge, R., Frappart, L., and Parkin, D.M. (2004). Accuracy of visual screening for cervical neoplasia: Results from an IARC multicentric study in India and Africa. *Int. J. Cancer* 110, 907–913.

Sellors, J.W., and Sankaranarayanan, R. (2003). *Colposcopy and Treatment Of Cervical Intraepithelial Neoplasia: A Beginners Manual.* Lyon: IARC Press.

Serraino, D., Carrieri, P., Pradier, C., Bidoli, E., Dorrucci, M., Ghetti, E., Schiesari, A., Zucconi, R., Pezzotti, P., Dellamonica, P., Franceschi, S., and Rezza, G. (1999). Risk of invasive cervical cancer among women with or at risk for HIV infection. *Int. J. Cancer* 82:334–337.

Shah, K.V. (1998). Do human papillomavirus infections cause oral cancer?. *J. Natl. Cancer Inst.* 90:1585–1586

Shin, H.R., Franceschi, S., Vaccarella, S., Roh, J.W., Ju, Y.H., Oh, J.K., Kong, H.J., Rha, S.H., Jung, S.I., Kim, J.I., Jung, K.Y., van Doorn, L.J., and Quint, W. (2004, Aug 1). Prevalence and determinants of genital infection with papillomavirus, in female and male university students in Busan, South Korea. *J. Infect. Dis.* 190(3):468–476.

Sigurdsson, K. (1995). Quality assurance in cervical cancer screening: The Icelandic experience 1964–1993. *Eur. J. Cancer* 31A:728–734.

Sigurdsson, K. (1999). The Icelandic and Nordic cervical screening programs: Trends in incidence and mortality rates through 1995. *Acta Obstet. Gynecol. Scand.* 78:478–485.

Smith, J.S., Muñoz, N., Herrero, R., Eluf-Neto, J., Ngelangel, C., Franceschi, S., Bosch, F.X., Walboomers, J.M., and Peeling, R.W. (2002). Evidence for Chlamydia trachomatis as an HPV cofactor in the etiology of invasive cervical cancer in Brazil and the Philippines. *J. Infect Dis.* 185:324–331.

Snijders, P.J.F., Steenbergen, R.D.M., Meijer, C.J.L.M., and Walboomers, J.M.M. (1997). Role of human papillomaviruses in cancer of the respiratory and upper digestive tract. *Clin. Dermatol.* 15:415–425.

Snijders, P.J., van den Brule, A.J., and Meijer, C.J. (2003). The clinical relevance of human papillomavirus testing: Relationship between analytical and clinical sensitivity. *J. Pathol.* 201:1–6.

Sobhani, I., Walker, F., Roudot-Thoraval, F., Abramowitz, L., Johanet, H., Henin, D., Delchier, J.C., and Soule, J.C. (2004). Anal carcinoma: Incidence and effect of cumulative infections. *AIDS.* 18:1561–1569.

Stjernsward, J., Eddy, D., Luthra, U.K., and Stanley, K. (1987). Plotting a new course for cervical cancer screening in developing countries. *World Health Forum.* 8:42–45.

Syrjanen, S.M. (1990). Basic concepts and practical applications of recombinant DNA techniques in detection of human papillomavirus (HPV) infection. *APMIS.* 98:95–110.

Taira, A.V., Neukermans, C.P., and Sanders, G.D. (2004). Evaluating Human Papillomavirus vaccination programs. *Emerging Infect. Dis.* 10:1915–1923. (See also http://www.cdc.gov/eid)

Thomas, D.B., Qin, Q., Kuypers, J., Kiviat, N., Ashley, R.L., Koetsawang, A., Ray, R.M., and Koetsawang, S. (2001). Human papillomaviruses and cervical cancer in Bangkok. II. Risk factors for in situ and invasive cervical carcinomas. *Am. J. Epidemiol.* 153:732–739.

Ursin, G., Peters, R.K., Henderson, B.E., d'Ablaing, G., III, Monroe, K.R., and Pike, M.C. (1994). Oral contraceptive use and adenocarcinoma of cervix. *Lancet* 344:1390–1394.

Vandelaer, J., Birmingham, M., Gasse, F., Kurian, M., Shaw, C., and, Garnier, S. (2003 Jul 28). Tetanus in developing countries: An update on the Maternal and Neonatal Tetanus Elimination Initiative. *Vaccine* 21:3442–3445.

van Houten, V.M.M., Snijders, P.J.F., van den Brekel, M.W.M., Kummer, J.A., Meijer, C.J.L.M., van Leeuwen, B., Denkers, F., Smeele, L.E., Snow, G.B., and Brakenhoff, R.H. (2001). Biological evidence that human papillomaviruses are etiologically involved in a subgroup of head and neck squamous cell carcinomas. *Int. J. Cancer* 93:232–235.

Vizcaino, A.P., Moreno, V., Bosch, F.X., Muñoz, N., Barros-Dios, X.M., and Parkin, D.M. (1998). International trends in the incidence of cervical cancer: I. Adenocarcinoma and adenosquamous cell carcinomas. *Int. J. Cancer* 75:536–545.

Walboomers, J.M., Jacobs, M.V., Manos, M.M., Bosch, F.X., Kummer, J.A., Shah, K.V., Snijders, P.J., Peto, J., Meijer, C.J., and Muñoz, N. Human papillomavirus is a necessary cause of invasive cervical cancer worldwide. (1999). *J. Pathol.* 189:12–19.

Winer, R.L., Lee, S.K., Hughes, J.P., Adam, D.E., Kiviat, N.B., and Koutsky, L.A. (2003). Genital human papillomavirus infection: Incidence and risk factors in a cohort of female university students. *Am. J. Epidemiol.* 157:218–226.

Woodman, C.B., Collins, S., Winter, H., Bailey, A., Ellis, J., Prior, P., Yates, M., Rollason, T.P., and Young, L.S. (2001). Natural history of cervical human papillomavirus infection in young women: A longitudinal cohort study. *Lancet* 357:1831–1836.

Yang, L., Li, L.D., Chen, T.D., and Parkin, D.M. (2003). Time Trends in Cancer Mortality in China:1987–1999. *Int. J. Cancer* 106:771–783.

zur Hausen, H. (2002). Papillomaviruses and cancer: From basic studies to clinical application. *Natl. Rev. Cancer* 2:342–350.

zur Hausen, H. (2000). Papillomaviruses causing cancer: Evasion from host-cell control in early events in carcinogenesis. *J. Natl. Cancer Inst.* 3. 92(9):690–698.

zur Hausen, H. (1976). Condylomata acuminata and human genital cancer. *Cancer Res.* 36(2 pt):794.

INDEX